Enzyme Kinetics

**METHODS OF
BIOCHEMICAL ANALYSIS**

Volume 53

Enzyme Kinetics

Rapid-Equilibrium Applications of Mathematica

Robert A. Alberty
Department of Chemistry
Massachusetts Institute of Technology
Cambridge, MA

WILEY

A JOHN WILEY & SONS, INC., PUBLICATION

Library of Congress Cataloging-in-Publication Data is available.

ISBN 978-0-470-63932-0

Printed in Singapore.

10 9 8 7 6 5 4 3 2 1

Contents

Preface

Chapter 1 Biochemical Thermodynamics   32 pages

Chapter 2 A = P   48 pages

Chapter 3 Ordered A + B → Products   50 pages

Chapter 4 Random A + B → Products   50 pages

Chapter 5 A + B = P + Q   38 pages

Chapter 6 A + B + C → Products   40 pages

Chapter 7 Ordered O + mR → Products   24 pages

Chapter 8 Random O + mR → Products   20 pages

Chapter 9 Inhibition and Activation of A → Products   48 pages

Chapter 10 Modification of A → Products   20 pages

Chapter 11 Inhibition, Activation, and Modification of A + B → Products

                                                             34 pages

Chapter 12 Systems of Enzyme-Catalyzed Reactions   32 pages

References   4 pages

Index   5 pages

Preface

Rapid-equilibrium enzyme kinetics has several advantages over steady-state enzyme kinetics. That is not to say that it is always applicable, but rapid-equilibrium enzyme kinetics yields simpler rate equations than steady-state enzyme kinetics. Since rapid-equilibrium rate equations are the simplest possible, they are the place to start. They are actually widely used. When rapid-equilibrium rate equations are applicable, Michaelis constants are equilibrium constants. For simple mechanisms, rapid-equilibrium rate equations are readily derived by hand, but, for more complicated mechanisms, they can be derived using a computer without having to write a computer program. This is especially useful when pH effects are involved, or there are many reactants, as for random A + B + C → products, or multiple forms of inhibition, activation, and modification. When the rapid-equilibrium assumption is applicable, all the reactions up to the rate-determining reaction are at equilibrium. The expression for the equilibrium concentration of the enzyme-substrate complex that yields products can be derived with a computer using Solve, which is an operation that solves a set of simultaneous polynomial equations. This operation is available in *Mathematica*^{R}, Maple^R, and MatLab^R. The input for Solve includes equilibrium expressions for an independent set of reactions and a conservation equation. The use of a computer has the additional advantage that the rapid-equilibrium rate equation is obtained in computer-readable form and does not have to be typed into the computer. Rapid-equilibrium rate equations can be derived with a computer for very complicated mechanisms.

Solve can also be used to calculate the values of the kinetic parameters in a rate equation from the minimum number of velocity measurements. This calculation can be made using steady-state rate equations as well as rapid-equilibrium rate equations. When rapid-equilibrium rate equations do not represent the experimental data, steady-state rate equations can be used, or empirical rate equations can be used.

There are two ways to organize a book on rapid-equilibrium kinetics. There could be chapters on deriving rate equations, estimating kinetic parameters, effects of pH, effects of inhibitors, etc. Or there could be chapters on various types of reactions with derivations of rate equations for different mechanisms, estimation of kinetic parameters, effects of pH, etc. This second option is followed here by treating different kinds of mechanisms separately so that a user can study various aspects of the kinetics of a certain type of enzyme-catalyzed reaction in a single chapter. The exceptions of this organizational plan are Chapter 1 on biochemical thermodynamics that is the foundation for the rapid-equilibrium assumption, the three chapters on inhibition, activation, and modification, and the last chapter on systems of enzyme-catalyzed reactions.

Biochemical thermodynamics is based on the Legendre transformed Gibbs energy G' that brings in the pH as an independent variable. This development has brought in new ideas about the use of thermodynamics in biochemistry. Enzyme kinetics necessarily contains biochemical thermodynamics because rate equations including the reverse reaction must yield the equilibrium composition at long times. When rate equations are derived on the assumption that there is a rate-determining reaction, the reactions before the rate-determining reaction are treated with biochemical thermodynamics. For example, the equilibrium composition of substrates, enzymatic sites, enzyme-substrate complexes, hydrogen ions, inhibitors, activators, and modifiers is calculated using an independent set of reactions. The relationship between biochemical thermodynamics and rapid-equilibrium enzyme kinetics is made evident by the Haldane relation that expresses the apparent equilibrium constant K' for the reaction that is catalyzed in terms of kinetic parameters in the rate equation. The existence of the Haldane relation means that the thermodynamic effects of temperature, pH, and ionic strength on equilibrium are all in the kinetic parameters. Of course biochemical thermodynamics is also built into transient kinetics, but that is not discussed in this book.

This book is concerned with six levels of understanding the kinetics of an enzyme-catalyzed reaction. (1) The first level is the understanding of the thermodynamics of the enzyme-catalyzed reaction. When an enzyme-catalyzed reaction occurs, the composition changes until equilibrium is reached. The equilibrium concentrations of the substrates can be calculated using the apparent equilibrium constant, which is a function of temperature, pH and ionic strength. The apparent equilibrium constant may also be a function of $[Mg^{2+}]$, for example. This thermodynamic information does not tell us anything about the mechanism or rate constants because the apparent equilibrium constant depends entirely on the thermodynamic properties of the substrates. (2) The second level of understanding is to be sure that the velocity at specified temperature, pH, and ionic strength is directly proportional to the total enzyme concentration. This is true for all the mechanisms discussed here. (3) The third level of understanding involves the determination of the rate equation at a particular temperature, pH, and ionic strength. There are two types of rapid-equilibrium rate equations: rate equations for the forward reaction and complete rate equations that include the reverse rate equation as well. In this book, when the catalyzed reaction is represented by A + B → products, only the forward reaction is discussed, and when the catalyzed reaction is represented by A + B = P + Q, the rate equation includes the parameters for the reverse reaction, as well as the forward reaction. This book is concerned with rapid-equilibrium rate equations, but there is the possibility that a steady-state rate equation is required. Even an empirical rate equation may be required. A complete rate equation can be tested by using the Haldane relation to calculate the apparent equilibrium constant at the experimental temperature, pH, and ionic strength. (4) The fourth level of understanding involves the study of the effects of pH on the kinetics. Even when the substrates do not have pKs, velocities will in general depend on the pH. These effects are determined by pKs of the enzymatic site, the enzyme-substrate complexes, and the substrates. It is possibile that one or more hydrogen ions may be consumed in the mechanism. When a steady-state rate equation is required, these effects become much more complicated, but when the rapid-equilibium assumption is applicable, all of these effects can be represented by pKs and the number of hydrogen ions consumed in the mechanism. When these effects can be determined quantitatively for both the forward and reverse reactions, the Haldane relation must yield the correct dependence of the apparent equilbrium constant on the pH and ionic strength. It is important to distinguish between the number n of hydrogen ions consumed in the mechanism and the change in binding of hydrogen ions determined by thermodynamic measurements. The number n of hydrogen ions consumed in a reaction in a mechanism is an integer, but the change in binding of hydrogen ions in an enzyme-catalyzed reaction depends on the pH. (5) When the rapid-equilibrium assumption applies, Michaelis constants are equilibrium constants, and so the study of the effect of temperature can yield standard transformed Gibbs energies of reaction, standard transformed enthalpies of reaction, and standard transformed entropies of reactions in the mechanism. (6) The sixth level of understanding involves inhibition, activation, and modification by binding at the catalyic site or other sites. Modifiers can be inhibiting or activating, but they provide additional pathways to products. This book deals with all six levels of understanding.

Mathematica is a wonderful application for rapid-equilibrium enzyme kinetics because Solve can be used to derive rapid-equilibrium rate equations and also to estimate the kinetic parameters using the minimum number of velocity measurements. *Mathematica* has the advantage of including a word processor so that a book with text and live equations can be written. In *Mathematica* the symbols for variables and kinetic parameters have to start with lower case letters because operations like Solve, TableForm, and Round start with capital letters. The names of programs also have to start with lower case letters. Names of reactants and functions cannot involve spaces, hyphens, dots or other mathematical symbols.

Each chapter in this book is a *Mathematica* notebook that includes text as well as calculations. These chapters include descriptions of the background for calculations, the calculations, and discussion of their significance. The CD at the back of the book contains the entire book in *Mathematica*. This CD can be downloaded into a personal computer with *Mathematica* installed, but it can also be read in a computer with *MathReader*, which is freely available from Wolfram Research, Inc. (100 Trade Center Drive, Champaign, IL 61820-7237, and www.wolfram.com). All the calculations in a chapter can be run by use of Evaluate/Evaluate Notebook. Programs and calculations can be copied and pasted in a new notebook. Calculations can be rerun with different input. The input to programs can be changed to apply these programs to other reactions and conditions.

It is not necessary to be a programmer to use the programs in this book. More programs and calculations on rapid-equilibrium enzyme kinetics are available at *MathSource*. The references are at the end of the book, and this includes web URLs and notebooks in *MathSource*.

This book builds on my previous book "Biochemical Thermodynamics: Applications of *Mathematica*", Wiley, Hoboken, NJ (2006). That book provides a great deal of information on the thermodynamics of enzyme-catalyzed reactions that is relevant to enzyme kinetics. For example, it gives the apparent equilibrium constants for 229 enzyme-catalyzed reactions at 298.15 K, 0.25 M ionic strength, and pHs 5-9. More apparent equilibrium constants can be calculated for similar reactions.

This is the second book on enzyme kinetics written in *Mathematica*. The first is P. J. Mulquiney and P. W. Kuchel, Modelling Metabolism with *Mathematica*, CRC Press, Boka Raton, Florida (2003). Their book uses steady-state rate equations, and it gives a program in *Mathematica* to derive steady-state rate equations. This program is based on the method described by A. Cornish-Bowden (Biochem. J. 165, 55-59 (1977)). They give steady-state rate equations for about 30 enzyme-catalyzed reactions.

I am indebted to Professor William Martin McClain, Wayne State University, for catching some errors in my treatment of thermodynamic cycles, which have been corrected. He has written an excellent book "Symmetry Theory in Molecular Physics with *Mathematica*," Springer, 2009.

I am indebted to the National Institutes of Health for support of research on which this book is based (5-RO1-GM4834812). At Wiley I am indebted to my Editor Anita Lekhwani.

Robert A. Alberty
Cambridge, Massachusetts

Use of Mathematica

This book is written in *Mathematica* 7.0. Even if you are not familiar with *Mathematica* (Wolfram Research, Inc. 100 World Center Drive, Champaign, IL 61820-7237 and www.wolfram.com) you should be able to read this book. The concepts and calculations in biochemical kinetics are explained in words in the textual parts of the book. And the results of calculations are discussed in words. When *Mathematica* is used to make tables and figures, explanatory titles are given.

Since *Mathematica* is a high level language that uses commands like Solve and D (for differentiate), you can see what mathematical operations are involved in a program. Everything that is involved in making these calculations is shown. When a calculation is made, a semicolon is often put at the end of the input so that the result is not shown there. When calculations are performed in you computer, the semicolons can be omitted to see what the result is. Semicolons are used in the book to save space.

Each chapter is a *Mathematica* notebook. When a chapter has been downloaded, it can be run by using Evaluation/Evaluate Notebook. This will take a few seconds or a minute to evaluate the operations. Quit *Mathematica* and restart before opening another chapter. Calculations and programs can be copied and pasted into a notebook of your own. Then the input can be changed to solve your problem.

The page of Contents shows that this book contains 12 Chapters that are each a *Mathematica* notebook. These 12 notebooks are selfsufficient in the sense that they can each be run alone. As mentioned above, only one of these notebooks should be open at one time. Pages are numbered within each of the notebooks.

The rate equations in this book can be used to calculate velocities for different sets of kinetic parameters and substrate concentrations. This is a good way to learn about the behavoir of a mechanism or to test the kinetic parameters that have been determined.

A number of books have been written to introduce *Mathematica* to new users. A list of books is available at the Wolfram Store. Wolfram offers free online seminars for new users: http://www.wolfram.com/services/education/seminars/.

Chapter 1 Biochemical Thermodynamics

1.1 Introduction

1.2 Chemical Thermodynamics

1.3 Transformed Thermodynamic Properties of Biochemical Reactants at a Specified pH

1.4 Calculation of the change in binding of hydrogen ions in a biochemical reaction at a specified pH by taking the derivative of $\log K'$ or $\Delta_r G'°$ with respect to pH

1.5 Calculation of the standard transformed Gibbs energies of formation of biochemical reactants

1.6 Calculation of the standard transformed Gibbs energy of hydrolysis of ATP to ADP and the apparent equilibrium constant K'

1.7 Calculation of the change in the binding of hydrogen ions $\Delta_r N_H$ in ATP + H_2O = ADP + P_i at 298.15 K

1.8 Calculation of the change in the binding of hydrogen ions in a biochemical reaction at a specified pH without information on standard Gibbs energies of formation of species

1.9 Another way to consider the pH dependence of the apparent equilibrium constant

- **1.9.1 Distributions of species of inorganic phosphate as functions of pH**

- **1.9.2 Distributions of species of ATP as functions of pH**

- **1.9.3 Distributions of species of ADP as functions of pH**

- **1.9.4 Calculations of the apparent equlibrium constant for ATP + H_2O = ADP + Pi as a function of pH**

1.10 Data on the thermodynamics of enzyme-catalyzed reactions in the literature and on the web

- **1.10.1 Experimental data on apparent equilibrium constants and heats of reactions**

- **1.10.2 Data on thermodynamic properties of species obtained chemically or from experimental measurements of apparent equilibrium constants and heats of reaction of enzyme-catalyzed reactions**

1.11 Discussion

1.1 Introduction

It is necessary to start a book on rapid-equilibrium enzyme kinetics with a chapter on biochemical thermodynamics for two reasons: (1) The complete rate equation for an enzyme-catalyzed reaction has to contain the biochemical thermodynamics of the reaction that is catalyzed. Setting the velocity in the complete rate equation equal to zero yields the Haldane relation for the apparent equilibrium constant K' in terms of the kinetic parameters at the specified temperature, pH and ionic strength. (2) The equilibrim concentration of the enzyme-substrate complex in the rate-determining reaction is calculated with a set of independent biochemical reactions. When an independent set of reactions is at equilibrium, other reactions that are the sums or differences of the reactions in the independent set are also at equilibrium.

In chemical thermodynamics, reactions are written in terms of species, and chemical reactions must balance the various atoms and electric charges. But in biochemical thermodynamics, reactions are written in terms of reactants (sums of species like ATP), and all atoms are balanced except for hydrogen atoms because the pH is held constant. Electric charges are also not balanced in a biochemical reaction. Conceptually, hydrogen ions are added or removed during the reaction so that the pH is held constant. In practice a buffer is used to hold the pH approximately constant.

This book is about rapid-equilibrium rate equations because they are simpler than steady-state rate equations. This is not to say that all enzyme-catalyzed reactions involve a rate-determining reaction, but simply to say that rapid-equilibrium rate equations are the place to start in the investigation of rate equations. When the kinetic data for an enzyme-catalyzed reaction cannot be represented by a rapid-equilibrium rate equation, steady-state rate equations or empirical rate equations have to be used.

The thermodynamics of enzyme-catalyzed reactions is based on chemical thermodynamics because it is the species of substrates, enzymatic sites, and enzyme-substrate complexes that react. Important new concepts are involved in biochemical thermodynamics because the pH is an independent variable like T and P; that is, the pH is chosen by the investigator. In chemical thermodynamics, on the other hand, the pH is calculated using the chemical equilibrium constants for an independent set of chemical reactions and the conservation of all atoms and electric charges.

To see how the specification of pH is handled in biochemical thermodynamics, it is necessary to consider the foundations of chemical thermodynamics. The most basic criterion for spontaneous change and equilibrium in chemical thermodynamics is provided by the entropy S because the second law specifies that the entropy of an isolated system can only increase and has its maximum value at equilibrium: thus $(dS)_{U,V} \geq 0$, where U is the internal energy and V is the volume. This criterion is only applicable to isolated systems because this is the only way to hold the internal energy and volume constant. The internal energy U provides the criterion for spontaneous change and equilibrium at constant volume and entropy: $(dU)_{V,S} \leq 0$. The internal energy decreases to a minimum when the system goes to equilibrium at constant V and S, but there is no way to hold the entropy constant. The enthalpy H is defined by the Legendre transform $H = U + PV$, and it provides the criterion for spontaneous change and equilibrium at specified P and S: $(dH)_{P,S} \leq 0$. Again there is no way to hold the entropy constant. A Legendre transform is the defintion of a new thermodynamic property by subtracting the product of two conjugate variables from an existing thermodynamic property [12,17,19]. Gibbs defined what we now call the Gibbs energy G by use of the Legendre transform $G = H - TS$. The Gibbs energy is so useful in chemistry because it provides the criterion for spontaneous change at specified temperature and pressure: $(dG)_{T,P} \leq 0$. Thus a chemical reaction at specified T and P can spontaneously go in the direction that decreases the Gibbs energy, and the Gibbs energy of a chemical reaction system is at its lowest value at equilibrium.

1.2 Chemical Thermodynamics

The fundamental equation for the Gibbs energy G of a chemical reaction system is given by [22]

$$dG = -SdT + VdP + \sum_{j=1}^{N} \mu_j \, d \, n_j \tag{1.2-1}$$

where μ_j is the chemical potential of species j, n_j is the amount of species j, and N is the number of different species. Equation 1.2-1 shows that T, P, and $\{n_j\}$ are the independent variables for the Gibbs energy of a chemical reaction system. For ideal solutions, the chemical potential of a species is given by

$$\mu_j = \mu_j^{\circ} + RT\ln[j] \tag{1.2-2}$$

where μ_j° is the standard chemical potential of species j and $[j]$ is the molar concentration of species j. When there is a single chemical reaction involving N species, the differential of the amount of species j is given by $dn_j = \nu_j \, d\xi$, where ν_j is the stoichiometric number for species j and ξ is the extent of the single reaction. In working with the fundamental equation, the chemical potential of a species is used, but, in working with experimental data, the symbol for the Gibbs energy of formation $\Delta_f G_j$ is used rather than μ_j. When there is a single chemical reaction in a system, equation 1.2-1 becomes

$$dG = -SdT + VdP + \sum_{j=1}^{N} \nu_j \, \Delta_f G_j \, d \, \xi \tag{1.2-3}$$

At constant temperature and pressure, the Gibbs energy of reaction is given by

$$\Delta_r G = dG/d\xi = \sum_{j=1}^{N} \nu_j \Delta_f G_j \tag{1.2-4}$$

Equation 1.2-2 for a chemical species can be written as

$$\Delta_f G_j = \Delta_f G_j^{\circ} + RT\ln[j] \tag{1.2-5}$$

where $\Delta_f G_j^{\circ}$ is the standard Gibbs energy of formation of species j. In chemical thermodynamics, this equation is used for ideal solutions and activity coefficients are introduced, but in biochemical thermodynamics this equation is used at the specified ionic strength and the Debye-Huckel equation is used to account for the effects of ionic strength. Substituting equation 1.2-5 into equation 1.2-4 yields

$$\Delta_r G = \Delta_r G° + RT \ln Q \qquad\qquad (1.2\text{-}6)$$

where $\Delta_r G°$ is the standard Gibbs energy of reaction $\sum v_j \Delta_f G_j°$, and Q is the reaction quotient. At equilibrium, $\Delta_r G = 0$ and Q becomes the chemical equilibrium constant K.

$$\Delta_r G° = -RT \ln K = \sum_{j=1}^{N} v_j \Delta_f G_j° \qquad\qquad (1.2\text{-}7)$$

Chemists have developed tables of values of $\Delta_f G_j°$ for species with respect to the elements. In other words, the $\Delta_f G_j°$ for elements are taken to be zero in a defined reference state. These tables can be used to calculate a number of chemical equilibrium constants K that are of biochemical interest, but the Natonal Bureau of Standards Tables [10] are limited to C_2.

More information about chemical thermodynamics is given in text books on chemical thermodynamics like Beattie and Oppenheim [8], and in my two books on biochemical thermodynamics [20,23].

The first publication that applied chemical thermodynamics to biochemical reactions was by Burton and Krebs [2], and the first table of thermodynamic properties was published by Burton in Krebs and Kornberg, Energy Transformations in Living Matter [4]. Burton recognized that the equilibrium constants for enzyme-catalyzed reactions together with the standard Gibbs energies $\Delta_f G_j°$ of species determined with chemical methods can yield $\Delta_f G_j°$ biochemical species. He made a table that could be used to calculate equilibrium constants of biochemical reactions that have not been studied. But he ran into problems with reactants like ATP that are sums of species at pH 7. Wilhoit [5] extended these tables, but ATP remained a problem. I became involved with ATP through electrophoresis, and my group determined acid dissociation constants and magnesium complex dissociation constants of ATP [1,3]. I worked on the thermodynamics of petroleum processing in the 1980-1990 period and learned that when the concentration of a species (like H^+) is held constant, the criterion for equilibrium is provided by a transformed Gibbs energy [13,15].

There is one more aspect of chemical thermodynamics that needs to be mentioned. If the equilibrium constants of a set of chemical reactions are known, the equilibrium composition can be calculated, but an independent set of chemical reactions must be used in the calculation. A set of chemical reactions is independent if no reaction in the set can be obtained by adding and subtracting reactions in the set. Linear algebra is needed for a complete discussion, but it leads to a simple equation: $N = C + R$, where N is the number of different species, C is the number of components, and R is the number of independent reactions [20,23]. The number of components in chemical thermodynamics is the number of elements involved.

1.3 Transformed Thermodynamic Properties of Biochemical Reactants at a Specified pH

To obtain the criterion for spontaneous change when the pH is held constant in addition to the temperature and presssure, it is necessary to use another Legendre transform. Two examples of Legendre transforms are $H = U + PV$ and $G = H - TS$ that have been discussed in Section 1.1. The conjugate variables required to introduce the pH as an independent variable are the amount of hydrogen atoms in the system $n_c(H)$, which is an extensive property, and the chemical potential of hydrogen ions $\mu(H^+)$, which is the intensive property corrresponding to the specified pH. The amount of the hydrogen component $n_c(H)$ is expressed in moles, and $\mu(H^+)$ is expressed kJ mol^{-1}. The product of conjugate variables is always an energy, and so $n_c(H)\mu(H^+)$ has the units of kJ, as does G. The relation between the chemical potential of hydrogen ions and the pH is

$$\mu(H^+) = \Delta_f G°(H^+) - RT\ln(10)\,pH \tag{1.3-1}$$

where $\Delta_f G°(H^+)$ is the standard Gibbs energy of formation of hydrogen ions and $R = 8.31451$ J mol^{-1}. Thus the transformed Gibbs energy G' of a thermodynamic system at a specified pH is defined by the Legendre transform

$$G' = G - n_c(H)\,\mu(H^+) = G - n_c(H)\{\Delta_f G°(H^+) - RT\ln(10)\,pH\} \tag{1.3-2}$$

The transformed Gibbs energy is needed in biochemical thermodynamics because it provides the criterion for spontaneous change and equilibrium at specified temperature, pressure, and pH: $(dG')_{T,P,pH} \le 0$. Thus a biochemical reaction at specified T, P, and pH can react spontaneously in the direction that decreases G', and G' has its minimum value at equilibrium.

The fundamental equation for the transformed Gibbs energy G' for a biochemical reaction system when N' reactants are present is given by [13,20,23]

$$dG' = -S'dT + VdP + \sum_{i=1}^{N'}\Delta_f G_i'\,dn_i' + RT\ln(10)n_c(H)dpH \tag{1.3-3}$$

where $\Delta_f G_i'$ is the transformed Gibbs energy of formation of reactant i (sum of species), and n_i' is the amount of reactant i. A number of steps are involved in deriving this equation. Equation 1.3-3 shows that there is a new type of term in the fundamental equation for the transformed Gibbs energy that is proportional to dpH.

When the reactants in a biochemical reaction system are involved in a single biochemical reaction, $dn_i' = v_i'd\xi'$, where v_i' is the stoichiometric number of reactant i in the biochemical reaction and $d\xi'$ is the differential of the extent of the biochemical reaction. The extent ξ' of a biochemical reaction is defined by $n_i' = (n_i')_0 + v_i'\xi'$, where $(n_i')_0$ the amount of reactant i when $\xi' = 0$. Replacing n_i' in equation 1.3-3 with this equation at constant T, P, and pH leads to the expression for the change in the transformed Gibbs energy in the biochemical reaction.

$$\Delta_r G' = \sum_{i=1}^{N'} \nu_i' \Delta_f G_i' \tag{1.3-4}$$

Thus equation 1.3-3 for a biochemical reaction system with a single biochemical reaction can be written as

$$dG' = -S' dT + V dP + \Delta_r G' d\xi' + RT\ln(10) n_c(H) dpH \tag{1.3-5}$$

Equation 1.3-5 can be used to derive the expression for the apparent equilibrium constant K' for a biochemical reaction at a specified pH. Substituting equation 1.3-4 in equation 1.3-5 yields

$$dG' = -S' dT + V dP + \sum_{i=1}^{N'} \nu_i' \Delta_f G_i' d\xi' + RT\ln(10) n_c(H) dpH \tag{1.3-6}$$

At specified T, P, and pH,

$$dG'/d\xi' = \sum_{i=1}^{N'} \nu_i' \Delta_f G_i' = \Delta_r G' \tag{1.3-7}$$

This is very much like equation 1.2-4 in chemical thermodynamics except that the i reactants are sums of species and ν_i' is the stoichiometric number of reactant i in the biochemical reaction at a specified pH. Substituting

$$\Delta_f G_i' = \Delta_f G_i'^\circ + RT\ln[i] \tag{1.3-8}$$

in equation 1.3-7 yields

$$\Delta_r G' = \sum_{i=1}^{N'} \nu_i' \Delta_f G'^\circ + RT\ln Q' = \Delta_r G'^\circ + RT\ln Q' \tag{1.3-9}$$

where $\Delta_r G'^\circ$ is the standard transformed Gibbs energy of reaction that is given by equation 1.3-8. Q' is the apparent reaction quotient. At equilibrium, $\Delta_r G' = 0$, and so

$$\Delta_r G'^\circ = -RT\ln K' \tag{1.3-10}$$

where K' is the apparent equilibrium constant for the biochemical reaction. The expression for K' is written in terms of reactant concentrations, except for H_2O.

Now we need to discuss the calculation of the standard transformed Gibbs energy of formation of a reactant $\Delta_f G_i'\,^\circ$ that was introduced in equation 1.3-8. The standard transformed Gibbs energies of formation $\Delta_f G_j'\,^\circ$ of the species that make up a reactant are given by

$$\Delta_f G_j'\,^\circ = \Delta_f G_j^\circ - N_H(j)\,\Delta_f G(H^+) \tag{1.3-11}$$

where $N_H(j)$ is the number of hydrogen atoms in species j. Since dilute solutions are assumed to be ideal, the Gibbs energy of species j is given by

$$\Delta_f G_j = \Delta_f G_j^\circ + RT\ln[j] \tag{1.3-12}$$

This equation can be applied to hydrogen ions at a specified pH.

$$\Delta_f G(H^+) = \Delta_f G^\circ(H^+) + RT\ln\left[10^{-pH}\right] \;\; = \Delta_f G^\circ(H^+) - RT\ln(10)\,pH \tag{1.3-13}$$

Substituting this equation into equation 1.3-11 yields

$$\Delta_f G_j'^\circ = \Delta_f G_j^\circ - N_H(j)\,(\Delta_f G^\circ(H^+) - RT\ln(10)pH) \tag{1.3-14}$$

When species have electric charges their standard thermodynamic properties need to be adjusted for the ionic strength I according to the extended Debye-Huckel theory.

$$\Delta_f G_j'\,^\circ(I) \;=\; \Delta_f G_j^\circ(I{=}0)\; -\; N_H(j)\,RT\ln(10)\,pH \;-\; RT\alpha\left(z_j^2 - N_H(j)\right)I^{1/2}\big/\left(1 + 1.6\,I^{1/2}\right) \tag{1.3-15}$$

In the ionic strength term, z_j is the electric charge of species j, and $\alpha = 1.17582$ $\mathrm{kg}^{1/2}\,\mathrm{mol}^{-1/2}$ at 298.15 K. This equation makes it possible to produce tables of standard transformed Gibbs energies of formation $\Delta_f G_j'\,^\circ$ of species at specified temperature, pH and ionic strength.

Now we can consider the relation between the standard transformed Gibbs energies of formation of the species and the standard transformed Gibbs energy of formation of the reactant at a specified temperature, pH, and ionic strength. Because of the entropy of mixing the species, the standard transformed Gibbs energy of formation is more negative than any of the species. The standard transformed Gibbs energy of formation of a reactant $\Delta_f G_i'\,^\circ$ is given in terms of the standard transformed Gibbs energies of formation $\Delta_f G_j'\,^\circ$ of the species (pseudoisomers) by [11]

$$\Delta_f G_i'^\circ = -\,RT\ln\sum_{j=1}^{N}\exp(-\Delta_f G_j'\,^\circ/RT) \tag{1.3-16}$$

where N is the number of different species in the pseudoisomer group. This summation is a partition function. The same result can be obtained by taking the mole-fraction-weighted average of the standard transformed Gibbs energies of formation of the species and adding a term for the transformed Gibbs energy of mixing. This equation made it possible to produce tables of standard transformed Gibbs energies of biochemical reactants at specified temperature, pH, and ionic strength. The equilibrium mole fractions r_j of the species in the reactant at a specified pH are given by

$$r_j = \exp[(\Delta_f G_i'^\circ - \Delta_f G_j'^\circ)/RT] \tag{1.3-17}$$

When the pH is specified, equation 1.2-7 is replaced with

$$\Delta_r G'^\circ = -RT\ln K' = \sum_{i=1}^{N'} \nu_i' \Delta_f G_i'^\circ \tag{1.3-18}$$

This makes it possible to produce tables of standard transformed Gibbs energies of reaction and tables of apparent equilibrium constants K' for biochemical reactions for which $\Delta_f G_i'^\circ$ (see equation 1.3-16) are known for all the reactants. It is important to notice that this does not require that K' has been measured for this reaction. More detailed derivations of these equations are given in references [20] and [23].

When the pH has been specified, the relation $N = C + R$ of chemical thermodynamics no longer applies; see the end of Section 1.2. It is replaced with $N' = C' + R'$, where N' is the number of reactants (sums of species), C' is the number of components, excluding hydrogen, and R' is the number of independent biochemical reactions [20,23]. The equilibrium composition of a system of biochemical reactions can be calculated using an independent set of biochemical reactions. A set of reactions is independent if no reaction in the set can be obtained by adding or subtracting reactions in the set. This relation for determining the number R' of independent reactions will be used often in this book.

1.4 Calculation of the change in binding of hydrogen ions in a biochemical reaction at a specified pH by taking the derivative of logK' or $\Delta_r G'^\circ$ with respect to pH

The fourth term in equation 1.3-5 shows how the transformed Gibbs energy of a biochemical reaction system with one reaction changes with the pH, but there is no direct way to determine G' for a reaction system. However, the terms on the right hand side of equation 1.3-5 are related by Maxwell equations. The derivative of $\Delta_r G'$ with respect to pH is equal to the derivative of $RT\ln(10)n_c(H)$ with respect to the extent of reaction ξ'.

$$\partial\Delta_r G'/\partial pH = \partial RT\ln(10)n_c(H)/\partial\xi' = RT\ln(10)\partial n_c(H)/\partial\xi' \tag{1.4-1}$$

Because of equation 1.3-9, the derivative $\partial\Delta_r G'/\partial pH$ is the same as $\partial\Delta_r G'^\circ/\partial pH$. Since $\Delta_r G'^\circ = -RT\ln K'$, equation 1.4-1 can be written as

$$\partial\ln K'/\partial pH = -\ln(10)\partial n_c(H)/\partial\xi' \tag{1.4-2}$$

This shows that when the apparent equilibrium constant depends on the pH, the amount of hydrogen atoms in the system changes when the reaction occurs. $\partial n_c(H)/\partial\xi'$ is the increase in the binding of hydrogen ions, and so it is represented by $\Delta_r N_H$.

$$\Delta_r N_H = -\partial\log K'/\partial pH \tag{1.4-3}$$

When the apparent equilibrium constant increases with the pH, $\Delta_r N_H$ is negative, and the amount of hydrogen atoms in the system decreases as the reaction occurs. This means that the reaction produces hydrogen ions. When the apparent equilibrium constant decreases with the pH, $\Delta_r N_H$ is positive, and the amount of hydrogen atoms in the system increases as the reaction occurs. This means that the reaction consumes hydrogen ions. This is an example of Le Chatelier's principle that if the conditions are changed on a system at equilibrium it will shift in the direction to oppose the change.

The change in binding of hydrogen ions in a biochemical reaction can also be calculated from the acid dissociation constants of the reactants using

$$\Delta_r N_H = \sum \nu_i' \overline{N}_H(i) \qquad\qquad (1.4\text{-}4)$$

where $\overline{N}_H(i)$ is the average number of hydrogen atoms bound by the species of reactant i. $\Delta_r N_H$ can be calculated for almost all enzyme-catalyzed reactions in the EC list [Web2] because the pKs of reactants can generally be estimated, and the number of hydrogen atoms in species are known from their structures [26].

When we get to kinetics, we will see that $\Delta_r N_H$ of thermodynamics is related to the consumption n of hydrogen ions in the rate-determining reaction [27]. When it is necessary to use the steady-state rate equation the relation between the consumption of hydrogen ions and the velocity is more complicated.

1.5 Calculation of the standard transformed Gibbs energies of formation of biochemical reactants

To calculate the standard transformed Gibbs energy of a biochemical reactant $\Delta_f G_i'^\circ$, it is necessary to use equation 1.3-15 to calculate the standard transformed Gibbs energy of formation of each species at the desired temperature, pH, and ionic strength. The standard transformed Gibbs energy of formation of a reactant is not a mole fraction average of the $\Delta_f G_j'^\circ$ of the species that make up the reactant because of the entropy of mixing. Equation 1.3-16 has to be used to calculate the standard transformed Gibbs energy of the sum of species (the reactant) at the desired temperature, pH and ionic strength. The database BasicBiochemData3 [*MathSource*3] contains $\{\Delta_f G^\circ, \Delta_f H^\circ, z_j, N_H\}$ for each species of 199 biochemical reactants. The energies are in kJ mol^{-1} at 298.15 K. In order to demonstrate the calculation of standard transformed Gibbs energies of reactants and the calculation of the apparent equilibrium constants of biochemical reactions, the species properties of the species of inorganic phosphate, ATP, ADP, and H_2O from BasicBiochemData3 are given here

The species properties for the hydrolysis of ATP are given by [14]:

```
pisp = {{-1096.1, -1299., -2, 1}, {-1137.3, -1302.6, -1, 2}};

atpsp =
   {{-2768.1, -3619.21, -4, 12}, {-2811.48, -3612.91, -3, 13}, {-2838.18, -3627.91, -2, 14}};

adpsp =
   {{-1906.13, -2626.54, -3, 12}, {-1947.1, -2620.94, -2, 13}, {-1971.98, -2638.54, -1, 14}};

h2osp = {{-237.19, -285.83, 0, 2}};
```

The following program derives the expression for the standard transformed Gibbs energy of formation of a reactant at 298.15 K.

```
calcdGmat[speciesmat_] :=
Module[{dGzero,dHzero, zi, nH, pHterm, isterm,gpfnsp},(*This program derives the
function of pH and ionic strength (is) that gives the standard transformed Gibbs
energy of formation of a reactant (sum of species) at 298.15 K.  The input speciesmat
is a matrix that gives the standard Gibbs energy of formation, the standard enthalpy
of formation, the electric charge, and the number of hydrogen atoms in each species.
There is a row in the matrix for each species of the reactant. gpfnsp is a list of
the functions for the species.  Energies are expressed in kJ mol^-1.*)
{dGzero,dHzero,zi,nH}=Transpose[speciesmat];
pHterm = nH*8.31451*.29815*Log[10^-pH];
isterm = 2.91482*((zi^2) - nH)*(is^.5)/(1 + 1.6*is^.5);
gpfnsp=dGzero - pHterm - isterm;
-8.31451*.29815*Log[Apply[Plus,Exp[-1*gpfnsp/(8.31451*.29815)]]]]]]
```

The function that expresses the standard transformed Gibbs energy of formation of inorganic phosphate at 298.15 K in the range pH 5 to 9 and ionic strength is calculated as follows:

piG = calcdGmat[pisp]

$$-2.47897 \, \text{Log}\left[e^{-0.403393\left(-1137.3+\frac{2.91482 \, is^{0.5}}{1+1.6 \, is^{0.5}}-4.95794 \, \text{Log}\left[10^{-pH}\right]\right)} + e^{-0.403393\left(-1096.1-\frac{8.74446 \, is^{0.5}}{1+1.6 \, is^{0.5}}-2.47897 \, \text{Log}\left[10^{-pH}\right]\right)}\right]$$

Table 1.1 Standard transformed Gibbs energies of formation $\Delta_f \, G'^o$ of inorganic phosphate in kJ mol^{-1} at 298.15 K.

TableForm[piG /. is → {0, .1, .25} /. pH → {5, 6, 7, 8, 9}, TableHeadings →
{{"I=0", "I=0.10", "I=0.25"}, {" pH 5", " pH 6", " pH 7", " pH 8", " pH 9"}}]

	pH 5	pH 6	pH 7	pH 8	pH 9
I=0	−1080.23	−1068.95	−1058.56	−1050.81	−1044.77
I=0.10	−1079.65	−1068.56	−1059.17	−1052.42	−1046.58
I=0.25	−1079.46	−1068.49	−1059.49	−1052.97	−1047.17

The function that expresses the standard transformed Gibbs energy of formation of ATP in the range pH 5 to 9 is calculated as follows:

atpG = calcdGmat[atpsp]

$$-2.47897 \, \text{Log}\left[e^{-0.403393\left(-2838.18+\frac{29.1482 \, is^{0.5}}{1+1.6 \, is^{0.5}}-34.7056 \, \text{Log}\left[10^{-pH}\right]\right)} + \right.$$
$$\left. e^{-0.403393\left(-2811.48+\frac{11.6593 \, is^{0.5}}{1+1.6 \, is^{0.5}}-32.2266 \, \text{Log}\left[10^{-pH}\right]\right)} + e^{-0.403393\left(-2768.1-\frac{11.6593 \, is^{0.5}}{1+1.6 \, is^{0.5}}-29.7477 \, \text{Log}\left[10^{-pH}\right]\right)}\right]$$

Table 1.2 Standard transformed Gibbs energies of formation $\Delta_f \, G'^o$ of ATP in kJ mol^{-1} at 298.15 K.

TableForm[atpG /. is → {0, .1, .25} /. pH → {5, 6, 7, 8, 9}, TableHeadings →
{{"I=0", "I=0.10", "I=0.25"}, {" pH 5", " pH 6", " pH 7", " pH 8", " pH 9"}}]

	pH 5	pH 6	pH 7	pH 8	pH 9
I=0	−2441.43	−2366.43	−2292.61	−2220.96	−2151.73
I=0.10	−2438.3	−2364.24	−2292.16	−2222.71	−2154.09
I=0.25	−2437.46	−2363.76	−2292.5	−2223.44	−2154.88

The function that expresses the standard transformed Gibbs energy of formation of ADP in the range pH 5 to 9 is calculated as follows:

adpG = calcdGmat[adpsp]

$$-2.47897 \, \text{Log}\left[e^{-0.403393\left(-1971.98+\frac{37.8927 \, is^{0.5}}{1+1.6 \, is^{0.5}}-34.7056 \, \text{Log}\left[10^{-pH}\right]\right)} + \right.$$
$$\left. e^{-0.403393\left(-1947.1+\frac{26.2334 \, is^{0.5}}{1+1.6 \, is^{0.5}}-32.2266 \, \text{Log}\left[10^{-pH}\right]\right)} + e^{-0.403393\left(-1906.13+\frac{8.74446 \, is^{0.5}}{1+1.6 \, is^{0.5}}-29.7477 \, \text{Log}\left[10^{-pH}\right]\right)}\right]$$

Table 1.3 Standard transformed Gibbs energies of formation $\Delta_f \, G'^o$ of ADP in kJ mol^{-1} at 298.15 K.

```
TableForm[adpG /. is → {0, .1, .25} /. pH → {5, 6, 7, 8, 9}, TableHeadings →
  {{"I=0", "I=0.10", "I=0.25"}, {"  pH 5", "  pH 6", "  pH 7", "  pH 8", "  pH 9"}}]
```

	pH 5	pH 6	pH 7	pH 8	pH 9
I=0	-1576.6	-1502.08	-1428.93	-1358.51	-1289.7
I=0.10	-1570.84	-1497.02	-1425.55	-1356.41	-1287.83
I=0.25	-1569.05	-1495.55	-1424.7	-1355.78	-1287.24

The function that expresses the standard transformed Gibbs energy of formation of H_2O in the range pH 5 to 9 is calculated as follows:

```
h2oG = calcdGmat[h2osp]
```

$$-2.47897 \, \text{Log}\left[e^{-0.403393\left(-237.19+\frac{5.82964\,is^{0.5}}{1+1.6\,is^{0.5}}-4.95794\,\text{Log}[10^{-pH}]\right)}\right]$$

Table 1.4 Standard transformed Gibbs energies of formation $\Delta_f G'^\circ$ of H_2O in kJ mol^{-1} at 298.15 K.

```
TableForm[h2oG /. is → {0, .1, .25} /. pH → {5, 6, 7, 8, 9}, TableHeadings →
  {{"I=0", "I=0.10", "I=0.25"}, {"  pH 5", "  pH 6", "  pH 7", "  pH 8", "  pH 9"}}]
```

	pH 5	pH 6	pH 7	pH 8	pH 9
I=0	-180.11	-168.693	-157.277	-145.861	-134.445
I=0.10	-178.885	-167.469	-156.053	-144.637	-133.221
I=0.25	-178.49	-167.074	-155.658	-144.242	-132.826

1.6 Calculation of the standard transformed Gibbs energy of hydrolysis of ATP to ADP and the apparent equilibrium constant K'

The biochemical reaction is

$$ATP + H_2O = ADP + P_i \tag{1.6-1}$$

For the hydrolysis of ATP, the change in standard transformed Gibbs energy of reaction is given by

$$\Delta_r G'^\circ = \Delta_f G'^\circ(ADP) + \Delta_f G'^\circ(P_i) - \Delta_f G'^\circ(ATP) - \Delta_f G'^\circ(H_2O) \tag{1.6-2}$$

The apparent equilibrium constant K' is given by

$$K' = \exp(-\Delta_r G'^\circ/RT) = \frac{[ADP][Pi]}{[ATP]} \tag{1.6-3}$$

The concentration of H_2O is omitted in the expression of the apparent equilibrium constant, but $\Delta_f G'^\circ(H_2O)$ is needed to calculate K' at the desired temperature, pH, and ionic strength. The biochemical reactant ATP is made up of three species (ATP^{4-}, $HATP^{3-}$, H_2ATP^{2-}) in the pH range 5 to 9.

The following program calculates the function of pH and ionic strength that yields the change in a standard thermodynamic property for a biochemical reaction at 298.15 K [23].

```
deriverxfn[eq_] :=
  Module[{function}, (*Derives the function of pH and ionic strength that gives the
     thermodynamic properties of a biochemical reaction typed in the form atpG+h2oG+de==
    adpG+piG.  Other suffixes can be used for H, S, and NH.*)
  function = Solve[eq, de]; function[[1, 1, 2]]]
```

```
atphydfnG = deriverxfn[atpG + h2oG + de == adpG + piG]
```

$$2.47897 \, \mathrm{Log}\left[e^{-0.403393\left(-237.19+\frac{5.82964\,\mathrm{is}^{0.5}}{1.+1.6\,\mathrm{is}^{0.5}}-4.95794\,\mathrm{Log}\left[10.^{-1.\,\mathrm{pH}}\right]\right)}\right] +$$

$$2.47897 \, \mathrm{Log}\left[e^{-0.403393\left(-2838.18+\frac{29.1482\,\mathrm{is}^{0.5}}{1.+1.6\,\mathrm{is}^{0.5}}-34.7056\,\mathrm{Log}\left[10.^{-1.\,\mathrm{pH}}\right]\right)} +$$

$$e^{-0.403393\left(-2811.48+\frac{11.6593\,\mathrm{is}^{0.5}}{1.+1.6\,\mathrm{is}^{0.5}}-32.2266\,\mathrm{Log}\left[10.^{-1.\,\mathrm{pH}}\right]\right)} + e^{-0.403393\left(-2768.1-\frac{11.6593\,\mathrm{is}^{0.5}}{1.+1.6\,\mathrm{is}^{0.5}}-29.7477\,\mathrm{Log}\left[10.^{-1.\,\mathrm{pH}}\right]\right)}\right] -$$

$$2.47897 \, \mathrm{Log}\left[e^{-0.403393\left(-1971.98+\frac{37.8927\,\mathrm{is}^{0.5}}{1.+1.6\,\mathrm{is}^{0.5}}-34.7056\,\mathrm{Log}\left[10.^{-1.\,\mathrm{pH}}\right]\right)} + e^{-0.403393\left(-1947.1+\frac{26.2334\,\mathrm{is}^{0.5}}{1.+1.6\,\mathrm{is}^{0.5}}-32.2266\,\mathrm{Log}\left[10.^{-1.\,\mathrm{pH}}\right]\right)} +$$

$$e^{-0.403393\left(-1906.13+\frac{8.74446\,\mathrm{is}^{0.5}}{1.+1.6\,\mathrm{is}^{0.5}}-29.7477\,\mathrm{Log}\left[10.^{-1.\,\mathrm{pH}}\right]\right)}\right] -$$

$$2.47897 \, \mathrm{Log}\left[e^{-0.403393\left(-1137.3+\frac{2.91482\,\mathrm{is}^{0.5}}{1.+1.6\,\mathrm{is}^{0.5}}-4.95794\,\mathrm{Log}\left[10.^{-1.\,\mathrm{pH}}\right]\right)} + e^{-0.403393\left(-1096.1-\frac{8.74446\,\mathrm{is}^{0.5}}{1.+1.6\,\mathrm{is}^{0.5}}-2.47897\,\mathrm{Log}\left[10.^{-1.\,\mathrm{pH}}\right]\right)}\right]$$

This function of the pH and ionic strength can be used to make tables or plots of $\Delta_r G'^{\circ}$ for the hydrolysis of ATP at 298.15 K.

Table 1.5 Standard transformed Gibbs energies of reaction in kJ mol^{-1} for ATP + H$_2$O = ADP + P$_i$ at 298.15 K

```
PaddedForm[TableForm[atphydfnG /. is → {0, .1, .25} /. pH → {5, 6, 7, 8, 9}, TableHeadings →
   {{"I=0", "I=0.10", "I=0.25"}, {"  pH 5", "  pH 6", "  pH 7", "  pH 8", "  pH 9"}}], {4, 2}]
```

	pH 5	pH 6	pH 7	pH 8	pH 9
I=0	−35.30	−35.91	−37.60	−42.50	−48.29
I=0.10	−33.30	−33.87	−36.50	−41.48	−47.10
I=0.25	−32.56	−33.22	−36.04	−41.07	−46.70

This calculation can be checked by using the values $\Delta_r G'^{\circ}$ at pH 7 and 0.25 M ionic strength for the four reactants.

```
-1059.49 - 1424.7 + 155.66 + 2292.5
```

```
-36.03
```

The corresponding apparent equilibrium constants are calculated as follows:

```
kprimeATPhyd = Exp[-atphydfnG / (8.31451 * .29815)]
```

$$e^{0.403393\left(-2.47897\,\mathrm{Log}\left[e^{-0.403393\left(-237.19+\frac{5.82964\,\mathrm{is}^{0.5}}{1.+1.6\,\mathrm{is}^{0.5}}-4.95794\,\mathrm{Log}\left[10.^{-1.\,\mathrm{pH}}\right]\right)}\right]-2.47897\,\mathrm{Log}\left[e^{-0.403393\left(-2838.18+\frac{29.1482\,\mathrm{is}^{0.5}}{1.+1.6\,\mathrm{is}^{0.5}}-34.7056\,\mathrm{Log}\left[10.^{-1.\,\mathrm{pH}}\right]\right)}+e^{-0.403393\left(-2811.48\right.}\right.}$$

Table 1.6 Apparent equilibrium constants K' at 298.15 K

```
PaddedForm[TableForm[kprimeATPhyd /. is → {0, .1, .25} /. pH → {5, 6, 7, 8, 9},
   TableHeadings → {{"I=0", "I=0.10", "I=0.25"},
      {"   pH 5", "   pH 6", "   pH 7", "   pH 8", "   pH 9"}}]]
```

	pH 5	pH 6	pH 7	pH 8	pH 9
I=0	1.52785×10^6	1.95781×10^6	3.87287×10^6	2.79036×10^7	2.88781×10^8
I=0.10	680808.	859076.	2.48023×10^6	1.8478×10^7	1.78325×10^8
I=0.25	506774.	659585.	2.05626×10^6	1.5698×10^7	1.51989×10^8

The apparent equilibrium constants for about 200 enzyme-catalyzed reactions are given in Chapter 12 of reference [23] and in BasicBiochemData3 [*MathSource*3]. $\Delta_r H'^{\circ}$ and $\Delta_r S'^{\circ}$ can also be calculated as functions of pH and ionic strength.

1.7 Calculation of the change in the binding of hydrogen ions $\Delta_r N_H$ in ATP + H_2O = ADP + P_i at 298.15 K

Equation 1.4-3 shows that the change in binding of hydrogen ions in the hydrolysis of ATP to ADP can be calculated using $\Delta_r N_H = -\partial \log[\text{kprimeATPhyd}]/\partial \text{pH}$.

Table 1.7 Change in the binding of hydrogen ions in ATP + H_2O = ADP + P_i at 298.15 K

```
PaddedForm[TableForm[
  -D[Log[10, kprimeATPhyd], pH] /. is → {0, .1, .25} /. pH → {5, 6, 7, 8, 9}, TableHeadings →
  {{"I=0", "I=0.10", "I=0.25"}, {"  pH 5", "  pH 6", "  pH 7", "  pH 8", "  pH 9"}}], {4, 2}]
```

	pH 5	pH 6	pH 7	pH 8	pH 9
I=0	−0.15	−0.12	−0.58	−1.01	−1.01
I=0.10	−0.05	−0.22	−0.72	−0.96	−1.00
I=0.25	−0.04	−0.25	−0.74	−0.96	−1.00

In Biochemical Thermodynamics: Applications of *Mathematica* [23] $\Delta_r N_H$ is calculated at 298.15 K and 0.25 M ionic strength for 229 enzyme-catalyzed reactions.

1.8 Calculation of the change in the binding of hydrogen ions in a biochemical reaction at a specified pH without information on standard Gibbs energies of formation of species

The calculation of $\Delta_r N_H$ without knowing the standard Gibbs energies of formation of the species in a reaction is a two-step process. First, the average number of hydrogen ions $\overline{N}_H$ in various reactants are calculated. Second, these average values are added and subtracted to obtain $\Delta_r N_H$ for various biochemical reactions. The calculation of the average binding of hydrogen ions by ATP as a function of pH and ionic strength is discussed on p. 29-32 in Biochemical Thermodynamics: Applications of *Mathematica* [23]. Since ATP is made up of three species in the range pH 5 to 9, its concentration is given by

$$[ATP] = [ATP^{4-}] + [HATP^{3-}] + [H_2 ATP^{2-}] \tag{1.8-1}$$

When the acid dissociations are at equilibrium, substituting the expressions for the two acid dissociation constants yields

$$[ATP] = [ATP^{4-}](1 + 10^{pK1ATP-pH} + 10^{pK1ATP+pK2ATP-2\,pH}) = [ATP^{4-}]*p \tag{1.8-2}$$

where $pK_{1\,ATP} = -\log K_{1\,ATP}$ is the highest pK that has to be considered in the range pH 5 to 9. The factor multiplying $\left[ATP^{4-}\right]$ is referred to as a binding polynomial and is represented here by p. Equation 1.8-2 shows that the equilibrium mole fraction r_1 for $\left[ATP^{4-}\right]$ is given by

$$r_1 = 1/p \tag{1.8-3}$$

The other two equilibrium mole fractions are given by

$$r_2 = 10^{pK1ATP-pH}/p \tag{1.8-4}$$

$$r_3 = 10^{pK1ATP+pK2ATP-2\,pH}/p \tag{1.8-5}$$

The pKs are at the desired temperature and ionic strength. The following calculations of the equilibrium concentrations of the species of ATP are for 298.15 K and 0.25 M ionic strength.

```
patp = 1 + 10^(pK1ATP - pH) + 10^(pK1ATP + pK2ATP - 2 * pH);
```

```
pK1ATP = 6.47;

pK2ATP = 3.83;
```

The equilibrium mole fractions of the three species of ATP are given by

```
r1 = 1 / patp;

r2 = 10 ^ (pK1ATP - pH) / patp;

r3 = 10 ^ (pK1ATP + pK2ATP - 2 * pH) / patp;
```

ATP^{4-} has 12 hydrogen atoms. $HATP^{3-}$ has 13 hydrogen atoms. H_2ATP^{2-} has 14 hydrogen atoms. The average number of hydrogen atoms in a biochemical reactant is given by

$$\overline{N}_H (i) = \sum r_j N_H(j) \tag{1.8-6}$$

The average number of hydrogen atoms in ATP is given by

```
nHATP = 12 * r1 + 13 * r2 + 14 * r3;

Plot[{nHATP}, {pH, 3, 9}, AxesLabel → {"pH", "N̄_H"}]
```

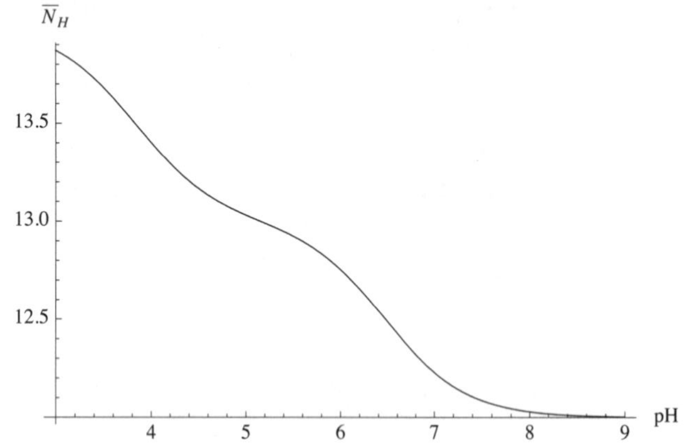

Fig. 1.1 Titration curve for ATP at 298.15 K and ionic strength 0.25 M.

In contrast with the way titration curves are usually plotted, this plot gives the total number of hydrogen atoms.

The following program is used to derive the function of pH and ionic strength that will yield the average number of hydrogen atoms in a reactant with two dissociable hydrogen atoms.

```
calcavnoH[pK1_, pK2_, nHbasicform_] :=
  Module[{p, r1, r2, r3}, (*This program derives the function of pH that gives the average
     number of hydrogen atoms in a reactant with three species.  The pKs are for the
     desired temperature and ionic strength.  nHbasicform is the number of hydrogen
     atoms in the species that dominates at the highest pH.  If there is a single pK,
     pK2 can be set equal to zero.  When there is a single species in the range pH 5 to 9,
     both pK1 and pK2 are set equal to zero.*)
    p = 1 + 10 ^ (pK1 - pH) + 10 ^ (pK1 + pK2 - 2 * pH);
    r1 = 1 / p;
  r2 = 10 ^ (pK1 - pH) / p;
  r3 = 10 ^ (pK1 + pK2 - 2 * pH) / p;
    nHbasicform * r1 + (nHbasicform + 1) * r2 + (nHbasicform + 2) * r3];
```

The function of pH that yields $\overline{N}_H$(ATP) is obtained as follows:

```
atpnhfun = calcavnoH[6.47, 3.83, 12]
```

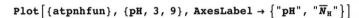

$$\frac{12}{1 + 10^{10.3-2\,pH} + 10^{6.47-pH}} + \frac{7\ 2^{11.3-2\,pH}\ 5^{10.3-2\,pH}}{1 + 10^{10.3-2\,pH} + 10^{6.47-pH}} + \frac{13\ 10^{6.47-pH}}{1 + 10^{10.3-2\,pH} + 10^{6.47-pH}}$$

Note that the function for ATP is named atpnhfun because the name atpNH is already used in BasicBiochemData3.nb [*MathSource*3] for the average number of hydrogen atoms in ATP calculated from its standard transformed Gibbs energy of formation.

```
Plot[{atpnhfun}, {pH, 3, 9}, AxesLabel → {"pH", "N̄_H"}]
```

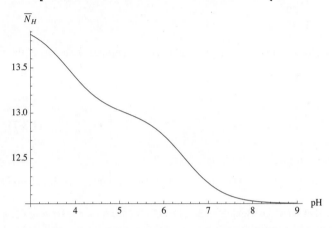

Fig. 1.2 This is the same plot as in Fig. 1.1 that was obtained without using the program calcavnoH.

The function of pH that yields $\overline{N}_H(ADP)$ is obtained as follows:

```
adpnhfun = calcavnoH[6.33, 3.79, 12];
```

The function of pH that yields $\overline{N}_H(AMP)$ is obtained as follows:

```
ampnhfun = calcavnoH[6.16, 3.71, 12];
```

These pKs in the ATP series were published by Alberty and Goldberg [14] in 1992.

The function of pH that yields $\overline{N}_H\ (P_i)$ is obtained as follows:

```
pinhfun = calcavnoH[6.65, 0, 1];
```

The second pK is set to zero to take it out of the calculation.

The function of pH that yields $\overline{N}_H\ (adenosine)$ is obtained as follows:

```
adenosinenhfun = calcavnoH[3.47, 0, 13];
```

The function of pH that yields $\overline{N}_H\ (H_2O)$ is obtained as follows:

```
h2onhfun = calcavnoH[0, 0, 2];
```

Since water does not have a pK in the range pH 5 to 9, both pK1 and pK2 are set equal to zero.

The average numbers of hydrogen atoms in these six reactants are summarized in Table 1.8.1.

```
listnames = {"ATP", "ADP", "AMP", "Adenosine", "Pi", "H₂O"};
```

Table 1.8 Average numbers of hydrogen atoms $\overline{N}_H$ in six reactants at 298.15 K and 0.25 M ionic strength

```
PaddedForm[TableForm[{atpnhfun, adpnhfun, ampnhfun, adenosinenhfun, pinhfun, h2onhfun} /.
    pH → {5, 6, 7, 8, 9}, TableHeadings →
    {listnames, {"  pH 5", "  pH 6", "  pH 7", "  pH 8", "  pH 9"}}] // N, {4, 2}]
```

	pH 5	pH 6	pH 7	pH 8	pH 9
ATP	13.03	12.75	12.23	12.03	12.00
ADP	13.01	12.69	12.18	12.02	12.00
AMP	12.98	12.60	12.13	12.01	12.00
Adenosine	13.03	13.00	13.00	13.00	13.00
Pi	1.98	1.82	1.31	1.04	1.00
H_2O	2.00	2.00	2.00	2.00	2.00

These numbers can be added and subtracted to obtain $\Delta_r N_H$ for a number of enzyme-catalyzed reactions, as shown in the next section.

The change in binding of hydrogen ions is given by

$$\Delta_r N_H = \sum \nu_i ' \overline{N}_H \text{ (i)}$$

(1.8-7)

where ν_i ' is the stoichiometric number for reactant i.

This type of equation applies to $\Delta_r G$ '°, $\Delta_r H$ '°, and $\Delta_r S$ '°. Therefore, correct stoichiometry for biochemical equations is very important.

It is convenient to rewrite the program trGibbsRxSummary (in BasicBiochemData3.nb [*MathSource*3]) to calculate $\Delta_r N_H$ for reactions that are typed in. The program chgNHbindSummary does not require information on the standard Gibbs energies of formation of species. The short program "round" rounds changes to two digits. The names of reactants are those used in *Mathematica* where names have to start with lower case letters and connot contain dashes, spaces or periods. The scientific names of reactants are given on p. 425 of Biochemical Thermodynamics: Applications of *Mathematica* [23].

```
round[vec_, params_:{6, 2}] :=(*When a list of numbers has more digits to the
right of the decimal point than you want, say 6, you can request 2 by using
round[vec,{6,2}],*)
  Flatten[Map[NumberForm[#1, params] & , {vec}, {2}]]

chgNHbindSummary[eq_, title_, reaction_, pHlist_] :=
  Module[{functiom, vectorNH}, (*When this program is given the equation for a biochemical
      reaction in the form acetaldehydenhfun+nadrednhfun+de==ethanolnhfun+nadoxnhfun,
    it calculates the change in binding of hydrogen ions in the reaction.  The
    temperature and ionic strength are the temperature and ionic strength for the
    pKs.  title_ is in the form "EC 1.1.1.1 Alcohol dehydrogenase".  reaction_
    is in the form "acetaldehyde+nadred=ethanol+nadox".*)
    function = Solve[eq, de];
    vectorNH = round[function[[1, 1, 2]] /. pH -> pHlist, {4, 2}];
    Print[title]; Print[reaction]; Print[vectorNH]]
```

The following calculations give $\Delta_r N_H$ at pHs 5, 6, 7, 8, and 9 at 298.15 K and 0.25 M ionic strength:

```
chgNHbindSummary[atpnhfun + h2onhfun + de == adpnhfun + pinhfun,
  "EC 3.6.1.3 Adenosinetriphosphatase", "atp+h2o=adp+pi", {5, 6, 7, 8, 9}]
```

EC 3.6.1.3 Adenosinetriphosphatase

atp+h2o=adp+pi

{-0.04, -0.25, -0.74, -0.96, -1.00}

This agrees with the preceding section, but it was obtained without information on Gibbs energies of formation of species.

1.9 Another way to consider the pH dependence of the apparent equilibrium constant for ATP + H$_2$O = ADP + Pi

It has been known for some time [16] that the pH dependence of the apparent equilibrium can be represented by

$$K' = \frac{[ADP][Pi]}{[ATP]} = K_{ref}10^{npH}f(pH) \tag{1.9-1}$$

where K_{ref} is the chemical equilibrium constant for a reference chemical reaction and n (positive integer if H$^+$ is produced and negative integer if H$^+$ is consumed) is the number of hydrogen ions in the reference reaction. The function $f(pH)$ introduces the effects of the pKs of the reactants. For the hydrolysis of ATP to ADP, the reference reaction can be taken to be

$$ATP^{4-} + H_2O = ADP^{3-} + HPO_4^{2-} + H^+ \qquad K_{ref1} = \frac{[ADP^{3-}][HPO_4^{2-}][H^+]}{[ATP^{4-}]} \tag{1.9-2}$$

When this is the reference reaction, the apparent equilibrium constant is given by $K' = K_{ref1}10^{pH}f(pH)$.

If the reference reaction is taken to be

$$ATP^{4-} + H_2O = ADP^{3-} + H_2PO_4^- \qquad K_{ref2} = \frac{[ADP^{3-}][H_2PO_4^{2-}]}{[ATP^{4-}]} \tag{1.9-3}$$

The pH dependence of K' is given by $K' = K_{ref2}f(pH)$. In biochemical thermodynamics the choice of reference reaction is arbitrary, but in rapid-equilibrium enzyme kinetics the rate is proportional to 10^{npH} where n is -1, -2, ... when one, two, ... hydrogen ions are consumed in the rate-determining reaction [27]. The following sections show how to calculate the distributions of species in Pi, ATP, and ADP, and how the apparent equilibrium constant for ATP + H$_2$O = ADP + Pi depends on pH.

■ **1.9.1 Distributions of species of inorganic phosphate as functions of pH**

Inorganic phosphate consists of 2 species in the range pH 5-9. The pK is pKpi = 6.65 at 298.15 K and ionic strength 0.25 M. The fractions of the two species in total phosphate as a function of pH can be calculated by hand, but it is convenient to use *Mathematica,* which is needed when there are more dissociable hydrogen ions. The fraction of a species can be calculated by using Solve[eqs,vars,elims] to derive the function of hydrogen ion concentration h that yields the equilibrium concentration for each species. In using Solve it is necesssary to use the hydrogen ion concentration, rather than the pH, because Solve was primarily developed to solve sets of simultaneous polynomial equations. In *Mathematica*, the initial letter for a property has to be a lower case letter, and so the following abbreviations are used to represent concentrations:

$[HPO_4^{2-}]$ = pi

$[H_2PO_4^-]$ = hpi

$[HPO_4^{2-}] + [H_2PO_4^-]$ = pi + hpi = pit

$[ATP^{4-}]$ = atp

$[HATP^{3-}]$ = hatp

$[H_2 ATP^{3-}]$ = h2atp

$[ATP^{4-}] + [HATP^{3-}] + [H_2 ATP^{2-}]$ = atp + hatp + h2atp = atpt

$[ADP^{3-}]$ = adp

$[HADP^{2-}]$ = hadp

$[H_2 ADP^-]$ = h2adp

$[ADP^{3-}] + [HADP^{2-}] + [H_2 ADP^-]$ = adp + hadp + h2adp = adpt

Calculation of the fraction of [Pi] that is [HPO$_4^{2-}$]

The dissociation constant K_1 for $H_2PO_4^- = H^+ + HPO_4^{2-}$ is represented in *Mathematica* with k1pi = h×pi/hpi. The following calculation yields the function of h for the equilibrium concentration of HPO_4^{2-}. The total concentration of inorganic phosphate $[Pi]_t$ is represented by pit.

```
Solve[{k1pi == h * pi / hpi, pit == pi + hpi}, {pi}, {hpi}]
```

$$\left\{\left\{pi \rightarrow \frac{k1pi\ pit}{h + k1pi}\right\}\right\}$$

The fraction pi/pit is

$$\frac{k1pi}{h + k1pi};$$

In calculating equilibrium concentrations of species, it is more convenient to use pHs and pKs. The ReplaceAll operation (/.x->) is used to make this change.

$$\frac{k1pi}{h + k1pi}\ /.\ h \rightarrow 10^{\char`\^}-pH\ /.\ k1pi \rightarrow 10^{\char`\^}-pKpi$$

$$\frac{10^{-pKpi}}{10^{-pH} + 10^{-pKpi}}$$

This is the expression for $[HPO_4^{2-}]$/[Pi]. To calculate this fraction as a function of pH at 298.15 K and 0.25 M ionic strength, pKpi is taken to be 6.65 [23].

```
  k1pi
————————  /. h → 10^-pH /. k1pi → 10^-pKpi /. pKpi → 6.65
h + k1pi
```

$$\frac{2.23872 \times 10^{-7}}{2.23872 \times 10^{-7} + 10^{-pH}}$$

Plot the fraction pi/pit as a function of pH.

```
plot1 = Plot[  2.2387211385683377`*^-7
             ——————————————————————————————————, {pH, 5, 9}, AxesLabel → {"pH", "[HPO₄²⁻]/[Pi]"}];
               2.2387211385683377`*^-7 + 10^-pH
```

This plot is in composite Fig. 1.3.

The reciprocal of the fraction $[\text{HPO}_4^{2-}]/[\text{Pi}]$ is given by

$$\frac{10^{-pH} + 10^{-pKpi}}{10^{-pKpi}} = 1 + 10^{-pH + pKpi} = [\text{Pi}]/[\text{HPO}_4^{2-}]$$

Thus $[\text{Pi}] = (1 + 10^{-pH + pKpi})[\text{HPO}_4^{2-}]$.

$[\text{Pi}]$ is of course always greater than $[\text{HPO}_4^{2-}]$. This equation will be used in Section 1.9.5.

Calculation of the fraction of [Pi] that is [H₂PO₄⁻]

The fraction of inorganic phosphate in the form $[\text{H}_2\text{PO}_4^-]$ is calculated as a function of pH as follows:

```
Solve[{k1pi == h * pi / hpi, pit == pi + hpi}, {hpi}, {pi}]
```

$$\left\{ \left\{ hpi \rightarrow \frac{h\,pit}{h + k1pi} \right\} \right\}$$

The fraction hpi/pit is

```
    h
————————;
h + k1pi
```

It is more convienent to use pHs and pKs.

```
    h
————————  /. h → 10^-pH /. k1pi → 10^-pKpi
h + k1pi
```

$$\frac{10^{-pH}}{10^{-pH} + 10^{-pKpi}}$$

This is the expression for $[\text{H}_2\text{PO}_4^-]/[\text{Pi}]$.

To calculate the fraction of H_2PO_4^- as a function of pH at 298.15 K and 0.25 M ionic strength, pKpi is taken to be 6.65.

```
    h
————————  /. h → 10^-pH /. k1pi → 10^-pKpi /. pKpi → 6.65
h + k1pi
```

$$\frac{10^{-pH}}{2.23872 \times 10^{-7} + 10^{-pH}}$$

The reciprocal $[\text{Pi}]/[\text{H}_2\text{PO}_4^-]$ is

$$1 \Big/ \frac{10^{-pH}}{10^{-pH} + 10^{-pKpi}}$$

$$10^{pH} \left(10^{-pH} + 10^{-pKpi} \right)$$

$$\texttt{Simplify}\left[1 \Big/ \frac{10^{-pH}}{10^{-pH} + 10^{-pKpi}} \right]$$

$$1 + 10^{pH-pKpi}$$

This is the expression for $[Pi]/[H_2PO_4^-]$, which is needed in Section 1.9.5. Notice that
$[Pi] = 10^{pH}(10^{-pH} + 10^{-pKi})[H_2PO_4^-] = \left(1 + 10^{pH-pKpi}\right)[H_2 PO_4^-]$
This shows that $[Pi] > [H_2 PO_4^-]$.

Plot the fraction hpi/pit as a function of pH.

$$\texttt{plot2} = \texttt{Plot}\left[\frac{10^{-pH}}{2.2387211385683377`*^-7 + 10^{-pH}}, \{pH, 5, 9\}, \texttt{AxesLabel} \rightarrow \{\texttt{"pH"}, \texttt{"[H}_2\texttt{PO}_4^-\texttt{]/[Pi]"}\} \right];$$

This plot is given in composite Fig. 1.3 (a) on the next page.

These two plots can be superimposed.

$$\texttt{plot3} = \texttt{Plot}\Big[\Big\{ \frac{2.2387211385683377`*^-7}{2.2387211385683377`*^-7 + 10^{-pH}}, \frac{10^{-pH}}{2.2387211385683377`*^-7 + 10^{-pH}} \Big\},$$

$$\{pH, 5, 9\}, \texttt{AxesLabel} \rightarrow \{\texttt{"pH"}, \texttt{"fractions"}\}, \texttt{PlotStyle} \rightarrow \{\texttt{Black}\}\Big];$$

This plot is given in composite Fig. 1.3.

These three plots can be shown together as follows:

```
GraphicsArray[{{plot1, plot2, plot3}}]
```

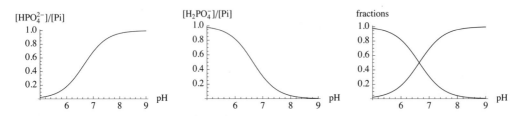

Fig. 1.3 (a) $\left[HPO_4^{2-}\right]/[Pi]$ versus pH at 298.15 K and ionic strength 0.25 M. (b) $\left[H_2 PO_4^-\right]\big/ [Pi]$ versus pH at 298.15 K and ionic strength 0.25 M. (c) Superposition of (a) and (b).

■ **1.9.2 Distributions of species of ATP as functions of pH**

ATP is composed of 3 species in the range pH 5-9. At 298.15 K 0.25 M ionic strength, pK1atp = 6.47 and pK2atp = 3.83

Calculation of the fraction of [ATP] that is [ATP $^{4-}$]

ATP has 2 acid dissociations and one conservation equation. First, the equilibrium concentration of ATP^{4-} is calculated:

```
Solve[{k1atp == h * atp / hatp, k2atp == h * hatp / h2atp, atpt == atp + hatp + h2atp},
 {atp}, {hatp, h2atp}]
```

$$\left\{\left\{atp \rightarrow \frac{atpt\ k1atp\ k2atp}{h^2 + h\ k2atp + k1atp\ k2atp}\right\}\right\}$$

The fraction (atp/atpt) = $[ATP^{4-}]/[ATP]$ is given by

$$\frac{k1atp\ k2atp}{h^2 + h\ k2atp + k1atp\ k2atp};$$

It is more convenient to use pHs and p*K*s.

$$\frac{k1atp\ k2atp}{h^2 + h\ k2atp + k1atp\ k2atp}\ /.\ h \rightarrow 10\hat{\ }-pH\ /.\ k1atp \rightarrow 10\hat{\ }-pK1atp\ /.\ k2atp \rightarrow 10\hat{\ }-pK2atp$$

$$\frac{10^{-pK1atp-pK2atp}}{10^{-2\ pH} + 10^{-pH-pK2atp} + 10^{-pK1atp-pK2atp}}$$

The fraction (atp/atpt) can be plotted versus pH.

$$\frac{10^{-pK1atp-pK2atp}}{10^{-2\ pH} + 10^{-pH-pK2atp} + 10^{-pK1atp-pK2atp}}\ /.\ pK1atp \rightarrow 6.47\ /.\ pK2atp \rightarrow 3.83$$

$$\frac{5.01187 \times 10^{-11}}{5.01187 \times 10^{-11} + 10^{-3.83-pH} + 10^{-2\ pH}}$$

```
plot1atp = Plot[ 10^-pK1atp-pK2atp
                ─────────────────────────────────────────────────── /. pK1atp → 6.47 /. pK2atp → 3.83,
                 10^-2 pH + 10^-pH-pK2atp + 10^-pK1atp-pK2atp

 {pH, 5, 9}, AxesLabel → {"pH", "[ATP^4-]/[ATP]"}, PlotLabel → "(a)"];
```

This plot is given later in Fig. 1.4.

The reciprocal of $[ATP^{4-}]/[ATP]$ is given by

$$1\Big/ \frac{10^{-pK1atp-pK2atp}}{10^{-2\ pH} + 10^{-pH-pK2atp} + 10^{-pK1atp-pK2atp}}$$

$$10^{pK1atp+pK2atp}\left(10^{-2\ pH} + 10^{-pH-pK2atp} + 10^{-pK1atp-pK2atp}\right)$$

$$\text{Simplify}\left[1\Big/ \frac{10^{-pK1atp-pK2atp}}{10^{-2\ pH} + 10^{-pH-pK2atp} + 10^{-pK1atp-pK2atp}}\right]$$

$$1 + 10^{-pH+pK1atp} + 10^{-2\ pH+pK1atp+pK2atp}$$

This is $[ATP]/[ATP^{4-}]$, and so the concentration of ATP is always greater than the concentration of ATP^{4-}.

Calculation of the fraction of [ATP] that is [HATP $^{3-}$]

```
Solve[{k1atp == h * atp / hatp, k2atp == h * hatp / h2atp, atpt == atp + hatp + h2atp},
 {hatp}, {atp, h2atp}]
```

$$\left\{\left\{hatp \rightarrow \frac{atpt\ h\ k2atp}{h^2 + h\ k2atp + k1atp\ k2atp}\right\}\right\}$$

The fraction (hatp/atpt) is given by

$$\frac{h\ k2atp}{h^2 + h\ k2atp + k1atp\ k2atp};$$

It is more convienent to use pHs and p*K*s.

$$\frac{h\ k2atp}{h^2 + h\ k2atp + k1atp\ k2atp}\ /.\ h \to 10^\wedge -pH\ /.\ k1atp \to 10^\wedge -pK1atp\ /.\ k2atp \to 10^\wedge -pK2atp$$

$$\frac{10^{-pH-pK2atp}}{10^{-2\,pH} + 10^{-pH-pK2atp} + 10^{-pK1atp-pK2atp}}$$

Plot the fraction hatp/atpt $= \left[HATP^{3-}\right] \big/ [ATP]$ as a function of pH

$$plot2atp = Plot\left[\frac{10^{-pH-pK2atp}}{10^{-2\,pH} + 10^{-pH-pK2atp} + 10^{-pK1atp-pK2atp}}\ /.\ pK1atp \to 6.47\ /.\ pK2atp \to 3.83,\right.$$

$$\left.\{pH, 5, 9\}, AxesLabel \to \{"pH", "[HATP^{3-}]/[ATP]"\}, PlotLabel \to " (b) "\right];$$

Calculation of the fraction of [ATP] that is [H_2 ATP $^{2-}$]

$$Solve[\{k1atp == h * atp / hatp, k2atp == h * hatp / h2atp, atpt == atp + hatp + h2atp\},$$
$$\{h2atp\}, \{atp, hatp\}]$$

$$\left\{\left\{h2atp \to \frac{atpt\ h^2}{h^2 + h\ k2atp + k1atp\ k2atp}\right\}\right\}$$

The fraction (h2atp/atpt) is given by

$$\frac{h^2}{h^2 + h\ k2atp + k1atp\ k2atp};$$

It is more convienent to use pHs and p*K*s.

$$\frac{h^2}{h^2 + h\ k2atp + k1atp\ k2atp}\ /.\ h \to 10^\wedge -pH\ /.\ k1atp \to 10^\wedge -pK1atp\ /.\ k2atp \to 10^\wedge -pK2atp$$

$$\frac{10^{-2\,pH}}{10^{-2\,pH} + 10^{-pH-pK2atp} + 10^{-pK1atp-pK2atp}}$$

Plot the fraction h2atp/atpt $= [H_2$ ATP $^{2-}]/[ATP]$ as a function of pH

$$plot3atp = Plot\left[\frac{10^{-2\,pH}}{10^{-2\,pH} + 10^{-pH-pK2atp} + 10^{-pK1atp-pK2atp}}\ /.\ pK1atp \to 6.47\ /.\ pK2atp \to 3.83,\right.$$

$$\left.\{pH, 5, 9\}, AxesLabel \to \{"pH", "[H_2\ ATP^{2-}] \big/ [ATP]"\}, PlotLabel \to " (c) "\right];$$

These three plots can be superimposed.

$$plot4atp = Plot\left[\left\{\frac{10^{-pK1atp-pK2atp}}{10^{-2\,pH} + 10^{-pH-pK2atp} + 10^{-pK1atp-pK2atp}},\ \frac{10^{-pH-pK2atp}}{10^{-2\,pH} + 10^{-pH-pK2atp} + 10^{-pK1atp-pK2atp}},\right.\right.$$

$$\left.\left.\frac{10^{-2\,pH}}{10^{-2\,pH} + 10^{-pH-pK2atp} + 10^{-pK1atp-pK2atp}}\right\}\ /.\ pK1atp \to 6.47\ /.\ pK2atp \to 3.83,\right.$$

$$\left.\{pH, 5, 9\}, AxesLabel \to \{"pH", "fractions"\}, PlotLabel \to " (d) "\right];$$

```
GraphicsArray[{{plot1atp, plot2atp}, {plot3atp, plot4atp}}]
```

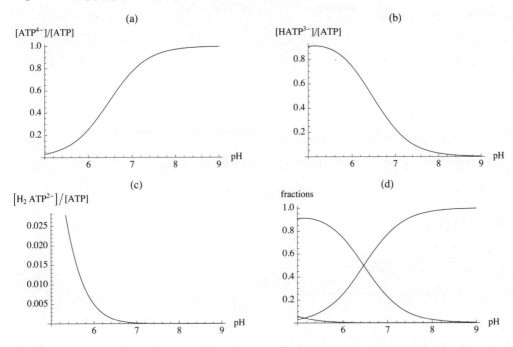

Fig. 1.4 Plots of fractions of [ATP] in (a) [ATP^{4-}], (b) [$HATP^{3-}$], (c) [$H_2 ATP^{2-}$], and (d) fractions at 298.15 K and ionic strength 0.25 M.

■ 1.9.3 Distributions of species of ADP as functions of pH

ADP is composed of 3 species in the range pH 5-9. At 298.15 K and 0.25 M ionic strength, pK1adp = 6.33 and pK2adp = 3.79.

Calculation of the fraction of [ADP] that is [ADP $^{3-}$]

ADP has 2 acid dissociations and one conservation equation. First, the equlibrium concentration of ADP^{3-} is calculated:

```
Solve[{k1adp == h * adp / hadp, k2adp == h * hadp / h2adp, adpt == adp + hadp + h2adp},
 {adp}, {hadp, h2adp}]
```

$$\left\{ \left\{ adp \to \frac{adpt\ k1adp\ k2adp}{h^2 + h\ k2adp + k1adp\ k2adp} \right\} \right\}$$

The fraction (adp/adpt) = [ADP^{4-}]/[ADP] is given by

$$\frac{k1adp\ k2adp}{h^2 + h\ k2adp + k1adp\ k2adp};$$

It is more convenient to use pHs and p*K*s.

$$\frac{k1adp\ k2adp}{h^2 + h\ k2adp + k1adp\ k2adp} \ /.\ h \to 10\hat{\ }-pH\ /.\ k1adp \to 10\hat{\ }-pK1adp\ /.\ k2adp \to 10\hat{\ }-pK2adp$$

$$\frac{10^{-pK1adp-pK2adp}}{10^{-2\,pH} + 10^{-pH-pK2adp} + 10^{-pK1adp-pK2adp}}$$

The fraction (adp/adpt) can be plotted versus pH.

$$\frac{10^{-\text{pK1adp-pK2adp}}}{10^{-2\,\text{pH}} + 10^{-\text{pH-pK2adp}} + 10^{-\text{pK1adp-pK2adp}}} \;/.\; \text{pK1adp} \to 6.33 \;/.\; \text{pK2adp} \to 3.79$$

$$\frac{7.58578 \times 10^{-11}}{7.58578 \times 10^{-11} + 10^{-3.79-\text{pH}} + 10^{-2\,\text{pH}}}$$

Plot the fraction adp/adpt as a function of pH

```
plot1adp = Plot[ 10^(-pK1adp-pK2adp) / (10^(-2 pH) + 10^(-pH-pK2adp) + 10^(-pK1adp-pK2adp)) /. pK1adp → 6.33 /. pK2adp → 3.79,

  {pH, 5, 9}, AxesLabel → {"pH", "[ADP³⁻]/[ADP]"}, PlotLabel → "(a)"];
```

Similar plots can be made for the equilibrium concentrations of hadp and h2adp.

The reciprocal of $[\text{ADP}^{3-}]/[\text{ADP}]$ is given by

$$1 \Big/ \frac{10^{-\text{pK1adp-pK2adp}}}{10^{-2\,\text{pH}} + 10^{-\text{pH-pK2adp}} + 10^{-\text{pK1adp-pK2adp}}}$$

$$10^{\text{pK1adp+pK2adp}} \left(10^{-2\,\text{pH}} + 10^{-\text{pH-pK2adp}} + 10^{-\text{pK1adp-pK2adp}}\right)$$

```
Simplify[1 / 10^(-pK1adp-pK2adp) / (10^(-2 pH) + 10^(-pH-pK2adp) + 10^(-pK1adp-pK2adp))]
```

$$1 + 10^{-\text{pH+pK1adp}} + 10^{-2\,\text{pH+pK1adp+pK2adp}}$$

This is $[\text{ADP}]/[\text{ADP}^{3-}]$, and so the concentration of ADP is always greater than the concentration of ADP^{3-}.

Calculation of the fraction of [ADP] that is [HADP $^{2-}$]

```
Solve[{k1adp == h * adp / hadp, k2adp == h * hadp / h2adp, adpt == adp + hadp + h2adp},
  {hadp}, {adp, h2adp}]
```

$$\left\{ \left\{ \text{hadp} \to \frac{\text{adpt h k2adp}}{h^2 + h\,\text{k2adp} + \text{k1adp k2adp}} \right\} \right\}$$

The fraction (hadp/adpt) is given by

$$\frac{h\,\text{k2adp}}{h^2 + h\,\text{k2adp} + \text{k1adp k2adp}};$$

It is more convenient to use pHs and pKs.

$$\frac{h\,\text{k2adp}}{h^2 + h\,\text{k2adp} + \text{k1adp k2adp}} \;/.\; h \to 10^{\wedge}-\text{pH} \;/.\; \text{k1adp} \to 10^{\wedge}-\text{pK1adp} \;/.\; \text{k2adp} \to 10^{\wedge}-\text{pK2adp}$$

$$\frac{10^{-\text{pH-pK2adp}}}{10^{-2\,\text{pH}} + 10^{-\text{pH-pK2adp}} + 10^{-\text{pK1adp-pK2adp}}}$$

Plot the fraction hadp/adpt = $\left[\text{HADP}^{2-}\right] \Big/ [\text{ADP}]$ as a function of pH

```
plot2adp = Plot[ 10^(-pH-pK2adp) / (10^(-2 pH) + 10^(-pH-pK2adp) + 10^(-pK1adp-pK2adp)) /. pK1adp → 6.33 /. pK2adp → 3.79,

  {pH, 5, 9}, AxesLabel → {"pH", "[HADP²⁻]/[ADP]"}, PlotLabel → "(b)"];
```

Calculation of the fraction of [ADP] that is [H$_2$ ADP $^-$]

```
Solve[{k1adp == h * adp / hadp, k2adp == h * hadp / h2adp, adpt == adp + hadp + h2adp},
  {h2adp}, {adp, hadp}]
```

$$\left\{\left\{\text{h2adp} \to \frac{\text{adpt } h^2}{h^2 + h \text{ k2adp} + \text{k1adp k2adp}}\right\}\right\}$$

The fraction (h2adp/adpt) is given by

$$\frac{h^2}{h^2 + h \text{ k2adp} + \text{k1adp k2adp}};$$

It is more convienent to use pHs and pKs.

$$\frac{h^2}{h^2 + h \text{ k2adp} + \text{k1adp k2adp}} \text{ /. } h \to 10\,\hat{}\,\text{-pH /. k1adp} \to 10\,\hat{}\,\text{-pK1adp /. k2adp} \to 10\,\hat{}\,\text{-pK2adp}$$

$$\frac{10^{-2\,\text{pH}}}{10^{-2\,\text{pH}} + 10^{-\text{pH-pK2adp}} + 10^{-\text{pK1adp-pK2adp}}}$$

Plot the fraction h2adp/adpt = [H$_2$ ADP $^-$]/[ADP] as a function of pH

```
plot3adp = Plot[
    10^(-2 pH)
  ----------------------------------------  /. pK1adp → 6.33 /. pK2adp → 3.79,
  10^(-2 pH) + 10^(-pH-pK2adp) + 10^(-pK1adp-pK2adp)

  {pH, 5, 9}, AxesLabel → {"pH", "[H₂ ADP⁻] / [ADP]"}, PlotLabel → "(c)"];
```

These three plots can be superimposed.

```
plot4adp = Plot[{
    10^(-pK1adp-pK2adp)                              10^(-pH-pK2adp)
  --------------------------------------------------, --------------------------------------------------,
  10^(-2 pH) + 10^(-pH-pK2adp) + 10^(-pK1adp-pK2adp)  10^(-2 pH) + 10^(-pH-pK2adp) + 10^(-pK1adp-pK2adp)

    10^(-2 pH)
  --------------------------------------------------} /. pK1adp → 6.47 /. pK2adp → 3.83,
  10^(-2 pH) + 10^(-pH-pK2adp) + 10^(-pK1adp-pK2adp)

  {pH, 5, 9}, AxesLabel → {"pH", "fractions"}, PlotLabel → "(d)"];
```

GraphicsArray[{{plot1adp, plot2adp}, {plot3adp, plot4adp}}]

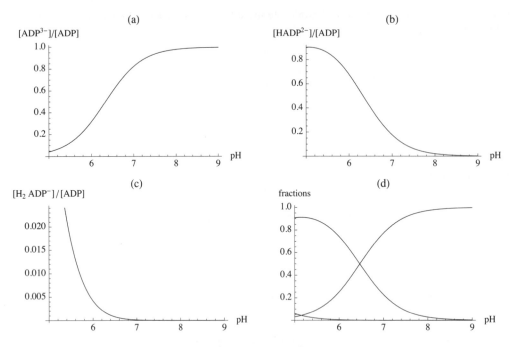

Fig. 1.5 Plots of fractions of [ADP] in (a) [ADP^{3-}], (b) [HADP^{2-}], (c) [H$_2$ADP$^-$], and (d) fractions at 298.15 K and ionic strength 0.25 M.

■ 1.9.4 Calculations of the apparent equlibrium constant for ATP + H$_2$O = ADP + Pi as a function of pH

The apparent equilibrium constant K' is given by [ADP][Pi]/[ATP].

adpt ∗ pit / atpt;

These 3 equilibrium expressions are used to replace adpt, pit, and atpt.

$$\left(1 + 10^{-pH+pK1adp} + 10^{-2\,pH+pK1adp+pK2adp}\right) * adp;$$

$$\left(1 + 10^{-pH+pKpi}\right) * pi;$$

$$\left(1 + 10^{-pH+pK1atp} + 10^{-2\,pH+pK1atp+pK2atp}\right) * atp;$$

These three expressions are used to replace atpt, adpt, and pit in the expression for the apparent equilibrium constant K'.

$$\textbf{(adpt ∗ pit / atpt) /. adpt -> } \left(1 + 10^{-pH+pK1adp} + 10^{-2\,pH+pK1adp+pK2adp}\right) \textbf{∗ adp /. pit -> } \left(1 + 10^{-pH+pKpi}\right) \textbf{∗ pi /.}$$
$$\textbf{atpt → } \left(1 + 10^{-pH+pK1atp} + 10^{-2\,pH+pK1atp+pK2atp}\right) \textbf{∗ atp}$$

$$\frac{\left(1 + 10^{-pH+pK1adp} + 10^{-2\,pH+pK1adp+pK2adp}\right)\left(1 + 10^{-pH+pKpi}\right)\,adp\,pi}{\left(1 + 10^{-pH+pK1atp} + 10^{-2\,pH+pK1atp+pK2atp}\right)\,atp}$$

The chemical reference reaction is ATP^{4-} + H$_2$O = ADP^{3-} + HPO$_4^{2-}$ + H$^+$ k_{ref1} = adp*pi*h/atp
Replace adp*pi/atp with kref1/h = kref1*10^pH. The apparent equilibrium constant is given by

$$\frac{\left(1 + 10^{-pH+pK1adp} + 10^{-2\,pH+pK1adp+pK2adp}\right)\left(1 + 10^{-pH+pKpi}\right) * \mathtt{kref1} * 10\,\hat{}\,\mathtt{pH}}{\left(1 + 10^{-pH+pK1atp} + 10^{-2\,pH+pK1atp+pK2atp}\right)}$$

$$\frac{10^{pH}\left(1 + 10^{-pH+pK1adp} + 10^{-2\,pH+pK1adp+pK2adp}\right)\left(1 + 10^{-pH+pKpi}\right)\mathtt{kref1}}{1 + 10^{-pH+pK1atp} + 10^{-2\,pH+pK1atp+pK2atp}}$$

$$\frac{\left(1 + 10^{-pH+pK1adp} + 10^{-2\,pH+pK1adp+pK2adp}\right)\left(1 + 10^{-pH+pKpi}\right) * \mathtt{kref1} * 10\,\hat{}\,\mathtt{pH}}{\left(1 + 10^{-pH+pK1atp} + 10^{-2\,pH+pK1atp+pK2atp}\right)}\ /.\ \mathtt{pK1adp} \to 6.33\ /.$$

$$\mathtt{pK2adp} \to 3.79\ /.\ \mathtt{pKpi} \to 6.65\ /.\ \mathtt{pK1atp} \to 6.47\ /.\ \mathtt{pK2atp} \to 3.83\ /.\ \mathtt{pH}\ \text{->}\ 7$$

$1.35578 \times 10^7\ \mathtt{kref1}$

The apparent equibrium constant K' for ATP + H_2O = ADP + Pi at 298.15 K, pH 7 and 0.25 M ionic strength (5,19) is

Exp[36.04 / (8.31451 * .29815)]

2.06015×10^6

Therefore, kref1 is given by

(2.06015 * 10^6) / (1.35578 * 10^7)

0.151953

The apparent equilibrium constant is given as a function of pH by

$$\frac{\left(1 + 10^{-pH+pK1adp} + 10^{-2\,pH+pK1adp+pK2adp}\right)\left(1 + 10^{-pH+pKpi}\right) * \mathtt{kref1} * 10\,\hat{}\,\mathtt{pH}}{\left(1 + 10^{-pH+pK1atp} + 10^{-2\,pH+pK1atp+pK2atp}\right)}\ /.\ \mathtt{pK1adp} \to 6.33\ /.$$

$$\mathtt{pK2adp} \to 3.79\ /.\ \mathtt{pKpi} \to 6.65\ /.\ \mathtt{pK1atp} \to 6.47\ /.\ \mathtt{pK2atp} \to 3.83\ /.\ \mathtt{kref1} \to .151953$$

$$\frac{0.151953\ 10^{pH}\left(1 + 10^{10.12-2\,pH} + 10^{6.33-pH}\right)\left(1 + 10^{6.65-pH}\right)}{1 + 10^{10.3-2\,pH} + 10^{6.47-pH}}$$

The apparent equilibrium constants at pHs 5, 6, 7, 8, and 9 are given by

$$\frac{\left(1 + 10^{-pH+pK1adp} + 10^{-2\,pH+pK1adp+pK2adp}\right)\left(1 + 10^{-pH+pKpi}\right) * \mathtt{kref1} * 10\,\hat{}\,\mathtt{pH}}{\left(1 + 10^{-pH+pK1atp} + 10^{-2\,pH+pK1atp+pK2atp}\right)}\ /.\ \mathtt{pK1adp} \to 6.33\ /.$$

$$\mathtt{pK2adp} \to 3.79\ /.\ \mathtt{pKpi} \to 6.65\ /.\ \mathtt{pK1atp} \to 6.47\ /.$$

$$\mathtt{pK2atp} \to 3.83\ /.\ \mathtt{kref1} \to .151953\ /.\ \mathtt{pH} \to \{5, 6, 7, 8, 9\}$$

$\left\{505\,886.,\ 659\,168.,\ 2.06015 \times 10^6,\ 1.57486 \times 10^7,\ 1.52508 \times 10^8\right\}$

The changes in the standard transformed Gibbs energies of reaction $\Delta_r G'^{\circ}$ are given by

$$-8.31451 * .29815 *$$

$$\mathtt{Log}\left[\frac{\left(1 + 10^{-pH+pK1adp} + 10^{-2\,pH+pK1adp+pK2adp}\right)\left(1 + 10^{-pH+pKpi}\right) * \mathtt{kref1} * 10\,\hat{}\,\mathtt{pH}}{\left(1 + 10^{-pH+pK1atp} + 10^{-2\,pH+pK1atp+pK2atp}\right)}\ /.\ \mathtt{pK1adp} \to 6.33\ /.\right.$$

$$\mathtt{pK2adp} \to 3.79\ /.\ \mathtt{pKpi} \to 6.65\ /.\ \mathtt{pK1atp} \to 6.47\ /.$$

$$\left.\mathtt{pK2atp} \to 3.83\ /.\ \mathtt{kref1} \to .151953\ /.\ \mathtt{pH} \to \{5, 6, 7, 8, 9\}\right]$$

$\{-32.559,\ -33.2151,\ -36.04,\ -41.0822,\ -46.7106\}$

These values are in agreement with BasicBiochemData3 [*MathSource*3, p. 227].

The expression for the apparent equilibrium constant for ATP + H_2O can be summarized by $K' = 10^{pH} K_{ref1} f(pH)$ where $f(pH)$ brings in the effects of the pKs of the reactants. However, the choice of a reference reaction is arbitrary. For example, the reference reaction can be taken to be $ATP^{4-} + H_2O = ADP^{3-} + H_2PO_4^-$ kref2 = adp*hpi/atp. In this case, pit is given as a function of pH by

```
(1 + 10 ^ (pH - pKpi)) * hpi
```

$$\left(1 + 10^{pH-pKpi}\right) hpi$$

Functions of pH are introduced.

```
(adpt * pit / atpt) /. adpt -> (1 + 10^-pH+pK1adp + 10^-2 pH+pK1adp+pK2adp) * adp /. pit -> (1 + 10^pH-pKpi) * hpi /.
   atpt → (1 + 10^-pH+pK1atp + 10^-2 pH+pK1atp+pK2atp) * atp
```

$$\frac{\left(1 + 10^{-pH+pK1adp} + 10^{-2\,pH+pK1adp+pK2adp}\right) \left(1 + 10^{pH-pKpi}\right) adp\ hpi}{\left(1 + 10^{-pH+pK1atp} + 10^{-2\,pH+pK1atp+pK2atp}\right) atp}$$

The expression for the chemical reference reaction is replaced with kref2.

$$\frac{\left(1 + 10^{-pH+pK1adp} + 10^{-2\,pH+pK1adp+pK2adp}\right) \left(1 + 10^{pH-pKpi}\right) kref2}{\left(1 + 10^{-pH+pK1atp} + 10^{-2\,pH+pK1atp+pK2atp}\right)}$$

$$\frac{\left(1 + 10^{-pH+pK1adp} + 10^{-2\,pH+pK1adp+pK2adp}\right) \left(1 + 10^{pH-pKpi}\right) kref2}{1 + 10^{-pH+pK1atp} + 10^{-2\,pH+pK1atp+pK2atp}}$$

The apparent equilibrium constant is expressed as a function of pH by

$$\frac{\left(1 + 10^{-pH+pK1adp} + 10^{-2\,pH+pK1adp+pK2adp}\right) \left(1 + 10^{pH-pKpi}\right) kref2}{\left(1 + 10^{-pH+pK1atp} + 10^{-2\,pH+pK1atp+pK2atp}\right)}\ /.\ pK1adp \to 6.33\ /.\ pK2adp \to 3.79\ /.$$
$$\quad pKpi \to 6.65\ /.\ pK1atp \to 6.47\ /.\ pK2atp \to 3.83$$

$$\frac{\left(1 + 10^{10.12-2\,pH} + 10^{6.33-pH}\right) \left(1 + 10^{-6.65+pH}\right) kref2}{1 + 10^{10.3-2\,pH} + 10^{6.47-pH}}$$

$$\frac{\left(1 + 10^{10.120000000000001`-2\,pH} + 10^{6.33`-pH}\right) \left(1 + 10^{-6.65`+pH}\right) kref2}{1 + 10^{10.3`-2\,pH} + 10^{6.47`-pH}}\ /.\ pH \to 7$$

```
3.03521 kref2
```

The apparent equribrium constant K' at 298.15 K, pH 7 and 0.25 M ionic strength is

```
Exp[36.04 / (8.31451 * .29815)]
```

2.06015×10^6

Therefore, kref2 is given by

```
(2.06015 * 10^6) / (3.03521)
```

```
678 750.
```

The apparent equilibrium constant is given as a function of pH by

$$\frac{\left(1 + 10^{10.120000000000001`-2\,\text{pH}} + 10^{6.33`-\text{pH}}\right)\left(1 + 10^{-6.65`+\text{pH}}\right)\text{kref2}}{1 + 10^{10.3`-2\,\text{pH}} + 10^{6.47`-\text{pH}}} \;/.\; \text{kref2} \rightarrow 678\,750$$

$$\frac{678\,750\left(1 + 10^{10.12-2\,\text{pH}} + 10^{6.33-\text{pH}}\right)\left(1 + 10^{-6.65+\text{pH}}\right)}{1 + 10^{10.3-2\,\text{pH}} + 10^{6.47-\text{pH}}}$$

$$\frac{\left(1 + 10^{10.120000000000001`-2\,\text{pH}} + 10^{6.33`-\text{pH}}\right)\left(1 + 10^{-6.65`+\text{pH}}\right)\text{kref2}}{1 + 10^{10.3`-2\,\text{pH}} + 10^{6.47`-\text{pH}}} \;/.\; \text{kref2} \rightarrow 678\,750 \;/.\; \text{pH} \rightarrow \{5, 6, 7, 8, 9\}$$

$$\left\{505\,886., 659\,169., 2.06015 \times 10^6, 1.57487 \times 10^7, 1.52508 \times 10^8\right\}$$

The standard transformed Gibbs energy of reaction $\Delta_r G'^\circ$ in kJ mol^{-1} are given by

```
-8.31451 * .29815 * Log[{505886.48270386615`,  659168.7875151065`,
     2.0601513062382448`*^6, 1.5748664776306028`*^7, 1.5250818400661832`*^8}]
```

$$\{-32.559, -33.2151, -36.04, -41.0822, -46.7106\}$$

The important point here is that although $K' = 10^{n\text{pH}}K_{ref}f(\text{pH})$, where n is zero or a positive or negative integer, the value of n is arbitrary in thermodynamics. In rapid-equilibrium enzyme kinetics the velocity is proportional to $10^{n\text{pH}}$, where n is the number of hydrogen ions consumed in the rate-determining reaction. When one hydrogen ion is consumed, $n = -1$.

1.10 Data on the thermodynamics of enzyme-catalyzed reactions in the literature and on the web

There are two types of thermodynamic data on enzyme-catalyzed reactions that are of interest in biochemical kinetics: (1) Experimental measurements of apparent equilibrium constants and heats of reaction. (2) Data on thermodynamic properties of species obtained chemically or from experimental measurements of apparent equilibrium constants and heats of reaction. This second type of data has the advantage that it can be used to calculate apparent equilibrium constants and heats of enzyme-catalyzed reactions under conditions that have not been studied experimentally. This second type of thermodynamic data can be used to calculate properties of biochemical reactions that are difficult to measure directly, but may be of interest in connection with enzyme kinetics. An example of this is checking the Haldane relation to calculate the apparent equilibrium constant with the kinetic parameters for the forward and reverse reactions. Another example is comparing the change in binding of hydrogen ions $\Delta_r N_H$ calculated using thermodynamics using equation 1.4-3 or 1.4-4 with the integer n for the number of hydrogen ions consumed in the rate-determining reaction in a rapid-equilibrium mechanism [27].

■ 1.10.1 Experimental data on apparent equilibrium constants and heats of reaction of enzyme-catalyzed reactions

The place to look for information on experimental measurements of apparent equilibrium costants and heats of reaction of enzyme-catalyzed reactions are the 10 compilations by Goldberg and Tewari of experimental data on apparent equilibrium constants and heats of reaction. These data are summarized on an interactive site at the National Institute Science and Technology [Web3]. The EC numbers [Web2] of the reactions are given, and the quality of the data is evaluated.

■ 1.10.2 Data on thermodynamic properties of species obtained chemically or from experimental measurements of apparent equilibrium constants and heats of enzyme-catalyzed reactions

At the time of Krebs [2,4], it was recognized that values of equilibrium constants on enzyme-catalyzed reactions and heats of reaction could be used to calculate standard Gibbs energies of formation and standard enthalpies of formation of species to augment the tables produced by chemists from the study of chemical reactions (2). This means that the most basic way to store information on apparent equlibrium constants and heats of biochemical reactions is to make tables of $\Delta_f G°$ and $\Delta_f H°$ of the species involved at 298.15 K and zero ionic strength. This makes it possible to calculate standard transformed Gibbs energies of formation $\Delta_f G_i'°$ and standard transformed enthalpies of formation $\Delta_f H_i'°$ of biochemical reactants at desired temperatures, pHs, and ionic strengths. These values can be used to calculate standard transformed Gibbs energies $\Delta_r G'°$ of reaction and standard transformed enthalpies $\Delta_r H'°$ of reaction. The crtical evaluations of apparent equilibrium constants and heats of reaction of enzyme catalyzed reactions by Goldberg and Tewari [Web3] have been used in developing BasicBiochemData3 [*MathSource*3] that gives $\Delta_f G°$ and $\Delta_f H°$ for species of199 biochemical reactants at 298.15 K in *Mathematica*. This list of species data can be extended by analyzing more data from the Goldberg-Tewari tables. As explained in Section 1.4, the fact that the pH is specified in biochemical thermodynamics makes it possible to calculate another reaction property that is not analogous to any property in chemical thermodynamics, and that is the change in binding of hydrogen ions $\Delta_r N_H$ in the enzyme-catalyzed reaction.

■ 1.10.3 Calculated values of standard transformed Gibbs energies of reaction, standard transformed enthalpies of reaction, apparent equilibrium constants, and changes in the binding of hydrogen ions

Species data on 199 reactants have been used to calculate these properties for about 300 biochemical reactions at 298.15 K and 0.25 M ionic strength [23]. For 94 of these reactants, the standard enthalpies of formation of the species are known at 298.15 K, and this makes it possible to calculate $\Delta_r G'°$, $\Delta_r H'°$, and $\Delta_r N_H$ at temperatures in the range 273.15 K to about 313.15 K. BasicBiochemData3 [*MathSource*3] gives the following functions in *Mathematica*:

(1) Functions of pH and ionic strength at 298.15 K for $\Delta_f G_i'°$ for 199 reactants, for example, atp.

(2) Functions of pH and ionic strength at 298.15 K for $\overline{N}_H$ for 199 reactants, for example, atpNH.

(3) Functions of temperature, pH and ionic strength for $\Delta_f G_i'°$ for 94 reactants, for example, atpGT.

(4) Functions of temperature, pH and ionic strength for $\Delta_f H_i'°$ for 94 reactants, for example, atpHT

(5) Functions of temperature, pH and ionic strength for $\Delta_f S_i'°$ for 94 reactants, for example, atpST

(6) Functions of temperature, pH and ionic strength for $\overline{N}_H$ for 94 reactants, for example, atpNHT

These 774 functions are not given here, but enough information is given to calculate the $\Delta_r G'°$ for ATP + H_2O = ADP + Pi at 273.15 K, 298.15 K, and 313.15 K, pHs 5, 6, 7, 8, and 9, and ionic strengths 0, 0.10, and 0.25 M. The following program derives the function of *T* (in Kelvin), pH, and ionic strength (is) that gives the standard transformed Gibbs energy of formation of a reactant (sum of species).

```
derivetrGibbsT[speciesmat_] := Module[{dGzero, dGzeroT, dHzero, zi, nH, gibbscoeff
, pHterm, isterm, gpfnsp}, (*This program derives the function of T (in Kelvin),
pH, and ionic strength (is) that gives the standard transformed Gibbs energy
of formation of a reactant (sum of species).  The input speciesmat is a matrix
that gives the standard Gibbs energy of formation in kJ mol^-1 at 298.15 K and
zero ionic strength, the standard enthalpy of formation in kJ mol^-1 at 298.15 K
and zero ionic strength, the electric charge, and the number of hydrogen atoms in
each species.  There is a row in the matrix for each species of the reactant.
gpfnsp is a list of the functions for the standard transformed Gibbs energies
of the species. The corresponding functions for other transformed properties
can be obtained by taking partial derivatives.  The standard transformed Gibbs
energy of formation of a reactant in kJ mol^-1 can be calculated at any temperature
in the range 273.15 K to 313.15 K, any pH in the range 5 to 9, and any ionic
strength in the range 0 to 0.35 M by use of ReplaceAll (/.).*)
{dGzero, dHzero, zi, nH} = Transpose[speciesmat];
gibbscoeff = (9.20483 * t) / 10^3 - (1.284668 * t^2) / 10^5 + (4.95199 * t^3) / 10^8;
dGzeroT = (dGzero * t) / 298.15 + dHzero * (1 - t / 298.15);
pHterm = (nH * 8.31451 * t * Log[10^ (-pH)]) / 1000;
istermG = (gibbscoeff * (zi^2 - nH) * is^0.5) / (1 + 1.6 * is^0.5);
gpfnsp = dGzeroT - pHterm - istermG;
- ((8.31451 * t * Log[Plus @@ (E^ (- (gpfnsp / ((8.31451 * t) / 1000))))]) / 1000)]
```

Since these properties are calculated here in Chapter 1, the functions are named atpGTch1, ...

atpGTch1 = derivetrGibbsT[atpsp]

$$-0.00831451\, t\, \mathrm{Log}\Big[e^{\frac{120.272\left(-3627.91\,(1-0.00335402\,t)-9.5193\,t+\frac{10\,is^{0.5}\left(0.00920483\,t-0.0000128467\,t^2+4.95199\times10^{-8}\,t^3\right)}{1+1.6\,is^{0.5}}-0.116403\,t\,\mathrm{Log}\left[10^{-pH}\right]\right)}{t}} +$$

$$e^{\frac{120.272\left(-3612.91\,(1-0.00335402\,t)-9.42975\,t+\frac{4\,is^{0.5}\left(0.00920483\,t-0.0000128467\,t^2+4.95199\times10^{-8}\,t^3\right)}{1+1.6\,is^{0.5}}-0.108089\,t\,\mathrm{Log}\left[10^{-pH}\right]\right)}{t}} +$$

$$e^{\frac{120.272\left(-3619.21\,(1-0.00335402\,t)-9.28425\,t+\frac{4\,is^{0.5}\left(0.00920483\,t-0.0000128467\,t^2+4.95199\times10^{-8}\,t^3\right)}{1+1.6\,is^{0.5}}-0.0997741\,t\,\mathrm{Log}\left[10^{-pH}\right]\right)}{t}} \Big]$$

adpGTch1 = derivetrGibbsT[adpsp]

$$-0.00831451\, t\, \mathrm{Log}\Big[e^{\frac{120.272\left(-2638.54\,(1-0.00335402\,t)-6.61405\,t+\frac{13\,is^{0.5}\left(0.00920483\,t-0.0000128467\,t^2+4.95199\times10^{-8}\,t^3\right)}{1+1.6\,is^{0.5}}-0.116403\,t\,\mathrm{Log}\left[10^{-pH}\right]\right)}{t}} +$$

$$e^{\frac{120.272\left(-2620.94\,(1-0.00335402\,t)-6.53061\,t+\frac{9\,is^{0.5}\left(0.00920483\,t-0.0000128467\,t^2+4.95199\times10^{-8}\,t^3\right)}{1+1.6\,is^{0.5}}-0.108089\,t\,\mathrm{Log}\left[10^{-pH}\right]\right)}{t}} +$$

$$e^{\frac{120.272\left(-2626.54\,(1-0.00335402\,t)-6.39319\,t+\frac{3\,is^{0.5}\left(0.00920483\,t-0.0000128467\,t^2+4.95199\times10^{-8}\,t^3\right)}{1+1.6\,is^{0.5}}-0.0997741\,t\,\mathrm{Log}\left[10^{-pH}\right]\right)}{t}} \Big]$$

piGTch1 = derivetrGibbsT[pisp]

$$-0.00831451\, t\, \mathrm{Log}\Big[e^{\frac{120.272\left(-1302.6\,(1-0.00335402\,t)-3.81452\,t+\frac{is^{0.5}\left(0.00920483\,t-0.0000128467\,t^2+4.95199\times10^{-8}\,t^3\right)}{1+1.6\,is^{0.5}}-0.016629\,t\,\mathrm{Log}\left[10^{-pH}\right]\right)}{t}} +$$

$$e^{\frac{120.272\left(-1299.\,(1-0.00335402\,t)-3.67634\,t+\frac{3\,is^{0.5}\left(0.00920483\,t-0.0000128467\,t^2+4.95199\times10^{-8}\,t^3\right)}{1+1.6\,is^{0.5}}-0.00831451\,t\,\mathrm{Log}\left[10^{-pH}\right]\right)}{t}} \Big]$$

h2oGTch1 = derivetrGibbsT[h2osp]

$$-0.00831451\, t\, \text{Log}\left[e^{-\frac{120.272\left(-285.83\,(1-0.00335402\,t)-0.795539\,t+\frac{2\,\text{is}^{0.5}\,\left(0.00920483\,t-0.0000128467\,t^2+4.95199\times10^{-8}\,t^3\right)}{1+1.6\,\text{is}^{0.5}}-0.016629\,t\,\text{Log}\left[10^{-\text{pH}}\right]\right)}{t}} \right]$$

Table 1.9 Standard transformed Gibbs energies of reaction for ATP + H_2O = ADP + Pi as a function of temperature, pH, and ionic strength

```
PaddedForm[
  TableForm[(piGTch1 + adpGTch1 - (atpGTch1 + h2oGTch1)) /. t → {273.15, 298.15, 313.15} /.
    pH → {5, 6, 7, 8, 9} /. is → {0, .1, .25},
  TableHeadings → {{"273.15 K", "298.15 K", "313.15 K"}, {"    pH 5", "    pH 6",
    "    pH 7", "    pH 8", "    pH 9"}, {"I=0", "I=0.10", "I=0.25"}}], 4]
```

		pH 5		pH 6		pH 7		pH 8		pH 9
	I=0	-34.29	I=0	-34.95	I=0	-36.5	I=0	-40.78	I=0	-45.98
273.15 K	I=0.10	-32.66	I=0.10	-33.17	I=0.10	-35.4	I=0.10	-39.79	I=0.10	-44.92
	I=0.25	-32.03	I=0.25	-32.58	I=0.25	-34.94	I=0.25	-39.43	I=0.25	-44.57
	I=0	-35.3	I=0	-35.91	I=0	-37.6	I=0	-42.5	I=0	-48.29
298.15 K	I=0.10	-33.3	I=0.10	-33.87	I=0.10	-36.5	I=0.10	-41.48	I=0.10	-47.1
	I=0.25	-32.56	I=0.25	-33.22	I=0.25	-36.04	I=0.25	-41.07	I=0.25	-46.7
	I=0	-35.9	I=0	-36.49	I=0	-38.27	I=0	-43.53	I=0	-49.68
313.15 K	I=0.10	-33.67	I=0.10	-34.28	I=0.10	-37.16	I=0.10	-42.48	I=0.10	-48.4
	I=0.25	-32.87	I=0.25	-33.59	I=0.25	-36.69	I=0.25	-42.05	I=0.25	-47.97

Tables like this can also be made for standard transformed enthalpies of reaction, standard transformed entropies of reaction, and changes in the binding of hydrogen ions in a reaction.

1.11 Discussion

This chapter on biochemical thermodynamics has been included because when there is a rate-determining reaction, all the reactions in the mechanism of catalysis prior to the rate-determining reaction are at equilibrium. When these reactions are at equilibrium, the composition can be calculated by use of a set of independent reactions and the conservation equation for enzymatic sites. Fortunately, this can be done using Solve in a personal computer, even for complicated mechanisms. Examples of complicated mechanisms are reactions like random A + B + C → products and mechanisms that include pKs of the enzymatic site, enzyme-substrate complexes, and the substrates.

Biochemical thermodynamics is different from chemical thermodynamics because the pH (and perhaps pMg) has to treated like temperature and pressure in chemical thermodynamics. In biochemical thermodynamics, T, P, and pH are independent variables set by the investigator. Therefore, the thermodynamic properties like equilibrium constants are functions of these variables. The way this is done is to use a Legendre transform to define a transformed Gibbs energy G'. There are corresponding transformed enthaplies H' and transformed entropies S'. When the pH is specified, it is necessary to deal with reactants like ATP, which are sums of species that are in equilibrium with each other. This is exactly what biochemists do when they write ATP + H_2O = ADP + Pi.

Chapter 2 A = P

2.1 Introduction to enzyme kinetics

2.2 A → Products

- 2.2.1 Derivation of the rapid - equilibrium rate equation
- 2.2.2 Estimation of kinetic parameters for A → products
- 2.2.3 Effects of pH on the velocity of A → products when the substrate does not have pKs
- 2.2.4 Effects of pH on the velocity of A → products when the substrate has two pKs

---

2.3 A = P

- 2.3.1 Derivation of the rapid - equilibrium rate equation using Solve
- 2.3.2 Estimation of kinetic parameters for A = P
- 2.3.3 Haldane relation for A = P
- 2.3.4 Effects of pH on the velocity for A = P
- 2.3.5 Haldane relation including pH effects for A = P

2.4 A → P when the apparent equilibrium constant is very large

- 2.4.1 Estimation of kinetic parameters when the apparent equilibrium constant is very large
- 2.4.2 Effects of pH on the velocity when the apparent equilibrium constant is very large
- 2.4.3 Haldane relation when the apparent equilibrium constant is very large

---

2.5 Appendix

2.6 Discussion

2.1 Introduction to enzyme kinetics

Enzyme kinetics deals with the rates of reactions catalyzed by enzymes, and there are three different types of studies: (1) transient- state kinetics, (2) steady-state kinetics, and (3) rapid-equilibrium kinetics. Transient-state kinetics deals with the very rapid changes that take place when solutions of enzyme and substrates are mixed or disturbed. This requires very special equipment to deal with very fast reactions, and it is not discussed here. When solutions of enzymes and substrates are mixed there is a very short induction period, followed by a nearly constant velocity, slowing velocity, and asymptotic approach to equilibrium. Henri (1902-1903) assumed that there was equilibrium between the free enzyme and the enzyme-substrate complexes, but experiments were pretty primitive at that time. The logarithmic scale of pH was not introduced until 1909 by Sorenson.

In 1913 Michaelis and Menten rediscovered the rate equation derived by Henri on the basis of simple chemical equilibrium principles and proposed the following mechanism:

$$E + A \rightleftharpoons EA \rightarrow E + P \tag{1}$$

They assumed that the first reaction was at equilibrium and the second reaction was a simple first-order reaction with rate equation k in sec^{-1} or min $^{-1}$. This mechanism led to the following expression for the velocity v of the catalyzed reaction in moles per liter or other unit of concentration:

$$v = \frac{k[E]_t}{1 + \frac{K_A}{[A]}} \tag{2}$$

$[E]_t$ is the total concentration of enzymatic sites, and K_A is an equilibrium constant. This is generally known as the Michaelis-Menten equation.

In 1925 Briggs and Haldane used the following more general rate equation:

$$E \quad + \quad A \quad \underset{k_2}{\overset{k_1}{\rightleftharpoons}} \quad X \xrightarrow{k_3} E + P \tag{3}$$

where X is an enzyme-substrate complex. Since $[E] + [X] = [E]_t$ and $[A] + [P] = [A]_0$, there are two independent rate equations for this mechanism.

$$\frac{d[X]}{dt} = k_1[E][A] - (k_2 + k_3)[X] \tag{4}$$

$$\frac{d[P]}{dt} = k_3[X] \tag{5}$$

Since enzymatic reactions are generally studied with enzyme concentrations (strictly speaking, molar concentrations of enzymatic sites) much lower than the concentrations of substrates, it is a good approximation to assume that after a very short transition period the enzymatic reaction is in a steady state in which $d[X]/dt = 0$. By introducing the equation for the conservation of enzymatic sites, $[E] = [E]_t - [X]$ and $d[A]/dt = 0$, into equation 4, we obtain

$$[X] = \frac{k_1[E]_0[A]}{k_1[A] + k_2 + k_3} \tag{6}$$

Substituting this expression in equation 5 yields

$$\frac{d[P]}{dt} = \frac{k_3[E]_t}{1 + (k_2 + k_3)/k_1[A]} \tag{7}$$

This is the steady-state rate equation for mechanism 3. The limiting velocity is given by $k_3[E]_t$ and the Michaelis constant is given by $(k_2 + k_3)/k_1$. The limiting velocity and the Michaelis constant can be determined by measuring two or more velocities $d[P]/dt$ at different substrate concentrations, but this does not provide enough information to calculate k_1 and k_2. Steady-state rate equations always involve more kinetic parameters than can be calculated from experimental velocities.

The concept of a steady-state can be applied to enzyme-catalyzed reactions with more reactants, and in 1956, E. L. King and C. Altman, J. Phys. Chem. 60, 1375-1378 published a method to derive steady-state rate equations for more complicated mechanisms using linear algebra. This method and nomenclature was further developed by W. W. Cleland, Biochem Biophys. Acta 67, 104-137 (1963), and it has been widely used.

In 1975 I. H. Segel published "Enzyme Kinetics: Behavoir and Analysis of Rapid-Equilibrium and Steady-State Enzyme Systems, Wiley". He pointed out that the general procedure of Henri-Michaelis-Menten that assumes equilibrium can be used to obtain velocity equations for more complicated reactions and mechanisms. He described the use of the rapid-equilibrium assumption as the simplest and most direct method to obtain rate equations for complex multiligand systems, and pointed out that "If the experimental data fit the velocity equation, then we have the simplest kinetic mechanism for the system. If the data do not fit, then we can proceed to more complex models and velocity equations." He also pointed out that "The major theories of allosteric enzymes are based on the rapid-equilibrium assumption."

2.2 A → Products

The kinetics for the forward reaction in an enzyme-catalyzed isomerization reaction or hydration reaction can be represented as A → P because the concentration of H_2O is omitted when reactions are studied in dilute aqueous solutions. The derivation of the rapid-equilibrium rate equation, the estimation of kinetic parameters, and the treatment of pH effects are discussed in textbooks on enzyme kinetics [6,21,25], but this chapter shows how *Mathematica* can be used to make derivations and calculations. The effects of pH can be very complicated, even for this simple reaction. The inclusion of the reverse reaction leads to the Haldane relation and product inhibition, even when the apparent equilibrium constant for the reaction is much greater than one.

■ **2.2.1 Derivation of the rapid - equilibrium rate equation**

The rapid-equilibrium rate equation for A → products is derived from one equilibrium equation and one conservation equation. The equilibrium constant for the dissociation of EA depends on the temperature, pH, and ionic strength.

The mechanism is

E + A = EA $K_A = [E][A]/[EA]$ kA = e*a/ea

EA → products   v = kf*ea = kf*et*(ea/et) = kfet*(ea/et)

In *Mathematica*, * is the multiply sign; an empty space is also a multiply sign. Symbols for mathematical quantities in *Mathematica* have to begin with lower case letters because operations like Plot and Table begin with upper case letters. Subscripts are avoided because they are more difficult to type. The product of kf and et is treated as a single kinetic parameter because their separate effects are not discussed here. The limiting velocity kfet is approached as [A] is raised to very high values. The total concentration of enzymatic sites $[E]_t$ is given by

$$[E]_t = [E] + [EA] = K_A[EA]/[A] + [EA] = [EA](1 + K_A/[A])$$

In *Mathematica* this is written as

et = e + ea = kA*ea/a + ea = ea (1 + kA/a)

ea/et = 1/(1 + kA/a)

The velocity v is given by

v = kfet/(1 + kA/a)

kfet is the limiting velocity in the forward direction at a specified pH. The limiting velocity kfet and the Michaelis constant kA are both functions of temperature, pH, and ionic strength. The expression for the velocity can be written in *Mathematica* as follows:

vA = kfet / (1 + kA / a)

$$\frac{kfet}{1 + \dfrac{kA}{a}}$$

Now any time we type vA in *Mathematica* we will get this equation.

vA

$$\frac{kfet}{1 + \dfrac{kA}{a}}$$

The velocity can be plotted versus [A] for specified values of kfet and kA. The values of kfet and kA can be assigned using the ReplaceAll operation (/. x ->). If kfet = 1 and kA = 10, the velocity can be plotted as a function of substrate concentration.

Plot[vA /. kfet → 1 /. kA → 10, {a, 0, 100}, PlotRange → {0, 1}, AxesLabel → {"[A]", "velocity"}]

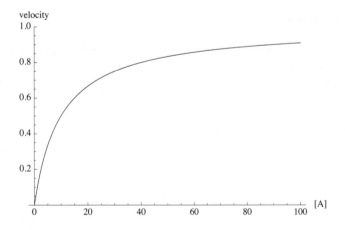

Fig. 2.1 Michaelis-Menten plot when kfet = 1 and kA = 10.

When the substrate concentration is very low compared with the Michaelis constant kA, kA/a is much greater than 1, and so the velocity is proportional to the substrate concentration: $v \approx$ (kfet/kA) a. When the substrate concentration is raised to high values, the velocity approaches kfet because the enzymatic sites are becoming saturated.

▪ 2.2.2 Estimation of kinetic parameters for A → products

Use of Lineweaver-Burk plots

The limiting velocities kfet and Michaelis constants kA for A → products can be calculated from experimental data by use of double reciprocal plots; these plots of 1/v versus 1/[A] are referred to as Lineweaver-Burk plots.

$$\frac{1}{v} = \frac{1}{kfet} + \frac{kA}{kfet[A]}$$

To plot velocity versus 1/a in *Mathematica*, it is necessary to replace a with 1/recipa using the ReplaceAll operation (/. x ->).

```
vA /. a → 1 / recipa
```

$$\frac{kfet}{1 + kA\ recipa}$$

The reciprocal velocity is given as a function of recipa by

```
1 / (vA /. a → 1 / recipa)
```

$$\frac{1 + kA\ recipa}{kfet}$$

The values of kfet and kA can be assigned using the ReplaceAll operation (/. x ->).

```
1 / (vA /. a → 1 / recipa) /. kfet -> 1 /. kA -> 10
```

```
1 + 10 recipa
```

```
Plot[1 / (vA /. a → 1 / recipa) /. kfet -> 1 /. kA -> 10,
  {recipa, 0, 1}, AxesLabel → {"1/[A]", "1/vA"}]
```

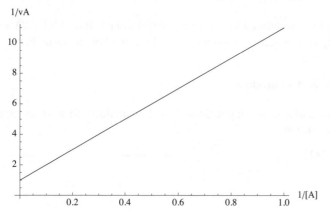

Fig. 1.2 Lineweaver-Burk plot when kfet = 1 and kA = 10.

The intercept is 1/kfet = 1 and the slope of 10 is equal to kA/kfet, so that kA = 10.

Use of two velocity measurements

Since there are 2 kinetic parameters for A → products, only 2 velocity measurements are required to estimate the 2 kinetic parameters. In 1979, Duggleby [9] suggested that the kinetic parameters for an enzyme-catalyzed reaction can be determined by measuring the velocities at as many sets of substrate concentrations as there are kinetic parameters. In 2008, Alberty [32] took advantage of the fact that Solve in *Mathematica* can be used to solve large sets of simultaneous polynomial equations to do this for more complicated mechanisms. For A → products, this can be done without a computer.

Two simultaneous rate equations for A → products at two different substrate concentrations are

v1 = kfet / (1 + kA / a1)

$$\frac{kfet}{1 + \frac{kA}{a1}}$$

v2 = kfet / (1 + kA / a2)

$$\frac{kfet}{1 + \frac{kA}{a2}}$$

To calculate the two kinetic parameters (vfexp and kA) from v1 and v2, these 2 rate equations are each solved for vfexp.

kfet = v1*(1 + kA/a1) = v2*(1 + kA/a2)

This equation can be written as

v1 + v1*kA/a1 = v2 + v2*kA/a2

Thus

$$v1 - v2 = kA*\left(\frac{v2}{a2} - \frac{v1}{a1}\right)$$

so that

$$kA = \frac{v1 - v2}{\left(\frac{v2}{a2} - \frac{v1}{a1}\right)}$$

When the Michaelis constant kA has been calculated from two velocities, kfet can be calculated using kfet = v1*(1 + kA/a). For A → products the calculation of 2 kinetic parameters using two velocity measurements can be carried out by hand, but a computer is needed when there is a larger number of kinetic parameters.

Use of Solve in *Mathematica* to derive the rate equation for A → products

The inputs in Solve[eqs,vars,elims] are equations, variables, and eliminations. For A → products, there is one equilibrium equation, one conservation equation, and e is to be eliminated.

Solve[{kA == e * a / ea, et == e + ea}, {ea}, {e}]

$$\left\{\left\{ea \to \frac{a\,et}{a + kA}\right\}\right\}$$

Note that double equal signs have to be used here. The rapid-equilibrium velocity is proportional to ea, and the proportionality constant is the rate constant kfexp. Thus the equation for the velocity vA can be obtained by replacing et with vfexp because kfet = kf*et.

$$vA = \frac{a\ kfet}{a + kA};$$

When a semicolon is put after an input, the output is not shown.

For the enzyme-catalyzed reaction A → products, only two velocity measurements are needed to calculate vfexp and kA. A computer is not need to do that, but a program is given here because programs are needed for more complicated mechanisms.

```
calc2kinpars[v1_, a1_, v2_, a2_] :=
  Module[{}, (*This program calculates 2 kinetic parameters
    from 2 experimental velocities for A → products*)
    Solve[{ (a1 kfet)/(a1 + kA) == v1, (a2 kfet)/(a2 + kA) == v2}, {kfet, kA}]]
```

To demonstrate this program, velocities are calculated at {a} = {100}, {1} when the kinetic parameters are kfet = 1 and kA = 10.

```
a kfet
------  /. kfet → 1 /. kA → 10 /. a → 100 // N
a + kA
```
0.909091

```
a kfet
------  /. kfet → 1 /. kA → 10 /. a → 1 // N
a + kA
```
0.0909091

```
calc2kinpars[.9091, 100, .09091, 1]
```
{{kfet → 1.00001, kA → 10.}}

These values are correct, but it is necessary to consider experimental errors in the velocity determinations. 5% errors are introduced in the velocities, one at a time.

```
calc2kinpars[1.05 * .9091, 100, .09091, 1]
```
{{kfet → 1.05588, kA → 10.6145}}

```
calc2kinpars[.9091, 100, 1.05 * .09091, 1]
```
{{kfet → 0.994747, kA → 9.42105}}

The effects of temperature on kfet and kA are discussed in Appendix 3 of this chapter.

■ **2.2.3 Effects of pH on the velocity of A → products when the substrate does not have pKs**

In the following mechanism for the forward reaction, $H_2 E^+$ and $H_2 EA^+$ each have 2 pKs.

$$E^-\qquad\qquad EA^-$$
$$\|\ pK_{E1}\qquad\quad \|\ pK_{EA1}$$
$$HE +\ \ A\ \ =\ HEA \rightarrow products\qquad K_{cHEA} = [HE][A]/[HEA]\qquad v = kfet*(hea/et)$$
$$\|\ pK_{E2}\qquad\quad \|\ pK_{EA2}$$
$$H_2 E^+\qquad\ H_2 EA^+$$

K_{cHEA} is a chemical equilibrium constant; that is, it is written in terms of concentrations of species. This mechanism involves 5 equilibrium expressions and a conservation equation for the 6 species of the enzyme. There are 8 reacting species and 3 components (E, A, and H). The mechanism is written in terms of chemical species, but H^+ is not shown explicitly. The number of species N is equal to the number of components C plus the number of independent chemical reactions R; $N = C + R$ is $8 = 3 + 5$ (see Section 1.2). The 5 independent equilibrium expressions in *Mathematica* form are

$$kHE = h*e/he = 10^{-pKE1}$$
$$kH2E = h*he/h2e = 10^{-pKE2}$$
$$kHEA = h*ea/hea = 10^{-pKEA1}$$
$$kH2EA = h*hea/h2ea = 10^{-pKEA2}$$
$$kcHEA = he*a/hea$$

The specified concentration of hydrogen ions is represented by h. In addition to the 5 independent reactions, enzymatic sites have to be conserved.

et = e + he + h2e + ea + hea + h2ea

The expression for the equilibrium concentration of HEA can be calculated using Solve[eqns,vars,elims].

```
Solve[{kHE == h * e / he, kH2E == h * he / h2e, kHEA == h * ea / hea, kH2EA == h * hea / h2ea,
    kcHEA == he * a / hea, et == e + he + h2e + ea + hea + h2ea}, {hea}, {e, he, h2e, ea, h2ea}]
```

$$\left\{\left\{hea \rightarrow (a\ et\ h\ kH2E\ kH2EA)\ /\ (a\ h^2\ kH2E + h^2\ kcHEA\ kH2EA +\right.\right.$$
$$\left.\left. a\ h\ kH2E\ kH2EA + h\ kcHEA\ kH2E\ kH2EA + kcHEA\ kH2E\ kH2EA\ kHE + a\ kH2E\ kH2EA\ kHEA)\right\}\right\}$$

The spaces in the output indicate multiplications.

As shown in the mechanism, the velocity is given by v = kfet*(hea/et). kfet = kf*et is treated as a single property here. Solve has provided the expression for hea (the equilibrium concentration of HEA), and so this expression needs to be divided by et and be multiplied by kfet to obtain the expression for the velocity. Thus, et in the Solve output is replaced with kfet to obtain the rapid-equilibrium velocity v6par for this mechanism.

```
v6par = (a kfet h kH2E kH2EA) / (a h² kH2E + h² kcHEA kH2EA +
    a h kH2E kH2EA + h kcHEA kH2E kH2EA + kcHEA kH2E kH2EA kHE + a kH2E kH2EA kHEA)
```

$$(a\ h\ kfet\ kH2E\ kH2EA)\ /\ (a\ h^2\ kH2E + h^2\ kcHEA\ kH2EA +$$
$$a\ h\ kH2E\ kH2EA + h\ kcHEA\ kH2E\ kH2EA + kcHEA\ kH2E\ kH2EA\ kHE + a\ kH2E\ kH2EA\ kHEA)$$

The rate equation was derived using h, rather than pH, because Solve was developed to solve sets of simultaneous polynomial equations and does not work on exponentials.

The following kinetic parameters are chosen arbitrarily for test calculations.

```
Grid[{{kHE, kH2E, kHEA, kH2EA, kcHEA, kfet}, {10^-8, 10^-6, 10^-7.5, 10^-5.5, 4, 1}}]
```

kHE	kH2E	kHEA	kH2EA	kcHEA	kfet
$\dfrac{1}{100\,000\,000}$	$\dfrac{1}{1\,000\,000}$	3.16228×10^{-8}	3.16228×10^{-6}	4	1

Calculation of 6 kinetic parameters from 6 velocities

The first step is to put the kinetic parameters into the rate equation using the ReplaceAll operation (/.x->).

vtest = v6par /. kHE → 10^-8 /. kH2E → 10^-6 /. kHEA → 10^-7.5 /. kH2EA → 10^-5.5 /. kcHEA → 4 /. kfet → 1

$$\frac{3.16228 \times 10^{-12} \, a \, h}{1.26491 \times 10^{-19} + 1. \times 10^{-19} \, a + 1.26491 \times 10^{-11} \, h + 3.16228 \times 10^{-12} \, a \, h + 0.0000126491 \, h^2 + \frac{a \, h^2}{1\,000\,000}}$$

Six velocities are calculated with high and low [A] at three pHs:

vtest /. h → 10^-9 /. a → 100

0.0302452

vtest /. h → 10^-9 /. a → 1

0.0130502

vtest /. h → 10^-7 /. a → 100

0.716409

vtest /. h → 10^-7 /. a → 1

0.162658

vtest /. h → 10^-5 /. a → 100

0.217133

vtest /. h → 10^-5 /. a → 1

0.0207601

These six velocities will be treated as experimental data and will be used to calculate the six pH-independent kinetic parametrs in two different ways:

(1) The first way is to write a program so that the six simultaneous equations for these velocities can be solved for the six kinetic parameters.

(2) The second way is to use the two measurements at each pH to calculate vfexp and kA at that pH. Then another program is used to estimate the pH-independent kinetic parameters.

These velocities will now be treated as experimental velocities to show that the kinetic parameters can be estimated by solving six simultaneous equations using Solve. The following program has been written to calculate the 6 kinetic parameters from 6 measurements of velocities.

```
calc6kinpars[v1_, h1_, a1_, v2_, h2_, a2_, v3_,
  h3_, a3_, v4_, h4_, a4_, v5_, h5_, a5_, v6_, h6_, a6_] := Module[{},
  (*This program calculates 6 kinetic parameters from 6 experimental velocities for A →
    products when E and EA each have 2pKs.  Velocities are measured at high pH,
    medium pH, and low pH at [A]=1 and 100.*)
  Solve[{ (a1 kfet h1 kH2E kH2EA) / (a1 h1² kH2E + h1² kcHEA kH2EA + a1 h1 kH2E kH2EA +
      h1 kcHEA kH2E kH2EA + kcHEA kH2E kH2EA kHE + a1 kH2E kH2EA kHEA) == v1,
    (a2 kfet h2 kH2E kH2EA) / (a2 h2² kH2E + h2² kcHEA kH2EA + a2 h2 kH2E kH2EA +
      h2 kcHEA kH2E kH2EA + kcHEA kH2E kH2EA kHE + a2 kH2E kH2EA kHEA) == v2,
    (a3 kfet h3 kH2E kH2EA) / (a3 h3² kH2E + h3² kcHEA kH2EA + a3 h3 kH2E kH2EA +
      h3 kcHEA kH2E kH2EA + kcHEA kH2E kH2EA kHE + a3 kH2E kH2EA kHEA) == v3,
    (a4 kfet h4 kH2E kH2EA) / (a4 h4² kH2E + h4² kcHEA kH2EA + a4 h4 kH2E kH2EA +
      h4 kcHEA kH2E kH2EA + kcHEA kH2E kH2EA kHE + a4 kH2E kH2EA kHEA) == v4,
    (a5 kfet h5 kH2E kH2EA) / (a5 h5² kH2E + h5² kcHEA kH2EA + a5 h5 kH2E kH2EA +
      h5 kcHEA kH2E kH2EA + kcHEA kH2E kH2EA kHE + a5 kH2E kH2EA kHEA) == v5,
    (a6 kfet h6 kH2E kH2EA) / (a6 h6² kH2E + h6² kcHEA kH2EA + a6 h6 kH2E kH2EA + h6 kcHEA kH2E kH2EA +
      kcHEA kH2E kH2EA kHE + a6 kH2E kH2EA kHEA) == v6}, {kHE, kH2E, kHEA, kH2EA, kcHEA, kfet}]]
```

The six velocities are used to calculate the six kinetic parameters assuming that there are no experimental errors in the measurements of the velocities.

```
calc6kinpars[.030245, 10^-9, 100, .013050, 10^-9, 1, .71641,
  10^-7, 100, .16266, 10^-7, 1, .21713, 10^-5, 100, .020760, 10^-5, 1]
```

$\{\{kfet \to 1., kHEA \to 3.16231 \times 10^{-8}, kHE \to 1.00005 \times 10^{-8},$
$kcHEA \to 3.99995, kH2E \to 9.99979 \times 10^{-7}, kH2EA \to 3.16221 \times 10^{-6}\}\}$

In order to make a table, the values of the kinetic parameters are extracted from the output.

```
line1 = {kfet, kHEA, kHE, kcHEA, kH2E, kH2EA} /. calc6kinpars[.030245, 10^-9, 100, .013050,
    10^-9, 1, .71641, 10^-7, 100, .16266, 10^-7, 1, .21713, 10^-5, 100, .020760, 10^-5, 1]
```

$\{\{1., 3.16231 \times 10^{-8}, 1.00005 \times 10^{-8}, 3.99995, 9.99979 \times 10^{-7}, 3.16221 \times 10^{-6}\}\}$

The acid dissociation constants are converted to pKs because it is easier to interpret pKs.

```
line1pK = {line1[[1, 1]], -Log[10, line1[[1, 2]]], -Log[10, line1[[1, 3]]],
  line1[[1, 4]], -Log[10, line1[[1, 5]]], -Log[10, line1[[1, 6]]]}
```

{1., 7.5, 7.99998, 3.99995, 6.00001, 5.50001}

Exactly correct values of kinetic parameters are obtained, but the effects of 5% errors in the velocity measurements, one by one, are calculated to indicate the effects of experimental errors.

Calculate effects of 5% errors in the six velocities, one at a time.

```
calc6kinpars[1.05 * .030245, 10^-9, 100, .013050, 10^-9, 1, .71641,
  10^-7, 100, .16266, 10^-7, 1, .21713, 10^-5, 100, .020760, 10^-5, 1]
```

$\{\{kfet \to 0.984036, kHEA \to 2.95372 \times 10^{-8}, kHE \to 1.04445 \times 10^{-8},$
$kcHEA \to 3.92011, kH2E \to 9.95523 \times 10^{-7}, kH2EA \to 3.22993 \times 10^{-6}\}\}$

```
line2 = {kfet, kHEA, kHE, kcHEA, kH2E, kH2EA} /.
  calc6kinpars[1.05 * .030245, 10^-9, 100, .013050, 10^-9, 1, .71641,
    10^-7, 100, .16266, 10^-7, 1, .21713, 10^-5, 100, .020760, 10^-5, 1]
```

$\{\{0.984036, 2.95372 \times 10^{-8}, 1.04445 \times 10^{-8}, 3.92011, 9.95523 \times 10^{-7}, 3.22993 \times 10^{-6}\}\}$

```
line2pK = {line2[[1, 1]], -Log[10, line2[[1, 2]]], -Log[10, line2[[1, 3]]],
   line2[[1, 4]], -Log[10, line2[[1, 5]]], -Log[10, line2[[1, 6]]]}
```

{0.984036, 7.52963, 7.98111, 3.92011, 6.00195, 5.49081}

A 5 % error in an acid dissociation constant corresponds with a 0.02 error in pK.

```
Log[10, 1.05]
```

0.0211893

Calculate the effect of a 5% error in 0.013050.

```
calc6kinpars[.030245, 10^-9, 100, 1.05 * .013050, 10^-9, 1, .71641,
   10^-7, 100, .16266, 10^-7, 1, .21713, 10^-5, 100, .020760, 10^-5, 1]
```

$\{\{kfet \to 1.00038, kHEA \to 3.16722 \times 10^{-8}, kHE \to 8.98511 \times 10^{-9},$
$\quad kcHEA \to 4.03907, kH2E \to 1.01032 \times 10^{-6}, kH2EA \to 3.16065 \times 10^{-6}\}\}$

```
line3 = {kfet, kHEA, kHE, kcHEA, kH2E, kH2EA} /.
   calc6kinpars[.030245, 10^-9, 100, 1.05 * .013050, 10^-9, 1, .71641,
     10^-7, 100, .16266, 10^-7, 1, .21713, 10^-5, 100, .020760, 10^-5, 1]
```

$\{\{1.00038, 3.16722 \times 10^{-8}, 8.98511 \times 10^{-9}, 4.03907, 1.01032 \times 10^{-6}, 3.16065 \times 10^{-6}\}\}$

```
line3pK = {line3[[1, 1]], -Log[10, line3[[1, 2]]], -Log[10, line3[[1, 3]]],
   line3[[1, 4]], -Log[10, line3[[1, 5]]], -Log[10, line3[[1, 6]]]}
```

{1.00038, 7.49932, 8.04648, 4.03907, 5.99554, 5.50022}

Calculate the effects of a 5% error in 0.71641:

```
calc6kinpars[.030245, 10^-9, 100, .013050, 10^-9, 1, 1.05 * .71641,
   10^-7, 100, .16266, 10^-7, 1, .21713, 10^-5, 100, .020760, 10^-5, 1]
```

$\{\{kfet \to 1.07355, kHEA \to 3.40225 \times 10^{-8}, kHE \to 9.81522 \times 10^{-9},$
$\quad kcHEA \to 4.36769, kH2E \to 1.01885 \times 10^{-6}, kH2EA \to 2.88311 \times 10^{-6}\}\}$

```
line4 = {kfet, kHEA, kHE, kcHEA, kH2E, kH2EA} /.
   calc6kinpars[.030245, 10^-9, 100, .013050, 10^-9, 1, 1.05 * .71641,
     10^-7, 100, .16266, 10^-7, 1, .21713, 10^-5, 100, .020760, 10^-5, 1]
```

$\{\{1.07355, 3.40225 \times 10^{-8}, 9.81522 \times 10^{-9}, 4.36769, 1.01885 \times 10^{-6}, 2.88311 \times 10^{-6}\}\}$

```
line4pK = {line4[[1, 1]], -Log[10, line4[[1, 2]]], -Log[10, line4[[1, 3]]],
   line4[[1, 4]], -Log[10, line4[[1, 5]]], -Log[10, line4[[1, 6]]]}
```

{1.07355, 7.46823, 8.0081, 4.36769, 5.99189, 5.54014}

Calculate the effects of a 5% error in 0.16266

```
calc6kinpars[.030245, 10^-9, 100, .013050, 10^-9, 1, .71641, 10^-7,
   100, 1.05 * .16266, 10^-7, 1, .21713, 10^-5, 100, .020760, 10^-5, 1]
```

$\{\{kfet \to 0.996995, kHEA \to 3.15249 \times 10^{-8}, kHE \to 1.0898 \times 10^{-8},$
$\quad kcHEA \to 3.68708, kH2E \to 9.17622 \times 10^{-7}, kH2EA \to 3.17478 \times 10^{-6}\}\}$

```
line5 = {kfet, kHEA, kHE, kcHEA, kH2E, kH2EA} /.
   calc6kinpars[.030245, 10^-9, 100, .013050, 10^-9, 1, .71641, 10^-7,
     100, 1.05 * .16266, 10^-7, 1, .21713, 10^-5, 100, .020760, 10^-5, 1]
```

$\{\{0.996995, 3.15249 \times 10^{-8}, 1.0898 \times 10^{-8}, 3.68708, 9.17622 \times 10^{-7}, 3.17478 \times 10^{-6}\}\}$

```
line5pK = {line5[[1, 1]], -Log[10, line5[[1, 2]]], -Log[10, line5[[1, 3]]],
   line5[[1, 4]], -Log[10, line5[[1, 5]]], -Log[10, line5[[1, 6]]]}
```

{0.996995, 7.50135, 7.96265, 3.68708, 6.03734, 5.49829}

Calculate the effects of a 5% error in 0.21713.

```
calc6kinpars[.030245, 10^-9, 100, .013050, 10^-9, 1, .71641, 10^-7,
   100, .16266, 10^-7, 1, 1.05 * .21713, 10^-5, 100, .020760, 10^-5, 1]
```

$\{\{kfet \to 0.997748, kHEA \to 3.15495 \times 10^{-8}, kHE \to 1.00067 \times 10^{-8},$
$\quad kcHEA \to 3.98867, kH2E \to 9.93854 \times 10^{-7}, kH2EA \to 3.41072 \times 10^{-6}\}\}$

```
line6 = {kfet, kHEA, kHE, kcHEA, kH2E, kH2EA} /.
   calc6kinpars[.030245, 10^-9, 100, .013050, 10^-9, 1, .71641, 10^-7,
      100, .16266, 10^-7, 1, 1.05 * .21713, 10^-5, 100, .020760, 10^-5, 1]
```

$\{\{0.997748, 3.15495 \times 10^{-8}, 1.00067 \times 10^{-8}, 3.98867, 9.93854 \times 10^{-7}, 3.41072 \times 10^{-6}\}\}$

```
line6pK = {line6[[1, 1]], -Log[10, line6[[1, 2]]], -Log[10, line6[[1, 3]]],
   line6[[1, 4]], -Log[10, line6[[1, 5]]], -Log[10, line6[[1, 6]]]}
```

{0.997748, 7.50101, 7.99971, 3.98867, 6.00268, 5.46715}

Calculate the effects of a 5% error in 0.020760

```
calc6kinpars[.030245, 10^-9, 100, .013050, 10^-9, 1, .71641, 10^-7,
   100, .16266, 10^-7, 1, .21713, 10^-5, 100, 1.05 * .020760, 10^-5, 1]
```

$\{\{kfet \to 1.00024, kHEA \to 3.16308 \times 10^{-8}, kHE \to 9.93589 \times 10^{-9},$
$\quad kcHEA \to 4.02454, kH2E \to 1.06841 \times 10^{-6}, kH2EA \to 3.13823 \times 10^{-6}\}\}$

```
line7 = {kfet, kHEA, kHE, kcHEA, kH2E, kH2EA} /.
   calc6kinpars[.030245, 10^-9, 100, .013050, 10^-9, 1, .71641, 10^-7,
      100, .16266, 10^-7, 1, .21713, 10^-5, 100, 1.05 * .020760, 10^-5, 1]
```

$\{\{1.00024, 3.16308 \times 10^{-8}, 9.93589 \times 10^{-9}, 4.02454, 1.06841 \times 10^{-6}, 3.13823 \times 10^{-6}\}\}$

```
line7pK = {line7[[1, 1]], -Log[10, line7[[1, 2]]], -Log[10, line7[[1, 3]]],
   line7[[1, 4]], -Log[10, line7[[1, 5]]], -Log[10, line7[[1, 6]]]}
```

{1.00024, 7.49989, 8.00279, 4.02454, 5.97126, 5.50331}

Table 2.1 Values of kinetic parameters for A $\to$ products when $H_2 E^+$ and $H_2 EA^+$ each have two pKs and velocities are measured at {h,a} = $\{10^{-9}, 100\}$, $\{10^{-9}, 1\}$, $\{10^{-7}, 100\}$, $\{10^{-7}, 1\}$, $\{10^{-5}, 100\}$, $\{10^{-5}, 1\}$,

```
TableForm[Round[{line1pK, line2pK, line3pK, line4pK, line5pK, line6pK, line7pK}, .01],
   TableHeadings -> {{"No errors", "1.05*v1", "1.05*v2", "1.05*v3", "1.05*v4", "1.05*v5",
      "1.05*v6", "1.05*v7"}, {"kfet", "pKHEA", "pKHE", "KcHEA", "pKH2E", "pKH2EA"}}]
```

	kfet	pKHEA	pKHE	KcHEA	pKH2E	pKH2EA
No errors	1.	7.5	8.	4.	6.	5.5
1.05*v1	0.98	7.53	7.98	3.92	6.	5.49
1.05*v2	1.	7.5	8.05	4.04	6.	5.5
1.05*v3	1.07	7.47	8.01	4.37	5.99	5.54
1.05*v4	1.	7.5	7.96	3.69	6.04	5.5
1.05*v5	1.	7.5	8.	3.99	6.	5.47
1.05*v6	1.	7.5	8.	4.02	5.97	5.5

This table shows that kfet, KcHEA, and four pKs can be estimated from six velocity measurement. More accurate values of the kinetic parameters can be obtained by using wider ranges of pH and substrate concentrations.

Expression of the rate equation in terms of pH and pKs

The rate equation was derived using h, rather than pH, because Solve was developed to solve sets of polynomial equations. But in enzyme kinetics it is more convenient to use pH as a variable, rather than h, and pKs, rather than acid dissociation constants. Therefore, the rate equation v6parpH is written in terms of pH and pKs.

```
v6parpH = v6par /. h → 10^-pH /. kHE → 10^-pKE1 /. kH2E → 10^-pKE2 /. kHEA → 10^-pKEA1 /.
   kH2EA → 10^-pKEA2
```

$$\left(10^{-pH-pKE2-pKEA2} \, a \, kfet\right) \Big/ \left(10^{-2\,pH-pKE2} \, a + 10^{-pH-pKE2-pKEA2} \, a + 10^{-pKE2-pKEA1-pKEA2} \, a + 10^{-2\,pH-pKEA2} \, kcHEA + 10^{-pH-pKE2-pKEA2} \, kcHEA + 10^{-pKE1-pKE2-pKEA2} \, kcHEA\right)$$

This shows how a rate equation in terms of [H$^+$] can be converted into a rate equation in terms of pH. The rate equation in terms of pH can be used to calculate velocities as functions of the concentration of A and pH when the six kinetic parameters are specified.

Now the acid dissociation constants are expressed as pKs:

```
Grid[{{{pKE1, pKE2, pKEA1, pKEA2, kcHEA, kfet}, {8, 6, 7.5, 5.5, 4, 1}}}]
```

pKE1	pKE2	pKEA1	pKEA2	kcHEA	kfet
8	6	7.5	5.5	4	1

Substitute these 6 kinetic parameters in the rate equation to obtain the following expression of the velocity in terms of pH and [A].

```
v6parpH1 = v6parpH /. pKE1 → 8 /. pKE2 → 6 /. pKEA1 → 7.5 /. pKEA2 → 5.5 /. kcHEA → 4 /. kfet → 1
```

$$\frac{10^{-11.5-pH} \, a}{1.26491 \times 10^{-19} + 2^{-3.5-2\,pH}\,5^{-5.5-2\,pH} + 2^{-9.5-pH}\,5^{-11.5-pH} + 1. \times 10^{-19} \, a + 10^{-6-2\,pH} \, a + 10^{-11.5-pH} \, a}$$

For example, the velocity as a function of [A] at pH 5 for the arbitrary kinetic parameters is given by

```
v6parpH1 /. pH → 5
```

$$\frac{3.16228 \times 10^{-17} \, a}{1.39153 \times 10^{-15} + 1.31723 \times 10^{-16} \, a}$$

To put this rate equation in the usual form, divide the numerator and denominator with a*1.31723×10^{-16}. The numerator becomes

```
(3.1622776601683796`*^-17 a) / (a * 1.317227766016838`*^-16)
```

```
0.240071
```

The denominator becomes 1 plus

```
(1.3915286615804937`*^-15) / (a * 1.317227766016838`*^-16)
```

$$\frac{10.5641}{a}$$

Thus the rapid-equilibrium rate equation at pH 5 is

$$0.24007068039043722` \Big/ \left(1 + \frac{10.564070219900799`}{a}\right)$$

$$\frac{0.240071}{1 + \frac{10.5641}{a}}$$

This shows that at pH 5 the limiting velocity kfet is 0.240 and the Michaelis constant K_A is 10.6.

Make Michaelis-Menten plots at pHs 5, 6, 7, and 8.

```
plot3 = Plot[v6parpH1 /. pH → 5, {a, 0, 50},
    PlotRange → {0, .2}, AxesLabel → {"[A]", "v"}, PlotLabel -> "pH 5"];

plot4 = Plot[v6parpH1 /. pH → 6, {a, 0, 50},
    PlotRange → {0, .8}, AxesLabel → {"[A]", "v"}, PlotLabel -> "pH 6"];

plot5 = Plot[v6parpH1 /. pH → 7, {a, 0, 50},
    PlotRange → {0, .8}, AxesLabel → {"[A]", "v"}, PlotLabel -> "pH 7"];

plot6 = Plot[v6parpH1 /. pH → 8, {a, 0, 50},
    PlotRange → {0, .25}, AxesLabel → {"[A]", "v"}, PlotLabel -> "pH 8"];

GraphicsArray[{{plot3, plot4}, {plot5, plot6}}]
```

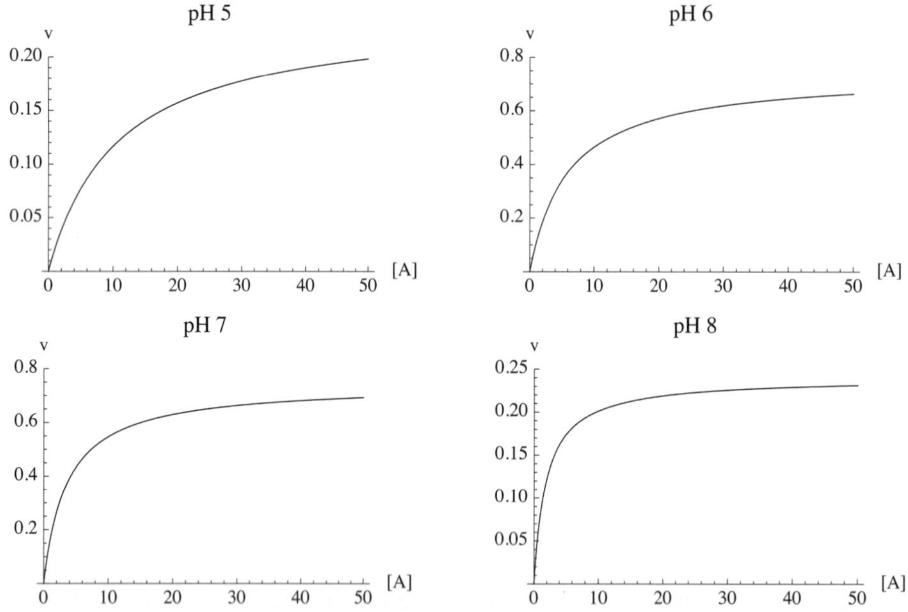

Fig. 2.3 Michaelis-Menten plots for A → products at four pHs. The underlying kinetic parameters are given above in the Grid table.

Very approximately these plots show that the Michaelis constants at pHs 5, 6, 7, and 8 are 10, 6, 4, and 2. Much higher [A] will have to be used to obtain the limiting velocity kfet.

Use calc2kinpars to estimate vfexp and kA at pHs 9, 7, and 5

When there are more reactants, mechanisms that take pKs into account become much more complicated and more velocities have to be determined. Therefore, it is fortunate that there is another way to determine the kinetic parameters. That way involves determining the pH-dependent kinetic parameters at three different pHs, as described in Section 2.3.2. When the enzymatic site and enzyme-substrate complexes each have two pKs, all the pH-independent kinetic parameters can be calculated by making plots of vfexp and certain ratios of kinetic parameters that are bell-shaped [33]. Bell-shaped plots of the type involved here are each characterized by three kinetic parameters that can be determined as described in Appendix 2 of this chapter.

The velocities are given under **Calculation of 6 kinetic parameters from 6 velocities** (in Section 2.2.3) makes it possible to calculate the two pH-dependent kinetic parameters (vfexp and kA) at three pHs.

At pH 9,

```
calc2kinpars[.03025, 100, .01305, 1]
```

$\{\{kfet \rightarrow 0.0306582, kA \rightarrow 1.34928\}\}$

At pH 7,

```
calc2kinpars[.7164, 100, .1627, 1]
```

$\{\{kfet \rightarrow 0.741903, kA \rightarrow 3.55995\}\}$

At pH 5

```
calc2kinpars[.2171, 100, .02076, 1]
```

$\{\{kfet \rightarrow 0.24003, kA \rightarrow 10.5622\}\}$

When the enzymatic site and the enzyme-substrate complex each has two pKs, the plots of vfexp and vfexp/kA versus pH are bell-shaped curves.

The values of vfexp at pHs 9, 7, and 5 make it possible to estimate three pH-independent kinetic parameters (kfet, k1, and k2).

```
calckfeth1h2[v1_, h1_, v2_, h2_, v3_, h3_] :=
 Module[{mat, kfet, k1, k2}, (*This program calculates kfet,k1,
   and k2 from 3 values of vfexp at 3 [H⁺].  The first velocity is at low [H⁺],
   the second velocity is at intermediate [H⁺],
   and the third velocity is at high [H⁺].  High is with respect to k1, and low is with
    respect to k2.  The acid dissociation constants in the output are expressed as pKs.*)
  mat = {kfet, k1, k2} /. Solve[{v1 == kfet / (1 + h1 / k2 + k1 / h1),
      v2 == kfet / (1 + h2 / k2 + k1 / h2),
      v3 == kfet / (1 + h3 / k2 + k1 / h3)}, {kfet, k1, k2}];
  kfet = mat[[1, 1]]; k1 = mat[[1, 2]]; k2 = mat[[1, 3]];
  {kfet, -Log[10, k1], -Log[10, k2]}]
```

```
two1 = calckfeth1h2[.03066, 10^-9, .7419, 10^-7, .2400, 10^-5]
```

$\{0.999897, 7.50015, 5.50011\}$

```
two2 = calckfeth1h2[1.05 * .03066, 10^-9, .7419, 10^-7, .2400, 10^-5]
```

$\{0.984301, 7.52908, 5.49112\}$

```
two3 = calckfeth1h2[.03066, 10^-9, 1.05 * .7419, 10^-7, .2400, 10^-5]
```

$\{1.06997, 7.46983, 5.53842\}$

```
two4 = calckfeth1h2[.03066, 10^-9, .7419, 10^-7, 1.05 * .2400, 10^-5]
```

$\{0.997877, 7.50105, 5.4708\}$

Table 2.2 Values of kinetic parameters for the effect of pH on V_{fexp} assuming 5% errors in the limiting velocities at pHs 9, 7, and 5

```
TableForm[Round[{two1, two2, two3, two4}, .01], TableHeadings →
  {{"No errors", "1.05*vfexp9", "1.05*vfexp7", "1.05*vfexp5"}, {"kfet", "pKHEA", "pKH2EA"}}]
```

	kfet	pKHEA	pKH2EA
No errors	1.	7.5	5.5
1.05*vfexp9	0.98	7.53	5.49
1.05*vfexp7	1.07	7.47	5.54
1.05*vfexp5	1.	7.5	5.47

The values of vfexp and kA calculated from measured velocities at the beginning of this section can be used to calculate the values of V_{fexp} and K_A from bell-shaped plots. The values of V_{fexp} and K_A at pHs 9, 7, and 6 are as follows:

At pH 9,

```
.03066 / 1.3493
```

0.0227229

At pH 7,

```
0.7419 / 3.560
```

0.208399

At pH 5,

```
.2400 / 10.56
```

0.0227273

These ratios can be put in calckfeth1h2 to obtain kfet/kcHEA, pK1ea, and pK2ea.

```
calckfeth1h2[.02272, 10^-9, .2084, 10^-7, .02273, 10^-5]
```

{0.250098, 7.9997, 6.00009}

The first value in the output is kfet/kcHEA, which is expected to be 1/4. The values of pK1e = 8.00 and pK2e = 6.00 are correct.

The effects of experimental errors in the values of V_{fexp}/K_A can be calculated by changing the three V_{fexp}/K_A by 5%, one at a time.

```
line1 = calckfeth1h2[.02272, 10^-9, .2084, 10^-7, .02273, 10^-5]
```

{0.250098, 7.9997, 6.00009}

```
line2 = calckfeth1h2[1.05 * .02272, 10^-9, .2084, 10^-7, .02273, 10^-5]
```

{0.248768, 8.02563, 5.99754}

```
line3 = calckfeth1h2[.02272, 10^-9, 1.05 * .2084, 10^-7, .02273, 10^-5]
```

{0.265586, 7.97109, 6.02871}

```
line4 = calckfeth1h2[.02272, 10^-9, .2084, 10^-7, 1.05 * .02273, 10^-5]
```

{0.248769, 8.00225, 5.97416}

Table 2.3 Values of kinetic parameters for the effect of pH on V_{fexp}/K_A assuming 5% errors in the experimentally determined ratios at pHs 9, 7, and 5

```
TableForm[Round[{line1, line2, line3, line4}, 0.01], TableHeadings →
   {{"No errors", "1.05*vfexp9", "1.05*vfexp7", "1.05*vfexp5"}, {"kfet/kA", "pKHE", "pKH2E"}}]
```

	kfet/kA	pKHE	pKH2E
No errors	0.25	8.	6.
1.05*vfexp9	0.25	8.03	6.
1.05*vfexp7	0.27	7.97	6.03
1.05*vfexp5	0.25	8.	5.97

The values of V_{fexp} and K_A can be plotted versus pH since the six pH-independent kinetic parameters can be calculated from six velocity measurements.

```
plot24a = Plot[1 / (1 + 10 ^ (5.5 - pH) + 10 ^ (-7.5 + pH)),
   {pH, 5, 9}, PlotRange → {0, 1}, AxesLabel → {"pH", "Vfexp"}];
```

```
plot24b = Plot[4 * (1 + 10 ^ (5.5 - pH) + 10 ^ (-7.5 + pH)) / (1 + 10 ^ (6 - pH) + 10 ^ (-8 + pH)),
   {pH, 5, 9}, PlotRange → {0, 15}, AxesLabel → {"pH", "KA"}];
```

```
GraphicsArray[{{plot24a, plot24b}}]
```

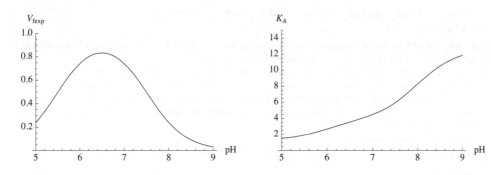

Fig. 2.4 (a) Plot of the limiting velocity versus pH. (b) Plot of the Michaelis constant versus pH.

▪ 2.2.4 Effects of pH on the velocity of A → products when the substrate has two pKs

In the following mechanism for the forward reaction, E, A, and EA each have 2 pKs.

$$
\begin{array}{lll}
E^- & A^- & HEA^- \\
\| \ pK_{E1} & \| \ pK_{A1} & \| \ pK_{EA1} \\
HE + \ HA & = H_2\,EA \rightarrow products & K_{H2EA} = [HE][EA]/[H_2\,EA] \qquad v = kf*et*h2ea/et = kfet*h2ea/et \\
\| \ pK_{E2} & \| \ pK_{A2} & \| \ pK_{EA2} \\
H_2\,E^+ & H_2\,A^+ & H_3\,EA^+
\end{array}
$$

To derive the rapid-equilibrium rate equation we will use a different approach from the preceding section; this is referred to as the pH factor approach. The total concentration of A is given by

$$[A] = [A^-] + [HA] + [H_2\,A^+] = K_{A1}[HA]/[H^+] + [HA] + [H^+][HA]/K_{A2}$$

where K_{A1} and K_{A2} are acid dissociation constants. This equation for [A] can be written in terms of pH and pKs (see Chapter 1):

$$[A] = [HA](1 + 10^{pH-pKA1} + 10^{pKA2-pH})$$

There are similar expressions for [E] and [EA]. The Michaelis constant for A is given by $K_A = [E][A]/[EA]$. (In biochemical thermodynamics this would be called an apparent equilibrium constant and have a prime on it because it is a function of pH.) Substituting the expression for [A] and the corresponding expressions for [E] and [EA] in the definition of K_A yields

$$K_A = K_{H2EA}(1 + 10^{pH-pKE1} + 10^{pKE2-pH})(1 + 10^{pH-pKA1} + 10^{pKA2-pH})/(1 + 10^{pH-pKEA1} + 10^{pKEA2-pH})$$

where K_{H2EA} is a chemical equilibrium constant.

The following factors make it possible to insert the pH dependencies of vfexp and kA into the rapid-equilibrium rate equation for A → P.

```
efactor = 1 + 10^pK2e-pH + 10^pH-pK1e;

eafactor = 1 + 10^pK2ea-pH + 10^pH-pK1ea;

afactor = 1 + 10^pK2a-pH + 10^pH-pK1a;
```

A *Mathematica* program can be written to use these factors to derive functions of pH for vfexp, kA, kAadj, vfexp/kAadj, and a*vfexp/(a + kA). The adjusted Michaelis constant is referred to as K_{Aadj} in Word and kaadj in *Mathematica*. Since A can be titrated, its pKs are considered to be known. The a factor can be taken out of the expression for the pH dependence of K_I.

Write a program to calculate the pH dependencies of vfexp, kA, and the velocity vA.

```
derAkinpH[pK1e_, pK2e_, pK1ea_, pK2ea_, pK1a_, pK2a_, kfEt_, kH2EA_] :=
Module[{eafactor, efactor,
    afactor, vfexp, kA, kAadj}, (*This program calculates 2 kinetic parameters (kA,vfexp) of
      the enzyme-catalyzed reaction A → products and vf as
      functions of pH. It also calculates the adjusted Michaelis constant for A,
    a ratio of kinetic parameters, and the expression for the initial reaction
      velocity vf in the absence of the product.  These calculations require 8
      inputs.  The output is a list of 5 functions, the last of which is the velocity.*)
    efactor = 1 + 10^pK2e-pH + 10^pH-pK1e;
    eafactor = 1 + 10^pK2ea-pH + 10^pH-pK1ea;
    afactor = 1 + 10^pK2a-pH + 10^pH-pK1a;
    vfexp = kfEt / eafactor;
    kA = kH2EA * efactor * afactor / eafactor;
    kAadj = kH2EA * efactor / eafactor;
    {vfexp, kA, kAadj, vfexp / kAadj, a * vfexp / (a + kA)}]
```

The outputs from this program are V_{fexp}, K_A, K_{Aadj}, V_{fexp}/K_{Aadj}, and v.

The following kinetic parameters are assigned arbitrarily:

```
Grid[{{pK1e, pK2e, pK1ea, pK2ea, pK1a, pK2a, kfEt, kH2EA}, {8, 6, 7.5, 5.5, 8, 5, 1, 4}}]
```

```
pK1e pK2e pK1ea pK2ea pK1a pK2a kfEt kH2EA
  8    6    7.5   5.5    8    5    1    4
```

```
derAkinpH[8, 6, 7.5, 5.5, 8, 5, 1, 4]
```

$$\left\{ \frac{1}{1 + 10^{5.5-pH} + 10^{-7.5+pH}}, \frac{4\left(1 + 10^{5-pH} + 10^{-8+pH}\right)\left(1 + 10^{6-pH} + 10^{-8+pH}\right)}{1 + 10^{5.5-pH} + 10^{-7.5+pH}}, \frac{4\left(1 + 10^{6-pH} + 10^{-8+pH}\right)}{1 + 10^{5.5-pH} + 10^{-7.5+pH}}, \right.$$

$$\left. \frac{1}{4\left(1 + 10^{6-pH} + 10^{-8+pH}\right)}, \frac{a}{\left(1 + 10^{5.5-pH} + 10^{-7.5+pH}\right)\left(\frac{4\left(1+10^{5-pH}+10^{-8+pH}\right)\left(1+10^{6-pH}+10^{-8+pH}\right)}{1+10^{5.5-pH}+10^{-7.5+pH}} + a\right)} \right\}$$

These kinetic parameters can be plotted as functions of pH.

```
plot25a =
  Plot[derAkinpH[8, 6, 7.5, 5.5, 8, 5, 1, 4][[1]], {pH, 5, 9}, AxesLabel → {"pH", "Vfexp"}];

plot25b = Plot[derAkinpH[8, 6, 7.5, 5.5, 8, 5, 1, 4][[2]],
    {pH, 5, 9}, AxesLabel → {"pH", "KA"}, PlotRange → {0, 25}];
```

```
plot25c = Plot[derAkinpH[8, 6, 7.5, 5.5, 8, 5, 1, 4][[3]],
    {pH, 5, 9}, AxesLabel → {"pH", "K_Aadj"}, PlotRange → {0, 12}];
```

```
plot25d = Plot[derAkinpH[8, 6, 7.5, 5.5, 8, 5, 1, 4][[4]],
    {pH, 5, 9}, AxesLabel → {"pH", "V_fexp / K_Aadj"}, PlotRange → {0, .25}];
```

```
GraphicsArray[{{plot25a, plot25b}, {plot25c, plot25d}}]
```

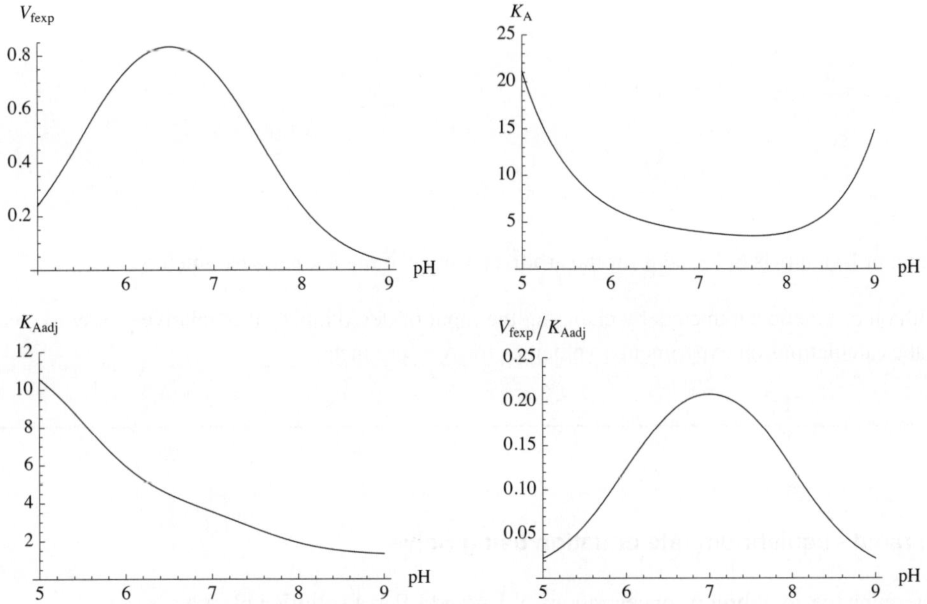

Fig. 2.5 Plots of kinetic parameters as functions of pH for arbitrary values of the 8 kinetic parameters. (a) V_{fexp}. (b) K_A. (c) K_{Aadj}. (d) V_{fexp}/K_{Aadj}.

If this was experimental data, kfET, pKE1 and pKE2 could be calculated from plot (a) (see Appendix 1). This plot makes it possible to obtain kfet, pK1e, and pK2e. The Michaelis constant K_A is a function of 7 kinetic parameters, and it would be difficult to calculate 7 parameters from this plot. pK1A and pK2A can be taken out by using K_{Aadj}.

The rate equation is expressed as a function of [A] and pH by

```
velAfpH = derAkinpH[8, 6, 7.5, 5.5, 8, 5, 1, 4][[5]]
```

$$\frac{a}{\left(1 + 10^{5.5-pH} + 10^{-7.5+pH}\right) \left(\frac{4\left(1+10^{5-pH}+10^{-8+pH}\right)\left(1+10^{6-pH}+10^{-8+pH}\right)}{1+10^{5.5-pH}+10^{-7.5+pH}} + a\right)}$$

This equation can be used to calculate plots of velocity versus [A] at desired pHs in the range pH 5 to 9.

```
Plot[velAfpH /. pH → {6, 7, 8}, {a, 0, 50}, AxesLabel → {"[A]", "velAfpH"}, PlotStyle → {Black}]
```

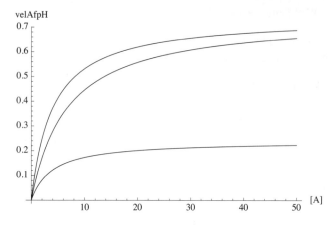

Fig. 2.6 Velocities as a function of [A] at pHs 6, 7, and 8 for the arbitrary values of the 8 kinetic parameters.

Plots can be made with other kinetic parameters by changing the input of derAkinpH. It is relatively easy to change the input of these programs to make calculations on experimental velocities for A → products.

2.3 A = P

■ 2.3.1 Derivation of the rapid - equilibrium rate equation using Solve

The equilibria that determine the equilibrium concentrations of EA and EP at a specified pH are

E + A = EA K_A = [E][A]/[EA]   kA = e*a/ea

E + P = EP K_P = [E][P]/[EP]   kP = e*p/ep

The total concentration of enzymatic sites $[E]_t$ is given by

$[E]_t$ = [E] + [EA] + [EP]   et = eq + ea + ep

It is not necessary to use *Mathematica* to derive the complete rate equation for A = P as described in Section 2.2.2, but *Mathematica* is used here as preparation for more complicated mechanisms where a computer is required. The equilibrium concentration of EA is obtained as follows:

```
Solve[{kA == e * a / ea, kP == e * p / ep, et == e + ea + ep}, {ea}, {e, ep}]
```

$$\left\{\left\{ea \rightarrow \frac{a \; et \; kP}{a \; kP + kA \; kP + kA \; p}\right\}\right\}$$

The equilibrium concentration of EP is obtained as follows :

```
Solve[{kA == e * a / ea, kP == e * p / ep, et == e + ea + ep}, {ep}, {e, ea}]
```

$$\left\{\left\{ep \rightarrow \frac{et \; kA \; p}{a \; kP + kA \; kP + kA \; p}\right\}\right\}$$

The complete mechanism is

E + A = EA K_A = [E][A]/[EA]

EA → products   vf = kf*ea = kf*et*(ea/et) = kfet*(ea/et) where kfet = kf*et

Therefore, divide ea by et and multiply by kfet to obtain vf.

E + P = EP K_P = [E][P]/[EP]

EP → products   vr = kr*ep = kret*(ep/et)

The velocity of A = B is obtained by multiplying [EA] by kfet and [EP] by kret.

$$\frac{a\ kfet\ kP}{a\ kP + kA\ kP + kA\ p} - \frac{kret\ kA\ p}{a\ kP + kA\ kP + kA\ p}$$

$$\frac{a\ kfet\ kP}{a\ kP + kA\ kP + kA\ p} - \frac{kA\ kret\ p}{a\ kP + kA\ kP + kA\ p}$$

Since the denominator terms are identical, the velocity vAP of A = P is given by

$$vAP = \frac{a\ kfet\ kP - kret\ kA\ p}{a\ kP + kA\ kP + kA\ p}$$

$$\frac{a\ kfet\ kP - kA\ kret\ p}{a\ kP + kA\ kP + kA\ p}$$

There are 4 kinetic parameters, and so the minimum number of velocity measurements to determine the values of the 4 kinetic parameters at a specified pH is 4. Various choices of pairs of substrate concentrations can be used.

▪ 2.3.2 Estimation of kinetic parameters for A = P

The estimation of the values of the 4 kinetic parameters can be tested by arbitrarily assigning values for the 4 kinetic parameters, calculating 4 velocities at chosen {[A],[P]}, and considering these velocities to be experimental data. The values of kinetic parameters for this test are chosen to be vfexp = 1, vrexp = 0.5, kA = 10, and kP = 20.

vAP /. kfet → 1 /. kret → 0.5 /. kA → 10 /. kP → 20

$$\frac{20\ a - 5.\ p}{200 + 20\ a + 10\ p}$$

The first pair of substrate concentrations {[A],[P]} is chosen to be {100,0} to obtain the predominant contribution to the determination of vfexp, the second pair is chosen to be {0,100} to obtain the predominant contribution to the determination of vrexp, the third pair is chosen to be {1,0} to obtain the predominant contribution to the determination of kA, and the fourth pair {0,1} is chosen to make the predominant contribution to the determination of kP.

vAP /. kfet → 1 /. kret → 0.5 /. kA → 10 /. kP → 20 /. a → 100 /. p → 0 // N

0.909091

vAP /. kfet → 1 /. kret → 0.5 /. kA → 10 /. kP → 20 /. a → 0 /. p → 100 // N

−0.416667

```
vAP /. kfet → 1 /. kret → 0.5 /. kA → 10 /. kP → 20 /. a → 1 /. p → 0 // N
```

0.0909091

```
vAP /. kfet → 1 /. kret → 0.5 /. kA → 10 /. kP → 20 /. a → 0 /. p → 1 // N
```

- 0.0238095

The input for the following program is

```
(*.90909,100,0,-.41667,0,100,.090909,1,0,-.023810,0,1*)
```

The (*...*) means that *Mathematica* will not operate on these numbers here.

```
calckinparsAP[v1_, a1_, p1_, v2_, a2_, p2_, v3_, a3_, p3_, v4_, a4_, p4_] :=
 Module[{}, (*This program calculates kfet, kA, kret,
    and kP from four experimental velocities for A = P at four pairs of
     substrate concentrations.  The first velocity is at high [A] and zero [B],
    the second velocity is at zero [A] and high [P], the third is at low [A] and zero [P],
    and the fourth velocity is at xero [A] and low [P]. High
     and low are with respect to the two Michaelis constants.*)
```

$$\text{Solve}\left[\left\{v1 == \frac{a1\,kfet\,kP - kret\,kA\,p1}{a1\,kP + kA\,kP + kA\,p1}, \; v2 == \frac{a2\,kfet\,kP - kret\,kA\,p2}{a2\,kP + kA\,kP + kA\,p2}, \right.\right.$$
$$\left.\left. v3 == \frac{a3\,kfet\,kP - kret\,kA\,p3}{a3\,kP + kA\,kP + kA\,p3}, \; v4 == \frac{a4\,kfet\,kP - kret\,kA\,p4}{a4\,kP + kA\,kP + kA\,p4}\right\}, \{kfet, kret, kA, kP\}\right]\right]$$

```
calckinparsAP[.90909, 100, 0, - .41667, 0, 100, .090909, 1, 0, - .023810, 0, 1]
```

{{kfet → 0.999999, kret → 0.500003, kA → 10., kP → 19.9997}}

```
line11 = {kfet, kret, kA, kP} /.
   calckinparsAP[.90909, 100, 0, - .41667, 0, 100, .090909, 1, 0, - .023810, 0, 1]
```

{{0.999999, 0.500003, 10., 19.9997}}

These values are correct, but it is necessary to take into account experimental errors in measured velocities and substrate concentrations. This is done by multiplyng the velocities, one by one, with 1.05. The effects of 5% errors in the substrate concentrations are easier to summarize; an error of 5% in a substrate concentration causes a 5% error in the corresponding Michaelis constant.

```
calckinparsAP[1.05 * .90909, 100, 0, - .41667, 0, 100, .090909, 1, 0, - .023810, 0, 1]
```

{{kfet → 1.05586, kret → 0.500003, kA → 10.6145, kP → 19.9997}}

```
line12 = {kfet, kret, kA, kP} /.
   calckinparsAP[1.05 * .90909, 100, 0, - .41667, 0, 100, .090909, 1, 0, - .023810, 0, 1]
```

{{1.05586, 0.500003, 10.6145, 19.9997}}

```
calckinparsAP[.90909, 100, 0, 1.05 * - .41667, 0, 100, .090909, 1, 0, - .023810, 0, 1]
```

{{kfet → 0.999999, kret → 0.530631, kA → 10., kP → 21.286}}

```
line13 = {kfet, kret, kA, kP} /.
   calckinparsAP[.90909, 100, 0, 1.05 * - .41667, 0, 100, .090909, 1, 0, - .023810, 0, 1]
```

{{0.999999, 0.530631, 10., 21.286}}

```
calckinparsAP[.90909, 100, 0, - .41667, 0, 100, 1.05 * .090909, 1, 0, - .023810, 0, 1]
```

{{kfet → 0.994736, kret → 0.500003, kA → 9.42105, kP → 19.9997}}

```
line14 = {kfet, kret, kA, kP} /.
   calckinparsAP[.90909, 100, 0, -.41667, 0, 100, 1.05 * .090909, 1, 0, -.023810, 0, 1]
```

$\{\{0.994736, 0.500003, 9.42105, 19.9997\}\}$

```
calckinparsAP[.90909, 100, 0, -.41667, 0, 100, .090909, 1, 0, 1.05 * -.023810, 0, 1]
```

$\{\{kfet \to 0.999999, kret \to 0.495003, kA \to 10., kP \to 18.7997\}\}$

```
line15 = {kfet, kret, kA, kP} /.
   calckinparsAP[.90909, 100, 0, -.41667, 0, 100, .090909, 1, 0, 1.05 * -.023810, 0, 1]
```

$\{\{0.999999, 0.495003, 10., 18.7997\}\}$

Table 2.4 Values of kinetic parameters for A = P obtained using {100,0}, {0,100}, {1,0}, and {0,1} and testing the effects of 5% errors in the measured velocities

```
TableForm[Round[{line11[[1]], line12[[1]], line13[[1]], line14[[1]], line15[[1]]}, 0.01],
   TableHeadings →
     {{"No errors", "1.05*v1", "1.05*v2", "1.05*v3", "1.05*v4"}, {"kfet", "kret", "kA", "kP"}}]
```

	kfet	kret	kA	kP
No errors	1.	0.5	10.	20.
1.05*v1	1.06	0.5	10.61	20.
1.05*v2	1.	0.53	10.	21.29
1.05*v3	0.99	0.5	9.42	20.
1.05*v4	1.	0.5	10.	18.8

■ 2.3.3 Haldane relation for A = P

The expression for the apparent equilibrium constant K' is obtained by setting the rate equal to zero at the end of Section 2.3.1. Thus at equilibrium,

$$\frac{kfet\, K_P}{kret\, K_A} = \frac{[P]}{[A]} = K'$$

If there are no experimental errors, the exact value of the apparent equilibrium constant is obtained.

$K' = kfet*kP/(kret*kA) = 1*20/(0.5*10) = 4$

Use Table 2.4 to show effects of experimental errors on the apparent equilibrium constant calculated using the Haldane relation. Rather than taking values from Table 2.2 and repeating this calculation, columns can be copied and multiplied and divided to calculate the values of K' when there are 5% errors in the measured velocities, one at a time.

$$\frac{\begin{matrix}1.` \\ 1.06` \\ 1.` \\ 0.99` \\ 1.`\end{matrix} \;\; *\begin{matrix}20.` \\ 20.` \\ 21.29` \\ 20.` \\ 18.8`\end{matrix}}{} \Bigg/ \left(\frac{\begin{matrix}0.5` \\ 0.5` \\ 0.53` \\ 0.5` \\ 0.5`\end{matrix} \;\; *\begin{matrix}10.` \\ 10.61` \\ 10.` \\ 9.42` \\ 10.`\end{matrix}}{}\right)$$

$\{\{4.\}, \{3.99623\}, \{4.01698\}, \{4.20382\}, \{3.76\}\}$

The frst value is obtained when there are no experimental errors, and the other four values show the effects of 5% errors in v_1, v_2, v_3, and v_4. This row of apparent equilibrium constants can be put in Table 2.4 as a column.

■ 2.3.4 Effects of pH on the velocity for A = P

A program can be written by using the factor method. Note that this section uses kfEt and krEt instead of kfet and kret

The program in Section 2.2.3 is expanded to include the reverse reaction.

```
derAPkinpH[pK1e_, pK2e_,
  pK1ea_, pK2ea_,
  pK1a_, pK2a_,
  pK1p_, pK2p_,
  pK1ep_, pK2ep_,
  kfEt_, krEt_,
  kH2EA_, kH2EP_] :=
Module[{eafactor, efactor, afactor, pfactor, epfactor, vfexp, vrexp, ka, kp, kaadj, kpadj},
  (*Calculates 4 pH-dependent kinetic parameters (kA, kP, vfexp, and vrexp) of
      the enzyme-catalyzed reaction A =
  P as functions of pH. It also calculates adjusted Michaelis constants for A and P,
  2 ratios of kinetic parameters, and the expression for the initial reaction velocity
    v.  The calculations require 14 inputs.  The output is a list of 9 functions,
  the last of which is the velocity for any pair of substrate concentrations.*)
  efactor = 1 + 10^(pK2e-pH) + 10^(pH-pK1e);
  eafactor = 1 + 10^(pK2ea-pH) + 10^(pH-pK1ea);
  afactor = 1 + 10^(pK2a-pH) + 10^(pH-pK1a);
  pfactor = 1 + 10^(pK2p-pH) + 10^(pH-pK1p);
  epfactor = 1 + 10^(pK2ep-pH) + 10^(pH-pK1ep);
  vfexp = kfEt / eafactor;
  vrexp = krEt / epfactor;
  ka = kH2EA * efactor * afactor / eafactor;
  kp = kH2EP * efactor * pfactor / epfactor;
  kaadj = kH2EA * efactor / eafactor;
  kpadj = kH2EP * efactor / epfactor;
  {vfexp, vrexp, ka, kp, kaadj, kpadj, vfexp / kaadj,
    vrexp / kpadj, (a * kp * vfexp - p * ka * vrexp) / (a * kp + ka * kp + p * kp)}]
```

These kinetic parameters can be plotted as functions of pH if arbitrary values are assigned to the various properties, as follows:

```
Grid[{{pK1e, pK2e, pK1ea, pK2ea, pK1a, pK2a, pK1p, pK2p,
  pK1ep, pK2ep, kfEt, krEt, kH2EA, kH2EP}, {8, 6, 7.5, 5.5, 8, 5, 9, 5, 7.5, 6.5, 1, .5, 4, 10}}]
```

pK1e	pK2e	pK1ea	pK2ea	pK1a	pK2a	pK1p	pK2p	pK1ep	pK2ep	kfEt	krEt	kH2EA	kH2EP
8	6	7.5	5.5	8	5	9	5	7.5	6.5	1	0.5	4	10

Note that pK1 > pK2. Taking pK1 > 10 or pK2 < 4 takes the pK out of the range of pH being considered.

```
derAPkinpH[8, 6, 7.5, 5.5, 8, 5, 9, 5, 7.5, 6.5, 1, .5, 4, 10]
```

$$
\left\{ \frac{1}{1 + 10^{5.5-pH} + 10^{-7.5+pH}}, \frac{0.5}{1 + 10^{6.5-pH} + 10^{-7.5+pH}}, \frac{4\left(1 + 10^{5-pH} + 10^{-8+pH}\right)\left(1 + 10^{6-pH} + 10^{-8+pH}\right)}{1 + 10^{5.5-pH} + 10^{-7.5+pH}}, \right.
$$

$$
\frac{10\left(1 + 10^{5-pH} + 10^{-9+pH}\right)\left(1 + 10^{6-pH} + 10^{-8+pH}\right)}{1 + 10^{6.5-pH} + 10^{-7.5+pH}}, \frac{4\left(1 + 10^{6-pH} + 10^{-8+pH}\right)}{1 + 10^{5.5-pH} + 10^{-7.5+pH}},
$$

$$
\frac{10\left(1 + 10^{6-pH} + 10^{-8+pH}\right)}{1 + 10^{6.5-pH} + 10^{-7.5+pH}}, \frac{1}{4\left(1 + 10^{6-pH} + 10^{-8+pH}\right)}, \frac{0.05}{1 + 10^{6-pH} + 10^{-8+pH}},
$$

$$
\left(\frac{10\left(1 + 10^{5-pH} + 10^{-9+pH}\right)\left(1 + 10^{6-pH} + 10^{-8+pH}\right)a}{\left(1 + 10^{5.5-pH} + 10^{-7.5+pH}\right)\left(1 + 10^{6.5-pH} + 10^{-7.5+pH}\right)} - \frac{2.\left(1 + 10^{5-pH} + 10^{-8+pH}\right)\left(1 + 10^{6-pH} + 10^{-8+pH}\right)p}{\left(1 + 10^{5.5-pH} + 10^{-7.5+pH}\right)\left(1 + 10^{6.5-pH} + 10^{-7.5+pH}\right)}\right) \Bigg/
$$

$$
\left(\frac{40\left(1 + 10^{5-pH} + 10^{-9+pH}\right)\left(1 + 10^{5-pH} + 10^{-8+pH}\right)\left(1 + 10^{6-pH} + 10^{-8+pH}\right)^2}{\left(1 + 10^{5.5-pH} + 10^{-7.5+pH}\right)\left(1 + 10^{6.5-pH} + 10^{-7.5+pH}\right)} + \right.
$$

$$
\left.\left. \frac{10\left(1 + 10^{5-pH} + 10^{-9+pH}\right)\left(1 + 10^{6-pH} + 10^{-8+pH}\right)a}{1 + 10^{6.5-pH} + 10^{-7.5+pH}} + \frac{10\left(1 + 10^{5-pH} + 10^{-9+pH}\right)\left(1 + 10^{6-pH} + 10^{-8+pH}\right)p}{1 + 10^{6.5-pH} + 10^{-7.5+pH}}\right)\right\}
$$

These 9 functions are numbered [[1]] to [[9]].

```
plot7a = Plot[derAPkinpH[8, 6, 7.5, 5.5, 8, 5, 9, 5, 7.5, 6.5, 1, .5, 4, 10][[1]],
    {pH, 5, 9}, AxesLabel → {"pH", "V_fexp"}];

plot7b = Plot[derAPkinpH[8, 6, 7.5, 5.5, 8, 5, 9, 5, 7.5, 6.5, 1, .5, 4, 10][[2]],
    {pH, 5, 9}, AxesLabel → {"pH", "V_rexp"}];

plot7c = Plot[derAPkinpH[8, 6, 7.5, 5.5, 8, 5, 9, 5, 7.5, 6.5, 1, .5, 4, 10][[3]],
    {pH, 5, 9}, AxesLabel → {"pH", "K_A"}, PlotRange → {0, 25}];

plot7d = Plot[derAPkinpH[8, 6, 7.5, 5.5, 8, 5, 9, 5, 7.5, 6.5, 1, .5, 4, 10][[4]],
    {pH, 5, 9}, AxesLabel → {"pH", "K_P"}, PlotRange → {4, 8}];
```

GraphicsArray[{{plot7a, plot7b}, {plot7c, plot7d}}]

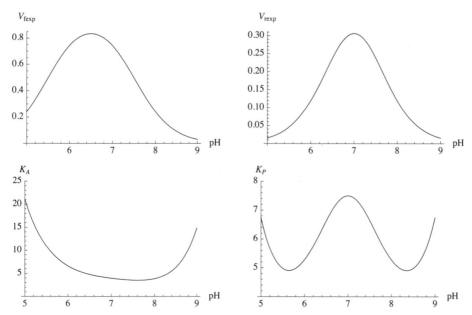

Fig. 2.7 Kinetic parameters at specified temperature, pH, and ionic strength as a function of pH for A = P. (a) V_{fexp}. Three kinetic parameters are readily calculated from this type of plot of experimental data, as described in the Appendix of this chapter. (b) V_{rexp}. (c) K_A. Seven kinetic parameters are involved in this plot, and so it is not practical to determine them from this plot. (d) K_P.

K_A and K_P can be adjusted for the pKs of these substrates to obtain adjusted properties K_{Aadj} and K_{Padj} that can be used to obtain properties that can be calculated from bell-shaped plots.

plot8a = Plot[derAPkinpH[8, 6, 7.5, 5.5, 8, 5, 9, 5, 7.5, 6.5, 1, .5, 4, 10][[5]],
    {pH, 5, 9}, AxesLabel → {"pH", "K_{Aadj}"}, PlotRange → {0, 12}];

This is a function of 5 parameters, but [[7]] (see Fig. 2.19) is a function of 3 parameters, and so all the kinetic parameters can be obtained from bell-shaped plots.

plot8b = Plot[derAPkinpH[8, 6, 7.5, 5.5, 8, 5, 9, 5, 7.5, 6.5, 1, .5, 4, 10][[6]],
    {pH, 5, 9}, AxesLabel → {"pH", "K_{Padj}"}, PlotRange → {0, 8}];

plot8c = Plot[derAPkinpH[8, 6, 7.5, 5.5, 8, 5, 9, 5, 7.5, 6.5, 1, .5, 4, 10][[7]],
    {pH, 5, 9}, AxesLabel → {"pH", "V_{fexp}/K_{Aadj}"}, PlotRange → {0, .3}];

plot8d = Plot[derAPkinpH[8, 6, 7.5, 5.5, 8, 5, 9, 5, 7.5, 6.5, 1, .5, 4, 10][[8]],
    {pH, 5, 9}, AxesLabel → {"pH", "V_{rexp}/K_{Padj}"}, PlotRange → {0, .05}];

```
GraphicsArray[{{plot8a, plot8b}, {plot8c, plot8d}}]
```

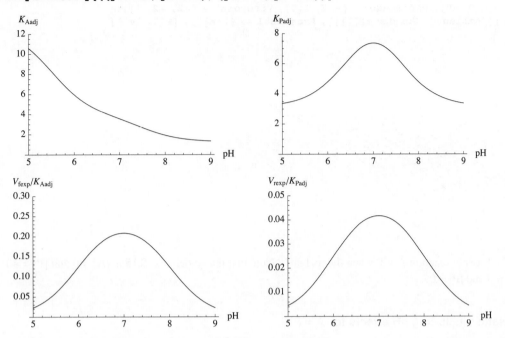

Fig. 2.8 Adjusted kinetic parameters at specified temperature, pH, and ionic strength as a function of pH for A = P. (a) K_{Aadj}. (b) K_{Padj}. This adjusted Michaelis constant is a function of 5 kinetic parameters. This looks like a bell-shaped plot, but it is not. V_{rexp} / K_{Padj} does yield a bell-shaped plot, and so all the kinetic parameters can be obtained from bell-shaped plots. (c) V_{fexp} / K_{Aadj}. (d) V_{rexp} / K_{Padj}.

The complete rate equation as a function of pH, [A], and [B] is given by

```
derAPkinpH[8, 6, 7.5, 5.5, 8, 5, 9, 5, 7.5, 6.5, 1, .5, 4, 10][[9]]
```

$$\left(\frac{10 \left(1 + 10^{5-pH} + 10^{-9+pH}\right) \left(1 + 10^{6-pH} + 10^{-8+pH}\right) a}{\left(1 + 10^{5.5-pH} + 10^{-7.5+pH}\right) \left(1 + 10^{6.5-pH} + 10^{-7.5+pH}\right)} - \frac{2. \left(1 + 10^{5-pH} + 10^{-8+pH}\right) \left(1 + 10^{6-pH} + 10^{-8+pH}\right) p}{\left(1 + 10^{5.5-pH} + 10^{-7.5+pH}\right) \left(1 + 10^{6.5-pH} + 10^{-7.5+pH}\right)} \right) \Bigg/$$

$$\left(\frac{40 \left(1 + 10^{5-pH} + 10^{-9+pH}\right) \left(1 + 10^{5-pH} + 10^{-8+pH}\right) \left(1 + 10^{6-pH} + 10^{-8+pH}\right)^2}{\left(1 + 10^{5.5-pH} + 10^{-7.5+pH}\right) \left(1 + 10^{6.5-pH} + 10^{-7.5+pH}\right)} + \right.$$

$$\left. \frac{10 \left(1 + 10^{5-pH} + 10^{-9+pH}\right) \left(1 + 10^{6-pH} + 10^{-8+pH}\right) a}{1 + 10^{6.5-pH} + 10^{-7.5+pH}} + \frac{10 \left(1 + 10^{5-pH} + 10^{-9+pH}\right) \left(1 + 10^{6-pH} + 10^{-8+pH}\right) p}{1 + 10^{6.5-pH} + 10^{-7.5+pH}} \right)$$

3 D plots of velocities

The velocity of the forward reaction A = P is a function of pH and {[A],[P]}. At pH 7, this velocity can be used to make a 3D plot versus [A] and [P]. Velocities for the forward reaction are positive, and velocities for the reverse reaction are negative. It is easier to see what is happening when the forward and reverse velocities are plotted separately.

```
Plot3D[Evaluate[derAPkinpH[8, 6, 7.5, 5.5, 8, 5, 9, 5, 7.5, 6.5, 1, .5, 4, 10][[9]]] /. pH → 7,
    {a, 0, 20}, {p, 0, 40}, PlotRange -> {-.2, 1.0}, ViewPoint → {-2, -2, 1},
    Lighting → {{"Ambient", GrayLevel[1]}}, AxesLabel -> {"[A]", "[P]", "v"}]
```

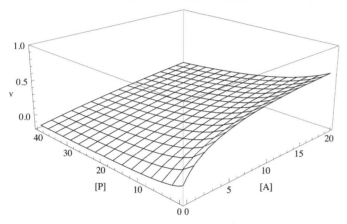

Fig. 2.9 Velocity of the forward reaction A = P when the pH is 7. Note that the velocity is -0.15 at [A] = 0 and [P] = 40. The velocity is zero at [A] = 0 and [P] = 0.

■ 2.3.5 Haldane relation including pH effects for A = P

Section 1.7 discusses the change in binding $\Delta_r N_H$ of hydrogen ions in a biochemical reaction. This is a thermodynamic property of the reaction that is catalyzed. $\Delta_r N_H$ can be calculated in two ways. When the apparent equilibrium constant K' of an enzyme-catalyzed reaction is known as a function of pH, the change in binding of hydrogen ions in an enzyme-catalyzed reaction can be calculated using (see equation 1.4-3).

$$\Delta_r N_H = -\mathrm{dlog}K'/\mathrm{dpH} \tag{8}$$

The values of $\Delta_r N_H$ at pHs 5, 6, 7, 8, and 9 have been calculated for about 200 enzyme-catalyzed reactions [23].

The second way to calculate $\Delta_r N_H$ only requires knowing the pKs of the reactants in the pH range of interest. The average number of hydrogen atoms in a biochemical reactant is given by [26, *MathSource*4]

$$\overline{N}_H (i) = \sum r_j N_H(j) \tag{9}$$

where r_j is the equilibrium mole fraction of species j in reactant i and $N_H(j)$ is the number of hydrogen atoms in species j. The change in binding of hydrogen ions is given by

$$\Delta_r N_H = \sum \nu_i' \overline{N}_H (i) \tag{10}$$

where ν_i' is the stoichiometric number for reactant i in the catalyzed reaction. The values of $\Delta_r N_H$ at pHs 5, 6, 7, 8, and 9 have been calculated [23,26] for about 200 enzyme-catalyzed reactions, and they are in agreement with the values calculated using equation 7. Values of $\Delta_r N_H$ at a desired pH can be calculated for most of the reactions in the EC list [Web2] because at least approximate pKs are known for most substrates.

An equation derived earlier in Section 2.2.3 indicates that when the velocity is equal to zero at equilibrium, a*kfet*kP is equal to p*kret*kA. The Haldane relation that relates the apparent equilibrium constant to the four kinetic parameters is given by

$$K' = \frac{p}{a} = \frac{\mathrm{kfet*kP}}{\mathrm{kret*kA}}$$

Substituting the functions of pH yields the following expression for the pH dependence of the apparent equilibrium constant:

```
derAPkinpH[8, 6, 7.5, 5.5, 8, 5, 9, 5, 7.5, 6.5, 1, .5, 4, 10][[1]] *
  derAPkinpH[8, 6, 7.5, 5.5, 8, 5, 9, 5, 7.5, 6.5, 1, .5, 4, 10][[4]] /
    (derAPkinpH[8, 6, 7.5, 5.5, 8, 5, 9, 5, 7.5, 6.5, 1, .5, 4, 10][[2]] *
      derAPkinpH[8, 6, 7.5, 5.5, 8, 5, 9, 5, 7.5, 6.5, 1, .5, 4, 10][[3]])
```

$$\frac{5. \left(1 + 10^{5-\text{pH}} + 10^{-9+\text{pH}}\right)}{1 + 10^{5-\text{pH}} + 10^{-8+\text{pH}}}$$

This shows that the pH dependence is due to the pKs of only the substrates.

$$\textbf{Plot}\left[\frac{\textbf{5.}^{\textasciigrave} \left(1 + 10^{5-\text{pH}} + 10^{-9+\text{pH}}\right)}{1 + 10^{5-\text{pH}} + 10^{-8+\text{pH}}}, \{\textbf{pH, 5, 9}\}, \textbf{AxesLabel} \rightarrow \{\textbf{"pH"}, \textbf{"K'"}\}\right]$$

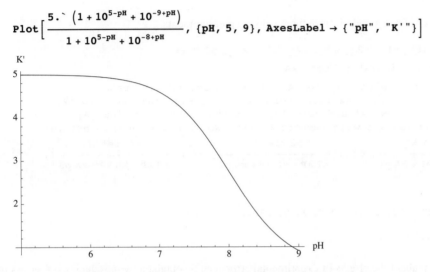

Fig. 2.10 Apparent equilibrium constant of A = P calculated from the pH-dependent kinetic parameters.

This section emphasizes the close relationship between biochemical thermodynamcs and enzyme kinetics. Since complete rate equations must include the thermodynamics of the enzyme-catalyzed reaction, the principles of biochemical thermodynamics have to be obeyed. The Haldane relation must yield the correct dependence of the apparent equilibrium constant on *T*, pH, and ionic strength.

2.4 A → P when the apparent equilibrium constant is very large

▪ 2.4.1 Estimation of kinetic parameters when the apparent equilibrium constant is very large

When kr is set equal to zero in the equation for the velocity of the forward reaction in Section 2.3.1, the velocity is given by

$$\textbf{vAlargeK} = \frac{\textbf{a kfet kP}}{\textbf{a kP + kA kP + kA p}}$$

$$\frac{\text{a kfet kP}}{\text{a kP + kA kP + kA p}}$$

The 3 kinetic parameters can be estimated by measuring 3 velocities;

vAlargeK /. kfet → 1 /. kA → 10 /. kP → 20 /. a → 100 /. p → 0 // N

0.909091

```
vAlargeK /. kfet → 1 /. kA → 10 /. kP → 20 /. a → 1 /. p → 0 // N
```

0.0909091

```
vAlargeK /. kfet → 1 /. kA → 10 /. kP → 20 /. a → 1 /. p → 100 // N
```

0.0163934

The first velocity has the predominate effect in determining kfet. The second velocity has the predominate effect in determining kA. The third velocity has the predominate effect in determining kP.

The following program calculates kfet, kA and kP from three velocity measurements.

$$\text{calckinparsAlargeK[v1\_, a1\_, p1\_, v2\_, a2\_, p2\_, v3\_, a3\_, p3\_] :=}$$
$$\text{Module}\Big[\{\}, \text{(*This program calculates kfet, kA,}$$

and kP from 3 experimental velocities for A = P at 3 pairs of substrate concentrations when kr = 0. The first velocity is at high [A] and zero [P], the second velocity is at low [A] and zero [P], and the third is at low [A] and high [P]. High and low are with respect to the two disociation constants.*)

$$\text{Solve}\Big[\Big\{v1 == \frac{a1\ kfet\ kP}{a1\ kP + kA\ kP + kA\ p1}, v2 == \frac{a2\ kfet\ kP}{a2\ kP + kA\ kP + kA\ p2}, v3 == \frac{a3\ kfet\ kP}{a3\ kP + kA\ kP + kA\ p3}\Big\},$$

$$\{kfet, kA, kP\}\Big]\Big]$$

```
calckinparsAlargeK[.90909, 100, 0, .090909, 1, 0, .016393, 1, 100]
```

$$\{\{kfet \to 0.999999, kP \to 19.9994, kA \to 10.\}\}$$

This yields exact values, but information about the effects of experimental errors can be obtained by introducing 5% errors in the velocities, one at a time.

■ 2.4.2 Effects of pH on the velocity when the apparent equilibrium constant is very large

Take the second term in the numerator out of the general expression for the velocity of A = P. This is the expression for the velocity when the apparent equilibrium constant is very large.

$$\text{vlargeK} = \frac{10\left(1 + 10^{5-pH} + 10^{-9+pH}\right)\left(1 + 10^{6-pH} + 10^{-8+pH}\right) a}{\left(1 + 10^{5.5`-pH} + 10^{-7.5`+pH}\right)\left(1 + 10^{6.5`-pH} + 10^{-7.5`+pH}\right)} \Big/$$

$$\left(\frac{40\left(1 + 10^{5-pH} + 10^{-9+pH}\right)\left(1 + 10^{5-pH} + 10^{-8+pH}\right)\left(1 + 10^{6-pH} + 10^{-8+pH}\right)^2}{\left(1 + 10^{5.5`-pH} + 10^{-7.5`+pH}\right)\left(1 + 10^{6.5`-pH} + 10^{-7.5`+pH}\right)} + \right.$$

$$\left. \frac{10\left(1 + 10^{5-pH} + 10^{-9+pH}\right)\left(1 + 10^{6-pH} + 10^{-8+pH}\right) a}{1 + 10^{6.5`-pH} + 10^{-7.5`+pH}} + \frac{10\left(1 + 10^{5-pH} + 10^{-9+pH}\right)\left(1 + 10^{6-pH} + 10^{-8+pH}\right) p}{1 + 10^{6.5`-pH} + 10^{-7.5`+pH}}\right);$$

This function of pH can be used to calculate the velocity at any [A], [P], and pH.

```
Plot3D[Evaluate[vlargeK /. pH → 7], {a, 0, 20},
 {p, 0, 40}, PlotRange -> {0, 1.0}, ViewPoint → {-2, -2, 1},
 Lighting → {{"Ambient", GrayLevel[1]}}, AxesLabel -> {"[A]", "[P]", "v"}]
```

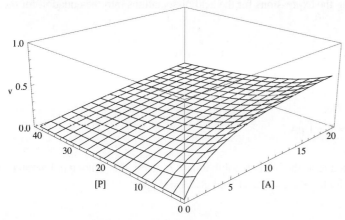

Fig. 2.11 This shows the competitive inhibition by P.

■ 2.4.3 Haldane relation when the apparent equilibrium constant is very large

Earlier in Section 2.3.5 it was shown that

$$K' = kfet*kP/(kret*kA)$$

When the apparent equilibrium constant is very large, it is not possible to estimate kret directly, but if kfet, kA, and kP can be estimated, and K' is known, kret can be calculated. It is not necessary that K' has been determined directly because the standard transformed Gibbs energies of formation of reactants have been tabulated in BasicBiochemicalData3 [*MathSource*1]. These properties of reactants can be used to calculate apparent equilibrium constants as functions of pH and ionic strength and 298.15 K for over 300 enzyme-catalyzed reactions, including reactions for which direct measurements of K' have not been made. Thus when the apparent equilibrium constant K' is known for A = B, kret can be estimated even when the apparent equilibrium constant is very large.

2.5 Appendix

■ Appendix 1 Titration curves for diprotic acids

Derivation of the binding polynomial

The objective of this section is to calculate the titration curve for $H_2 A$. The acid dissociations of a diprotic acid are given by

$$HA^- = H^+ + A^{2-} \qquad 10^{-pK1A} = [H^+][A^{2-}]/[HA^-]$$
$$H_2 A = H^+ + HA^- \qquad 10^{-pK2A} = [H^+][HA^-]/[H_2A]$$

The total concentration of A is given by $[A] = [A^{2-}] + [HA^-] + [H_2A]$. The equilibrium concentrations of A^{2-}, HA^-, and H_2A as functions of pH can be calculated as follows:

The expression for $[A^{2-}]$ can be obtained by substituting the expressions for the acid dissociations into the equation for the total concentration of A.

$$[A] = [A^{2-}](1 + [HA^-]/[A^{2-}] + [H_2A]/[A^{2-}]) = [A^{2-}](1 + 10^{pK1A-pH} + 10^{pK1A+pK2A-2pH})$$

The fraction of A that is A^{2-} is

$$r(A^{2-}) = [A^{2-}]/[A] = 1/(1 + 10^{pK1A-pH} + 10^{pK1A+pK2A-2pH}) = 1/pA$$

where $pA = 1 + 10^{pK1A-pH} + 10^{pK1A+pK2A-2pH}$ is referred to as the binding polynomial. Actually this is not a polynomial, but it can be converted to a polynomial by substituting pH = -logh, pK1A = -logk1, and pK2A = -logk2.

The fraction of A that is HA^- is

$$r(HA^-) = [HA^-]/[A] = (10^{pK1A-pH})/(1 + 10^{pK1A-pH} + 10^{pK1A+pK2A-2pH}) = (10^{pK1A-pH})/pA$$

The fraction of A that is H_2A is

$$r(H_2A) = [H_2A]/[A] = (10^{pK1A+pK2A-2pH})/(1 + 10^{pK1A-pH} + 10^{pK1A+pK2A-2pH}) = (10^{pK1A+pK2A-2pH})/pA$$

These three mole fractions add up to one as they must.

These three mole fractions are put into *Mathematica* as follows:

$$\mathbf{pA = 1 + 10^{pK1A-pH} + 10^{pK1A+pK2A-2pH}}$$

$$1 + 10^{-pH+pK1A} + 10^{-2pH+pK1A+pK2A}$$

The equilibrium mole fractions of the three species of A (r1 is $r(A^{2-})$, r2 is $r(HA^-)$, and r3 is $r(A^{2-})$) are given in *Mathematica* by

$$\mathbf{r1 = 1/pA}$$

$$\frac{1}{1 + 10^{-pH+pK1A} + 10^{-2pH+pK1A+pK2A}}$$

$$\mathbf{r2 = \left(10^{pK1A-pH}\right)/pA}$$

$$\frac{10^{-pH+pK1A}}{1 + 10^{-pH+pK1A} + 10^{-2pH+pK1A+pK2A}}$$

$$\mathbf{r3 = 10^{-2pH+pK1A+pK2A}/pA}$$

$$\frac{10^{-2pH+pK1A+pK2A}}{1 + 10^{-pH+pK1A} + 10^{-2pH+pK1A+pK2A}}$$

Derivation of the titration curve for H_2A, which is assumed to have pK1A = 7.00 and pK2A = 6.00

■ **A1.2**

r1 /. pK1A → 7.00 /. pK2A → 6.00

$$\frac{1}{1 + 10^{13.-2\,pH} + 10^{7.-pH}}$$

r2 /. pK1A → 7.00 /. pK2A → 6.00

$$\frac{10^{7.-pH}}{1 + 10^{13.-2\,pH} + 10^{7.-pH}}$$

r3 /. pK1A → 7.00 /. pK2A → 6.00

$$\frac{10^{13.-2\,pH}}{1 + 10^{13.-2\,pH} + 10^{7.-pH}}$$

The equilibrium mole fractions of these three species can be plotted as a function of pH.

```
Plot[{r1, r2, r3} /. pK1A → 7.00 /. pK2A → 6.00,
  {pH, 3, 9}, AxesLabel → {"pH", "rⱼ"}, PlotStyle → {Black}]
```

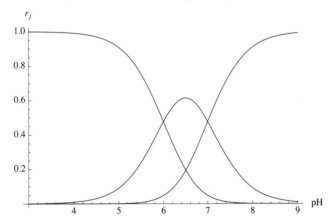

Fig. 2.12 Equilibrium mole fractions of H_2A, HA^-, and A^{2-} as functions of pH.

The average number $\overline{N}_H$ of hydrogen ions bound by A bound is given as a function of pH by $\overline{N}_H = r_2 + 2r_3$.

$$\overline{N}_H = \left(10^{pK1A-pH} + 2*10^{-2\,pH+pK1A+pK2A}\right) \big/ pA$$

$$\frac{2^{1-2\,pH+pK1A+pK2A}\,5^{-2\,pH+pK1A+pK2A} + 10^{-pH+pK1A}}{1 + 10^{-pH+pK1A} + 10^{-2\,pH+pK1A+pK2A}}$$

$\overline{N}_H$ /. pK1A → 7.00 /. pK2A → 6.00

$$\frac{2^{14.-2\,pH}\,5^{13.-2\,pH} + 10^{7.-pH}}{1 + 10^{13.-2\,pH} + 10^{7.-pH}}$$

Plot $\overline{N}_H$ as a function of pH.

```
Plot[N̄_H /. pK1A → 7.00 /. pK2A → 6.00, {pH, 3, 9}, AxesLabel → {"pH", "N̄_H"}]
```

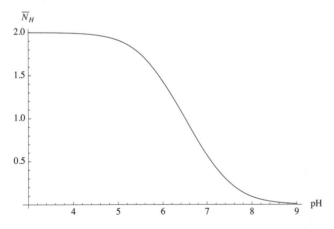

Fig. 2.13 Average number of hydrogen ions bound by A as a function of pH when pK1A = 7 and pK2A = 6.

Slope of the titration curve as a function of pH

More information about the shape of the titration curve can be obtained by calculating the slope as a function of pH. The derivative of $\overline{N}_H$ with respect to pH is obtained as follows:

$$D\left[\overline{N}_H, pH\right]$$

$$\frac{-2^{2-2\,pH+pK1A+pK2A}\,5^{-2\,pH+pK1A+pK2A}\,Log[2] - 2^{2-2\,pH+pK1A+pK2A}\,5^{-2\,pH+pK1A+pK2A}\,Log[5] - 10^{-pH+pK1A}\,Log[10]}{1 + 10^{-pH+pK1A} + 10^{-2\,pH+pK1A+pK2A}} -$$

$$\left(\left(2^{1-2\,pH+pK1A+pK2A}\,5^{-2\,pH+pK1A+pK2A} + 10^{-pH+pK1A}\right)\left(-2^{1-2\,pH+pK1A+pK2A}\,5^{-2\,pH+pK1A+pK2A}\,Log[10] - 10^{-pH+pK1A}\,Log[10]\right)\right) \Big/$$

$$\left(1 + 10^{-pH+pK1A} + 10^{-2\,pH+pK1A+pK2A}\right)^2$$

$$D\left[\overline{N}_H \;/. \; pK1A \to 7.00 \;/. \; pK2A \to 6.00, \; pH\right]$$

$$\frac{-2^{15.-2\,pH}\,5^{13.-2\,pH}\,Log[2] - 2^{15.-2\,pH}\,5^{13.-2\,pH}\,Log[5] - 10^{7.-pH}\,Log[10]}{1 + 10^{13.-2\,pH} + 10^{7.-pH}} -$$

$$\frac{\left(2^{14.-2\,pH}\,5^{13.-2\,pH} + 10^{7.-pH}\right)\left(-2^{14.-2\,pH}\,5^{13.-2\,pH}\,Log[10] - 10^{7.-pH}\,Log[10]\right)}{\left(1 + 10^{13.-2\,pH} + 10^{7.-pH}\right)^2}$$

$$\mathtt{Plot}\left[\mathtt{Evaluate}\left[D\left[\overline{N}_H, pH\right] \;/. \; pK1A \to 7.00 \;/. \; pK2A \to 6.00\right], \; \{pH, 3, 9\}, \; \mathtt{AxesLabel} \to \left\{\text{"pH"}, \text{"d}\overline{N}_H\text{/dpH"}\right\}\right]$$

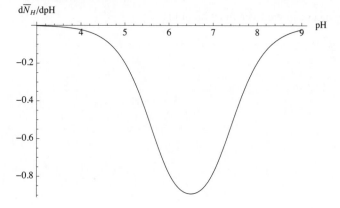

Fig. 2.14 d $\overline{N}_H$ / dpH as a function of pH when pK1A = 7 and pK2A = 6.

The slope can be calculated of various values of pK2, but pK2 < pK1.

```
Plot[Evaluate[D[N̄_H /. pK1A → 7.00 /. pK2A → {5.0, 5.5, 6.0, 6.5, 7.0}, pH]],
  {pH, 3, 9}, AxesLabel → {"pH", "dN̄_H/dpH"}, PlotStyle → {Black}]
```

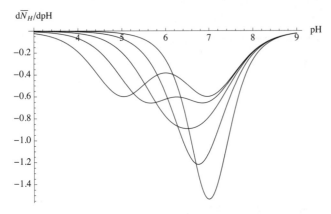

Fig. 2.15 $d\,\overline{N}_H\,/\,dpH$ as a function of pH for pK1A = 7.0 and pK2A = {5.0,5.5,6.0,6.5,7.0}

The plot for pK2 = 7 is the plot for a diprotic acid of the type HCO_2-$(CH_2)_n$-CO_2H, where *n* is large enough to make the two carboxyl groups independent.

The equilibrium mole fractions of the three species can be plotted as a function of pH.

```
Plot[Evaluate[{r1, r2, r3} /. pK1A → 7.00 /. pK2A → 7.0],
  {pH, 3, 9}, AxesLabel → {"pH", "r_j"}, PlotStyle → {Black}]
```

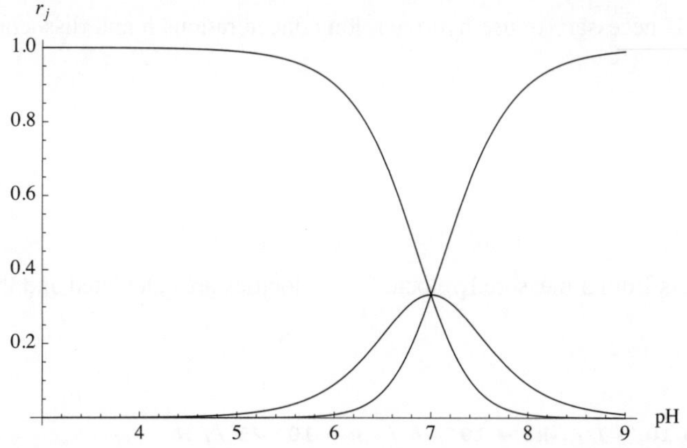

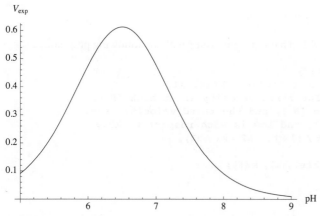

Fig. 2.16 Equilibrium mole fractions of H_2A, HA^-, and A^{2-} as functions of pH when pK1A = pK2A = 7.0. Notice that when the acid dissociations are equivalent and independent, the three species have the same concentration at pH = 7.

▪ Appendix 2 Calculation of kinetic parameters from bell - shaped plots

When $H_2 EA^+$ has two pKs, V_{exp} is given by

$$kfet/(1 + 10^{pK2ea-pH} + 10^{pH-pK1ea})$$

Solve can be used to calculate the three kinetic parameters.

In the example that is discussed here, kfet = 1, pK1ea = 7, and pK2ea = 6.

```
Plot[kfet / (1 + 10^(pK2 - pH) + 10^(pH - pK1)) /. kfet → 1 /. pK1 → 7 /. pK2 → 6,
  {pH, 5, 9}, AxesLabel → {"pH", "V_exp"}]
```

Fig. 2.17 Plot of V_{exp} versus pH.

Since there are three kinetic parmeters (kfet, pK1ea, and pK2ea), their values can be estimated from three velocity measurements.

Solve cannot handle these exponential functions, and so it is necessary to use hydrogen ion concentrations h and dissociation constants k1 and k2.

```
kfet / (1 + h / k2 + k1 / h)
```

$$\frac{kfet}{1 + \frac{k1}{h} + \frac{h}{k2}}$$

To demonstrate the estimation of 3 kinetic parameters from 3 measured velocities, 3 velocities are calculated and then are assumed to be experimental data.

When kfet = 1, k1 = 10^-7, and k2 = 10^-6,

```
kfet / (1 + h / k2 + k1 / h) /. kfet → 1 /. k1 → 10^-7 /. k2 → 10^-6 /. h → 10^-5 // N
```

0.0908265

```
kfet / (1 + h / k2 + k1 / h) /. kfet → 1 /. k1 → 10^-7 /. k2 → 10^-6 /. h → 10^-7 // N
```

0.47619

```
kfet / (1 + h / k2 + k1 / h) /. kfet → 1 /. k1 → 10^-7 /. k2 → 10^-6 /. h → 10^-8 // N
```

0.0908265

```
calckfeth1h2[v1_, h1_, v2_, h2_, v3_, h3_] := Module[{}, (*This program calculates kfet,
   k1, and k2 from 3 velocities at 3 [H⁺].  The first velocity is at high [H⁺],
   the second velocity is at intermediate [H⁺], and the third velocity is at
   low [H⁺].  High is with respect to k1, and low is with respect to k2.*)
   Solve[{v1 == kfet / (1 + h1 / k2 + k1 / h1),
     v2 == kfet / (1 + h2 / k2 + k1 / h2),
     v3 == kfet / (1 + h3 / k2 + k1 / h3)}, {kfet, k1, k2}]]
```

```
calckfeth1h2[.090827, 10^-5, .47619, 10^-7, .090827, 10^-8]
```

$$\{\{kfet → 0.99999, k1 → 9.99984 \times 10^{-8}, k2 → 1.00002 \times 10^{-6}\}\}$$

It is more convenient to have the program output pK1 and pK2. That is accomplished with the following program :

```
calckfetK3[v1_, h1_, v2_, h2_, v3_, h3_] :=
  Module[{mat, kfet, k1, k2}, (*This program calculates kfet, k1,
   and k2 from 3 velocities at 3 [H⁺].  The first velocity is at high [H⁺],
   the second velocity is at intermediate [H⁺], and the third velocity is at
   low [H⁺].  High is with respect to k1, and low is with respect to k2.*)
  mat = {kfet, k1, k2} /. Solve[{v1 == kfet / (1 + h1 / k2 + k1 / h1),
      v2 == kfet / (1 + h2 / k2 + k1 / h2),
      v3 == kfet / (1 + h3 / k2 + k1 / h3)}, {kfet, k1, k2}];
  kfet = mat[[1, 1]];
  k1 = mat[[1, 2]];
  k2 = mat[[1, 3]];
  {kfet, -Log[10, k1], -Log[10, k2]}]
```

The input to this program involvs hydrogen ion concentrations, but the program outputs pKs.

```
calckfetK3[.090827, 10^-5, .47619, 10^-7, .090827, 10^-8]
```

{0.99999, 7.00001, 5.99999}

```
line16 = calckfetK3[.090827, 10^-5, .47619, 10^-7, .090827, 10^-8]
```

{0.99999, 7.00001, 5.99999}

Put in 5% errors in the 3 velocities, one by one,

```
calckfetK3[1.05 * .090827, 10^-5, .47619, 10^-7, .090827, 10^-8]
```

{0.994193, 7.00276, 5.97381}

```
line17 = calckfetK3[1.05 * .090827, 10^-5, .47619, 10^-7, .090827, 10^-8]
```

{0.994193, 7.00276, 5.97381}

```
calckfetK3[.090827, 10^-5, 1.05 * .47619, 10^-7, .090827, 10^-8]
```

{1.12655, 6.9434, 6.0566}

```
line18 = calckfetK3[.090827, 10^-5, 1.05 * .47619, 10^-7, .090827, 10^-8]
```

{1.12655, 6.9434, 6.0566}

```
calckfetK3[.090827, 10^-5, .47619, 10^-7, 1.05 * .090827, 10^-8]
```

{0.944372, 7.05095, 5.9726}

```
line19 = calckfetK3[.090827, 10^-5, .47619, 10^-7, 1.05 * .090827, 10^-8]
```

{0.944372, 7.05095, 5.9726}

Table 2.5 Values of kinetic parameters for the effect of pH on V_{fexp} obtained using velocities at 3 pHs and testing the effects of 5% errors in the measured velocities

```
TableForm[Round[{line16, line17, line18, line19}, 0.01],
  TableHeadings → {{"No errors", "1.05*v1", "1.05*v2", "1.05*v3"}, {"kfet", "pK1", "pK2"}}]
```

	kfet	pK1	pK2
No errors	1.	7.	6.
1.05*v1	0.99	7.	5.97
1.05*v2	1.13	6.94	6.06
1.05*v3	0.94	7.05	5.97

The input requires hydrogen ion concentrations, but output yields pKs because that makes it easier to see the effects of errors in the measured velocities. This method can also be appled to V_{fexp}/K_{IA} because this also yields a bell-shaped plot.

▪ Appendix 3 Effect of temperature on kinetic properties

Since the Michaelis constants in rapid-equilibrium rate equations are equilibrium constants, they obey thermodynamic relations that define standard transformed thermodynamic properties like $\Delta_r G'^\circ$, $\Delta_r H'^\circ$, and $\Delta_r S'^\circ$. There is a difference between Michaelis constants and dissociation constants determined by direct study of the binding of small molecules by proteins. Michaelis constants are dissociation constants of substrates from enzyme-substrate complexes, but the direct thermodynamic study of the binding of small molecules by a protein may include binding at sites in addition to the enzymatic sites.

The rapid-equilibrium rate equation for A → products at specified T, pH, and ionic strength is

```
vA = a kfet / (a + kA)
```

$$\frac{a\,kfet}{a + kA}$$

The objective of this Appendix is to show what can be learned about the thermodynamics of substrate binding by measuring kfet and kA at two temperatures. To make test calculations, arbitrary values of kfet and kA at 298.15 K and 273.15 K are chosen. The pH and ionic strength are held constant for the estimation of kinetic parameters at the two temperatures. Molar concentrations are used.

Table 2.6 Assumed experimental data at two temperatures.

```
TableForm[{{1, 0.02}, {.3, 0.005}}, TableHeadings → {{"298.15 K", "273.15 K"}, {"kfet", "kA"}}]
```

	kfet	kA
298.15 K	1	0.02
273.15 K	0.3	0.005

These kinetic parameters can be estimated by measuring two velocities at each temperature, as shown in this chapter.

At 298.15 K, the rate equation is given by

```
vA /. kfet → 1 /. kA → .020
```

$$\frac{a}{0.02 + a}$$

At 273.15 K, the rate equation is given by

```
vA /. kfet → .3 /. kA → .005
```

$$\frac{0.3\,a}{0.005 + a}$$

The velocities at these two temperatures can be plotted as functions of [A].

```
Plot[{vA /. kfet → 1 /. kA → .020, vA /. kfet → .3 /. kA → .005},
 {a, 0, .2}, AxesLabel → {"[A]", "vA"}, PlotRange → {0, 1}, PlotStyle → {Black}]
```

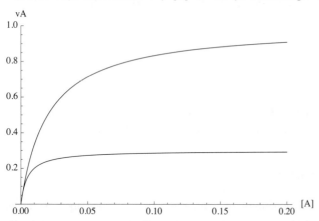

Fig. 2.18 Michelis-Menten plots of vA at two temperatures calculated with the parameters in Table 2.6. The upper curve is for 298.15 K, and the lower curve is for 273.15 K.

This shows that A is bound more strongly at the catalytic site at low temperatures, but the limiting velocity is reduced at the lower temperature.

It is also useful to plot [EA]/[E]$_t$ as a function of [A] at the two temperatures.

ea/et = 1/(1 + kA/a)

```
Plot[{1 / (1 + .02 / a), 1 / (1 + .005 / a)}, {a, 0, .1},
   PlotRange → {0, 1}, AxesLabel → {"[A]", "([EA]/[E]ₜ)"}, PlotStyle → {Black}]
```

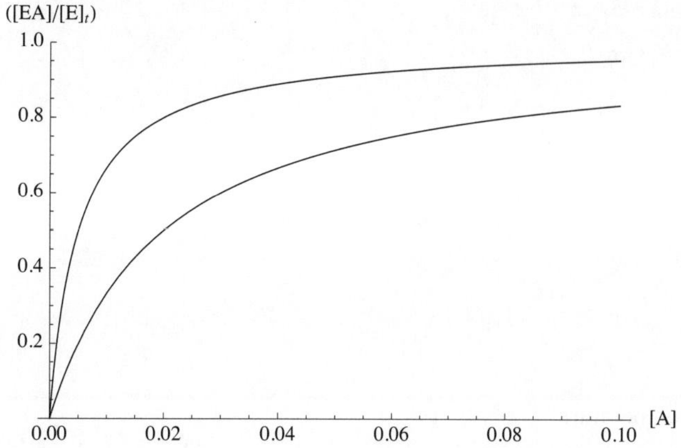

Fig. 2.19 Fraction of enzymatic sites occupied. The upper curve is for 273.15 K, and the lower curve is for 298.15 K.

Expression of the limiting velocity as a function of temperature

In chemical kinetics, it is generally found that plots of the logarithm of a rate constant versus $1/T$ are linear. This can be tested experimentally with kfet. When plots of the logarithms of rate constants versus $1/T$ are linear, the activation energy actE can be calculated using

ln(kfet2/kfet1) = (actE/R)(t2-t1)/t1t2   (A3.1)

Solving for actE yields

actE = 8.3145*ln(kfet2/kfet1)*(t2*t1)/(t2-t1)   (A3.2)

Natural logarithms in *Mathematica* are calculated using Log.

Inserting the parameters in Table 2.6 in equation A3.2 in *Mathematica* yields the activation energy actE in J mol^{-1} for kfet.

```
8.3145 * Log[.3] * (273.15 * 298.15) / (-25)

32 609.9
```

Note that kfet1 is taken to be 1.00 at 298.15 K and kfet2 is taken to be 0.30 at 273.15 K.

A general equation for kfet as a function of T can be derived using

kfet = const*E^-32610/RT

The basis for matural logarithms is represented by E in *Mathematica*.

At 298.15 K,

1 = const*E^(-32610/(8.3145*298.15))

The constant is given by

```
E^(32 610 / (8.3145 * 298.15))

516 415.
```

The value of kfet at 298.15 K is given by

```
516 415 * E ^ (- 32 610 / (8.3145 * 298.15))
```

```
0.999999
```

The value of kfet at 273.15 K is given by

```
516 415 * E ^ (- 32 610 / (8.3145 * 273.15))
```

```
0.299998
```

At temperature t, kfet is given by kfett:

```
kfett = 516 415 * E ^ (- 32 610 / (8.3145 * t))
```

$$516\,415\; e^{-3922.06/t}$$

The limiting velocity kfet can be plotted versus the temperature.

```
Plot[kfett, {t, 273.15, 298.15}, PlotRange -> {0, 1}, AxesLabel → {"T/K", "kfet"}]
```

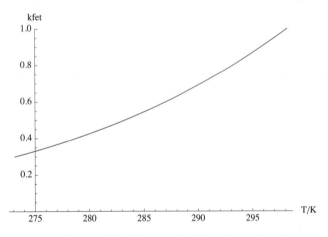

Fig. 2.20 The limiting velocity kfet as a function of temperature.

Knowing the Michaelis constants at two temperatures makes it possible to calculate the standard transformed enthalpy of reaction $\Delta_r H'^\circ$ and the standard transformed entropy of reaction $\Delta_r S'^\circ$.

Calculation of the standard transformed Gibbs energy of reaction for EA = E + A at two temperatures

Since the Michaelis constants in rapid-equilibrium rate equations are equilibrium constants, they can be used to calculate the standard transformed Gibbs energies of reaction $\Delta_r G'^\circ$ for EA = E + A at the two temperatures at the specified pH and ionic strength.

$$\Delta_r G'^\circ = -RT\ln K' \qquad\qquad\qquad\qquad (A.3.3)$$

The assumed values of K' are given in Table 2.6. At 298.15 K, the standard transformed Gibbs energy of reaction is given in J mol^{-1} by

```
- 8.3145 * 298.15 * Log[.02]
```

```
9697.78
```

At 273.15 K, the standard transformed Gibbs energy of reaction is given in J mol^{-1} by

```
- 8.3145 * 273.15 * Log[.005]
```

```
12 033.
```

Calculation of the standard transformed enthalpy of reaction for EA = E + A

When the standard transformed enthalpy of a biochemical reaction is independent of temperature, it can be calculated from the ratio of apparent equilibrium constants using

$$\ln(K_2'/K_1') = \Delta_r H'^\circ (T_2 - T_1)/RT_1 T_2 \tag{A3.4}$$

Another way to say that $\Delta_r H'^\circ$ is independent of temperature is to say that the standard transformed heat capacity at constant pressure $\Delta_r C_P'^\circ$ is equal to zero.

Thus the standard transformed enthalpy of reaction is calculated using

$$\Delta_r H'^\circ = \ln(K_2'/K_1')RT_1 T_2/(T_2 - T_1) \tag{A3.5}$$

If T_1 is taken to be the higher temperature, $\Delta_r H'^\circ$ in J mol^{-1} is given by

Log[.005 / .02] * 8.3145 * 298.15 * 273.15 / (-25)

37548.1

The fact that $\Delta_r H'^\circ$ is positive shows that heat is absorbed in the reaction EA = E + A. This is an endothermic reaction, and the dissociation increases when the temperature is raised.

Calculation of the standard transformed entropy of reaction for EA = E + A

Since

$$\Delta_r G'^\circ = \Delta_r H'^\circ - T\Delta_r S'^\circ \tag{A3.6}$$

The standard transformed entropy of reaction is given by

$$\Delta_r S'^\circ = (\Delta_r H'^\circ - \Delta_r G'^\circ)/T \tag{A3.7}$$

Using the values of $\Delta_r G'^\circ$ and $\Delta_r H'^\circ$ calculated from the assumed experimental data at 298.15 K, the standard transformed entropy of reaction $\Delta_r S'^\circ$ in J K^{-1} mol^{-1} is given by

(37 548 – 9698) / 298.15

93.4094

This is the standard transformed entropy of reaction at 298.15 K.

At 273.15 K, the standard transformed entropy of reaction is given by

(37 548 – 12 033) / 273.15

93.4102

This is the standard transformed entropy of reaction at 273.15 K.

The standard transformed entropy of reaction is independent of temperature as expected. The fact that the standard transformed entropy of reaction is positive shows that E + A is more random than EA.

Table 2.7 Standard transformed thermodynamic properties for EA = E + A calculated from the assumed experimental data in Table 2.6

```
TableForm[{{9698, 37548, 93.41, 0.02, 1}, {12033, 37548, 93.41, 0.005, 0.3}},
  TableHeadings → {{"298.15 K", "273.15 K"}, {G, H, S, K', kfet}}]
```

	G	H	S	K'	kfet
298.15 K	9698	37548	93.41	0.02	1
273.15 K	12033	37548	93.41	0.005	0.3

in this table G is $\Delta_r G'^\circ$, H is $\Delta_r H'^\circ$, and S is $\Delta_r S'^\circ$.

The apparent equilibrium constant at 298.15 K can be calculated using $\Delta_r G'^\circ$ in this table.

$$K' = e^{-\Delta_r G'^\circ/RT} \tag{A3.8}$$

```
E^(-9698 / (8.3145 * 298.15))
```

0.0199982

The apparent equilibrium constant K' at 273.15 K is given by

```
E^(-12033 / (8.3145 * 273.15))
```

0.00500009

Calculation of the Michaelis constant as a function of *T*

In order to compare the relative contributions of $\Delta_r H'^\circ$ and $\Delta_r S'^\circ$ to the value of the apparent equilibrium constant it is useful to express the apparent equilibrium constant as the product of an enthalpy factor and an entropy factor

$$K' = e^{-\Delta_r H'^\circ/RT} \, e^{\Delta_r S'^\circ/R} \tag{A3.9}$$

At 298.15 K, the enthalpy factor is given by

```
E^(-37548 / (8.3145 * 298.15))
```

2.64183×10^{-7}

At 273.15 K, the enthalpy factor is given by

```
E^(-37548 / (8.3145 * 273.15))
```

6.60459×10^{-8}

At both temperatures the factor for the standard transformed entropy of reaction is given by

```
E^(93.41 / 8.314)
```

75755.5

At 298.15 K, the apparent equilibrium constant K' calculated using equation A3.9 is given by

```
(2.64183 * 10^-8) * 75755
```

0.00200132

At 273.15 K, the apparent equilibrium constant K' calculated using equation A3.9 is given by

```
(6.60459 * 10^-8) * 75755
```

0.00500331

These apparent equilibrium constants are in agreement with Table 2.7.

Calculation of the apparent equilibrium constant as a function of temperature

Substituting $\Delta_r H'^{\circ} = 37548$ and $\Delta_r S'^{\circ} = 93.41$ in equation A.3.9 yields the following expression for the apparent equilibrium constant at temperatre t.

```
(E^(-37548 / (8.3145 * t))) * E^(93.41 / 8.3145)
```

$75\,704.3\, e^{-4515.97/t}$

```
(E^(-37548 / (8.3145 * t))) * E^(93.41 / 8.3145) /. t → 298.15
```

0.0199998

```
(E^(-37548 / (8.3145 * t))) * E^(93.41 / 8.3145) /. t → 273.15
```

0.00499996

The apparent equilibrium constant is given as a function of temperature by

```
kprimet = (E^(-37548 / (8.3145 * t))) * E^(93.41 / 8.3145)
```

$75\,704.3\, e^{-4515.97/t}$

This expression for the apparent equilibrium constant can be used to plot K' versus T.

```
Plot[kprimet, {t, 273.15, 298.15}, PlotRange → {0, .02}, AxesLabel → {"T/K", "K'"}]
```

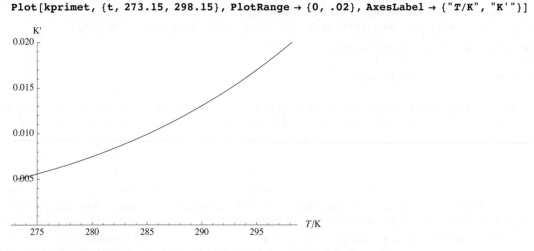

Fig. 2.21 The apparent equilibrium constant for EA = E + A as a function of temperature.

Plot kfet/kA as a function of temperature

The values of kfet/kA at 298.15 K and 273.25 K are calculated as follows:

```
(516415 * E^(-32610 / (8.3145 * t))) /
  (E^(-37548 / (8.3145 * t)) * E^(93.41 / 8.3145)) /. t → 298.15
```

50.0005

```
(516415 * E^(-32610 / (8.3145 * t))) /
  (E^(-37548 / (8.3145 * t)) * E^(93.41 / 8.3145)) /. t → 273.15
```

60.0002

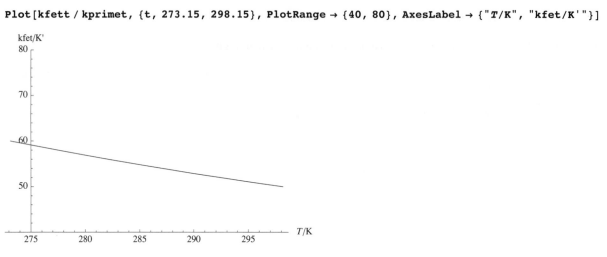

```
Plot[kfett / kprimet, {t, 273.15, 298.15}, PlotRange → {40, 80}, AxesLabel → {"T/K", "kfet/K'"}]
```

Fig. 2.22 kfet/K_A as a function of temperature

This ratio is referred to as a "specificity constant."

It is convenient to talk about strong binding and weak binding, but it is more precise to talk about apparent equilibrium constants or standard transformed Gibbs energies of reaction. These three ways to discuss binding are contrasted in Table 2.8.

Table 2.8 Three ways to discuss binding of a substrate by an enzyme

```
Grid[{{"Strong binding", "Neutral", "Weak binding"},
  {"K'<1", "K'=1", "K'>1"}, {"positive Δᵣ G' °", "Δᵣ G' °=0", "negative Δᵣ G' °"}}]
```

```
Strong binding  Neutral   Weak binding
    K'<1          K'=1         K'>1
positive Δᵣ G' ° Δᵣ G' °=0 negative Δᵣ G' °
```

A big dissociation constant indicates weak binding. The bigger the dissociation constant, the higher the concentrations of [E] and [A] at equilibrium. A small dissociation constant indicates strong binding. The smaller the dissociation constant, the lower the concentrations of [E] and [A] at equilibrium. The transition from weak and strong is at $K' = 1$. The dissociation constants of 0.02 and 0.005 can both be characterized as strong binding, but 0.005 represents stronger binding than 0.02.

These calculations are based on the determination of two kinetic parameters at two temperatures. This can be done with four velocity determinations. The two limiting velocities make it possible to calculate the activation energy for kfet. The two values of the Michaelis constant makes it possible to calculate the standard transformed Gibbs energies for the dissociation reactions. The temperature dependency of the Michaelis constant make it possible to calculate the standard transformed enthalpy. Knowledge of the standard transformed Gibbs energy of reaction and the standard transformed enthalpy of reaction makes it possible to calculate the standard transformed entropy of reaction. It is of interest to see the relative contributions of the standard transformed enthalpy and standard transformed entropy to Michaelis constants.

2.6 Discussion

The simplest enzyme-catalyzed reaction turns out not to be so simple after all. The effects of pH are especially complicated when the substrates have pKs in the pH range of interest. More complications will be introduced in Chapters 9-11 on inhibition, activation, and modification of A → Products.

This chapter has shown how Solve can be used to derive rapid-equilibrium rate equations for four mechanisms for A = P.

$$vA = \frac{a \, kfet}{a + kA};$$

$$v6par = (a\ kfet\ h\ kH2E\ kH2EA) \big/ \big(a\ h^2\ kH2E + h^2\ kcHEA\ kH2EA +$$
$$a\ h\ kH2E\ kH2EA + h\ kcHEA\ kH2E\ kH2EA + kcHEA\ kH2E\ kH2EA\ kHE + a\ kH2E\ kH2EA\ kHEA\big);$$

$$vAP = \frac{a\ kfet\ kP - kret\ kA\ p}{a\ kP + kA\ kP + kA\ p};$$

$$vAlargeK = \frac{a\ kfet\ kP}{a\ kP + kA\ kP + kA\ p};$$

These four rate equations can be used to calculate velocities at specified concentrations of A and P for arbitrarily chosen values of kinetic parameters. They can also be used to write programs to estimate the values of kinetic parameters from the minimum number of measured velocities. This has been illustrated by writing and using the following programs.

calc6kinpars   Section 2.2.3
calckfeth1h2   Section 2.2.3
derAkinpH   Section 2.2.4
calckinparsAP   Section 2.3.2
derAPkinpH   Section 2.3.4
calckinparsAlarge K   Section 2.4.1

Chapter 3 Ordered A + B → Products

3.1 Derivation of the rapid-equilibrium rate equation for ordered A + B → products

3.2 Estimation of the kinetic parameters for ordered A + B → products

- 3.2.1 Use of Solve to estimate the kinetic parameters from the minimum number of velocity measurements for ordered A + B → products
- 3.2.2 Calculation of the effects of changes in the kinetic parameters on the velocities

3.3 Effects of pH on the kinetics of ordered A + B → products

- 3.3.1 Mechanism for ordered A + B → products when the enzymatic site and enzyme-substrate complexes each have two pKs
- 3.3.2 Derivation of the rate equation involving hydrogen ion concentrations for ordered A + B → products when the enzymatic site and enzyme-substrate complexes each have two pKs
- 3.3.3 Estimation of nine kinetic parameters from nine velocities using simpvelordAB
- 3.3.4 Derivation of the dependencies of V_{fexp}, K_{IA}, and K_B on the hydrogen ion concentration
- 3.3.5 The program calckinparsordABsimp can be used to determine when a pK is missing

3.4 Effects of pH on the kinetics of ordered A + B → products when one hydrogen ion is consumed in the rate-determining reaction

3.5 Effects of pH on the kinetics of ordered A + B → products when n hydrogen ions are consumed in the rate-determining reaction

- 3.5.1 Mechanism and rate equation for ordered A + B → products when the enzymatic site and the enzyme-substrate complexes each have 2 pKs and n hydrogen ions are consumed
- 3.5.2 Determination of the number of hydrogen ions consumed from V_{fexp}

3.6 Appendix on effects of temperature on the kinetic properties for ordered A + B → products

3.7 Discussion

3.1 Derivation of the rapid-equilibrium rate equation for ordered A + B → products

There are many articles in which rapid-equilibrium rate equations have been used, and these applications have been discussed in Segel [6], Cornish-Bowden [21] and Cook and Cleland [25]. The derivations of these rate equations have generally regarded as simple because the reactions up to the rate-determining reaction are at equilibrium, but when effects of pKs, the consumption of hydrogen ions, more reactants, inhibitors, activators, and modifiers are involved, the derivations are not simple and are more readily made by the use of a computer. Using a computer to derive the expression for the velocity has the additional advantage that the rate equation does not have to be typed into the computer. *Mathematica* is very convenient for deriving the rapid-equilibrium rate equation from the mechanism of an enzyme-catalyzed reaction. In this ordered mechanism, the equations are written in their usual form, and then in *Mathematica*.

$$E + A = EA \qquad K_{IA} = [E][A]/[EA] \qquad kIA = e*a/ea \tag{1}$$

$$EA + B = EAB \qquad K_B = [EA][B]/[EAB] \qquad kB = ea*b/eab \tag{2}$$

$$\begin{array}{l} k_f \\ EAB \rightarrow products \end{array} \qquad v = k_f[EAB] \qquad v= kf*eab = kf*et*(eab/et) = kfet*(eab/et) \tag{3}$$

The separate effects of the rate constant kf and total concentration of enzymatic sites et are not discussed, and so kfet is used to represent the limiting velocity.

In using biochemical thermodynamics to obtain the expression for [EAB] at equilibrium, it is important to be careful about the number of equilibrium expressions that are used. The reactions used to derive a rapid-equilibrium rate equation must be independent; that is, none of them can be obtained by adding or subtracting other reactions in the mechanism. There is a simple test of independence of a set of reactions, and that is $N' = C' + R'$, where N' is the number of reactants, C' is the number of components, and R' is the number of independent reactions (see Section 1.3). It is important to understand that the equilibrium composition that is calculated using thermodynamics satisfies all possible equilibrium constant expressions that can be obtained by adding and subtracting reactions in the set of independent reactions with N' reactants.

To calculate the equilibrium concentration eab, there are two equilibrium equations and one conservation equation. The number N' of reactants is 5 (a, b, ea, eab, and e), the number C' of components is 3 (a, b, and e), and the number R' of independent reactions is 2, so that $N' = C' + R'$ is 5 = 3 + 2.

Solve[eqs,vars,elims] in *Mathematica* can be used to derive the expression for the equilibrium concentration eab. This operation solves for the variables when the elims are eliminated.

```
Solve[{kIA == e * a / ea, kB == ea * b / eab, et == e + ea + eab}, {eab}, {e, ea}]
```

$$\left\{ \left\{ eab \rightarrow \frac{a\,b\,et}{a\,b + a\,kB + kB\,kIA} \right\} \right\}$$

Note that

eab/et = 1/(1+kB/b+kBkIA/ab) so that v = kfet/(1+kB/b+kBkIA/ab)

When the expression for eab derived using Solve is multiplied by kf, kf*et can be replaced by kfet, the limiting velocity in the forward direction. The velocity v of the forward reaction in this case is named vordAB.

```
vordAB = a b kfet / (a b + a kB + kB kIA);
```

An advantage of the derivation of the rate equation in *Mathematica* is that the rate equation is in computer-readable form. The number of kinetic parameters is equal to the number of terms in the denominator of the rapid-equilibrium rate equation.

This rate equation can be used in two alternative forms:

(a) Trial values of the kinetic parameters can be inserted by use of the ReplaceAll operation (/.x->).

vordAB /. kfet → 1 /. kIA → 5 /. kB → 20

$$\frac{a\,b}{100 + 20\,a + a\,b}$$

vordAB can be used to plot the velocity versus [A] at constant [B] or versus the concentration of [B] at constant [A]. It can be used to make 3D plots of velocity versus [A] and [B].

(b) The other form of the rate equation is obtained by substitution of specific values of a and b.

vordAB /. a → 100 /. b → 80

$$\frac{8000\,\text{kfet}}{8000 + 100\,\text{kB} + \text{kB kIA}}$$

This form of the rate equation is a function of the values of the kinetic parameters.

vordAB can be used to plot v versus [A] at constant [B] and versus [B] at constant [A]

```
Plot[vordAB /. kfet → 1 /. kIA → 5 /. kB → 20 /. b → {5, 20, 80},
  {a, 0, 100}, AxesLabel → {"[A]", "v"}]
```

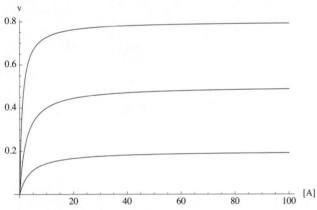

Fig. 3.1 Rapid-equilibrium velocities at [B] = {5,20,80} as functions of [A].

```
Plot[vordAB /. kfet → 1 /. kIA → 5 /. kB → 20 /. a → {1, 5, 50},
  {b, 0, 100}, AxesLabel → {"[B]", "v"}]
```

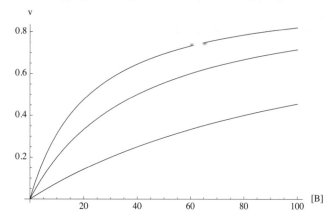

Fig. 3.2 Rapid-equilibrium velocities at [A] = {1,5,50} as functions of [B].

Rate equation vordAB can be rearranged to the Michaelis form

> **vordAB**
>
> $$\frac{a\,b\,kfet}{a\,b + a\,kB + kB\,kIA}$$

Divide the denominator by a*b, one term at a time.

> **{a b, a kB, kB kIA} / (a * b)**
>
> $$\left\{1, \ \frac{kB}{b}, \ \frac{kB\,kIA}{a\,b}\right\}$$

This yields the following rate equation.

> $$\mathbf{kfet} \Big/ \left(1 + \frac{\mathbf{kB}}{\mathbf{b}} + \frac{\mathbf{kB\,kIA}}{\mathbf{a\,b}}\right)$$
>
> $$\frac{kfet}{1 + \frac{kB}{b} + \frac{kB\,kIA}{a\,b}}$$

Take kB/b out of two terms in denominator.

> **kfet / (1 + (kB / b) * (1 + kIA / a))**
>
> $$\frac{kfet}{1 + \frac{kB\left(1 + \frac{kIA}{a}\right)}{b}}$$

This is the usual form for the velocity for ordered A + B → products.

Plot reciprocal velocity versus 1/[A] at specified [B]

Lineweaver-Burk plots are readily constructed using vordAB /. kfet → 1 /. kIA → 5 /. kB → 20.

```
vordAB /. kfet → 1 /. kIA → 5 /. kB → 20
```

$$\frac{a\,b}{100 + 20\,a + a\,b}$$

Make reciprocal v:

```
1 / vordAB
```

$$\frac{a\,b + a\,kB + kB\,kIA}{a\,b\,kfet}$$

Introduce reciprocal a that is defined by a = 1/reca.

```
(1 / vordAB) /. a → 1 / reca
```

$$\frac{\left(kB\,kIA + \frac{b}{reca} + \frac{kB}{reca}\right)reca}{b\,kfet}$$

Plot of reciprocal velocity versus 1/[A] at specified [B]

The expressions for reciprocal velocity at [B] = {5,20,50} are given by

```
(1 / vordAB) /. a → 1 / reca /. b → {5, 20, 80}
```

$$\left\{\frac{\left(kB\,kIA + \frac{5}{reca} + \frac{kB}{reca}\right)reca}{5\,kfet}, \frac{\left(kB\,kIA + \frac{20}{reca} + \frac{kB}{reca}\right)reca}{20\,kfet}, \frac{\left(kB\,kIA + \frac{80}{reca} + \frac{kB}{reca}\right)reca}{80\,kfet}\right\}$$

Arbitrary values of kfet, kIA, and kB can be inserted.

```
(1 / vordAB) /. a → 1 / reca /. kfet → 1 /. kIA → 5 /. kB → 20 /. b → {5, 20, 80}
```

$$\left\{\frac{1}{5}\left(100 + \frac{25}{reca}\right)reca, \frac{1}{20}\left(100 + \frac{40}{reca}\right)reca, \frac{1}{80}\left(100 + \frac{100}{reca}\right)reca\right\}$$

```
Plot[(1 / vordAB) /. a → 1 / reca /. kfet → 1 /. kIA → 5 /. kB → 20 /. b → {5, 20, 80},
  {reca, -.5, 1}, AxesLabel → {"1/[A]", "1/v"}, PlotRange → {0, 15}]
```

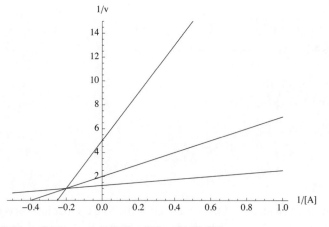

Fig. 3.3 Plot of 1/v versus 1/[A] at [B] = {5,20,80}.

Plot reciprocal velocity versus 1/[B] at specified [A]

Plots of 1/v versus 1/[B] can be made at several [A] as follows:

```
(1 / vordAB) /. b → 1 / recb /. kfet → 1 /. kIA → 5 /. kB → 20 /. a → {1, 5, 50}
```

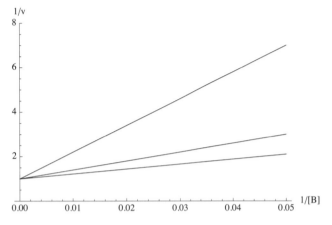

```
Plot[(1 / vordAB) /. b → 1 / recb /. kfet → 1 /. kIA → 5 /. kB → 20 /. a → {1, 5, 50},
  {recb, 0, .05}, AxesLabel → {"1/[B]", "1/v"}, PlotRange → {0, 8}]
```

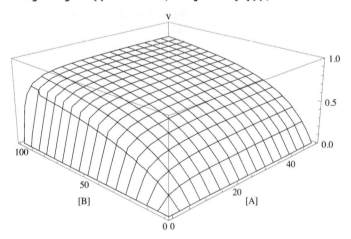

Fig. 3.4 Plot of 1/v versus 1/[B] at [A] = {1,5,50}.

Three dimensional plots

Since vordAB is a function of {[A],[B]} at specified values of the kinetic parameters, v can be plotted versus [A] and [B].

```
Plot3D[vordAB /. kfet → 1 /. kIA → 5 /. kB → 20, {a, 0, 50}, {b, 0, 100},
  PlotRange → {0, 1}, ViewPoint → {-2, -2, 1}, AxesLabel → {"[A]", "[B]", ""},
  Lighting -> {{"Ambient", GrayLevel[1]}}, PlotLabel → "v"]
```

Fig. 3.5 3D plot of the velocity for ordered A + B -> products when kfet = 1, kIA = 5 and kB = 20. The contour lines at constant [B] are Michaelis-Menten plots for A. The contour lines at constant [A] are Michaelis-Menten plots for B. Various scales can be used on the A and B axes.

More information about the shape of this surface can be obtained by taking partial derivatives of v with respect to [A] and [B], and the mixed second derivative with respect to [A] and [B].

```
plotA2 = Plot3D[Evaluate[vordAB] /. kfet → 1 /. kIA → 5 /. kB → 20,
    {a, .0001, 15}, {b, .0001, 60}, PlotRange → {0, 1}, ViewPoint → {-2, -2, 1},
    AxesLabel → {"[A]", "[B]", ""}, Lighting -> {{"Ambient", GrayLevel[1]}}, PlotLabel → "v"];

plotB2 = Plot3D[Evaluate[5 * D[vordAB /. kfet → 1 /. kIA → 5 /. kB → 20, a]], {a, .0001, 15},
    {b, .0001, 60}, PlotRange → {0, 1.5}, ViewPoint → {-2, -2, 1}, AxesLabel → {"[A]", "[B]", ""},
    Lighting -> {{"Ambient", GrayLevel[1]}}, PlotLabel → "10dv/d[A]"];

plotC2 = Plot3D[Evaluate[30 * D[vordAB /. kfet → 1 /. kIA → 5 /. kB → 20, b]], {a, .0001, 15},
    {b, .0001, 60}, PlotRange → {0, 1.3}, ViewPoint → {-2, -2, 1}, AxesLabel → {"[A]", "[B]", ""},
    Lighting -> {{"Ambient", GrayLevel[1]}}, PlotLabel → "10dv/d[B]"];

plotD2 = Plot3D[Evaluate[1000 * D[vordAB /. kfet → 1 /. kIA → 5 /. kB → 20, a, b]], {a, .0001, 15},
    {b, .0001, 60}, PlotRange → {-1, 7}, ViewPoint → {-2, -2, 1}, AxesLabel → {"[A]", "[B]", ""},
    Lighting -> {{"Ambient", GrayLevel[1]}}, PlotLabel → "1000d(dv/d[A]/d[B]"];

figlord2 = Show[GraphicsArray[{{plotA2, plotB2}, {plotC2, plotD2}}]]
```

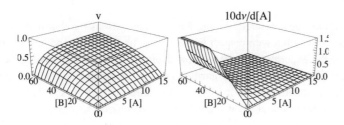

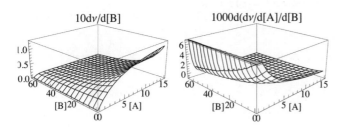

Fig. 3.6 (a) v versus [A] and [B] for the ordered mechanism for A + B → products with kfet = 1.00, K_{IA} = 5.00, and K_B = 20.00. (b) 10dv/d[A] for the ordered mechanism. (c) 10dv/d[B] for the ordered mechanism. (d) 1000d(dv/d[A]/d[B]) for the ordered mechanism.

The plot of 10dv/d[A] shows that the velocity is most sensitive to [A] at high [B] and low [A]. The plot of 10dv/d[B] shows that the velocity is most sensitive to [B] at high [A] and low [B]. The plot of 1000d(dv/d[A]/d[B]) shows that the velocity is most sensitive to [A] and [B] together at low [A] and low [B]; this plot will be important in considering the random A + B → products in the next chapter.

3.2 Estimation of the kinetic parameters for ordered A + B → products

The use of Lineweaver-Burk plots for the determination of kinetic parameters is well known [6,21,25]. One problem with this method is that it becomes much more complicated when the number of reactants is increased, as with the five rapid-equilibrium rate equations for A + B + C → products, A + mB → products, and when there are pKs or inhibitors, activators, or modifiers. In 1979 Duggleby [9] suggested that the kinetic parameters in a rate equation can be calculated by solving simultaneous equations for the minimum number of velocity measurements. *Mathematica*, Maple, and MatLab have solve programs that can be used to solve sets of complicated polynomial equations. *Mathematica* has been used [32] to write a program that calculates the 3 kinetic parameters for rapid-equilibrium ordered A + B → products at a specified pH. The effects of 5% errors in the three velocity measurements are readily calculated.

- ### 3.2.1 Use of Solve to estimate the kinetic parameters from the minimum number of velocity measurements for ordered A + B → products

Since there are 3 kinetic parameters, 3 velocities have to be measured. Figure 3.6 provides reasons for choosing {a,b} = {high a,high b},{low a,high b}, and {low a,low b}. High means high with respect to the Michaelis constants, and low is low with respect to the Michaelis constants, but limitations of the experimental methods for determining velocities have to be considered. Notice that the first measured velocity makes the most important contribution to the determination of kfet, the second velocity makes the most important contribution to the determination of kIA, and the third velocity makes the most important contribution to the determination of kB. As a trial, three velocities are calculated for specified kfet, kIA, and kB (see Section 3.1) by use of the ReplaceAll operation (/.x->). The following pairs {a,b} of substrate concentrations are used for test calculations: {100,100}, {1,100}, and {100,1}.

```
vordAB /. kfet → 1 /. kIA → 5 /. kB → 20 /. a → 100 /. b → 100 // N
```

```
0.826446
```

```
vordAB /. kfet → 1 /. kIA → 5 /. kB → 20 /. a → 1 /. b → 100 // N
```

```
0.454545
```

```
vordAB /. kfet → 1 /. kIA → 5 /. kB → 20 /. a → 100 /. b → 1 // N
```

```
0.0454545
```

The following program was written in *Mathematica* to estimate the values of kfet, kA, and kB from three measured velocities at three pairs of substrate concentrations on the assumption that the mechanism is ordered, with A being bound first.

```
calckinparsordAB[v1_, a1_, b1_, v2_, a2_, b2_, v3_, a3_, b3_] :=
 Module[{}, (*This program calculates kfet, kIA,
  and kB from three experimental velocities for A + B → products at three pairs
    of substrate concentrations on the assumption that the mechanism is ordered,
  with A bound first.  The first velocity is at high [A] and high [B],
    the second velocity is at low [A] and high [B],
    and the third velocity is at high [A] and low [B].*)
  Solve[{v1 == a1 * b1 * kfet / (a1 * b1 + a1 * kB + kIA * kB),
    v2 == a2 * b2 * kfet / (a2 * b2 + a2 * kB + kIA * kB),
    v3 == a3 * b3 * kfet / (a3 * b3 + a3 * kB + kIA * kB)}, {kfet, kIA, kB}]]
```

This program is tested by using it to estimate the kinetic parameters using 3 velocities calculated using vordAB.

Calculation of kinetic parameters from velocities at {100, 100}, {1, 100}, and {100, 1}

The program calckinparsordAB is used to calculate the kinetic parameters from the 3 velocities calculated above (now considered to be measured velocities) at {[A],[B]} taken to be {100,100}, {1,100}, and {100,1}.

```
calckinparsordAB[.826446, 100, 100, .4544545, 1, 100, .0454545, 100, 1]
```

```
{{kfet → 1., kIA → 5.00233, kB → 19.9996}}
```

line11 is used to obtain a list of kinetic parameters that can be used to make a table.

```
line11 =
 {kfet, kIA, kB} /. calckinparsordAB[.826446, 100, 100, .4544545, 1, 100, .0454545, 100, 1]
```

```
{{1., 5.00233, 19.9996}}
```

The values of the 3 kinetic parameters calculated using calckinparsordAB are correct, but the question is "How sensitive are the kinetic parameters to experimental errors in the measured velocities?" Introduce 5% errors in the velocities, one at a time.

```
calckinparsordAB[1.05 * .826446, 100, 100, .4544545, 1, 100, .0454545, 100, 1]
```

```
{{kfet → 1.0618, kIA → 5.29334, kB → 21.2355}}
```

```
line12 =
 {kfet, kIA, kB} /. calckinparsordAB[1.05 * .826446, 100, 100, .4544545, 1, 100, .0454545, 100, 1]
```

```
{{1.0618, 5.29334, 21.2355}}
```

```
calckinparsordAB[.826446, 100, 100, 1.05 * .4544545, 1, 100, .0454545, 100, 1]
```

```
{{kfet → 1., kIA → 4.44956, kB → 20.1054}}
```

```
line13 =
 {kfet, kIA, kB} /. calckinparsordAB[.826446, 100, 100, 1.05 * .4544545, 1, 100, .0454545, 100, 1]
```

```
{{1., 4.44956, 20.1054}}
```

```
calckinparsordAB[.826446, 100, 100, .4544545, 1, 100, 1.05 * .0454545, 100, 1]
```

```
{{kfet → 0.989529, kIA → 5.28179, kB → 18.743}}
```

```
line14 =
 {kfet, kIA, kB} /. calckinparsordAB[.826446, 100, 100, .4544545, 1, 100, 1.05 * .0454545, 100, 1]
```

```
{{0.989529, 5.28179, 18.743}}
```

Table 3.1 Values of kinetic parameters obtained using {100,100}, {1,100}, and {100,1} and testing the effects of 5% errors in the measured velocities, one at a time.

```
TableForm[Round[{line11[[1]], line12[[1]], line13[[1]], line14[[1]]}, 0.01],
 TableHeadings → {{"No errors", "1.05*v1", "1.05*v2", "1.05*v3"}, {"kf[E]t", "KIA", "KB"}}]
```

	$k_f[E]_t$	K_{IA}	K_B
No errors	1.	5.	20.
1.05*v1	1.06	5.29	21.24
1.05*v2	1.	4.45	20.11
1.05*v3	0.99	5.28	18.74

The effects of experimental errors in substrate concentrations can be calculated in the same way, but their effects are more easily summarized: An error of 5% in the concentration of a substrate causes an error of 5% in the corresponding Michaelis constant.

Calculation of kinetic parameters from velocities at {150,150}, {0.1,150}, {150,0.1}

This shows the effects of widening the range of substrate equations. The 3 velocities are calculated as follows:

```
vordAB /. kfet → 1 /. kIA → 5 /. kB → 20 /. a → 150 /. b → 150 // N
```

```
0.878906
```

```
vordAB /. kfet → 1 /. kIA → 5 /. kB → 20 /. a → .1 /. b → 150 // N
```

0.128205

```
vordAB /. kfet → 1 /. kIA → 5 /. kB → 20 /. a → 150 /. b → .1 // N
```

0.00481541

This velocity is quite low, but it can be determined by increasing the enzyme concentration.

Estimation of the kinetic parameters.

```
calckinparsordAB[.878906, 150, 150, .128205, .1, 150, .00481541, 150, .1]
```

{{kfet → 1., kIA → 5.00001, kB → 20.}}

```
line21 =
 {kfet, kIA, kB} /. calckinparsordAB[.878906, 150, 150, .128205, .1, 150, .00481541, 150, .1]
```

{{1., 5.00001, 20.}}

Introduce 5% errors in the 3 velocities, one at a time.

```
calckinparsordAB[1.05 * .878906, 150, 150, .128205, .1, 150, .00481541, 150, .1]
```

{{kfet → 1.05732, kIA → 5.04067, kB → 21.1465}}

```
line22 = {kfet, kIA, kB} /.
   calckinparsordAB[1.05 * .878906, 150, 150, .128205, .1, 150, .00481541, 150, .1]
```

{{1.05732, 5.04067, 21.1465}}

```
calckinparsordAB[.878906, 150, 150, 1.05 * .128205, .1, 150, .00481541, 150, .1]
```

{{kfet → 1., kIA → 4.71249, kB → 20.0372}}

```
line23 = {kfet, kIA, kB} /.
   calckinparsordAB[.878906, 150, 150, 1.05 * .128205, .1, 150, .00481541, 150, .1]
```

{{1., 4.71249, 20.0372}}

```
calckinparsordAB[.878906, 150, 150, .128205, .1, 150, 1.05 * .00481541, 150, .1]
```

{{kfet → 0.993446, kIA → 5.26027, kB → 18.8859}}

```
line24 = {kfet, kIA, kB} /.
   calckinparsordAB[.878906, 150, 150, .128205, .1, 150, 1.05 * .00481541, 150, .1]
```

{{0.993446, 5.26027, 18.8859}}

Table 3.2 Values of kinetic parameters obtained using {150,150}, {0.1,150}, and {150,0.1} and testing the effects of 5% errors in the measured velocities

```
TableForm[Round[{line21[[1]], line22[[1]], line23[[1]], line24[[1]]}, 0.01],
 TableHeadings → {{"No errors", "1.05*v1", "1.05*v2", "1.05*v3"}, {"k_f[E]_t", "K_IA", "K_B"}}]
```

	$k_f[E]_t$	K_{IA}	K_B
No errors	1.	5.	20.
1.05*v1	1.06	5.04	21.15
1.05*v2	1.	4.71	20.04
1.05*v3	0.99	5.26	18.89

The errors are a little smaller than for {100,100},{1,100},{100,1} in Table 3.1.

Calculation of kinetic parameters from velocities at {20, 20}, {5, 20}, and {20,5}

Try a narrower range of substrate concentrations.

```
vordAB /. kfet → 1 /. kIA → 5 /. kB → 20 /. a → 20 /. b → 20 // N
```

0.444444

```
vordAB /. kfet → 1 /. kIA → 5 /. kB → 20 /. a → 5 /. b → 20 // N
```

0.333333

```
vordAB /. kfet → 1 /. kIA → 5 /. kB → 20 /. a → 20 /. b → 5 // N
```

0.166667

Calculate the kinetic parameters.

```
calckinparsordAB[.444444, 20, 20, .333333, 5, 20, .166667, 20, 5]
```

{{kfet → 0.999993, kIA → 5.00003, kB → 19.9998}}

```
line31 = {kfet, kIA, kB} /. calckinparsordAB[.444444, 20, 20, .333333, 5, 20, .166667, 20, 5]
```

{{0.999993, 5.00003, 19.9998}}

This is correct, as expected, but now calculate the effects of 5% errors in the measured velocities, one at a time.

```
calckinparsordAB[1.05 * .444444, 20, 20, .333333, 5, 20, .166667, 20, 5]
```

{{kfet → 1.16666, kIA → 5.71432, kB → 23.333}}

```
line32 =
 {kfet, kIA, kB} /. calckinparsordAB[1.05 * .444444, 20, 20, .333333, 5, 20, .166667, 20, 5]
```

{{1.16666, 5.71432, 23.333}}

```
calckinparsordAB[.444444, 20, 20, 1.05 * .333333, 5, 20, .166667, 20, 5]
```

{{kfet → 0.999993, kIA → 3.86366, kB → 20.9521}}

```
line33 =
 {kfet, kIA, kB} /. calckinparsordAB[.444444, 20, 20, 1.05 * .333333, 5, 20, .166667, 20, 5]
```

{{0.999993, 3.86366, 20.9521}}

```
calckinparsordAB[.444444, 20, 20, .333333, 5, 20, 1.05 * .166667, 20, 5]
```

{{kfet → 0.913038, kIA → 5.52635, kB → 16.5215}}

```
line34 =
 {kfet, kIA, kB} /. calckinparsordAB[.444444, 20, 20, .333333, 5, 20, 1.05 * .166667, 20, 5]
```

```
{{0.913038, 5.52635, 16.5215}}
```

Table 3.3 Values of kinetic parameters obtained using {20,20}, {5,20}, and {20,5} and testing the effects of 5% errors in the measured velocities

```
TableForm[Round[{line31[[1]], line32[[1]], line33[[1]], line34[[1]]}, 0.01],
 TableHeadings → {{"No errors", "1.05*v1", "1.05*v2", "1.05*v3"}, {"kf[E]t", "KIA", "KB"}}]
```

	$k_f[E]_t$	K_{IA}	K_B
No errors	1.	5.	20.
1.05*v1	1.17	5.71	23.33
1.05*v2	1.	3.86	20.95
1.05*v3	0.91	5.53	16.52

These errors are larger than in the preceding two cases, and so it is clear that the widest practical range of substrate concentrations should be used. The calculations in this section emphasize the importance of velocity measurements at low substrate concentrations. Thus the accuracy of the estimation of kinetic parameters is very dependent on th analytical equipment used. Replicate measurements can be used to improve the accuracy of the estimation of the kinetic paremeters.

■ 3.2.2 Calculation of the effects of changes in the kinetic parameters on the velocities

It is of interest to plot v, dv/dkfet, dv/kIA, and dv/kB versus [A] and [B] in a 3D plot to see how sensitive v is to the values of these 3 kinetic parameters. 3D plots of v, dv/d[A], dv/d[B], and d(dv/d[A])/d[B] were given earlier in Fig. 3.6. 3D plots of v, dv/dkfet, dv/dkIA, and dv/dkB are now calculated for Fig. 3.7.

```
plotA3 = Plot3D[Evaluate[vordAB] /. kfet → 1 /. kIA → 5 /. kB → 20,
   {a, .0001, 15}, {b, .0001, 60}, PlotRange → {0, 1}, ViewPoint → {-2, -2, 1},
   AxesLabel → {"[A]", "[B]", ""}, Lighting -> {{"Ambient", GrayLevel[1]}}, PlotLabel → "v"];
```

```
vordAB
```

$$\frac{a\, b\, kfet}{a\, b + a\, kB + kB\, kIA}$$

The derivative of the velocity with respect to kfet is

```
D[vordAB, kfet]
```

$$\frac{a\, b}{a\, b + a\, kB + kB\, kIA}$$

```
D[vordAB, kfet] /. kIA → 5 /. kB → 20
```

$$\frac{a\, b}{100 + 20\, a + a\, b}$$

```
plotB3 = Plot3D[D[vordAB, kfet] /. kIA → 5 /. kB → 20, {a, .0001, 15}, {b, .0001, 60},
   PlotRange → {0, 1.}, ViewPoint → {-2, -2, 1}, AxesLabel → {"[A]", "[B]", ""},
   Lighting -> {{"Ambient", GrayLevel[1]}}, PlotLabel → "dv/dkfet"];
```

The second plot in Fig. 3.7 shows that changing the value of kfet has the largest effect on v at high [A] and high [B].

```
D[vordAB, kIA]
```

$$-\frac{a\, b\, kB\, kfet}{(a\, b + a\, kB + kB\, kIA)^2}$$

```
D[vordAB, kIA] /. kfet → 1 /. kIA → 5 /. kB → 20
```

$$-\frac{20\,a\,b}{(100 + 20\,a + a\,b)^2}$$

```
plotC3 = Plot3D[Evaluate[D[vordAB, kIA]] /. kfet → 1 /. kIA → 5 /. kB → 20,
    {a, .0001, 15}, {b, .0001, 60}, PlotRange → {0, -0.04},
    ViewPoint → {-2, -2, 1}, AxesLabel → {"[A]", "[B]", ""},
    Lighting -> {{"Ambient", GrayLevel[1]}}, PlotLabel → "dv/dkIA"];
```

The third plot in Fig. 3.7 shows that v is most sensitive to kIA at low [A] and low [B].

```
plotD3 =
  Plot3D[Evaluate[D[vordAB, kB]] /. kfet → 1 /. kIA → 5 /. kB → 20, {a, .0001, 15}, {b, .0001, 60},
    (*PlotRange→{0,-0.04},*)ViewPoint → {-2, -2, 1}, AxesLabel → {"[A]", "[B]", ""},
    Lighting -> {{"Ambient", GrayLevel[1]}}, PlotLabel → "dv/dkB"];
```

This plot shows that changing the value of kB has the largest effect on v at low [A] and low [B].

```
figlord3 = GraphicsArray[{{plotA3, plotB3}, {plotC3, plotD3}}]
```

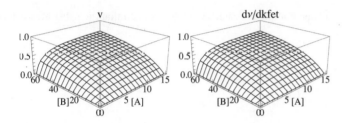

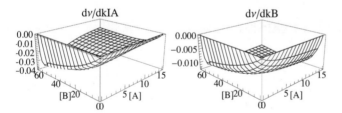

Fig. 3.7 Sensitivities of the velocity for ordered A + B → products with kfet = 1.00, K_{IA} = 5.00, and K_B = 20.00 to the values of the kinetic parameters. (a) v versus [A] and [B] . (b) dv/dkfet. (c) dv/dkIA. (d) dv/dkB.

This figure supports the recommendation to use {high a,high b}, {low a, high b}, and {low a,low b} to obtain the most accurate values of the kinetic parameters.

3.3 Effects of pH on the kinetics of ordered A + B → products

The effects of pH on the rapid-equilibrium velocity of an enzyme-catalyzed reaction are a consequence of the pKs of the substrates, enzymatic site, enzyme-substrate complexes, and the consumption of hydrogen ions in the rate-determining reaction. In investigating these effects, the rapid-equilibrium method is especially useful because a large number of chemical reactions have to be taken into account. The effects of pH on the kinetics of A + B → products has been discussed in the literature [33,35].

There are two ways to write the expression for the velocity of ordered A + B → products when the enzymatic site and the enzyme-substrate complexes each have two pKs. The factor method, which is discussed first, utilizes pKs and expresses the velocity as a function of pH, as well as [A] and [B] (Sections 3.3.2 and 3.3.3). The second method utilizes acid dissociation constants and expresses the velocity in terms of [H$^+$], as well as [A] and [B]. These two forms of the rate equation yield the same velocities. The advantage of the second method is that it can be used to write a computer program to estimate the kinetic parameters using the minimum number of velocity measurements. Rate equations derived using the factor method cannot be used to estimate kinetic parameters using Solve because Solve has been developed to solve sets of polynomial equations.

■ 3.3.1 Mechanism for ordered A + B → products when the enzymatic site and enzyme-substrate complexes each have two pKs

It is assumed that the enzymatic site and enzyme-substrate complexes each have two pKs and that the substrates A and B do not have pKs in the pH range of interest.

$$
\begin{array}{ll}
E^- & EA^- \\
\| \ pK1e & \| \ pK1ea \\
HE \ + \ A = HEA & \\
\| \ pK2e & \| \ pK2ea \\
H_2 E^+ & H_2 EA^+
\end{array}
\qquad K_{HEA} = [HE][A]/[HEA] \qquad kcHEA = he*a/hea \qquad (4)
$$

$$
\begin{array}{ll}
EA^- & EAB^- \\
\| & \| \ pK1eab \\
HEA \ + \ B = HEAB & \\
\| & \| \ pK2eab \\
H_2 EA^+ & H_2 EAB^+
\end{array}
\qquad K_{HEAB} = [HEA][B]/[HEAB] \qquad kcHEAB = hea*b/heab \qquad (5)
$$

$$
\begin{array}{l}
EAB^- \\
\| \ k_f \\
HEAB \ \to \ products \\
\| \\
H_2 EAB^+
\end{array}
\qquad v = k_f[HEAB] = kf*et*heab/et = kfet*heab/et \qquad (6)
$$

The chemical equilibrium constants K_{HEA} and K_{HEAB} are always obeyed. kfet is used for $k_f[E]_t$ because the relative contributions of k_f and $[E]_t$ are not discussed here. Notice that this mechanism is written in terms of chemical species, whereas mechanism 1-3 was written in terms of reactants (sums of species).

The equilibrium constants involved in this mechanism are defined as follows:

$kHE = h*e/he = 10^{-pK1e}$

$kH2E = h*he/h2e = 10^{-pK2e}$

$kHEA = h*ea/hea = 10^{-pK1ea}$

$kH2EA = h*hea/h2ea = 10^{-pK2ea}$

$kHEAB = h*eab/heab = 10^{-pK1eab}$

$kH2EAB = h*heab/h2eab = 10^{-pK2eab}$

$kcHEA = he*a/hea$

$kcHEAB = hea*b/heab$

The specified concentration of hydrogen ions is represented by h. The total concentration of enzymatic sites $[E]_t$ = et is given by

et = e + he + h2e + ea + hea + h2ea + eab + heab + h2eab

The rapid-equilibrium rate equation can be derived using the factor method (see Section 2.1.4) or using Solve. The equations for the factor method are discussed first.

The following *Mathematica* program has been written to input these 6 pKs, 2 chemical equilibrium constants, and $k_f[E]_t$ = kfet in the calculation of the pH dependencies of V_{fexp}, K_{IA}, K_B, V_f/K_B, and $V_f/K_{IA}K_B$. The factor approach is used; see Section 2.1.4. The **derordABzero** program also derives the equation for the velocity v of the forward reaction as a function of [A], [B], and pH.

```
derordABzero[pK1e_, pK2e_, pK1ea_, pK2ea_, pK1eab_, pK2eab_, kfet_, kHEA_, kHEAB_] :=
Module[{eafactor, efactor, eabfactor, vfexp, kia, kb, v},
   (*This program derives the pH dependencies of the kinetic parameters of
      the forward enzyme-catalyzed reaction ordered A+B = products when no
      hydrogen ions are consumed. The output is a list of 5 functions of pH: vfexp,
   kia, kb, vf/kb, and vf/kiakb.  v is a function of [A] = a, [B]= b, and pH.*)
   efactor = 1 + 10^(pK2e-pH) + 10^(pH-pK1e);
   eafactor = 1 + 10^(pK2ea-pH) + 10^(pH-pK1ea);
   eabfactor = 1 + 10^(pK2eab-pH) + 10^(pH-pK1eab);
   vfexp = kfet / eabfactor;
   kia = kHEA * efactor / eafactor;
   kb = kHEAB * eafactor / eabfactor;
   v = vfexp / (1 + (kb / b) * (1 + (kia / a)));
   {vfexp, kia, kb, vfexp / kb, vfexp / (kia * kb), v}]
```

When the symbols for the parameters are used as input, the most general forms of the kinetic parameters, two ratios, and v are obtained.

```
derordABzero[pK1e, pK2e, pK1ea, pK2ea, pK1eab, pK2eab, kfet, kHEA, kHEAB]
```

$$\left\{ \frac{kfet}{1 + 10^{pH-pK1eab} + 10^{-pH+pK2eab}}, \frac{\left(1 + 10^{pH-pK1e} + 10^{-pH+pK2e}\right) kHEA}{1 + 10^{pH-pK1ea} + 10^{-pH+pK2ea}}, \frac{\left(1 + 10^{pH-pK1ea} + 10^{-pH+pK2ea}\right) kHEAB}{1 + 10^{pH-pK1eab} + 10^{-pH+pK2eab}}, \right.$$

$$\frac{kfet}{\left(1 + 10^{pH-pK1ea} + 10^{-pH+pK2ea}\right) kHEAB}, \frac{kfet}{\left(1 + 10^{pH-pK1e} + 10^{-pH+pK2e}\right) kHEA \, kHEAB},$$

$$\left. \frac{kfet}{\left(1 + 10^{pH-pK1eab} + 10^{-pH+pK2eab}\right)\left(1 + \frac{\left(1+10^{pH-pK1ea}+10^{-pH+pK2ea}\right)\left(1+\frac{\left(1+10^{pH-pK1e}+10^{-pH+pK2e}\right) kHEA}{\left(1+10^{pH-pK1ea}+10^{-pH+pK2ea}\right) a}\right) kHEAB}{\left(1+10^{pH-pK1eab}+10^{-pH+pK2eab}\right) b}\right)} \right\}$$

The last function in the list gives the velocity vel.

```
vel =
```

$$\text{kfet} \Big/ \left(\left(1 + 10^{pH-pK1eab} + 10^{-pH+pK2eab}\right) \left(1 + \frac{\left(1 + 10^{pH-pK1ea} + 10^{-pH+pK2ea}\right) \left(1 + \frac{\left(1+10^{pH-pK1e}+10^{-pH+pK2e}\right)\text{kHEA}}{\left(1+10^{pH-pK1ea}+10^{-pH+pK2ea}\right)a}\right)\text{kHEAB}}{\left(1 + 10^{pH-pK1eab} + 10^{-pH+pK2eab}\right)b}\right)\right)$$

$$\frac{\text{kfet}}{\left(1 + 10^{pH-pK1eab} + 10^{-pH+pK2eab}\right)\left(1 + \frac{\left(1+10^{pH-pK1ea}+10^{-pH+pK2ea}\right)\left(1+\frac{\left(1+10^{pH-pK1e}+10^{-pH+pK2e}\right)\text{kHEA}}{\left(1+10^{pH-pK1ea}+10^{-pH+pK2ea}\right)a}\right)\text{kHEAB}}{\left(1+10^{pH-pK1eab}+10^{-pH+pK2eab}\right)b}\right)}$$

The expression for V_{fexp} is readily obtained by setting [A] and [B] equal to infinity.

```
vel /. a → ∞ /. b → ∞
```

$$\frac{\text{kfet}}{1 + 10^{pH-pK1eab} + 10^{-pH+pK2eab}}$$

When the pH, [A], and [B] are specified, the velocity is given by

```
vel /. pH → 7 /. a → 100 /. b → 100
```

$$\frac{\text{kfet}}{\left(1 + 10^{7-pK1eab} + 10^{-7+pK2eab}\right)\left(1 + \frac{\left(1+10^{7-pK1ea}+10^{-7+pK2ea}\right)\left(1+\frac{\left(1+10^{7-pK1e}+10^{-7+pK2e}\right)\text{kHEA}}{100\left(1+10^{7-pK1ea}+10^{-7+pK2ea}\right)}\right)\text{kHEAB}}{100\left(1+10^{7-pK1eab}+10^{-7+pK2eab}\right)}\right)}$$

This rate equation cannot be used in Solve to obtain the values of the kinetic parameters from the minimum number of velocity measurements because it is not a polynomial equation. It can be converted into a polynomial equation, but that is not done here.

The following values for kinetic parameters are chosen arbitrarily for test calculations.

```
Grid[{{pK1e, pK2e, pK1ea, pK2ea, pK1eab, pK2eab, kfet, kHEA, kHEAB}, {7, 6, 8, 6, 8, 7, 2, 1, 2}}]
```

pK1e	pK2e	pK1ea	pK2ea	pK1eab	pK2eab	kfet	kHEA	kHEAB
7	6	8	6	8	7	2	1	2

The pH dependencies of vfexp, kia, kb, vfexp/kb, vfexp/(kia*kb), and v are obtained as by substituting these kinetic parameters in the expressions for the kinetic parameters and v.

```
vordABzero = derordABzero[7, 6, 8, 6, 8, 7, 2, 1, 2]
```

$$\left\{ \frac{2}{1 + 10^{7-pH} + 10^{-8+pH}}, \; \frac{1 + 10^{6-pH} + 10^{-7+pH}}{1 + 10^{6-pH} + 10^{-8+pH}}, \; \frac{2\left(1 + 10^{6-pH} + 10^{-8+pH}\right)}{1 + 10^{7-pH} + 10^{-8+pH}}, \; \frac{1}{1 + 10^{6-pH} + 10^{-8+pH}}, \right.$$

$$\left. \frac{1}{1 + 10^{6-pH} + 10^{-7+pH}}, \; \frac{2}{\left(1 + 10^{7-pH} + 10^{-8+pH}\right)\left(1 + \frac{2\left(1+10^{6-pH}+10^{-8+pH}\right)\left(1+\frac{1+10^{6-pH}+10^{-7+pH}}{\left(1+10^{6-pH}+10^{-8+pH}\right)a}\right)}{\left(1+10^{7-pH}+10^{-8+pH}\right)b}\right)} \right\}$$

Notice that the expression for the velocity includes a and b.

The pH dependencies of the kinetic parameters can be plotted as follows:

```
plot1fr = Plot[Evaluate[vordABzero[[1]]],
    {pH, 5, 9}, AxesLabel → {"pH", "V_fexp"}, DisplayFunction → Identity];
```

```
plot2fr = Plot[Evaluate[vordABzero[[2]]],
    {pH, 5, 9}, AxesLabel → {"pH", "K_IA"}, DisplayFunction → Identity];

plot3fr = Plot[Evaluate[vordABzero[[3]]],
    {pH, 5, 9}, AxesLabel → {"pH", "K_B"}, DisplayFunction → Identity];

plot4fr = Plot[Evaluate[vordABzero[[4]]],
    {pH, 5, 9}, AxesLabel → {"pH", "V_fexp/K_B"}, DisplayFunction → Identity];

plot5fr = Plot[Evaluate[vordABzero[[5]]], {pH, 5, 9},
    AxesLabel → {"pH", "V_fexp/K_IA K_B"}, DisplayFunction → Identity];

fig3jpc = GraphicsGrid[{{plot1fr, plot2fr}, {plot3fr, plot4fr}, {plot5fr}}]
```

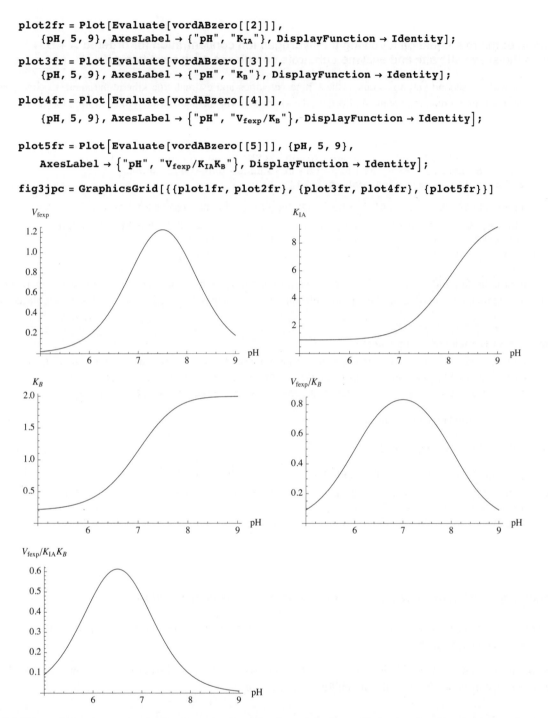

Fig. 3.8 Plots of kinetic parameters for the ordered mechanism of the forward reaction A + B → products when no hydrogen ions are consumed. The pKs, chemical equilibrium constants and kfet are given by Grid. These plots can be used to calculate the kinetic parameters [33].

The rate equation vordABzero[[6]] can be used to calculate velocities at desired pHs and substrate concentrations when the kinetic parameters have been specified.

This section provides a second way to calculate these nine velocities and estimate the kinetic parameters.

3.3.2 Derivation of the rate equation involving the hydrogen ion concentration for ordered A + B → products when the enzymatic site and enzyme-substrate complexes each have two p*K*s

This section provides a second way to calculate these nine velocities and estimate the kinetic parameters. Since the reactions up to the rate-determining reaction are at equilibrium, the expression for the equilibrium concentration of HEAB can be derived using Solve[eqns,vars,elims].

```
Solve[{kHE == h * e / he, kH2E == h * he / h2e, kHEA == h * ea / hea, kH2EA == h * hea / h2ea,
  kHEAB == h * eab / heab, kH2EAB == h * heab / h2eab, kcHEA == he * a / hea,
  kcHEAB == hea * b / heab, et == e + he + h2e + ea + hea + h2ea + eab + heab + h2eab},
 {heab}, {e, he, h2e, ea, hea, h2ea, eab, h2eab}]
```

$$\{\{heab \rightarrow (a\,b\,et\,h\,kH2E\,kH2EA\,kH2EAB)\,/\,(a\,b\,h^2\,kH2E\,kH2EA + a\,h^2\,kcHEAB\,kH2E\,kH2EAB +$$
$$h^2\,kcHEA\,kcHEAB\,kH2EA\,kH2EAB + a\,b\,h\,kH2E\,kH2EA\,kH2EAB + a\,h\,kcHEAB\,kH2E\,kH2EA\,kH2EAB +$$
$$h\,kcHEA\,kcHEAB\,kH2E\,kH2EA\,kH2EAB + kcHEA\,kcHEAB\,kH2E\,kH2EA\,kH2EAB\,kHE +$$
$$a\,kcHEAB\,kH2E\,kH2EA\,kH2EAB\,kHEA + a\,b\,kH2E\,kH2EA\,kH2EAB\,kHEAB)\}\}$$

The spaces in the output indicate multiplications. The expression for heab has 9 terms in the denominator because there are 9 kinetic parameters. To obtain the expression for the rapid-equilibrium velocity velordAB in terms of a and b, et is replaced with kfet.

$$velordAB = (a\,b\,kfet\,h\,kH2E\,kH2EA\,kH2EAB)\,/$$
$$(a\,b\,h^2\,kH2E\,kH2EA + a\,h^2\,kcHEAB\,kH2E\,kH2EAB + h^2\,kcHEA\,kcHEAB\,kH2EA\,kH2EAB + a\,b\,h\,kH2E\,kH2EA\,kH2EAB +$$
$$a\,h\,kcHEAB\,kH2E\,kH2EA\,kH2EAB + h\,kcHEA\,kcHEAB\,kH2E\,kH2EA\,kH2EAB + kcHEA\,kcHEAB\,kH2E\,$$
$$kH2EA\,kH2EAB\,kHE + a\,kcHEAB\,kH2E\,kH2EA\,kH2EAB\,kHEA + a\,b\,kH2E\,kH2EA\,kH2EAB\,kHEAB);$$

This rate equation can be simplified by using Simplify.

```
simpvelordAB = Simplify[velordAB]
```

$$(a\,b\,h\,kfet\,kH2E\,kH2EA\,kH2EAB)\,/\,(kcHEA\,kcHEAB\,kH2EA\,kH2EAB\,(h^2 + h\,kH2E + kH2E\,kHE) +$$
$$a\,kH2E\,(kcHEAB\,kH2EAB\,(h^2 + h\,kH2EA + kH2EA\,kHEA) + b\,kH2EA\,(h^2 + h\,kH2EAB + kH2EAB\,kHEAB)))$$

This rate equation is equivalent to the rapid-equilibrium rate equations in the literature [32,33], as shown below. This rate equation is used to calculate nine velocities that will be treated here like experimental data to estimate the nine kinetic parameters using Solve.

3.3.3 Estimation of nine kinetic parameters from nine velocities using simpvelordAB

Rate equation simpvelordAB is used to calculate nine velocities that will be treated here like experimental data to estimate the nine kinetic parameters using Solve.

The equilibrium constants for the acid dissociations can be expressed as p*K*s or as dissociation constants k. The following values for the kinetic parameters are arbitrarily chosen for test calculations:

```
Grid[{{pK1e, pK2e, pK1ea, pK2ea, pK1eab, pK2eab, kfet, kcHEA, kcHEAB},
  {7, 6, 7.5, 6.5, 8, 6, 2, 5, 20}}]
```

pK1e	pK2e	pK1ea	pK2ea	pK1eab	pK2eab	kfet	kcHEA	kcHEAB
7	6	7.5	6.5	8	6	2	5	20

```
Grid[{{kHE, kH2E, kHEA, kH2EA, kHEAB, kH2EAB, kfet, kcHEA, kcHEAB},
  {10^-7, 10^-6, 10^-7.5, 10^-6.5, 10^-8, 10^-6, 2, 5, 20}}] // N
```

kHE	kH2E	kHEA	kH2EA	kHEAB	kH2EAB	kfet	kcHEA	kcHEAB
$1.\times10^{-7}$	$1.\times10^{-6}$	3.16228×10^{-8}	3.16228×10^{-7}	$1.\times10^{-8}$	$1.\times10^{-6}$	2.	5.	20.

These are different valus of kinetic parameters than those in the previous Grid.

The arbitrary parameters can be inserted into simpvtestordAB2 to obtain the expression for the velocity as a function of [A], [B], and [H$^+$]. Notice that pK1 > pK2. The derivations of rapid-equilibrium rate equations and estimation of kinetic parameters are carried out with hydrogen ion concentrations h and dissociation constants k because Solve was developed for polynomial equations. But tables of results of calculations are given in terms of pH and pKs because they are easier to interpret.

These arbitrary kinetic parameters are substituted in the rate equation to obtain the equation vtestordAB2 that can be used to calculate velocities for chosen values of {h,a,b}.

```
vtestordAB2 =
  simpvelordAB /. kHE → 10^-7 /. kH2E → 10^-6 /. kHEA -> 10^-7.5 /. kH2EA → 10^-6.5 /.
    kHEAB → 10^-8 /. kH2EAB → 10^-6 /. kfet → 2 /. kcHEA → 5 /. kcHEAB → 20
```

$$\left(6.32456 \times 10^{-19}\, a\, b\, h\right) \Big/ \left(3.16228 \times 10^{-11} \left(\frac{1}{10\,000\,000\,000\,000} + \frac{h}{1\,000\,000} + h^2 \right) + \right.$$

$$\left. \frac{a\left(\frac{1. \times 10^{-14} + 3.16228 \times 10^{-7}\, h + h^2}{50\,000} + 3.16228 \times 10^{-7}\, b\left(\frac{1}{100\,000\,000\,000\,000} + \frac{h}{1\,000\,000} + h^2 \right)\right)}{1\,000\,000} \right)$$

Since there are nine kinetic parameters, nine velocities have to be measured. The following velocities can be measured at pHs 6, 7, and 8: {a,b} = {200,200}, {1,200}, {200,1}.

```
vtestordAB2 /. h → 10^-8 /. a → 200 /. b → 200 // N
```

0.814029

```
vtestordAB2 /. h → 10^-8 /. a → 1 /. b → 200 // N
```

0.252067

```
vtestordAB2 /. h → 10^-8 /. a → 200 /. b → 1 // N
```

0.0218835

```
vtestordAB2 /. h → 10^-7 /. a → 200 /. b → 200 // N
```

1.46146

```
vtestordAB2 /. h → 10^-7 /. a → 1 /. b → 200 // N
```

0.828759

```
vtestordAB2 /. h → 10^-7 /. a → 200 /. b → 1 // N
```

0.0573081

```
vtestordAB2 /. h → 10^-6 /. a → 200 /. b → 200 // N
```

0.821477

```
vtestordAB2 /. h → 10^-6 /. a → 1 /. b → 200 // N
```

0.574813

```
vtestordAB2 /. h → 10^-6 /. a → 200 /. b → 1 // N
```

0.0230049

These nine velocities are used to calculate the nine kinetic parameters. Only four digits are retained because experimental errors of the order of 5% will be considered.

The following program has been written to calculate nine kinetic parameters from nine measured velocities:

```
calckinparsordABsimp[v1_, h1_, a1_, b1_, v2_, h2_, a2_,
  b2_, v3_, h3_, a3_, b3_, v4_, h4_, a4_, b4_, v5_, h5_, a5_, b5_, v6_, h6_,
  a6_, b6_, v7_, h7_, a7_, b7_, v8_, h8_, a8_, b8_, v9_, h9_, a9_, b9_] :=
 Module[{}, (*This program calculates 9 kinetic parameters from 9 experimental
       velocities for A + B → products at 9 triplets of [H⁺], [A],
       and [B] on the assumption that the mechanism is ordered, with A bound
       first.  Velocities are measured at {high [A],high [B]}, {low [A], high[B]},
     {high [A], low [B]} at low [H⁺], intermediate [H⁺], and at high [H⁺].*)
  Solve[{ (a1 b1 h1 kfet kH2E kH2EA kH2EAB) /
     (kcHEA kcHEAB kH2EA kH2EAB (h1² + h1 kH2E + kH2E kHE) + a1 kH2E
        (kcHEAB kH2EAB (h1² + h1 kH2EA + kH2EA kHEA) + b1 kH2EA (h1² + h1 kH2EAB + kH2EAB kHEAB))) ==
    v1, (a2 b2 h2 kfet kH2E kH2EA kH2EAB) / (kcHEA kcHEAB kH2EA kH2EAB (h2² + h2 kH2E + kH2E kHE) +
       a2 kH2E (kcHEAB kH2EAB (h2² + h2 kH2EA + kH2EA kHEA) +
         b2 kH2EA (h2² + h2 kH2EAB + kH2EAB kHEAB))) == v2,
    (a3 b3 h3 kfet kH2E kH2EA kH2EAB) / (kcHEA kcHEAB kH2EA kH2EAB (h3² + h3 kH2E + kH2E kHE) + a3 kH2E
        (kcHEAB kH2EAB (h3² + h3 kH2EA + kH2EA kHEA) + b3 kH2EA (h3² + h3 kH2EAB + kH2EAB kHEAB))) ==
    v3, (a4 b4 h4 kfet kH2E kH2EA kH2EAB) / (kcHEA kcHEAB kH2EA kH2EAB (h4² + h4 kH2E + kH2E kHE) +
       a4 kH2E (kcHEAB kH2EAB (h4² + h4 kH2EA + kH2EA kHEA) +
         b4 kH2EA (h4² + h4 kH2EAB + kH2EAB kHEAB))) == v4,
    (a5 b5 h5 kfet kH2E kH2EA kH2EAB) / (kcHEA kcHEAB kH2EA kH2EAB (h5² + h5 kH2E + kH2E kHE) + a5 kH2E
        (kcHEAB kH2EAB (h5² + h5 kH2EA + kH2EA kHEA) + b5 kH2EA (h5² + h5 kH2EAB + kH2EAB kHEAB))) ==
    v5, (a6 b6 h6 kfet kH2E kH2EA kH2EAB) / (kcHEA kcHEAB kH2EA kH2EAB (h6² + h6 kH2E + kH2E kHE) +
       a6 kH2E (kcHEAB kH2EAB (h6² + h6 kH2EA + kH2EA kHEA) +
         b6 kH2EA (h6² + h6 kH2EAB + kH2EAB kHEAB))) == v6,
    (a7 b7 h7 kfet kH2E kH2EA kH2EAB) / (kcHEA kcHEAB kH2EA kH2EAB (h7² + h7 kH2E + kH2E kHE) + a7 kH2E
        (kcHEAB kH2EAB (h7² + h7 kH2EA + kH2EA kHEA) + b7 kH2EA (h7² + h7 kH2EAB + kH2EAB kHEAB))) ==
    v7, (a8 b8 h8 kfet kH2E kH2EA kH2EAB) / (kcHEA kcHEAB kH2EA kH2EAB (h8² + h8 kH2E + kH2E kHE) +
       a8 kH2E (kcHEAB kH2EAB (h8² + h8 kH2EA + kH2EA kHEA) +
         b8 kH2EA (h8² + h8 kH2EAB + kH2EAB kHEAB))) == v8,
    (a9 b9 h9 kfet kH2E kH2EA kH2EAB) / (kcHEA kcHEAB kH2EA kH2EAB (h9² + h9 kH2E + kH2E kHE) + a9 kH2E
        (kcHEAB kH2EAB (h9² + h9 kH2EA + kH2EA kHEA) + b9 kH2EA (h9² + h9 kH2EAB + kH2EAB kHEAB))) ==
    v9}, {kHE, kH2E, kHEA, kH2EA, kHEAB, kH2EAB, kfet, kcHEA, kcHEAB}]]
```

This calculation was made using velocities at pHs 6, 7, and 8.

```
calckinparsordABsimp[.8140, 10^-8, 200, 200, .2521, 10^-8, 1, 200, .02188,
  10^-8, 200, 1, 1.4615, 10^-7, 200, 200, .8288, 10^-7, 1, 200, .05731, 10^-7,
  200, 1, .8215, 10^-6, 200, 200, .5748, 10^-6, 1, 200, .02300, 10^-6, 200, 1]
```

$\{\{kfet \to 2.00004, kHEAB \to 1.00006 \times 10^{-8}, kHEA \to 3.16422 \times 10^{-8}, kHE \to 9.99731 \times 10^{-8}, kcHEA \to 5.0016,$
$kcHEAB \to 19.9949, kH2E \to 9.99756 \times 10^{-7}, kH2EA \to 3.1602 \times 10^{-7}, kH2EAB \to 1.00012 \times 10^{-6}\}\}$

```
line1 = {kfet, kHEAB, kHEA, kHE, kcHEA, kcHEAB, kH2E, kH2EA, kH2EAB} /.
  calckinparsordABsimp[.8140, 10^-8, 200, 200, .2521, 10^-8, 1, 200, .02188,
    10^-8, 200, 1, 1.4615, 10^-7, 200, 200, .8288, 10^-7, 1, 200, .05731, 10^-7,
    200, 1, .8215, 10^-6, 200, 200, .5748, 10^-6, 1, 200, .02300, 10^-6, 200, 1]
```

$$\left\{\left\{2.00004, 1.00006 \times 10^{-8}, 3.16422 \times 10^{-8}, 9.99731 \times 10^{-8},\right.\right.$$
$$\left.\left.5.0016, 19.9949, 9.99756 \times 10^{-7}, 3.1602 \times 10^{-7}, 1.00012 \times 10^{-6}\right\}\right\}$$

These values are correct, but it is necessary to test their sensitivity to experimental errors in the measured velocities.

```
calckinparsordABsimp[1.05 * .8140, 10^-8, 200, 200, .2521, 10^-8, 1, 200, .02188,
  10^-8, 200, 1, 1.4615, 10^-7, 200, 200, .8288, 10^-7, 1, 200, .05731, 10^-7,
  200, 1, .8215, 10^-6, 200, 200, .5748, 10^-6, 1, 200, .02300, 10^-6, 200, 1]
```

$$\left\{\left\{\text{kfet} \to 1.97143, \text{kHEAB} \to 8.55662 \times 10^{-9},\right.\right.$$
$$\text{kHEA} \to 3.16422 \times 10^{-8}, \text{kHE} \to 1.05681 \times 10^{-7}, \text{kcHEA} \to 4.85639, \text{kcHEAB} \to 19.7088,$$
$$\left.\left.\text{kH2E} \to 9.45774 \times 10^{-7}, \text{kH2EA} \to 3.1602 \times 10^{-7}, \text{kH2EAB} \to 1.02821 \times 10^{-6}\right\}\right\}$$

```
line2 = {kfet, kHEAB, kHEA, kHE, kcHEA, kcHEAB, kH2E, kH2EA, kH2EAB} /.
  calckinparsordABsimp[1.05 * .8140, 10^-8, 200, 200, .2521, 10^-8, 1, 200, .02188,
    10^-8, 200, 1, 1.4615, 10^-7, 200, 200, .8288, 10^-7, 1, 200, .05731, 10^-7,
    200, 1, .8215, 10^-6, 200, 200, .5748, 10^-6, 1, 200, .02300, 10^-6, 200, 1]
```

$$\left\{\left\{1.97143, 8.55662 \times 10^{-9}, 3.16422 \times 10^{-8}, 1.05681 \times 10^{-7},\right.\right.$$
$$\left.\left.4.85639, 19.7088, 9.45774 \times 10^{-7}, 3.1602 \times 10^{-7}, 1.02821 \times 10^{-6}\right\}\right\}$$

```
calckinparsordABsimp[.8140, 10^-8, 200, 200, 1.05 * .2521, 10^-8, 1, 200, .02188,
  10^-8, 200, 1, 1.4615, 10^-7, 200, 200, .8288, 10^-7, 1, 200, .05731, 10^-7,
  200, 1, .8215, 10^-6, 200, 200, .5748, 10^-6, 1, 200, .02300, 10^-6, 200, 1]
```

$$\left\{\left\{\text{kfet} \to 2.00004, \text{kHEAB} \to 1.00006 \times 10^{-8},\right.\right.$$
$$\text{kHEA} \to 3.19301 \times 10^{-8}, \text{kHE} \to 8.36127 \times 10^{-8}, \text{kcHEA} \to 5.48332, \text{kcHEAB} \to 19.948,$$
$$\left.\left.\text{kH2E} \to 1.19532 \times 10^{-6}, \text{kH2EA} \to 3.15067 \times 10^{-7}, \text{kH2EAB} \to 1.00012 \times 10^{-6}\right\}\right\}$$

```
line3 = {kfet, kHEAB, kHEA, kHE, kcHEA, kcHEAB, kH2E, kH2EA, kH2EAB} /.
  calckinparsordABsimp[.8140, 10^-8, 200, 200, 1.05 * .2521, 10^-8, 1, 200, .02188,
    10^-8, 200, 1, 1.4615, 10^-7, 200, 200, .8288, 10^-7, 1, 200, .05731, 10^-7,
    200, 1, .8215, 10^-6, 200, 200, .5748, 10^-6, 1, 200, .02300, 10^-6, 200, 1]
```

$$\left\{\left\{2.00004, 1.00006 \times 10^{-8}, 3.19301 \times 10^{-8}, 8.36127 \times 10^{-8},\right.\right.$$
$$\left.\left.5.48332, 19.948, 1.19532 \times 10^{-6}, 3.15067 \times 10^{-7}, 1.00012 \times 10^{-6}\right\}\right\}$$

```
calckinparsordABsimp[.8140, 10^-8, 200, 200, .2521, 10^-8, 1, 200, 1.05 * .02188,
  10^-8, 200, 1, 1.4615, 10^-7, 200, 200, .8288, 10^-7, 1, 200, .05731, 10^-7,
  200, 1, .8215, 10^-6, 200, 200, .5748, 10^-6, 1, 200, .02300, 10^-6, 200, 1]
```

$$\left\{\left\{\text{kfet} \to 2.00546, \text{kHEAB} \to 1.02738 \times 10^{-8},\right.\right.$$
$$\text{kHEA} \to 2.84189 \times 10^{-8}, \text{kHE} \to 9.99731 \times 10^{-8}, \text{kcHEA} \to 4.87005, \text{kcHEAB} \to 20.5906,$$
$$\left.\left.\text{kH2E} \to 9.99756 \times 10^{-7}, \text{kH2EA} \to 3.27095 \times 10^{-7}, \text{kH2EAB} \to 9.94974 \times 10^{-7}\right\}\right\}$$

```
line4 = {kfet, kHEAB, kHEA, kHE, kcHEA, kcHEAB, kH2E, kH2EA, kH2EAB} /.
  calckinparsordABsimp[.8140, 10^-8, 200, 200, .2521, 10^-8, 1, 200, 1.05 * .02188,
    10^-8, 200, 1, 1.4615, 10^-7, 200, 200, .8288, 10^-7, 1, 200, .05731, 10^-7,
    200, 1, .8215, 10^-6, 200, 200, .5748, 10^-6, 1, 200, .02300, 10^-6, 200, 1]
```

$$\left\{\left\{2.00546, 1.02738 \times 10^{-8}, 2.84189 \times 10^{-8}, 9.99731 \times 10^{-8},\right.\right.$$
$$\left.\left.4.87005, 20.5906, 9.99756 \times 10^{-7}, 3.27095 \times 10^{-7}, 9.94974 \times 10^{-7}\right\}\right\}$$

```
calckinparsordABsimp[.8140, 10^-8, 200, 200, .2521, 10^-8, 1, 200, .02188,
  10^-8, 200, 1, 1.05 * 1.4615, 10^-7, 200, 200, .8288, 10^-7, 1, 200, .05731, 10^-7,
  200, 1, .8215, 10^-6, 200, 200, .5748, 10^-6, 1, 200, .02300, 10^-6, 200, 1]
```

$$\{\{kfet \to 2.1779, \, kHEAB \to 1.17704 \times 10^{-8}, \, kHEA \to 3.16422 \times 10^{-8}, \, kHE \to 8.45478 \times 10^{-8}, \, kcHEA \to 5.81845,$$
$$kcHEAB \to 21.773, \, kH2E \to 1.38731 \times 10^{-6}, \, kH2EA \to 3.1602 \times 10^{-7}, \, kH2EAB \to 8.49729 \times 10^{-7}\}\}$$

```
line5 = {kfet, kHEAB, kHEA, kHE, kcHEA, kcHEAB, kH2E, kH2EA, kH2EAB} /.
  calckinparsordABsimp[.8140, 10^-8, 200, 200, .2521, 10^-8, 1, 200, .02188, 10^-8,
    200, 1, 1.05 * 1.4615, 10^-7, 200, 200, .8288, 10^-7, 1, 200, .05731, 10^-7,
    200, 1, .8215, 10^-6, 200, 200, .5748, 10^-6, 1, 200, .02300, 10^-6, 200, 1]
```

$$\{\{2.1779, \, 1.17704 \times 10^{-8}, \, 3.16422 \times 10^{-8}, \, 8.45478 \times 10^{-8},$$
$$5.81845, \, 21.773, \, 1.38731 \times 10^{-6}, \, 3.1602 \times 10^{-7}, \, 8.49729 \times 10^{-7}\}\}$$

```
calckinparsordABsimp[.8140, 10^-8, 200, 200, .2521, 10^-8, 1, 200, .02188,
  10^-8, 200, 1, 1.4615, 10^-7, 200, 200, 1.05 * .8288, 10^-7, 1, 200, .05731, 10^-7,
  200, 1, .8215, 10^-6, 200, 200, .5748, 10^-6, 1, 200, .02300, 10^-6, 200, 1]
```

$$\{\{kfet \to 2.00004, \, kHEAB \to 1.00006 \times 10^{-8},$$
$$kHEA \to 3.13451 \times 10^{-8}, \, kHE \to 1.44416 \times 10^{-7}, \, kcHEA \to 3.53569, \, kcHEAB \to 20.1389,$$
$$kH2E \to 5.53921 \times 10^{-7}, \, kH2EA \to 3.19015 \times 10^{-7}, \, kH2EAB \to 1.00012 \times 10^{-6}\}\}$$

```
line6 = {kfet, kHEAB, kHEA, kHE, kcHEA, kcHEAB, kH2E, kH2EA, kH2EAB} /.
  calckinparsordABsimp[.8140, 10^-8, 200, 200, .2521, 10^-8, 1, 200, .02188, 10^-8,
    200, 1, 1.4615, 10^-7, 200, 200, 1.05 * .8288, 10^-7, 1, 200, .05731, 10^-7,
    200, 1, .8215, 10^-6, 200, 200, .5748, 10^-6, 1, 200, .02300, 10^-6, 200, 1]
```

$$\{\{2.00004, \, 1.00006 \times 10^{-8}, \, 3.13451 \times 10^{-8}, \, 1.44416 \times 10^{-7},$$
$$3.53569, \, 20.1389, \, 5.53921 \times 10^{-7}, \, 3.19015 \times 10^{-7}, \, 1.00012 \times 10^{-6}\}\}$$

```
calckinparsordABsimp[.8140, 10^-8, 200, 200, .2521, 10^-8, 1, 200, .02188,
  10^-8, 200, 1, 1.4615, 10^-7, 200, 200, .8288, 10^-7, 1, 200, 1.05 * .05731, 10^-7,
  200, 1, .8215, 10^-6, 200, 200, .5748, 10^-6, 1, 200, .02300, 10^-6, 200, 1]
```

$$\{\{kfet \to 1.97943, \, kHEAB \to 9.79551 \times 10^{-9},$$
$$kHEA \to 3.64722 \times 10^{-8}, \, kHE \to 9.99731 \times 10^{-8}, \, kcHEA \to 5.58311, \, kcHEAB \to 17.7277,$$
$$kH2E \to 9.99756 \times 10^{-7}, \, kH2EA \to 2.7417 \times 10^{-7}, \, kH2EAB \to 1.02106 \times 10^{-6}\}\}$$

```
line7 = {kfet, kHEAB, kHEA, kHE, kcHEA, kcHEAB, kH2E, kH2EA, kH2EAB} /.
  calckinparsordABsimp[.8140, 10^-8, 200, 200, .2521, 10^-8, 1, 200, .02188, 10^-8,
    200, 1, 1.4615, 10^-7, 200, 200, .8288, 10^-7, 1, 200, 1.05 * .05731, 10^-7,
    200, 1, .8215, 10^-6, 200, 200, .5748, 10^-6, 1, 200, .02300, 10^-6, 200, 1]
```

$$\{\{1.97943, \, 9.79551 \times 10^{-9}, \, 3.64722 \times 10^{-8}, \, 9.99731 \times 10^{-8},$$
$$5.58311, \, 17.7277, \, 9.99756 \times 10^{-7}, \, 2.7417 \times 10^{-7}, \, 1.02106 \times 10^{-6}\}\}$$

```
calckinparsordABsimp[.8140, 10^-8, 200, 200, .2521, 10^-8, 1, 200, .02188,
  10^-8, 200, 1, 1.4615, 10^-7, 200, 200, .8288, 10^-7, 1, 200, .05731, 10^-7, 200,
  1, 1.05 * .8215, 10^-6, 200, 200, .5748, 10^-6, 1, 200, .02300, 10^-6, 200, 1]
```

$$\{\{kfet \to 1.97168, \, kHEAB \to 9.72987 \times 10^{-9},$$
$$kHEA \to 3.16422 \times 10^{-8}, \, kHE \to 1.03204 \times 10^{-7}, \, kcHEA \to 4.85771, \, kcHEAB \to 19.7113,$$
$$kH2E \to 7.69738 \times 10^{-7}, \, kH2EA \to 3.1602 \times 10^{-7}, \, kH2EAB \to 1.16715 \times 10^{-6}\}\}$$

```
line8 = {kfet, kHEAB, kHEA, kHE, kcHEA, kcHEAB, kH2E, kH2EA, kH2EAB} /.
   calckinparsordABsimp[.8140, 10^-8, 200, 200, .2521, 10^-8, 1, 200, .02188,
     10^-8, 200, 1, 1.4615, 10^-7, 200, 200, .8288, 10^-7, 1, 200, .05731, 10^-7, 200,
     1, 1.05 * .8215, 10^-6, 200, 200, .5748, 10^-6, 1, 200, .02300, 10^-6, 200, 1]
```

$$\{\{1.97168, \ 9.72987 \times 10^{-9}, \ 3.16422 \times 10^{-8}, \ 1.03204 \times 10^{-7},$$
$$4.85771, \ 19.7113, \ 7.69738 \times 10^{-7}, \ 3.1602 \times 10^{-7}, \ 1.16715 \times 10^{-6}\}\}$$

```
calckinparsordABsimp[.8140, 10^-8, 200, 200, .2521, 10^-8, 1, 200, .02188,
   10^-8, 200, 1, 1.4615, 10^-7, 200, 200, .8288, 10^-7, 1, 200, .05731, 10^-7, 200,
   1, .8215, 10^-6, 200, 200, 1.05 * .5748, 10^-6, 1, 200, .02300, 10^-6, 200, 1]
```

$$\{\{kfet \to 2.00004, \ kHEAB \to 1.00006 \times 10^{-8}, \ kHEA \to 3.16841 \times 10^{-8}, \ kHE \to 9.5666 \times 10^{-8}, \ kcHEA \to 5.2126,$$
$$kcHEAB \to 19.9743, \ kH2E \to 1.66187 \times 10^{-6}, \ kH2EA \to 3.14766 \times 10^{-7}, \ kH2EAB \to 1.00012 \times 10^{-6}\}\}$$

```
line9 = {kfet, kHEAB, kHEA, kHE, kcHEA, kcHEAB, kH2E, kH2EA, kH2EAB} /.
   calckinparsordABsimp[.8140, 10^-8, 200, 200, .2521, 10^-8, 1, 200, .02188,
     10^-8, 200, 1, 1.4615, 10^-7, 200, 200, .8288, 10^-7, 1, 200, .05731, 10^-7, 200,
     1, .8215, 10^-6, 200, 200, 1.05 * .5748, 10^-6, 1, 200, .02300, 10^-6, 200, 1]
```

$$\{\{2.00004, \ 1.00006 \times 10^{-8}, \ 3.16841 \times 10^{-8}, \ 9.5666 \times 10^{-8},$$
$$5.2126, \ 19.9743, \ 1.66187 \times 10^{-6}, \ 3.14766 \times 10^{-7}, \ 1.00012 \times 10^{-6}\}\}$$

```
calckinparsordABsimp[.8140, 10^-8, 200, 200, .2521, 10^-8, 1, 200, .02188,
   10^-8, 200, 1, 1.4615, 10^-7, 200, 200, .8288, 10^-7, 1, 200, .05731, 10^-7, 200,
   1, .8215, 10^-6, 200, 200, .5748, 10^-6, 1, 200, 1.05 * .02300, 10^-6, 200, 1]
```

$$\{\{kfet \to 2.0052, \ kHEAB \to 1.00498 \times 10^{-8}, \ kHEA \to 3.06217 \times 10^{-8}, \ kHE \to 9.99731 \times 10^{-8}, \ kcHEA \to 4.87629,$$
$$kcHEAB \to 20.5615, \ kH2E \to 9.99756 \times 10^{-7}, \ kH2EA \to 3.49977 \times 10^{-7}, \ kH2EAB \to 9.74781 \times 10^{-7}\}\}$$

```
line10 = {kfet, kHEAB, kHEA, kHE, kcHEA, kcHEAB, kH2E, kH2EA, kH2EAB} /.
   calckinparsordABsimp[.8140, 10^-8, 200, 200, .2521, 10^-8, 1, 200, .02188,
     10^-8, 200, 1, 1.4615, 10^-7, 200, 200, .8288, 10^-7, 1, 200, .05731, 10^-7, 200,
     1, .8215, 10^-6, 200, 200, .5748, 10^-6, 1, 200, 1.05 * .02300, 10^-6, 200, 1]
```

$$\{\{2.0052, \ 1.00498 \times 10^{-8}, \ 3.06217 \times 10^{-8}, \ 9.99731 \times 10^{-8},$$
$$4.87629, \ 20.5615, \ 9.99756 \times 10^{-7}, \ 3.49977 \times 10^{-7}, \ 9.74781 \times 10^{-7}\}\}$$

Table 3.4 Estimation of nine kinetic parameters from nine measured velocities using {a,b} = {200,200}, {1,200}, {200,1} at pHs 6, 7, and 8

```
TableForm[
  Round[-Log[10, {line1[[1]], line2[[1]], line3[[1]], line4[[1]], line5[[1]], line6[[1]],
     line7[[1]], line8[[1]], line9[[1]], line10[[1]]}], .01],
  TableHeadings → {{"No errors", "1.05*v1", "1.05*v2", "1.05*v3", "1.05*v4", "1.05*v5",
     "1.05*v6", "1.05*v7", "1.05*v8", "1.05*v9", "1.05*v10"}, {"-log(kfet)", "pK1eab",
     "pK1ea", "pK1e", "-log(kcHEA)", "-log(kcHEAB)", "pK2e", "pK2ea", "pK2eab"}}]
```

	$-\log(kfet)$	pK1eab	pK1ea	pK1e	$-\log(kcHEA)$	$-\log(kcHEAB)$	pK2e	pK2ea	pK2eab
No errors	-0.3	8.	7.5	7.	-0.7	-1.3	6.	6.5	6.
1.05*v1	-0.29	8.07	7.5	6.98	-0.69	-1.29	6.02	6.5	5.99
1.05*v2	-0.3	8.	7.5	7.08	-0.74	-1.3	5.92	6.5	6.
1.05*v3	-0.3	7.99	7.55	7.	-0.69	-1.31	6.	6.49	6.
1.05*v4	-0.34	7.93	7.5	7.07	-0.76	-1.34	5.86	6.5	6.07
1.05*v5	-0.3	8.	7.5	6.84	-0.55	-1.3	6.26	6.5	6.
1.05*v6	-0.3	8.01	7.44	7.	-0.75	-1.25	6.	6.56	5.99
1.05*v7	-0.29	8.01	7.5	6.99	-0.69	-1.29	6.11	6.5	5.93
1.05*v8	-0.3	8.	7.5	7.02	-0.72	-1.3	5.78	6.5	6.
1.05*v9	-0.3	8.	7.51	7.	-0.69	-1.31	6.	6.46	6.01

The negative logarithms have been used as entries in this table because it is easier to interpret pKs. Notice that -Log[10, kfet] is

> **-Log[10, 2] // N**
>
> - 0.30103

-Log[10,kcHEA] is

> **-Log[10, 5] // N**
>
> - 0.69897

-Log[10,kcHEAB] is

> **-Log[10, 20] // N**
>
> - 1.30103

■ **3.3.4 Derivation of the dependencies of V_{fexp}, K_{IA}, and K_B on the hydrogen ion concentration**

> **simpvelordAB**
>
> $(a\,b\,h\,kfet\,kH2E\,kH2EA\,kH2EAB) \big/ \big(kcHEA\,kcHEAB\,kH2EA\,kH2EAB\,(h^2 + h\,kH2E + kH2E\,kHE) + a\,kH2E\,(kcHEAB\,kH2EAB\,(h^2 + h\,kH2EA + kH2EA\,kHEA) + b\,kH2EA\,(h^2 + h\,kH2EAB + kH2EAB\,kHEAB))\big)$

This equation can be compared with vordAB in Section 3.1.

> **vordAB**
>
> $$\frac{a\,b\,kfet}{a\,b + a\,kB + kB\,kIA}$$

To obtain the pH dependencies of the limiting velocity and the two Michaelis constants, it is more convenient to write this expression for the velocity as kfet/(1 + kB/b + kBkIA/a b). To make it clearer that there are 3 terms in the denominator of simpvelordAB, it is written as

> **$(a\,b\,h\,kfet\,kH2E\,kH2EA\,kH2EAB) \big/ \big(kcHEA\,kcHEAB\,kH2EA\,kH2EAB\,(h^2 + h\,kH2E + kH2E\,kHE) +$ $a\,kH2E\,kcHEAB\,kH2EAB\,(h^2 + h\,kH2EA + kH2EA\,kHEA) + a\,b\,kH2E\,kH2EA\,(h^2 + h\,kH2EAB + kH2EAB\,kHEAB)\big)$**

> $(a\,b\,h\,kfet\,kH2E\,kH2EA\,kH2EAB) \big/ \big(kcHEA\,kcHEAB\,kH2EA\,kH2EAB\,(h^2 + h\,kH2E + kH2E\,kHE) +$ $a\,kcHEAB\,kH2E\,kH2EAB\,(h^2 + h\,kH2EA + kH2EA\,kHEA) + a\,b\,kH2E\,kH2EA\,(h^2 + h\,kH2EAB + kH2EAB\,kHEAB)\big)$

The expression for V_{fexp} is derived first. The ab term in the denominator is

> **$\big(a\,b\,kH2E\,kH2EA\,(h^2 + h\,kH2EAB + kH2EAB\,kHEAB)\big)$**

> $a\,b\,kH2E\,kH2EA\,(h^2 + h\,kH2EAB + kH2EAB\,kHEAB)$

Divide the numerator by this term

> **$(a\,b\,h\,kfet\,kH2E\,kH2EA\,kH2EAB) \big/ \big(a\,b\,kH2E\,kH2EA\,(h^2 + h\,kH2EAB + kH2EAB\,kHEAB)\big)$**

> $$\frac{h\,kfet\,kH2EAB}{h^2 + h\,kH2EAB + kH2EAB\,kHEAB}$$

Divide the denominator of this rate equation by the coefficient of kfet in the numerator, one term at a time.

$$\{h^2, \ h \ kH2EAB, \ kH2EAB \ kHEAB\} \ / \ (h \ kH2EAB)$$

$$\left\{ \frac{h}{kH2EAB}, \ 1, \ \frac{kHEAB}{h} \right\}$$

This shows that V_{fexp} is given as a function of the hydrogen ion concentration by

$$\texttt{kfet} \ / \ (1 + \texttt{kHEAB} \ / \ h + h \ / \ \texttt{kH2EAB})$$

$$\frac{kfet}{1 + \dfrac{h}{kH2EAB} + \dfrac{kHEAB}{h}}$$

This can be expressed in terms of pH as follows:

$$\texttt{kfet} \ / \ (1 + \texttt{kHEAB} \ / \ h + h \ / \ \texttt{kH2EAB}) \ /. \ h \rightarrow 10\,\char`\^\,-pH \ /. \ \texttt{kHEAB} \rightarrow 10\,\char`\^\,-pK1eab \ /. \ \texttt{kH2EAB} \rightarrow 10\,\char`\^\,-pK2eab$$

$$\frac{kfet}{1 + 10^{pH-pK1eab} + 10^{-pH+pK2eab}}$$

This is the same as the expression in the first output of derordABzero in Section 3.3.1.

This process can be continued to obtain the expressions for K_{B} and K_{IA} from simpvelordAB. Since the numerator of simpvelordAB was divided by $\left(a \ b \ kH2E \ kH2EA \ \left(h^2 + h \ kH2EAB + kH2EAB \ kHEAB\right)\right)$, the terms in the denominator have to be divided by this factor, one term at a time.

$$\{ \left(\texttt{kcHEA} \ \texttt{kcHEAB} \ \texttt{kH2EA} \ \texttt{kH2EAB} \ \left(h^2 + h \ \texttt{kH2E} + \texttt{kH2E} \ \texttt{kHE}\right)\right),$$
$$\left(a \ \texttt{kH2E} \ \texttt{kcHEAB} \ \texttt{kH2EAB} \ \left(h^2 + h \ \texttt{kH2EA} + \texttt{kH2EA} \ \texttt{kHEA}\right)\right), \ \left(a \ b \ \texttt{kH2E} \ \texttt{kH2EA} \ \left(h^2 + h \ \texttt{kH2EAB} + \texttt{kH2EAB} \ \texttt{kHEAB}\right)\right)\} \ /$$
$$\left(a \ b \ \texttt{kH2E} \ \texttt{kH2EA} \ \left(h^2 + h \ \texttt{kH2EAB} + \texttt{kH2EAB} \ \texttt{kHEAB}\right)\right)$$

$$\left\{ \frac{\texttt{kcHEA} \ \texttt{kcHEAB} \ \texttt{kH2EAB} \ \left(h^2 + h \ \texttt{kH2E} + \texttt{kH2E} \ \texttt{kHE}\right)}{a \ b \ \texttt{kH2E} \ \left(h^2 + h \ \texttt{kH2EAB} + \texttt{kH2EAB} \ \texttt{kHEAB}\right)}, \ \frac{\texttt{kcHEAB} \ \texttt{kH2EAB} \ \left(h^2 + h \ \texttt{kH2EA} + \texttt{kH2EA} \ \texttt{kHEA}\right)}{b \ \texttt{kH2EA} \ \left(h^2 + h \ \texttt{kH2EAB} + \texttt{kH2EAB} \ \texttt{kHEAB}\right)}, \ 1 \right\}$$

This shows that K_{B} is given by

$$\frac{\texttt{kcHEAB} \ \texttt{kH2EAB} \ \left(h^2 + h \ \texttt{kH2EA} + \texttt{kH2EA} \ \texttt{kHEA}\right)}{\texttt{kH2EA} \ \left(h^2 + h \ \texttt{kH2EAB} + \texttt{kH2EAB} \ \texttt{kHEAB}\right)}$$

$$\frac{\texttt{kcHEAB} \ \texttt{kH2EAB} \ \left(h^2 + h \ \texttt{kH2EA} + \texttt{kH2EA} \ \texttt{kHEA}\right)}{\texttt{kH2EA} \ \left(h^2 + h \ \texttt{kH2EAB} + \texttt{kH2EAB} \ \texttt{kHEAB}\right)}$$

This equation can be written as

$$\frac{\texttt{kcHEAB} \ \left(h^2 \ / \ \texttt{kH2EA} + h + \texttt{kHEA}\right)}{\left(h^2 \ / \ \texttt{kH2EAB} + h + \texttt{kHEAB}\right)}$$

$$\frac{\texttt{kcHEAB} \ \left(h + \dfrac{h^2}{kH2EA} + kHEA\right)}{h + \dfrac{h^2}{kH2EAB} + kHEAB}$$

Dividing the numerator and the denominator by h yields the pH dependence of K_{B}.

$$\text{kcHEAB (h / kH2EA + 1 + kHEA / h)} \over \text{(h / kH2EAB + 1 + kHEAB / h)}$$

$$\frac{\text{kcHEAB} \left(1 + \frac{h}{\text{kH2EA}} + \frac{\text{kHEA}}{h}\right)}{1 + \frac{h}{\text{kH2EAB}} + \frac{\text{kHEAB}}{h}}$$

This equation can be written in terms of pH and pKs.

$$\frac{\text{kcHEAB (h / kH2EA + 1 + kHEA / h)}}{\text{(h / kH2EAB + 1 + kHEAB / h)}} \;\; /. \; h \to 10 \char`^ -pH \; /. \; \text{kHEA} \to 10 \char`^ -pK1ea \; /. \; \text{kH2EA} \to 10 \char`^ -pK2ea \; /.$$
$$\text{kHEAB} \to 10 \char`^ -pK1eab \; /. \; \text{kH2EAB} \to 10 \char`^ -pK2eab$$

$$\frac{\left(1 + 10^{\text{pH}-\text{pK1ea}} + 10^{-\text{pH}+\text{pK2ea}}\right) \text{kcHEAB}}{1 + 10^{\text{pH}-\text{pK1eab}} + 10^{-\text{pH}+\text{pK2eab}}}$$

This is the same as the expression for K_B in Section 3.3.1.

The third term in the denominator yields kBkIA/a b.

$$\frac{\text{kcHEA kcHEAB kH2EAB} \left(h^2 + h\, \text{kH2E} + \text{kH2E kHE}\right)}{\text{a b kH2E} \left(h^2 + h\, \text{kH2EAB} + \text{kH2EAB kHEAB}\right)} ;$$

kIA is given by kB a b denominator term/kB

$$\frac{\text{kcHEAB (h / kH2EA + 1 + kHEA / h)}}{\text{(h / kH2EAB + 1 + kHEAB / h)}} \Big/ \frac{\text{kcHEA kcHEAB kH2EAB} \left(h^2 + h\, \text{kH2E} + \text{kH2E kHE}\right)}{\text{kH2E} \left(h^2 + h\, \text{kH2EAB} + \text{kH2EAB kHEAB}\right)}$$

$$\frac{\text{kH2E} \left(1 + \frac{h}{\text{kH2EA}} + \frac{\text{kHEA}}{h}\right) \left(h^2 + h\, \text{kH2EAB} + \text{kH2EAB kHEAB}\right)}{\text{kcHEA kH2EAB} \left(h^2 + h\, \text{kH2E} + \text{kH2E kHE}\right) \left(1 + \frac{h}{\text{kH2EAB}} + \frac{\text{kHEAB}}{h}\right)}$$

We expect

$$\frac{\left(1 + 10^{\text{pH}-\text{pK1e}} + 10^{-\text{pH}+\text{pK2e}}\right) \text{kHEA}}{1 + 10^{\text{pH}-\text{pK1ea}} + 10^{-\text{pH}+\text{pK2ea}}} ;$$

Go back to the terms in the denominator and divide the term for kB kIA/ab by the term for kB/b to get kIA/a.

$$\frac{\text{kcHEA kcHEAB kH2EAB} \left(h^2 + h\, \text{kH2E} + \text{kH2E kHE}\right)}{\text{a b kH2E} \left(h^2 + h\, \text{kH2EAB} + \text{kH2EAB kHEAB}\right)} \Big/ \frac{\text{kcHEAB kH2EAB} \left(h^2 + h\, \text{kH2EA} + \text{kH2EA kHEA}\right)}{\text{b kH2EA} \left(h^2 + h\, \text{kH2EAB} + \text{kH2EAB kHEAB}\right)}$$

$$\frac{\text{kcHEA kH2EA} \left(h^2 + h\, \text{kH2E} + \text{kH2E kHE}\right)}{\text{a kH2E} \left(h^2 + h\, \text{kH2EA} + \text{kH2EA kHEA}\right)}$$

Since this is kIA/a, multiply by a and divide numerator with kH2E and denominator kH2EA.

$$\frac{\text{kcHEA} \left(h^2 \big/ \text{kH2E} + h + \text{kHE}\right)}{\left(h^2 \big/ \text{kH2EA} + h + \text{kHEA}\right)}$$

$$\frac{\text{kcHEA} \left(h + \frac{h^2}{\text{kH2E}} + \text{kHE}\right)}{h + \frac{h^2}{\text{kH2EA}} + \text{kHEA}}$$

Divide numerator and denominator by h.

$$\frac{\text{kcHEA (h / kH2E + 1 + kHE / h)}}{\text{(h / kH2EA + 1 + kHEA / h)}}$$

$$\frac{\text{kcHEA} \left(1 + \dfrac{h}{kH2E} + \dfrac{kHE}{h}\right)}{1 + \dfrac{h}{kH2EA} + \dfrac{kHEA}{h}}$$

This can be expressed in terms of p*K*s and pH as follows:

$$\frac{\text{kcHEA (h / kH2E + 1 + kHE / h)}}{\text{(h / kH2EA + 1 + kHEA / h)}} \;\; /. \; h \to 10\text{^-pH} \; /. \; kHEA \to 10\text{^-pK1ea} \; /. \; kH2EA \to 10\text{^-pK2ea} \; /.$$
$$kHE \to 10\text{^-pK1e} \; /. \; kH2E \to 10\text{^-pK2e}$$

$$\frac{\left(1 + 10^{pH-pK1e} + 10^{-pH+pK2e}\right) \text{kcHEA}}{1 + 10^{pH-pK1ea} + 10^{-pH+pK2ea}}$$

This is the same as in Section 3.3.1.

■ 3.3.5 The program calckinparsordABsimp can be used to determine when a p*K* is missing

Assume species EA$^-$ is missing in the mechanism; or, in other words, assume that p*K*1ea and kHEA are missing. The corresponding rate equation can be derived as follows:

```
Solve[{kHE == h * e / he, kH2E == h * he / h2e, kH2EA == h * hea / h2ea, kHEAB == h * eab / heab,
   kH2EAB == h * heab / h2eab, kcHEA == he * a / hea, kcHEAB == hea * b / heab,
   et == e + he + h2e + hea + h2ea + eab + heab + h2eab}, {heab}, {e, he, h2e, hea, h2ea, eab, h2eab}]
```

$$\Big\{\Big\{\text{heab} \to \text{(a b et h kH2E kH2EA kH2EAB)} \Big/$$
$$\Big(\text{a b h}^2 \text{ kH2E kH2EA + a h}^2 \text{ kcHEAB kH2E kH2EAB + h}^2 \text{ kcHEA kcHEAB kH2EA kH2EAB +}$$
$$\text{a b h kH2E kH2EA kH2EAB + a h kcHEAB kH2E kH2EA kH2EAB + h kcHEA kcHEAB kH2E kH2EA kH2EAB +}$$
$$\text{kcHEA kcHEAB kH2E kH2EA kH2EAB kHE + a b kH2E kH2EA kH2EAB kHEAB}\Big)\Big\}\Big\}$$

```
velordABea = (a b kfet h kH2E kH2EA kH2EAB) /
```
$$\Big(\text{a b h}^2 \text{ kH2E kH2EA + a h}^2 \text{ kcHEAB kH2E kH2EAB + h}^2 \text{ kcHEA kcHEAB kH2EA kH2EAB +}$$
$$\text{a b h kH2E kH2EA kH2EAB + a h kcHEAB kH2E kH2EA kH2EAB + h kcHEA kcHEAB kH2E kH2EA kH2EAB +}$$
$$\text{kcHEA kcHEAB kH2E kH2EA kH2EAB kHE + a b kH2E kH2EA kH2EAB kHEAB}\Big)$$

$$\text{(a b h kfet kH2E kH2EA kH2EAB)} \Big/$$
$$\Big(\text{a b h}^2 \text{ kH2E kH2EA + a h}^2 \text{ kcHEAB kH2E kH2EAB + h}^2 \text{ kcHEA kcHEAB kH2EA kH2EAB +}$$
$$\text{a b h kH2E kH2EA kH2EAB + a h kcHEAB kH2E kH2EA kH2EAB + h kcHEA kcHEAB kH2E kH2EA kH2EAB +}$$
$$\text{kcHEA kcHEAB kH2E kH2EA kH2EAB kHE + a b kH2E kH2EA kH2EAB kHEAB}\Big)$$

```
simpvelordABea = Simplify[velordABea]
```

$$\text{(a b h kfet kH2E kH2EA kH2EAB)} \Big/ \Big(\text{kcHEA kcHEAB kH2E kH2EA kH2EAB} \left(h^2 + h\ kH2E + kH2E\ kHE\right) +$$
$$\text{a kH2E} \left(h\ kcHEAB\ (h + kH2EA)\ kH2EAB + b\ kH2EA\ \left(h^2 + h\ kH2EAB + kH2EAB\ kHEAB\right)\right)\Big)$$

The rate equation with the arbitrarily specified kinetic parameters is

```
vtestordAB2ea =
 simpvelordABea /. kHE → 10^-7 /. kH2E → 10^-6 /. kH2EA → 10^-6.5 /. kHEAB → 10^-8 /.
   kH2EAB → 10^-6 /. kfet → 2 /. kcHEA → 5 /. kcHEAB → 20
```

$$\left(6.32456 \times 10^{-19}\, a\, b\, h\right) \Bigg/ \left(3.16228 \times 10^{-11}\left(\frac{1}{10\,000\,000\,000\,000} + \frac{h}{1\,000\,000} + h^2\right) + \right.$$

$$\left. \frac{a\left(\frac{h\left(3.16228\times10^{-7}+h\right)}{50\,000} + 3.16228 \times 10^{-7}\, b\left(\frac{1}{100\,000\,000\,000\,000} + \frac{h}{1\,000\,000} + h^2\right)\right)}{1\,000\,000} \right)$$

This equation can be used to calculate 9 velocities under the same conditions as Table 3.3.

```
vtestordAB2ea /. h → 10^-8 /. a → 200 /. b → 200 // N
```

0.934279

```
vtestordAB2ea /. h → 10^-8 /. a → 1 /. b → 200 // N
```

0.262531

```
vtestordAB2ea /. h → 10^-8 /. a → 200 /. b → 1 // N
```

0.0710544

```
vtestordAB2ea /. h → 10^-7 /. a → 200 /. b → 200 // N
```

1.49603

```
vtestordAB2ea /. h → 10^-7 /. a → 1 /. b → 200 // N
```

0.839764

```
vtestordAB2ea /. h → 10^-7 /. a → 200 /. b → 1 // N
```

0.0699923

```
vtestordAB2ea /. h → 10^-6 /. a → 200 /. b → 200 // N
```

0.822545

```
vtestordAB2ea /. h → 10^-6 /. a → 1 /. b → 200 // N
```

0.575336

```
vtestordAB2ea /. h → 10^-6 /. a → 200 /. b → 1 // N
```

0.0231735

Estimation of kinetic parameters.

```
calckinparsordABsimp[0.93428, 10^-8, 200, 200, 0.26253, 10^-8, 1, 200, 0.071054,
  10^-8, 200, 1, 1.4960, 10^-7, 200, 200, 0.83976, 10^-7, 1, 200, 0.069992, 10^-7,
  200, 1, 0.82255, 10^-6, 200, 200, 0.57534, 10^-6, 1, 200, 0.023174, 10^-6, 200, 1]
```

$\{\{$ kfet → 1.99994, kHEAB → 9.99934×10^{-9},

  kHEA → -1.11325×10^{-13}, kHE → 1.00005×10^{-7}, kcHEA → 4.9997, kcHEAB → 19.9997,

  kH2E → 9.99934×10^{-7}, kH2EA → 3.16245×10^{-7}, kH2EAB → $1.00007 \times 10^{-6}\}\}$

Notice that kHEA is negative, which is impossible physically. The other 8 parameters are all correct, as shown by the Grid table used before. This is the way the computer program shows that the mechanism does not include EHA = EA$^-$ + H$^+$.

```
Grid[{{kHE, kH2E, kHEA, kH2EA, kHEAB, kH2EAB, kfet, kcHEA, kcHEAB},
   {10^-7, 10^-6, 10^-7.5, 10^-6.5, 10^-8, 10^-6, 2, 5, 20}}] // N
```

kHE	kH2E	kHEA	kH2EA	kHEAB	kH2EAB	kfet	kcHEA	kcHEAB
$1. \times 10^{-7}$	$1. \times 10^{-6}$	3.16228×10^{-8}	3.16228×10^{-7}	$1. \times 10^{-8}$	$1. \times 10^{-6}$	2.	5.	20.

Calculation of the effects of 5% errors.

```
line21 = {kfet, kHEAB, kHEA, kHE, kcHEA, kcHEAB, kH2E, kH2EA, kH2EAB} /.
   calckinparsordABsimp[0.93428, 10^-8, 200, 200, 0.26253, 10^-8, 1, 200, 0.071054,
      10^-8, 200, 1, 1.4960, 10^-7, 200, 200, 0.83976, 10^-7, 1, 200, 0.069992, 10^-7,
      200, 1, 0.82255, 10^-6, 200, 200, 0.57534, 10^-6, 1, 200, 0.023174, 10^-6, 200, 1]
```

$\{\{1.99994, 9.99934 \times 10^{-9}, -1.11325 \times 10^{-13}, 1.00005 \times 10^{-7},$
   $4.9997, 19.9997, 9.99934 \times 10^{-7}, 3.16245 \times 10^{-7}, 1.00007 \times 10^{-6}\}\}$

```
calckinparsordABsimp[1.05 * 0.93428, 10^-8, 200, 200, 0.26253, 10^-8, 1, 200, 0.071054,
   10^-8, 200, 1, 1.4960, 10^-7, 200, 200, 0.83976, 10^-7, 1, 200, 0.069992, 10^-7,
   200, 1, 0.82255, 10^-6, 200, 200, 0.57534, 10^-6, 1, 200, 0.023174, 10^-6, 200, 1]
```

$\{\{kfet \rightarrow 1.97496, kHEAB \rightarrow 8.73902 \times 10^{-9},$
   $kHEA \rightarrow -1.11325 \times 10^{-13}, kHE \rightarrow 1.0496 \times 10^{-7}, kcHEA \rightarrow 4.87322, kcHEAB \rightarrow 19.7499,$
   $kH2E \rightarrow 9.5273 \times 10^{-7}, kH2EA \rightarrow 3.16245 \times 10^{-7}, kH2EAB \rightarrow 1.0245 \times 10^{-6}\}\}$

```
line22 = {kfet, kHEAB, kHEA, kHE, kcHEA, kcHEAB, kH2E, kH2EA, kH2EAB} /.
   calckinparsordABsimp[1.05 * 0.93428, 10^-8, 200, 200, 0.26253, 10^-8, 1, 200, 0.071054,
      10^-8, 200, 1, 1.4960, 10^-7, 200, 200, 0.83976, 10^-7, 1, 200, 0.069992, 10^-7,
      200, 1, 0.82255, 10^-6, 200, 200, 0.57534, 10^-6, 1, 200, 0.023174, 10^-6, 200, 1]
```

$\{\{1.97496, 8.73902 \times 10^{-9}, -1.11325 \times 10^{-13}, 1.0496 \times 10^{-7},$
   $4.87322, 19.7499, 9.5273 \times 10^{-7}, 3.16245 \times 10^{-7}, 1.0245 \times 10^{-6}\}\}$

```
calckinparsordABsimp[0.93428, 10^-8, 200, 200, 1.05 * 0.26253, 10^-8, 1, 200, 0.071054,
   10^-8, 200, 1, 1.4960, 10^-7, 200, 200, 0.83976, 10^-7, 1, 200, 0.069992, 10^-7,
   200, 1, 0.82255, 10^-6, 200, 200, 0.57534, 10^-6, 1, 200, 0.023174, 10^-6, 200, 1]
```

$\{\{kfet \rightarrow 1.99994, kHEAB \rightarrow 9.99934 \times 10^{-9}, kHEA \rightarrow 2.04944 \times 10^{-10}, kHE \rightarrow 8.4237 \times 10^{-8}, kcHEA \rightarrow 5.4621,$
   $kcHEAB \rightarrow 19.9547, kH2E \rightarrow 1.1871 \times 10^{-6}, kH2EA \rightarrow 3.1533 \times 10^{-7}, kH2EAB \rightarrow 1.00007 \times 10^{-6}\}\}$

```
line23 = {kfet, kHEAB, kHEA, kHE, kcHEA, kcHEAB, kH2E, kH2EA, kH2EAB} /.
   calckinparsordABsimp[0.93428, 10^-8, 200, 200, 1.05 * 0.26253, 10^-8, 1, 200, 0.071054,
      10^-8, 200, 1, 1.4960, 10^-7, 200, 200, 0.83976, 10^-7, 1, 200, 0.069992, 10^-7,
      200, 1, 0.82255, 10^-6, 200, 200, 0.57534, 10^-6, 1, 200, 0.023174, 10^-6, 200, 1]
```

$\{\{1.99994, 9.99934 \times 10^{-9}, 2.04944 \times 10^{-10}, 8.4237 \times 10^{-8},$
   $5.4621, 19.9547, 1.1871 \times 10^{-6}, 3.1533 \times 10^{-7}, 1.00007 \times 10^{-6}\}\}$

```
calckinparsordABsimp[0.93428, 10^-8, 200, 200, 0.26253, 10^-8, 1, 200, 1.05 * 0.071054,
   10^-8, 200, 1, 1.4960, 10^-7, 200, 200, 0.83976, 10^-7, 1, 200, 0.069992, 10^-7,
   200, 1, 0.82255, 10^-6, 200, 200, 0.57534, 10^-6, 1, 200, 0.023174, 10^-6, 200, 1]
```

$\{\{kfet \rightarrow 2.0016, kHEAB \rightarrow 1.00833 \times 10^{-8},$
   $kHEA \rightarrow -7.49809 \times 10^{-10}, kHE \rightarrow 1.00005 \times 10^{-7}, kcHEA \rightarrow 4.95847, kcHEAB \rightarrow 20.1828,$
   $kH2E \rightarrow 9.99934 \times 10^{-7}, kH2EA \rightarrow 3.19639 \times 10^{-7}, kH2EAB \rightarrow 9.98482 \times 10^{-7}\}\}$

```
line24 = {kfet, kHEAB, kHEA, kHE, kcHEA, kcHEAB, kH2E, kH2EA, kH2EAB} /.
    calckinparsordABsimp[0.93428, 10^-8, 200, 200, 0.26253, 10^-8, 1, 200, 1.05 * 0.071054,
      10^-8, 200, 1, 1.4960, 10^-7, 200, 200, 0.83976, 10^-7, 1, 200, 0.069992, 10^-7,
      200, 1, 0.82255, 10^-6, 200, 200, 0.57534, 10^-6, 1, 200, 0.023174, 10^-6, 200, 1]
```

$$\{\{2.0016, 1.00833 \times 10^{-8}, -7.49809 \times 10^{-10}, 1.00005 \times 10^{-7},$$
$$4.95847, 20.1828, 9.99934 \times 10^{-7}, 3.19639 \times 10^{-7}, 9.98482 \times 10^{-7}\}\}$$

```
calckinparsordABsimp[0.93428, 10^-8, 200, 200, 0.26253, 10^-8, 1, 200, 0.071054, 10^-8,
  200, 1, 1.05 * 1.4960, 10^-7, 200, 200, 0.83976, 10^-7, 1, 200, 0.069992, 10^-7,
  200, 1, 0.82255, 10^-6, 200, 200, 0.57534, 10^-6, 1, 200, 0.023174, 10^-6, 200, 1]
```

$$\{\{kfet \to 2.17332, kHEAB \to 1.17246 \times 10^{-8},$$
$$kHEA \to -1.11325 \times 10^{-13}, kHE \to 8.48809 \times 10^{-8}, kcHEA \to 5.79748, kcHEAB \to 21.7336,$$
$$kH2E \to 1.37703 \times 10^{-6}, kH2EA \to 3.16245 \times 10^{-7}, kH2EAB \to 8.52912 \times 10^{-7}\}\}$$

```
line25 = {kfet, kHEAB, kHEA, kHE, kcHEA, kcHEAB, kH2E, kH2EA, kH2EAB} /.
    calckinparsordABsimp[0.93428, 10^-8, 200, 200, 0.26253, 10^-8, 1, 200, 0.071054, 10^-8,
      200, 1, 1.05 * 1.4960, 10^-7, 200, 200, 0.83976, 10^-7, 1, 200, 0.069992, 10^-7,
      200, 1, 0.82255, 10^-6, 200, 200, 0.57534, 10^-6, 1, 200, 0.023174, 10^-6, 200, 1]
```

$$\{\{2.17332, 1.17246 \times 10^{-8}, -1.11325 \times 10^{-13}, 8.48809 \times 10^{-8},$$
$$5.79748, 21.7336, 1.37703 \times 10^{-6}, 3.16245 \times 10^{-7}, 8.52912 \times 10^{-7}\}\}$$

```
calckinparsordABsimp[0.93428, 10^-8, 200, 200, 0.26253, 10^-8, 1, 200, 0.071054, 10^-8,
  200, 1, 1.4960, 10^-7, 200, 200, 1.05 * 0.83976, 10^-7, 1, 200, 0.069992, 10^-7,
  200, 1, 0.82255, 10^-6, 200, 200, 0.57534, 10^-6, 1, 200, 0.023174, 10^-6, 200, 1]
```

$$\{\{kfet \to 1.99994, kHEAB \to 9.99934 \times 10^{-9},$$
$$kHEA \to -6.99714 \times 10^{-11}, kHE \to 1.43655 \times 10^{-7}, kcHEA \to 3.55323, kcHEAB \to 20.1419,$$
$$kH2E \to 5.58512 \times 10^{-7}, kH2EA \to 3.19203 \times 10^{-7}, kH2EAB \to 1.00007 \times 10^{-6}\}\}$$

```
line26 = {kfet, kHEAB, kHEA, kHE, kcHEA, kcHEAB, kH2E, kH2EA, kH2EAB} /.
    calckinparsordABsimp[0.93428, 10^-8, 200, 200, 0.26253, 10^-8, 1, 200, 0.071054, 10^-8,
      200, 1, 1.4960, 10^-7, 200, 200, 1.05 * 0.83976, 10^-7, 1, 200, 0.069992, 10^-7,
      200, 1, 0.82255, 10^-6, 200, 200, 0.57534, 10^-6, 1, 200, 0.023174, 10^-6, 200, 1]
```

$$\{\{1.99994, 9.99934 \times 10^{-9}, -6.99714 \times 10^{-11}, 1.43655 \times 10^{-7},$$
$$3.55323, 20.1419, 5.58512 \times 10^{-7}, 3.19203 \times 10^{-7}, 1.00007 \times 10^{-6}\}\}$$

```
calckinparsordABsimp[0.93428, 10^-8, 200, 200, 0.26253, 10^-8, 1, 200, 0.071054, 10^-8,
  200, 1, 1.4960, 10^-7, 200, 200, 0.83976, 10^-7, 1, 200, 1.05 * 0.069992, 10^-7,
  200, 1, 0.82255, 10^-6, 200, 200, 0.57534, 10^-6, 1, 200, 0.023174, 10^-6, 200, 1]
```

$$\{\{kfet \to 1.98303, kHEAB \to 9.83111 \times 10^{-9},$$
$$kHEA \to 9.22699 \times 10^{-10}, kHE \to 1.00005 \times 10^{-7}, kcHEA \to 5.4657, kcHEAB \to 18.1399,$$
$$kH2E \to 9.99934 \times 10^{-7}, kH2EA \to 2.81761 \times 10^{-7}, kH2EAB \to 1.01718 \times 10^{-6}\}\}$$

```
line27 = {kfet, kHEAB, kHEA, kHE, kcHEA, kcHEAB, kH2E, kH2EA, kH2EAB} /.
    calckinparsordABsimp[0.93428, 10^-8, 200, 200, 0.26253, 10^-8, 1, 200, 0.071054, 10^-8,
      200, 1, 1.4960, 10^-7, 200, 200, 0.83976, 10^-7, 1, 200, 1.05 * 0.069992, 10^-7,
      200, 1, 0.82255, 10^-6, 200, 200, 0.57534, 10^-6, 1, 200, 0.023174, 10^-6, 200, 1]
```

$$\{\{1.98303, 9.83111 \times 10^{-9}, 9.22699 \times 10^{-10}, 1.00005 \times 10^{-7},$$
$$5.4657, 18.1399, 9.99934 \times 10^{-7}, 2.81761 \times 10^{-7}, 1.01718 \times 10^{-6}\}\}$$

```
calckinparsordABsimp[0.93428, 10^-8, 200, 200, 0.26253, 10^-8, 1, 200, 0.071054, 10^-8,
 200, 1, 1.4960, 10^-7, 200, 200, 0.83976, 10^-7, 1, 200, 0.069992, 10^-7, 200, 1,
 1.05 * 0.82255, 10^-6, 200, 200, 0.57534, 10^-6, 1, 200, 0.023174, 10^-6, 200, 1]
```

$\{\{kfet \to 1.97161, kHEAB \to 9.72898 \times 10^{-9},$
$kHEA \to -1.11325 \times 10^{-13}, kHE \to 1.03232 \times 10^{-7}, kcHEA \to 4.85604, kcHEAB \to 19.7165,$
$kH2E \to 7.70065 \times 10^{-7}, kH2EA \to 3.16245 \times 10^{-7}, kH2EAB \to 1.16683 \times 10^{-6}\}\}$

```
line28 = {kfet, kHEAB, kHEA, kHE, kcHEA, kcHEAB, kH2E, kH2EA, kH2EAB} /.
  calckinparsordABsimp[0.93428, 10^-8, 200, 200, 0.26253, 10^-8, 1, 200, 0.071054, 10^-8,
   200, 1, 1.4960, 10^-7, 200, 200, 0.83976, 10^-7, 1, 200, 0.069992, 10^-7, 200, 1,
   1.05 * 0.82255, 10^-6, 200, 200, 0.57534, 10^-6, 1, 200, 0.023174, 10^-6, 200, 1]
```

$\{\{1.97161, 9.72898 \times 10^{-9}, -1.11325 \times 10^{-13}, 1.03232 \times 10^{-7},$
$4.85604, 19.7165, 7.70065 \times 10^{-7}, 3.16245 \times 10^{-7}, 1.16683 \times 10^{-6}\}\}$

```
calckinparsordABsimp[0.93428, 10^-8, 200, 200, 0.26253, 10^-8, 1, 200, 0.071054, 10^-8,
 200, 1, 1.4960, 10^-7, 200, 200, 0.83976, 10^-7, 1, 200, 0.069992, 10^-7, 200, 1,
 0.82255, 10^-6, 200, 200, 1.05 * 0.57534, 10^-6, 1, 200, 0.023174, 10^-6, 200, 1]
```

$\{\{kfet \to 1.99994, kHEAB \to 9.99934 \times 10^{-9},$
$kHEA \to 9.23386 \times 10^{-12}, kHE \to 9.57001 \times 10^{-8}, kcHEA \to 5.21044, kcHEAB \to 19.9792,$
$kH2E \to 1.66143 \times 10^{-6}, kH2EA \to 3.14991 \times 10^{-7}, kH2EAB \to 1.00007 \times 10^{-6}\}\}$

```
line29 = {kfet, kHEAB, kHEA, kHE, kcHEA, kcHEAB, kH2E, kH2EA, kH2EAB} /.
  calckinparsordABsimp[0.93428, 10^-8, 200, 200, 0.26253, 10^-8, 1, 200, 0.071054, 10^-8,
   200, 1, 1.4960, 10^-7, 200, 200, 0.83976, 10^-7, 1, 200, 0.069992, 10^-7, 200, 1,
   0.82255, 10^-6, 200, 200, 1.05 * 0.57534, 10^-6, 1, 200, 0.023174, 10^-6, 200, 1]
```

$\{\{1.99994, 9.99934 \times 10^{-9}, 9.23386 \times 10^{-12}, 9.57001 \times 10^{-8},$
$5.21044, 19.9792, 1.66143 \times 10^{-6}, 3.14991 \times 10^{-7}, 1.00007 \times 10^{-6}\}\}$

```
calckinparsordABsimp[0.93428, 10^-8, 200, 200, 0.26253, 10^-8, 1, 200, 0.071054, 10^-8,
 200, 1, 1.4960, 10^-7, 200, 200, 0.83976, 10^-7, 1, 200, 0.069992, 10^-7, 200, 1,
 0.82255, 10^-6, 200, 200, 0.57534, 10^-6, 1, 200, 1.05 * 0.023174, 10^-6, 200, 1]
```

$\{\{kfet \to 2.00505, kHEAB \to 1.00481 \times 10^{-8},$
$kHEA \to -2.26123 \times 10^{-10}, kHE \to 1.00005 \times 10^{-7}, kcHEA \to 4.8754, kcHEAB \to 20.5621,$
$kH2E \to 9.99934 \times 10^{-7}, kH2EA \to 3.4996 \times 10^{-7}, kH2EAB \to 9.74921 \times 10^{-7}\}\}$

```
line30 = {kfet, kHEAB, kHEA, kHE, kcHEA, kcHEAB, kH2E, kH2EA, kH2EAB} /.
  calckinparsordABsimp[0.93428, 10^-8, 200, 200, 0.26253, 10^-8, 1, 200, 0.071054, 10^-8,
   200, 1, 1.4960, 10^-7, 200, 200, 0.83976, 10^-7, 1, 200, 0.069992, 10^-7, 200, 1,
   0.82255, 10^-6, 200, 200, 0.57534, 10^-6, 1, 200, 1.05 * 0.023174, 10^-6, 200, 1]
```

$\{\{2.00505, 1.00481 \times 10^{-8}, -2.26123 \times 10^{-10}, 1.00005 \times 10^{-7},$
$4.8754, 20.5621, 9.99934 \times 10^{-7}, 3.4996 \times 10^{-7}, 9.74921 \times 10^{-7}\}\}$

Table 3.5 Application of calckinparsordABsimp to data for a mechanism without HEA = EA$^-$ + H$^+$, that is without kHEA or pK1ea.

The following pairs of substrate concentrations have been used: {200,200}, {1,200}, and {200,1}.

```
TableForm[
  Round[-Log[10, {line21[[1]], line22[[1]], line23[[1]], line24[[1]], line25[[1]], line26[[1]],
      line27[[1]], line28[[1]], line29[[1]], line30[[1]]}], .01],
  TableHeadings → {{"No errors", "1.05*v1", "1.05*v2", "1.05*v3", "1.05*v4", "1.05*v5",
      "1.05*v6", "1.05*v7", "1.05*v8", "1.05*v9", "1.05*v10"}, {"-log(kfet)", "pK1eab",
      "pK1ea", "pK1e", "-log(kcHEA)", "-log(kcHEAB)", "pK2e", "pK2ea", "pK2eab"}}]
```

	-log(kfet)	pK1eab	pK1ea	pK1e	-log(kcHEA)	-log(kcHEAB)	pK2e	pK2ea	pK2eab
No errors	-0.3	8.	12.95 - 1.36 i	7.	-0.7	-1.3	6.	6.5	6.
1.05*v1	-0.3	8.06	12.95 - 1.36 i	6.98	-0.69	-1.3	6.02	6.5	5.99
1.05*v2	-0.3	8.	9.69	7.07	-0.74	-1.3	5.93	6.5	6.
1.05*v3	-0.3	8.	9.13 - 1.36 i	7.	-0.7	-1.3	6.	6.5	6.
1.05*v4	-0.34	7.93	12.95 - 1.36 i	7.07	-0.76	-1.34	5.86	6.5	6.07
1.05*v5	-0.3	8.	10.16 - 1.36 i	6.84	-0.55	-1.3	6.25	6.5	6.
1.05*v6	-0.3	8.01	9.03	7.	-0.74	-1.26	6.	6.55	5.99
1.05*v7	-0.29	8.01	12.95 - 1.36 i	6.99	-0.69	-1.29	6.11	6.5	5.93
1.05*v8	-0.3	8.	11.03	7.02	-0.72	-1.3	5.78	6.5	6.
1.05*v9	-0.3	8.	9.65 - 1.36 i	7.	-0.69	-1.31	6.	6.46	6.01

The negative logarithms have been used as entries because it is easier to interpret pKs. The imaginary terms are obtained for pK1ea because

```
Log[10, -1]
```

$$\frac{i \pi}{\text{Log}[10]}$$

There are negative values for pK1ea 7 times out of 10. The pKs 9.69, 9.03, and 11.03 are outside the range that can be determined when the velocity measurements are limited to the pH 6 to 8 range.

This calculation is an example of the statement earler [32] that "When a reaction is studied for the first time, there is an advantage in using the most general mechanism available for that type of reaction."

3.4 Effects of pH on the kinetics of ordered A + B → products when a hydrogen ion is consumed in the rate-determining reaction

There are two different ways that hydrogen ions are involved in the thermodynamics and rapid-equilibrium kinetics of reactions at a specified pH. In addition to the effects of pKs, an integer number of hydrogen ions may be consumed in the rate-determining reaction. This is often the case for reactions involving NAD. Up to 8 hydrogen ions may be consumed in reductase reactions [23,26,*MathSource*4]. When a single hydrogen ion is consumed, the last reaction in the mechanism in Section 3.2 is replaced with

EAB$^-$

  ‖ k$_f$

H$^+$ + HEAB → products $v = k_f$[H$^+$][HEAB]   v = kfet*h*eab/et

  ‖

H$_2$ EAB$^+$

There is a simple change in the expression for the velocity. If one hydrogen ion is consumed, kfet is replaced with h*kfet. If the rate equation is expressed in terms of the pH, kfet is replaced with 10^{-pH}*kfet. Since this introduces large numbers, it is convenient to use 10^{-pH+7}*kfet = $10^{-(pH-7)}$*kfet. This is equivalent to using $h/10^{-7}$*kfet. $V_{fexp} = 10^{n(pH-7)}$kfet×eab/et, where $n = 0, -1, -2,....$ The values of n are negative because that is the convention for reactants in chemical reactions.

When a single hydrogen ion is consumed, the rate equation is given by
simpvelordABh = h 10^7 simpvelordAB.

```
simpvelordABh =
 (a b h^2 kfet 10^7 kH2E kH2EA kH2EAB) / (kcHEA kcHEAB kH2EA kH2EAB (h^2 + h kH2E + kH2E kHE) +
    a kH2E (kcHEAB kH2EAB (h^2 + h kH2EA + kH2EA kHEA) + b kH2EA (h^2 + h kH2EAB + kH2EAB kHEAB)))
```

$$\left(10\,000\,000\ a\ b\ h^2\ kfet\ kH2E\ kH2EA\ kH2EAB\right) / \left(kcHEA\ kcHEAB\ kH2EA\ kH2EAB\ \left(h^2 + h\ kH2E + kH2E\ kHE\right) + a\ kH2E\ \left(kcHEAB\ kH2EAB\ \left(h^2 + h\ kH2EA + kH2EA\ kHEA\right) + b\ kH2EA\ \left(h^2 + h\ kH2EAB + kH2EAB\ kHEAB\right)\right)\right)$$

This rate equation is used to write a program to estimate the 9 kinetic parameters from 9 measured velocities.

```
calckinparsordABsimph[v1_, h1_, a1_, b1_, v2_, h2_, a2_,
   b2_, v3_, h3_, a3_, b3_, v4_, h4_, a4_, b4_, v5_, h5_, a5_, b5_, v6_, h6_,
   a6_, b6_, v7_, h7_, a7_, b7_, v8_, h8_, a8_, b8_, v9_, h9_, a9_, b9_] :=
 Module[{}, (*A single hydrogen ion is consumed in the rate-
       determining reaction.  This program calculates 9 kinetic parameters from
         9 experimental velocities for A + B → products at 9 triplets of [H⁺], [
        A], and [B] on the assumption that the mechanism is ordered, with A bound
         first.  Velocities are measured at {high [A],high [B]}, {low [A], high[B]},
        {high [A], low [B]} at low [H⁺], intermediate [H⁺], and at high [H⁺].*)
  Solve[{
```

$$\frac{a1\ b1\ h1^2\ kfet\ 10^{\wedge}7\ kH2E\ kH2EA\ kH2EAB}{\left(kcHEA\ kcHEAB\ kH2EA\ kH2EAB\ \left(h1^2 + h1\ kH2E + kH2E\ kHE\right) + a1\ kH2E\ \left(kcHEAB\ kH2EAB\ \left(h1^2 + h1\ kH2EA + kH2EA\ kHEA\right) + b1\ kH2EA\ \left(h1^2 + h1\ kH2EAB + kH2EAB\ kHEAB\right)\right)\right)} ==$$

v1,

$$\frac{a2\ b2\ h2^2\ kfet\ 10^{\wedge}7\ kH2E\ kH2EA\ kH2EAB}{\left(kcHEA\ kcHEAB\ kH2EA\ kH2EAB\ \left(h2^2 + h2\ kH2E + kH2E\ kHE\right) + a2\ kH2E\ \left(kcHEAB\ kH2EAB\ \left(h2^2 + h2\ kH2EA + kH2EA\ kHEA\right) + b2\ kH2EA\ \left(h2^2 + h2\ kH2EAB + kH2EAB\ kHEAB\right)\right)\right)} ==$$

v2,

$$\frac{a3\ b3\ h3^2\ kfet\ 10^{\wedge}7\ kH2E\ kH2EA\ kH2EAB}{\left(kcHEA\ kcHEAB\ kH2EA\ kH2EAB\ \left(h3^2 + h3\ kH2E + kH2E\ kHE\right) + a3\ kH2E\ \left(kcHEAB\ kH2EAB\ \left(h3^2 + h3\ kH2EA + kH2EA\ kHEA\right) + b3\ kH2EA\ \left(h3^2 + h3\ kH2EAB + kH2EAB\ kHEAB\right)\right)\right)} ==$$

v3,

$$\frac{a4\ b4\ h4^{\wedge}2\ kfet\ 10^{\wedge}7\ kH2E\ kH2EA\ kH2EAB}{\left(kcHEA\ kcHEAB\ kH2EA\ kH2EAB\ \left(h4^2 + h4\ kH2E + kH2E\ kHE\right) + a4\ kH2E\ \left(kcHEAB\ kH2EAB\ \left(h4^2 + h4\ kH2EA + kH2EA\ kHEA\right) + b4\ kH2EA\ \left(h4^2 + h4\ kH2EAB + kH2EAB\ kHEAB\right)\right)\right)} ==$$

v4,

$$\frac{a5\ b5\ h5^{\wedge}2\ kfet\ 10^{\wedge}7\ kH2E\ kH2EA\ kH2EAB}{\left(kcHEA\ kcHEAB\ kH2EA\ kH2EAB\ \left(h5^2 + h5\ kH2E + kH2E\ kHE\right) + a5\ kH2E\ \left(kcHEAB\ kH2EAB\ \left(h5^2 + h5\ kH2EA + kH2EA\ kHEA\right) + b5\ kH2EA\ \left(h5^2 + h5\ kH2EAB + kH2EAB\ kHEAB\right)\right)\right)} ==$$

v5,

$$\frac{a6\ b6\ h6^{\wedge}2\ kfet\ 10^{\wedge}7\ kH2E\ kH2EA\ kH2EAB}{\left(kcHEA\ kcHEAB\ kH2EA\ kH2EAB\ \left(h6^2 + h6\ kH2E + kH2E\ kHE\right) + a6\ kH2E\ \left(kcHEAB\ kH2EAB\ \left(h6^2 + h6\ kH2EA + kH2EA\ kHEA\right) + b6\ kH2EA\ \left(h6^2 + h6\ kH2EAB + kH2EAB\ kHEAB\right)\right)\right)} ==$$

v6,

$$\frac{a7\ b7\ h7^{\wedge}2\ kfet\ 10^{\wedge}7\ kH2E\ kH2EA\ kH2EAB}{\left(kcHEA\ kcHEAB\ kH2EA\ kH2EAB\ \left(h7^2 + h7\ kH2E + kH2E\ kHE\right) + a7\ kH2E\ \left(kcHEAB\ kH2EAB\ \left(h7^2 + h7\ kH2EA + kH2EA\ kHEA\right) + b7\ kH2EA\ \left(h7^2 + h7\ kH2EAB + kH2EAB\ kHEAB\right)\right)\right)} ==$$

v7,

$$\frac{a8\ b8\ h8^{\wedge}2\ kfet\ 10^{\wedge}7\ kH2E\ kH2EA\ kH2EAB}{\left(kcHEA\ kcHEAB\ kH2EA\ kH2EAB\ \left(h8^2 + h8\ kH2E + kH2E\ kHE\right) + a8\ kH2E\ \left(kcHEAB\ kH2EAB\ \left(h8^2 + h8\ kH2EA + kH2EA\ kHEA\right) + b8\ kH2EA\ \left(h8^2 + h8\ kH2EAB + kH2EAB\ kHEAB\right)\right)\right)} ==$$

v8,

$$\frac{a9\ b9\ h9^{\wedge}2\ kfet\ 10^{\wedge}7\ kH2E\ kH2EA\ kH2EAB}{\left(kcHEA\ kcHEAB\ kH2EA\ kH2EAB\ \left(h9^2 + h9\ kH2E + kH2E\ kHE\right) + a9\ kH2E\ \left(kcHEAB\ kH2EAB\ \left(h9^2 + h9\ kH2EA + kH2EA\ kHEA\right) + b9\ kH2EA\ \left(h9^2 + h9\ kH2EAB + kH2EAB\ kHEAB\right)\right)\right)} ==$$

```
  v9}, {kHE, kH2E, kHEA, kH2EA, kHEAB, kH2EAB, kfet, kcHEA, kcHEAB}]]
```

The same arbitrary values of the kinetic parameters are used.

```
Grid[{{pK1e, pK2e, pK1ea, pK2ea, pK1eab, pK2eab, kfet, kcHEA, kcHEAB},
  {7, 6, 7.5, 6.5, 8, 6, 2, 5, 20}}]
```

pK1e	pK2e	pK1ea	pK2ea	pK1eab	pK2eab	kfet	kcHEA	kcHEAB
7	6	7.5	6.5	8	6	2	5	20

```
Grid[{{kHE, kH2E, kHEA, kH2EA, kHEAB, kH2EAB, kfet, kcHEA, kcHEAB},
  {10^-7, 10^-6, 10^-7.5, 10^-6.5, 10^-8, 10^-6, 2, 5, 20}}] // N
```

kHE	kH2E	kHEA	kH2EA	kHEAB	kH2EAB	kfet	kcHEA	kcHEAB
$1.\times 10^{-7}$	$1.\times 10^{-6}$	3.16228×10^{-8}	3.16228×10^{-7}	$1.\times 10^{-8}$	$1.\times 10^{-6}$	2.	5.	20.

The arbitrary parameters can be inserted into simpvtestordAB2 to obtain the expression for the velocity as a function of [A], [B], and [H$^+$].

```
simpvtestordABh =
  simpvelordABh /. kHE → 1. * 10^-7 /. kH2E → 10^-6 /. kHEA → 10^-7.5 /. kH2EA → 10^-6.5 /.
    kHEAB → 1. * 10^-8 /. kH2EAB → 1. * 10^-6 /. kfet → 2. /. kcHEA → 5. /. kcHEAB → 20.
```

$$\left(6.32456 \times 10^{-12} \, a\, b\, h^2\right) \Bigg/ \left(3.16228 \times 10^{-11} \left(1. \times 10^{-13} + \frac{h}{1\,000\,000} + h^2\right) + \right.$$

$$\left. \frac{a\left(0.00002\left(1. \times 10^{-14} + 3.16228 \times 10^{-7}\, h + h^2\right) + 3.16228 \times 10^{-7}\, b\left(1. \times 10^{-14} + 1. \times 10^{-6}\, h + h^2\right)\right)}{1\,000\,000}\right)$$

This rate equation can be used to calculate 9 velocities for 9 chosen triplets of concentrations {h,a,b} at 3 pHs, when these velocities are treated as experimental data.

Calculate the rate equation at pH 6 and rearrange it to obtain the kinetic parameters.

```
simpvtestordABh /. h → 10^-6
```

$$\frac{6.32456 \times 10^{-24} \, a\, b}{6.64078 \times 10^{-23} + \dfrac{a\left(2.65246\times10^{-17} + 6.35618\times10^{-19}\, b\right)}{1\,000\,000}}$$

```
Simplify[simpvtestordABh /. h → 10^-6]
```

$$\frac{6.32456 \times 10^{-24} \, a\, b}{6.64078 \times 10^{-23} + 2.65246 \times 10^{-23}\, a + 6.35618 \times 10^{-25}\, a\, b}$$

Divide the numerator and denominator by the coefficient of the a b term in the denominator.

```
(6.324555320336758`*^-24 a b) / (6.356178096938442`*^-25 a b) /
 ((6.640783086353597`*^-23) / (6.356178096938442`*^-25 a b) +
  (2.6524555320336755`*^-23 a / (6.356178096938442`*^-25 a b)) +
  (6.356178096938442`*^-25 a b) / (6.356178096938442`*^-25 a b))
```

$$\frac{9.95025}{1. + \dfrac{41.7304}{b} + \dfrac{104.478}{a\, b}}$$

This is the same as obtained before, except that the numerator is 10 times larger.

```
104.5 / 41.73
```

```
2.50419
```

The rate equation can be written as

$$\frac{9.950248756218906`}{1.` + \dfrac{41.73035260467724`}{b}\left(1 + \dfrac{2.504}{a}\right)}$$

$$\frac{9.95025}{1. + \dfrac{41.7304\left(1 + \dfrac{2.504}{a}\right)}{b}}$$

Calculate the rate equation at pH 7 and rearrange it to obtain the kinetic parameters.

```
simpvtestordABh /. h → 10^-7
```

$$\frac{6.32456 \times 10^{-26} \text{ a b}}{6.64078 \times 10^{-24} + \frac{\text{a} \left(1.03246\times10^{-18}+3.79473\times10^{-20} \text{ b}\right)}{1\,000\,000}}$$

```
Simplify[simpvtestordABh /. h → 10^-7]
```

$$\frac{6.32456 \times 10^{-26} \text{ a b}}{6.64078 \times 10^{-24} + 1.03246 \times 10^{-24} \text{ a} + 3.79473 \times 10^{-26} \text{ a b}}$$

Divide the numerator and denominator by the coefficient of the a b term in the denominator.

```
(6.324555320336758`*^-26 a b) / (3.7947331922020547`*^-26 a b) /
 ((6.640783086353596`*^-24) / (3.7947331922020547`*^-26 a b) +
  (1.032455532033676`*^-24 a) / (3.7947331922020547`*^-26 a b) +
  (3.7947331922020547`*^-26 a b) / (3.7947331922020547`*^-26 a b))
```

$$\frac{1.66667}{1. + \frac{27.2076}{b} + \frac{175.}{a\,b}}$$

This is the same as obtained before.

Calculate the rate equation at pH 8 and rearrange it to obtain the kinetic parameters.

```
simpvtestordABh /. h → 10^-8
```

$$\frac{6.32456 \times 10^{-28} \text{ a b}}{3.48167 \times 10^{-24} + \frac{\text{a} \left(2.65246\times10^{-19}+6.35618\times10^{-21} \text{ b}\right)}{1\,000\,000}}$$

```
Simplify[simpvtestordABh /. h → 10^-8]
```

$$\frac{6.32456 \times 10^{-28} \text{ a b}}{3.48167 \times 10^{-24} + 2.65246 \times 10^{-25} \text{ a} + 6.35618 \times 10^{-27} \text{ a b}}$$

Divide the numerator and denominator by the coefficient of the ab term in the denominator.

```
(6.324555320336758`*^-28 a b) / (6.356178096938441`*^-27 a b) /
 ((3.4816677038453854`*^-24) / (6.356178096938441`*^-27 a b) +
  (2.6524555320336756`*^-25 a) / (6.356178096938441`*^-27 a b) +
  (6.356178096938441`*^-27 a b) / (6.356178096938441`*^-27 a b))
```

$$\frac{0.0995025}{1. + \frac{41.7304}{b} + \frac{547.761}{a\,b}}$$

This is the same as obtained before, but V_{fexp} is 10 times smaller.

Table 3.6 Values of the pH-dependent kinetic parameters when a single hydrogen ion is consumed by the rate-determining reaction

```
TableForm[{{9.950, 25.03, 41.73}, {1.667, 6.431, 27.21}, {.09950, 13.13, 41.73}},
 TableHeadings → {{"pH 6", "pH 7", "pH 8"}, {"Vfexp", "KIA", "KB"}}]
```

	V_{fexp}	K_{IA}	K_B
pH 6	9.95	25.03	41.73
pH 7	1.667	6.431	27.21
pH 8	0.0995	13.13	41.73

Calculate approximate values for limiting velocities V_{fexp} at pHs 5, 6, 7, 8, and 9.

```
simpvtestordABh /. h → {10^-5, 10^-6, 10^-7, 10^-8, 10^-9} /. a → 1000 /. b → 1000 // N
```

{17.1606, 9.5507, 1.62225, 0.0954664, 0.00171473}

```
simpvtestordABh /. h → {10^-5, 10^-6, 10^-7, 10^-8, 10^-9} /. a → 10 000 /. b → 10 000 // N
```

{18.0729, 9.90889, 1.66214, 0.0990884, 0.00180728}

Neither of these calculations yields exact values of V_{fexp} bcause a → ∞ and b → ∞ are required.

Plot these limiting velocities V_{fexp} versus pH in comparison with the limiting velocities when a hydrogen ion is not consumed in the rate-determining reaction.

```
Plot[{simpvtestordABh /. a → 1000 /. b → 1000 /. h → 10^-pH,
   simpvtestordABh /. a → 10 000 /. b → 10 000 /. h → 10^-pH,
   vtestordAB2 /. a → 10 000 /. b → 10 000 /. h → 10^-pH},
 {pH, 5, 9}, AxesLabel → {"pH", "approx. V_fexp"}, PlotStyle → {Black}]
```

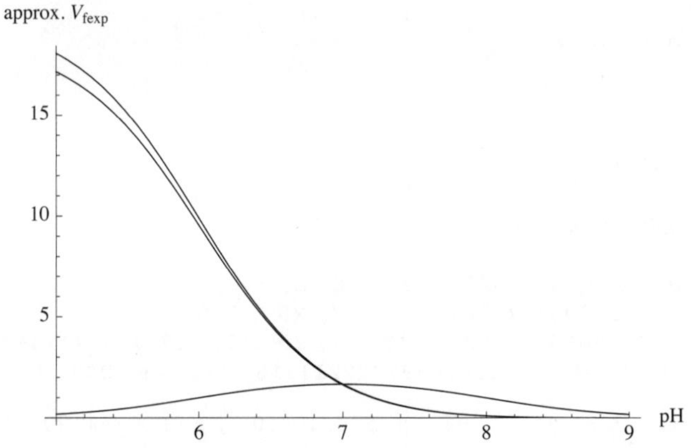

Fig. 3.9 Plots of approximate V_{fexp} when a single hydrogen ion is consumed in the rate-determining reaction. The bell-shaped curve is obtained when no hydrogen ions are consumed.

The curves cross at pH 7 because h/10^-7 is unity. This makes it easier to compare the $n = 0$ and $n = -1$ cases. Since this package is based on kfet which does not deal with the magnitudes of kf or et separately, making the curves cross at pH 7 is arbitrary.

Estimation of kinetic parameters when one hydrogen ion is consumed

The following 9 velocities are used to calculate the 9 kinetic parameters when one hydrogen ion is consumed in the rate-determining reaction.

```
simpvtestordABh /. h → 10^-8 /. a → 200 /. b → 200 // N
```

0.0814029

```
simpvtestordABh /. h → 10^-8 /. a → 1 /. b → 200 // N
```

0.0252067

```
simpvtestordABh /. h → 10^-8 /. a → 200 /. b → 1 // N
```

0.00218835

These velocities at pH 8 are 10 - fold smaller than those calculated earlier.

```
simpvtestordABh /. h → 10^-7 /. a → 200 /. b → 200 // N
```

1.46146

```
simpvtestordABh /. h → 10^-7 /. a → 1 /. b → 200 // N
```

0.828759

```
simpvtestordABh /. h → 10^-7 /. a → 200 /. b → 1 // N
```

0.0573081

These velocities at pH 7 agree with those calculated earlier.

```
simpvtestordABh /. h → 10^-6 /. a → 200 /. b → 200 // N
```

8.21477

```
simpvtestordABh /. h → 10^-6 /. a → 1 /. b → 200 // N
```

5.74813

```
simpvtestordABh /. h → 10^-6 /. a → 200 /. b → 1 // N
```

0.230049

These velocities at pH 6 are 10 times bigger than those calculated earlier.

Calculation of 9 kinetic parameters when one hydrogen ion is consumed

```
calckinparsordABsimph[0.081403, 10^-8, 200, 200, 0.0252067, 10^-8, 1, 200,
  0.002188351196304686`, 10^-8, 200, 1, 1.4614588957329162`, 10^-7, 200, 200,
  0.8287594262196729`, 10^-7, 1, 200, 0.057308050643247024`, 10^-7, 200, 1, 8.214766717544927`,
  10^-6, 200, 200, 5.74813393983966`, 10^-6, 1, 200, 0.23004897732211316`, 10^-6, 200, 1]
```

$\{\{$ kfet → 2., kHEAB → 9.99996 × 10^{-9}, kHEA → 3.16228 × 10^{-8}, kHE → 1.00001 × 10^{-7}, kcHEA → 4.99999,

kcHEAB → 20., kH2E → 9.99995 × 10^{-7}, kH2EA → 3.16228 × 10^{-7}, kH2EAB → 1. × 10^{-6} $\}\}$

These kinetic parameters are all correct, but it is important to check the effects of 5% errors in the measured velocities, one at a time.

```
line31 = {kfet, kHEAB, kHEA, kHE, kcHEA, kcHEAB, kH2E, kH2EA, kH2EAB} /.
  calckinparsordABsimph[0.081403, 10^-8, 200, 200, 0.0252067, 10^-8, 1,
    200, 0.002188351196304686`, 10^-8, 200, 1, 1.4614588957329162`, 10^-7,
    200, 200, 0.8287594262196729`, 10^-7, 1, 200, 0.057308050643247024`,
    10^-7, 200, 1, 8.214766717544927`, 10^-6, 200, 200, 5.74813393983966`,
    10^-6, 1, 200, 0.23004897732211316`, 10^-6, 200, 1]
```

$\{\{$2., 9.99996 × 10^{-9}, 3.16228 × 10^{-8},

  1.00001 × 10^{-7}, 4.99999, 20., 9.99995 × 10^{-7}, 3.16228 × 10^{-7}, 1. × 10^{-6} $\}\}$

```
calckinparsordABsimph[1.05 * 0.081403, 10^-8, 200, 200, 0.0252067, 10^-8, 1,
  200, 0.002188351196304686`, 10^-8, 200, 1, 1.4614588957329162`, 10^-7, 200, 200,
  0.8287594262196729`, 10^-7, 1, 200, 0.057308050643247024`, 10^-7, 200, 1, 8.214766717544927`,
  10^-6, 200, 200, 5.74813393983966`, 10^-6, 1, 200, 0.23004897732211316`, 10^-6, 200, 1]
```

$\{\{$kfet → 1.97138, kHEAB → 8.55607 × 10^{-9},

  kHEA → 3.16228 × 10^{-8}, kHE → 1.05709 × 10^{-7}, kcHEA → 4.85482, kcHEAB → 19.7138,

  kH2E → 9.45993 × 10^{-7}, kH2EA → 3.16228 × 10^{-7}, kH2EAB → 1.02808 × 10^{-6} $\}\}$

```
line32 = {kfet, kHEAB, kHEA, kHE, kcHEA, kcHEAB, kH2E, kH2EA, kH2EAB} /.
   calckinparsordABsimph[1.05 * 0.081403, 10^-8, 200, 200, 0.0252067, 10^-8,
   1, 200, 0.002188351196304686`, 10^-8, 200, 1, 1.4614588957329162`, 10^-7,
   200, 200, 0.8287594262196729`, 10^-7, 1, 200, 0.057308050643247024`,
   10^-7, 200, 1, 8.214766717544927`, 10^-6, 200, 200,
   5.74813393983966`, 10^-6, 1, 200, 0.23004897732211316`, 10^-6, 200, 1]
```

$$\{\{1.97138, 8.55607 \times 10^{-9}, 3.16228 \times 10^{-8}, 1.05709 \times 10^{-7},$$
$$4.85482, 19.7138, 9.45993 \times 10^{-7}, 3.16228 \times 10^{-7}, 1.02808 \times 10^{-6}\}\}$$

```
calckinparsordABsimph[0.081403, 10^-8, 200, 200, 1.05 * 0.0252067, 10^-8, 1,
   200, 0.002188351196304686`, 10^-8, 200, 1, 1.4614588957329162`, 10^-7, 200, 200,
   0.8287594262196729`, 10^-7, 1, 200, 0.057308050643247024`, 10^-7, 200, 1, 8.214766717544927`,
   10^-6, 200, 200, 5.74813393983966`, 10^-6, 1, 200, 0.23004897732211316`, 10^-6, 200, 1]
```

$$\{\{kfet \to 2., kHEAB \to 9.99996 \times 10^{-9}, kHEA \to 3.19107 \times 10^{-8}, kHE \to 8.36353 \times 10^{-8}, kcHEA \to 5.48163,$$
$$kcHEAB \to 19.9531, kH2E \to 1.19567 \times 10^{-6}, kH2EA \to 3.15274 \times 10^{-7}, kH2EAB \to 1. \times 10^{-6}\}\}$$

```
line33 = {kfet, kHEAB, kHEA, kHE, kcHEA, kcHEAB, kH2E, kH2EA, kH2EAB} /.
   calckinparsordABsimph[0.081403, 10^-8, 200, 200, 1.05 * 0.0252067, 10^-8,
   1, 200, 0.002188351196304686`, 10^-8, 200, 1, 1.4614588957329162`, 10^-7,
   200, 200, 0.8287594262196729`, 10^-7, 1, 200, 0.057308050643247024`,
   10^-7, 200, 1, 8.214766717544927`, 10^-6, 200, 200,
   5.74813393983966`, 10^-6, 1, 200, 0.23004897732211316`, 10^-6, 200, 1]
```

$$\{\{2., 9.99996 \times 10^{-9}, 3.19107 \times 10^{-8}, 8.36353 \times 10^{-8},$$
$$5.48163, 19.9531, 1.19567 \times 10^{-6}, 3.15274 \times 10^{-7}, 1. \times 10^{-6}\}\}$$

```
calckinparsordABsimph[0.081403, 10^-8, 200, 200, 0.0252067, 10^-8, 1, 200,
   1.05 * 0.002188351196304686`, 10^-8, 200, 1, 1.4614588957329162`, 10^-7, 200, 200,
   0.8287594262196729`, 10^-7, 1, 200, 0.057308050643247024`, 10^-7, 200, 1, 8.214766717544927`,
   10^-6, 200, 200, 5.74813393983966`, 10^-6, 1, 200, 0.23004897732211316`, 10^-6, 200, 1]
```

$$\{\{kfet \to 2.00541, kHEAB \to 1.02731 \times 10^{-8},$$
$$kHEA \to 2.84014 \times 10^{-8}, kHE \to 1.00001 \times 10^{-7}, kcHEA \to 4.86854, kcHEAB \to 20.5956,$$
$$kH2E \to 9.99995 \times 10^{-7}, kH2EA \to 3.27306 \times 10^{-7}, kH2EAB \to 9.94859 \times 10^{-7}\}\}$$

```
line34 = {kfet, kHEAB, kHEA, kHE, kcHEA, kcHEAB, kH2E, kH2EA, kH2EAB} /.
   calckinparsordABsimph[0.081403, 10^-8, 200, 200, 0.0252067,
   10^-8, 1, 200, 1.05 * 0.002188351196304686`, 10^-8, 200, 1,
   1.4614588957329162`, 10^-7, 200, 200, 0.8287594262196729`, 10^-7, 1, 200,
   0.057308050643247024`, 10^-7, 200, 1, 8.214766717544927`, 10^-6, 200, 200,
   5.74813393983966`, 10^-6, 1, 200, 0.23004897732211316`, 10^-6, 200, 1]
```

$$\{\{2.00541, 1.02731 \times 10^{-8}, 2.84014 \times 10^{-8}, 1.00001 \times 10^{-7},$$
$$4.86854, 20.5956, 9.99995 \times 10^{-7}, 3.27306 \times 10^{-7}, 9.94859 \times 10^{-7}\}\}$$

```
calckinparsordABsimph[0.081403, 10^-8, 200, 200, 0.0252067, 10^-8, 1, 200,
   0.002188351196304686`, 10^-8, 200, 1, 1.05 * 1.4614588957329162`, 10^-7, 200, 200,
   0.8287594262196729`, 10^-7, 1, 200, 0.057308050643247024`, 10^-7, 200, 1, 8.214766717544927`,
   10^-6, 200, 200, 5.74813393983966`, 10^-6, 1, 200, 0.23004897732211316`, 10^-6, 200, 1]
```

$$\{\{kfet \to 2.17785, kHEAB \to 1.17697 \times 10^{-8},$$
$$kHEA \to 3.16228 \times 10^{-8}, kHE \to 8.45704 \times 10^{-8}, kcHEA \to 5.81664, kcHEAB \to 21.7785,$$
$$kH2E \to 1.38774 \times 10^{-6}, kH2EA \to 3.16228 \times 10^{-7}, kH2EAB \to 8.49637 \times 10^{-7}\}\}$$

```
line35 = {kfet, kHEAB, kHEA, kHE, kcHEA, kcHEAB, kH2E, kH2EA, kH2EAB} /.
   calckinparsordABsimph[0.081403, 10^-8, 200, 200, 0.0252067, 10^-8, 1, 200,
     0.002188351196304686`, 10^-8, 200, 1, 1.05 * 1.4614588957329162`, 10^-7,
     200, 200, 0.8287594262196729`, 10^-7, 1, 200, 0.057308050643247024`,
     10^-7, 200, 1, 8.214766717544927`, 10^-6, 200, 200,
     5.74813393983966`, 10^-6, 1, 200, 0.23004897732211316`, 10^-6, 200, 1]
```

$$\left\{\left\{2.17785,\ 1.17697 \times 10^{-8},\ 3.16228 \times 10^{-8},\ 8.45704 \times 10^{-8},\right.\right.$$
$$\left.\left.5.81664,\ 21.7785,\ 1.38774 \times 10^{-6},\ 3.16228 \times 10^{-7},\ 8.49637 \times 10^{-7}\right\}\right\}$$

```
calckinparsordABsimph[0.081403, 10^-8, 200, 200, 0.0252067,
   10^-8, 1, 200, 0.002188351196304686`, 10^-8, 200, 1, 1.4614588957329162`,
   10^-7, 200, 200, 1.05 * 0.8287594262196729`, 10^-7, 1, 200,
   0.057308050643247024`, 10^-7, 200, 1, 8.214766717544927`, 10^-6, 200, 200,
   5.74813393983966`, 10^-6, 1, 200, 0.23004897732211316`, 10^-6, 200, 1]
```

$$\left\{\left\{\text{kfet} \to 2.,\ \text{kHEAB} \to 9.99996 \times 10^{-9},\ \text{kHEA} \to 3.13259 \times 10^{-8},\ \text{kHE} \to 1.4446 \times 10^{-7},\ \text{kcHEA} \to 3.53443,\right.\right.$$
$$\left.\left.\text{kcHEAB} \to 20.144,\ \text{kH2E} \to 5.53992 \times 10^{-7},\ \text{kH2EA} \to 3.19224 \times 10^{-7},\ \text{kH2EAB} \to 1. \times 10^{-6}\right\}\right\}$$

```
line36 = {kfet, kHEAB, kHEA, kHE, kcHEA, kcHEAB, kH2E, kH2EA, kH2EAB} /.
   calckinparsordABsimph[0.081403, 10^-8, 200, 200, 0.0252067, 10^-8,
     1, 200, 0.002188351196304686`, 10^-8, 200, 1, 1.4614588957329162`,
     10^-7, 200, 200, 1.05 * 0.8287594262196729`, 10^-7, 1, 200,
     0.057308050643247024`, 10^-7, 200, 1, 8.214766717544927`, 10^-6, 200, 200,
     5.74813393983966`, 10^-6, 1, 200, 0.23004897732211316`, 10^-6, 200, 1]
```

$$\left\{\left\{2.,\ 9.99996 \times 10^{-9},\ 3.13259 \times 10^{-8},\ 1.4446 \times 10^{-7},\right.\right.$$
$$\left.\left.3.53443,\ 20.144,\ 5.53992 \times 10^{-7},\ 3.19224 \times 10^{-7},\ 1. \times 10^{-6}\right\}\right\}$$

```
calckinparsordABsimph[0.081403, 10^-8, 200, 200,
   0.0252067, 10^-8, 1, 200, 0.002188351196304686`, 10^-8, 200, 1,
   1.4614588957329162`, 10^-7, 200, 200, 0.8287594262196729`, 10^-7, 1, 200,
   1.05 * 0.057308050643247024`, 10^-7, 200, 1, 8.214766717544927`, 10^-6, 200,
   200, 5.74813393983966`, 10^-6, 1, 200, 0.23004897732211316`, 10^-6, 200, 1]
```

$$\left\{\left\{\text{kfet} \to 1.97939,\ \text{kHEAB} \to 9.79487 \times 10^{-9},\right.\right.$$
$$\text{kHEA} \to 3.64492 \times 10^{-8},\ \text{kHE} \to 1.00001 \times 10^{-7},\ \text{kcHEA} \to 5.58115,\ \text{kcHEAB} \to 17.7327,$$
$$\left.\left.\text{kH2E} \to 9.99995 \times 10^{-7},\ \text{kH2EA} \to 2.74354 \times 10^{-7},\ \text{kH2EAB} \to 1.02094 \times 10^{-6}\right\}\right\}$$

```
line37 = {kfet, kHEAB, kHEA, kHE, kcHEA, kcHEAB, kH2E, kH2EA, kH2EAB} /.
   calckinparsordABsimph[0.081403, 10^-8, 200, 200, 0.0252067, 10^-8, 1,
     200, 0.002188351196304686`, 10^-8, 200, 1, 1.4614588957329162`, 10^-7,
     200, 200, 0.8287594262196729`, 10^-7, 1, 200, 1.05 * 0.057308050643247024`,
     10^-7, 200, 1, 8.214766717544927`, 10^-6, 200, 200,
     5.74813393983966`, 10^-6, 1, 200, 0.23004897732211316`, 10^-6, 200, 1]
```

$$\left\{\left\{1.97939,\ 9.79487 \times 10^{-9},\ 3.64492 \times 10^{-8},\ 1.00001 \times 10^{-7},\right.\right.$$
$$\left.\left.5.58115,\ 17.7327,\ 9.99995 \times 10^{-7},\ 2.74354 \times 10^{-7},\ 1.02094 \times 10^{-6}\right\}\right\}$$

```
calckinparsordABsimph[0.081403, 10^-8, 200, 200, 0.0252067, 10^-8,
   1, 200, 0.002188351196304686`, 10^-8, 200, 1, 1.4614588957329162`, 10^-7,
   200, 200, 0.8287594262196729`, 10^-7, 1, 200, 0.057308050643247024`,
   10^-7, 200, 1, 1.05 * 8.214766717544927`, 10^-6, 200, 200,
   5.74813393983966`, 10^-6, 1, 200, 0.23004897732211316`, 10^-6, 200, 1]
```

$$\left\{\left\{\text{kfet} \to 1.97164,\ \text{kHEAB} \to 9.72923 \times 10^{-9},\right.\right.$$
$$\text{kHEA} \to 3.16228 \times 10^{-8},\ \text{kHE} \to 1.03232 \times 10^{-7},\ \text{kcHEA} \to 4.85614,\ \text{kcHEAB} \to 19.7164,$$
$$\left.\left.\text{kH2E} \to 7.6987 \times 10^{-7},\ \text{kH2EA} \to 3.16228 \times 10^{-7},\ \text{kH2EAB} \to 1.167 \times 10^{-6}\right\}\right\}$$

```
line38 = {kfet, kHEAB, kHEA, kHE, kcHEA, kcHEAB, kH2E, kH2EA, kH2EAB} /.
    calckinparsordABsimph[0.081403, 10^-8, 200, 200, 0.0252067, 10^-8, 1,
    200, 0.002188351196304686`, 10^-8, 200, 1, 1.4614588957329162`, 10^-7,
    200, 200, 0.8287594262196729`, 10^-7, 1, 200, 0.057308050643247024`,
    10^-7, 200, 1, 1.05 * 8.214766717544927`, 10^-6, 200, 200,
    5.74813393983966`, 10^-6, 1, 200, 0.23004897732211316`, 10^-6, 200, 1]
```

$$\left\{ \left\{ 1.97164, 9.72923 \times 10^{-9}, 3.16228 \times 10^{-8}, 1.03232 \times 10^{-7}, \right. \right.$$
$$\left. \left. 4.85614, 19.7164, 7.6987 \times 10^{-7}, 3.16228 \times 10^{-7}, 1.167 \times 10^{-6} \right\} \right\}$$

```
calckinparsordABsimph[0.081403, 10^-8, 200, 200, 0.0252067, 10^-8, 1, 200,
    0.002188351196304686`, 10^-8, 200, 1, 1.4614588957329162`, 10^-7, 200, 200,
    0.8287594262196729`, 10^-7, 1, 200, 0.057308050643247024`, 10^-7, 200, 1, 8.214766717544927`,
    10^-6, 200, 200, 1.05 * 5.74813393983966`, 10^-6, 1, 200, 0.23004897732211316`, 10^-6, 200, 1]
```

$$\left\{ \left\{ kfet \to 2., kHEAB \to 9.99996 \times 10^{-9}, kHEA \to 3.16647 \times 10^{-8}, kHE \to 9.56923 \times 10^{-8}, kcHEA \to 5.21092, \right. \right.$$
$$\left. \left. kcHEAB \to 19.9794, kH2E \to 1.66252 \times 10^{-6}, kH2EA \to 3.14972 \times 10^{-7}, kH2EAB \to 1. \times 10^{-6} \right\} \right\}$$

```
line39 = {kfet, kHEAB, kHEA, kHE, kcHEA, kcHEAB, kH2E, kH2EA, kH2EAB} /.
    calckinparsordABsimph[0.081403, 10^-8, 200, 200, 0.0252067, 10^-8, 1, 200,
    0.002188351196304686`, 10^-8, 200, 1, 1.4614588957329162`, 10^-7, 200,
    200, 0.8287594262196729`, 10^-7, 1, 200, 0.057308050643247024`, 10^-7,
    200, 1, 8.214766717544927`, 10^-6, 200, 200, 1.05 * 5.74813393983966`,
    10^-6, 1, 200, 0.23004897732211316`, 10^-6, 200, 1]
```

$$\left\{ \left\{ 2., 9.99996 \times 10^{-9}, 3.16647 \times 10^{-8}, 9.56923 \times 10^{-8}, \right. \right.$$
$$\left. \left. 5.21092, 19.9794, 1.66252 \times 10^{-6}, 3.14972 \times 10^{-7}, 1. \times 10^{-6} \right\} \right\}$$

```
calckinparsordABsimph[0.081403, 10^-8, 200, 200, 0.0252067, 10^-8, 1, 200,
    0.002188351196304686`, 10^-8, 200, 1, 1.4614588957329162`, 10^-7, 200, 200,
    0.8287594262196729`, 10^-7, 1, 200, 0.057308050643247024`, 10^-7, 200, 1, 8.214766717544927`,
    10^-6, 200, 200, 5.74813393983966`, 10^-6, 1, 200, 1.05 * 0.23004897732211316`, 10^-6, 200, 1]
```

$$\left\{ \left\{ kfet \to 2.00515, kHEAB \to 1.00491 \times 10^{-8}, \right. \right.$$
$$kHEA \to 3.06033 \times 10^{-8}, kHE \to 1.00001 \times 10^{-7}, kcHEA \to 4.87478, kcHEAB \to 20.5665,$$
$$\left. \left. kH2E \to 9.99995 \times 10^{-7}, kH2EA \to 3.50207 \times 10^{-7}, kH2EAB \to 9.74675 \times 10^{-7} \right\} \right\}$$

```
line40 = {kfet, kHEAB, kHEA, kHE, kcHEA, kcHEAB, kH2E, kH2EA, kH2EAB} /.
    calckinparsordABsimph[0.081403, 10^-8, 200, 200, 0.0252067, 10^-8, 1,
    200, 0.002188351196304686`, 10^-8, 200, 1, 1.4614588957329162`, 10^-7,
    200, 200, 0.8287594262196729`, 10^-7, 1, 200, 0.057308050643247024`,
    10^-7, 200, 1, 8.214766717544927`, 10^-6, 200, 200, 5.74813393983966`,
    10^-6, 1, 200, 1.05 * 0.23004897732211316`, 10^-6, 200, 1]
```

$$\left\{ \left\{ 2.00515, 1.00491 \times 10^{-8}, 3.06033 \times 10^{-8}, 1.00001 \times 10^{-7}, \right. \right.$$
$$\left. \left. 4.87478, 20.5665, 9.99995 \times 10^{-7}, 3.50207 \times 10^{-7}, 9.74675 \times 10^{-7} \right\} \right\}$$

Table 3.7 Effects of 5% errors in measured velocities on the estimation of the 9 kinetic parameters of ordered A + B → products when a single hydrogen ion is consumed in the rate-determining reaction

```
TableForm[
  Round[-Log[10, {line31[[1]], line32[[1]], line33[[1]], line34[[1]], line35[[1]], line36[[1]],
      line37[[1]], line38[[1]], line39[[1]], line40[[1]]}], .01],
    TableHeadings → {{"No errors", "1.05*v1", "1.05*v2", "1.05*v3", "1.05*v4", "1.05*v5",
      "1.05*v6", "1.05*v7", "1.05*v8", "1.05*v9", "1.05*v10"}, {"-log(kfet)", "pK1eab",
      "pK1ea", "pK1e", "-log(kcHEA)", "-log(kcHEAB)", "pK2e", "pK2ea", "pK2eab"}}]
```

	-log(kfet)	pK1eab	pK1ea	pK1e	-log(kcHEA)	-log(kcHEAB)	pK2e	pK2ea	pK2eab
No errors	-0.3	8.	7.5	7.	-0.7	-1.3	6.	6.5	6.
1.05*v1	-0.29	8.07	7.5	6.98	-0.69	-1.29	6.02	6.5	5.99
1.05*v2	-0.3	8.	7.5	7.08	-0.74	-1.3	5.92	6.5	6.
1.05*v3	-0.3	7.99	7.55	7.	-0.69	-1.31	6.	6.49	6.
1.05*v4	-0.34	7.93	7.5	7.07	-0.76	-1.34	5.86	6.5	6.07
1.05*v5	-0.3	8.	7.5	6.84	-0.55	-1.3	6.26	6.5	6.
1.05*v6	-0.3	8.01	7.44	7.	-0.75	-1.25	6.	6.56	5.99
1.05*v7	-0.29	8.01	7.5	6.99	-0.69	-1.29	6.11	6.5	5.93
1.05*v8	-0.3	8.	7.5	7.02	-0.72	-1.3	5.78	6.5	6.
1.05*v9	-0.3	8.	7.51	7.	-0.69	-1.31	6.	6.46	6.01

This table is the same as Table 3.4 for $n = 0$. When a hydrogen ion is consumed in the rate-determining reaction there is no change in the kinetic parameters, but there is a $[H^+]$ in the rate equation.

{h,a,b} was taken to be {10^-8,200,200}, {10^-8,1,200}, {10^-8,200,1}, {10^-7,200,200}, {10^-7,1,200}, {10^-7,200,1}, {10^-6,200,200}, {10^-6,1,200}, {10^-6,200,1}.

3.5 Effects of pH on the kinetics of ordered A + B → products when *n* hydrogen ions are consumed in the rate-determining reaction

■ **3.5.1 Mechanism and rate equation for ordered A + B → products when the enzymatic site and the enzyme-substrate complexes each have 2 p*K*s and *n* hydrogen ions are consumed**

When *n* hydrogen ions are consumed, the last reaction in the mechanism in Section 3.2 is replaced with

$$
\begin{array}{l}
\text{EAB}^- \\
\quad \| \quad k_f \\
n\text{H}^+ + \text{HEAB} \;\rightarrow\; \text{products} \qquad v = k_f[\text{H}^+]^n[\text{HEAB}] = k_f[\text{H}^+]^n[\text{E}]_t * [\text{HEAB}]/[\text{E}]_t \\
\quad \| \\
\text{H}_2\,\text{EAB}^+
\end{array}
$$

Strictly speaking, the number of hydrogen ions consumed should be represented by |n| as discussed in Section 3.4 and the Appendix in this chapter. This is needed to obtain a thermodynamically correct expression for the apparent equilibrium constant using the Haldane relation, but in deriving rate equations in *Mathematica*, it is more convenient to use *n* for the number of hydrogen ions consumed.

Review of rate equations for A + B → products when E, EA, and EAB each have two p*K*s

Section 3.3.4 gives the following rate equation when no hydrogen ions are consumed:

simpvelordAB

$$
(a\,b\,h\,\text{kfet}\,\text{kH2E}\,\text{kH2EA}\,\text{kH2EAB}) \big/ \big(\text{kcHEA}\,\text{kcHEAB}\,\text{kH2EA}\,\text{kH2EAB}\,(h^2 + h\,\text{kH2E} + \text{kH2E}\,\text{kHE}) +
$$
$$
a\,\text{kH2E}\,\big(\text{kcHEAB}\,\text{kH2EAB}\,(h^2 + h\,\text{kH2EA} + \text{kH2EA}\,\text{kHEA}) + b\,\text{kH2EA}\,(h^2 + h\,\text{kH2EAB} + \text{kH2EAB}\,\text{kHEAB})\big)\big)
$$

Section 3.4.1 gives the following rate equation when one hydrogen ion is consumed:

simpvelordABh = h * 10^7 * simpvelordAB

$$\left(10\,000\,000\;a\;b\;h^2\;kfet\;kH2E\;kH2EA\;kH2EAB\right) \big/ \left(kcHEA\;kcHEAB\;kH2EA\;kH2EAB\;\left(h^2 + h\;kH2E + kH2E\;kHE\right) + \right.$$
$$\left. a\;kH2E\;\left(kcHEAB\;kH2EAB\;\left(h^2 + h\;kH2EA + kH2EA\;kHEA\right) + b\;kH2EA\;\left(h^2 + h\;kH2EAB + kH2EAB\;kHEAB\right)\right)\right)$$

Besides multiplying simpvelordAB by h, it is multiplied by 10^7 to avoid very high velocities. Since the contributions of k_f and et are not consdered separately, the 10^7 is arbitrary.

When two hydrogen ions are consumed in the rate-determining reaction, the following rate equation is used:

simpvelordABh2 = h^2 * 10^14 * simpvelordAB

$$\left(100\,000\,000\,000\,000\;a\;b\;h^3\;kfet\;kH2E\;kH2EA\;kH2EAB\right) \big/$$
$$\left(kcHEA\;kcHEAB\;kH2EA\;kH2EAB\;\left(h^2 + h\;kH2E + kH2E\;kHE\right) + \right.$$
$$\left. a\;kH2E\;\left(kcHEAB\;kH2EAB\;\left(h^2 + h\;kH2EA + kH2EA\;kHEA\right) + b\;kH2EA\;\left(h^2 + h\;kH2EAB + kH2EAB\;kHEAB\right)\right)\right)$$

These equations can be generalized to the consumption of *n* hydrogen ions by use of

simpvelordABhn = 10^(n * 7) * h^n * simpvelordAB

$$\left(10^{7\,n}\;a\;b\;h^{1+n}\;kfet\;kH2E\;kH2EA\;kH2EAB\right) \big/ \left(kcHEA\;kcHEAB\;kH2EA\;kH2EAB\;\left(h^2 + h\;kH2E + kH2E\;kHE\right) + \right.$$
$$\left. a\;kH2E\;\left(kcHEAB\;kH2EAB\;\left(h^2 + h\;kH2EA + kH2EA\;kHEA\right) + b\;kH2EA\;\left(h^2 + h\;kH2EAB + kH2EAB\;kHEAB\right)\right)\right)$$

This can be tested by choosing $n = 0, 1, 2,$ and 3.

simpvelordABhn /. n → 0

$$\left(a\;b\;h\;kfet\;kH2E\;kH2EA\;kH2EAB\right) \big/ \left(kcHEA\;kcHEAB\;kH2EA\;kH2EAB\;\left(h^2 + h\;kH2E + kH2E\;kHE\right) + \right.$$
$$\left. a\;kH2E\;\left(kcHEAB\;kH2EAB\;\left(h^2 + h\;kH2EA + kH2EA\;kHEA\right) + b\;kH2EA\;\left(h^2 + h\;kH2EAB + kH2EAB\;kHEAB\right)\right)\right)$$

simpvelordABhn /. n → 1

$$\left(10\,000\,000\;a\;b\;h^2\;kfet\;kH2E\;kH2EA\;kH2EAB\right) \big/ \left(kcHEA\;kcHEAB\;kH2EA\;kH2EAB\;\left(h^2 + h\;kH2E + kH2E\;kHE\right) + \right.$$
$$\left. a\;kH2E\;\left(kcHEAB\;kH2EAB\;\left(h^2 + h\;kH2EA + kH2EA\;kHEA\right) + b\;kH2EA\;\left(h^2 + h\;kH2EAB + kH2EAB\;kHEAB\right)\right)\right)$$

simpvelordABhn /. n → 2

$$\left(100\,000\,000\,000\,000\;a\;b\;h^3\;kfet\;kH2E\;kH2EA\;kH2EAB\right) \big/$$
$$\left(kcHEA\;kcHEAB\;kH2EA\;kH2EAB\;\left(h^2 + h\;kH2E + kH2E\;kHE\right) + \right.$$
$$\left. a\;kH2E\;\left(kcHEAB\;kH2EAB\;\left(h^2 + h\;kH2EA + kH2EA\;kHEA\right) + b\;kH2EA\;\left(h^2 + h\;kH2EAB + kH2EAB\;kHEAB\right)\right)\right)$$

simpvelordABhn /. n → 3

$$\left(1\,000\,000\,000\,000\,000\,000\,000\;a\;b\;h^4\;kfet\;kH2E\;kH2EA\;kH2EAB\right) \big/$$
$$\left(kcHEA\;kcHEAB\;kH2EA\;kH2EAB\;\left(h^2 + h\;kH2E + kH2E\;kHE\right) + \right.$$
$$\left. a\;kH2E\;\left(kcHEAB\;kH2EAB\;\left(h^2 + h\;kH2EA + kH2EA\;kHEA\right) + b\;kH2EA\;\left(h^2 + h\;kH2EAB + kH2EAB\;kHEAB\right)\right)\right)$$

■ 3.5.2 Determination of the number of hydrogen ions consumed from V_{fexp}

The consumption of hydrogen ions in the rate-determining reaction does not compete with the other reactions in the mechanism. The pKs determine the distribution of the various species, but the consumption of hydrogen ions determines the rate that EAB is converted to products. The effect of pH on V_{fexp} is dramatic. Thus the determination of velocities at very high substrate concentrations provides an opportunity to determine n. As discussed in Section 3.4.1, the pH-dependent limiting velocity is given by $V_{fexp} = 10^{n(pH-7)}$kfet×eab/et, where $n = 0, -1, -2, ...$ At high concentrations of substrates, the pH-dependent limiting velocity approaches

```
kfet * 10^(n (pH - 7)) / (1 + 10^(pH - pK1eab) + 10^(-pH + pK2eab))
```

$$\frac{10^{n\,(-7+pH)}\,\text{kfet}}{1 + 10^{pH-pK1eab} + 10^{-pH+pK2eab}}$$

If kfet = 1, pK1eab = 8, pK2eab = 6, and $n = 0, -1, -2, -3, -4$.

```
kfet * 10^(n (pH - 7)) / (1 + 10^(pH - pK1eab) + 10^(-pH + pK2eab)) /. kfet → 1 /. pK1eab → 8 /.
    pK2eab → 6 /. n → {0, -1, -2, -3, -4}
```

$$\left\{ \frac{1}{1 + 10^{6-pH} + 10^{-8+pH}}, \frac{10^{7-pH}}{1 + 10^{6-pH} + 10^{-8+pH}}, \frac{10^{-2\,(-7+pH)}}{1 + 10^{6-pH} + 10^{-8+pH}}, \frac{10^{-3\,(-7+pH)}}{1 + 10^{6-pH} + 10^{-8+pH}}, \frac{10^{-4\,(-7+pH)}}{1 + 10^{6-pH} + 10^{-8+pH}} \right\}$$

```
Plot[{ 1/(1 + 10^(6-pH) + 10^(-8+pH)), 10^(7-pH)/(1 + 10^(6-pH) + 10^(-8+pH)),
    10^(-2 (-7+pH))/(1 + 10^(6-pH) + 10^(-8+pH)), 10^(-3 (-7+pH))/(1 + 10^(6-pH) + 10^(-8+pH)), 10^(-4 (-7+pH))/(1 + 10^(6-pH) + 10^(-8+pH))}, {pH, 5, 9},

    PlotRange → {0, 10}, PlotStyle → {Black}, AxesLabel -> {"pH", "Vfexp"}]
```

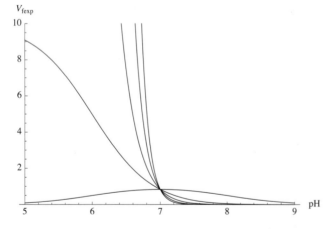

Fig. 3.10 V_{fexp} when $n = 0, -1, -2, -3$, and -4. The kinetic parameters are kfet = 1, pK1eab = 8, and pK2eab = 6. These plots can be determined experimentally by use of very high concentrations of A and B without the need to determine the Michaelis constants. Since these pH effects are so large, it is more useful to plot logV_{fexp}.

```
Plot[Log[10,
```

$$\left\{ \frac{1}{1 + 10^{6-pH} + 10^{-8+pH}} , \frac{10^{7-pH}}{1 + 10^{6-pH} + 10^{-8+pH}} , \frac{10^{-2\,(-7+pH)}}{1 + 10^{6-pH} + 10^{-8+pH}} , \frac{10^{-3\,(-7+pH)}}{1 + 10^{6-pH} + 10^{-8+pH}} , \frac{10^{-4\,(-7+pH)}}{1 + 10^{6-pH} + 10^{-8+pH}} \right\} ,$$

```
{pH, 5, 9}, PlotRange → {-1, 5}, PlotStyle → {Black}, AxesLabel -> {"pH", "Log[10,V_fexp]"}]
```

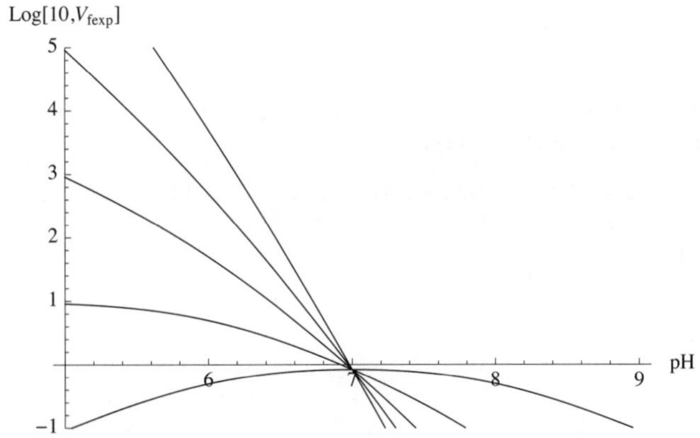

Fig. 3.11 $\text{Log}\big[10, V_{\text{fexp}}\big]$ when $n = 0, -1, -2, -3,$ and -4.

For some reductase reactions [29] the change in binding of hydrogen ions $\Delta_r N_H$ is as large as 8; this indicates that n may be this large.

3.6 Appendix on effects of temperature on the kinetic parameters for ordered A + B → products

The effects of temperature on the kinetic parameters for A → products have been discussed in Appendix 3 of Chapter 2. That discussion is extended here to ordered A + B → products using the mechanism in Section 3.1. The rapid-equilibrium rate equation at specified temperature, pH and ionic strength is given by

$$\text{vordAB} = \frac{a\,b\,\text{kfet}}{a\,b + a\,\text{kB} + \text{kB}\,\text{kIA}} ;$$

To make test calculations, assumed values of the kinetic parameters at 298.15 K and 273.15 K are used. They are given in the following table.

Table 3.8 Assumed experimental parameters at two temperatures for ordered A + B → products

```
TableForm[{{1, .005, .020}, {.3, .001, .010}},
   TableHeadings → {{"298.15 K", "273.15 K"}, {"kfet", "kIA", "kB"}}]
```

	kfet	kIA	kB
298.15 K	1	0.005	0.02
273.15 K	0.3	0.001	0.01

The Michaelis constants are expressed as molar concentrations.

At 298.15 K, the rate equation is given by

```
vordAB /. kfet → 1 /. kIA → .005 /. kB → .02
```

$$\frac{a\,b}{0.0001 + 0.02\,a + a\,b}$$

At 273.15 K, the rate equation is given by

```
vordAB /. kfet → .3 /. kIA → .001 /. kB → .010
```

$$\frac{0.3\, a\, b}{0.00001 + 0.01\, a + a\, b}$$

The velocities at these two temperatures can be plotted as functions of [A] and [B].

```
plotA298 = Plot3D[Evaluate[vordAB] /. kfet → 1 /. kIA → .005 /. kB → .02, {a, .00001, .02},
    {b, .00001, .05}, PlotRange → {0, 1}, ViewPoint → {-2, -2, 1}, AxesLabel → {"[A]", "[B]", ""},
    Lighting -> {{"Ambient", GrayLevel[1]}}, PlotLabel → "v298"];
```

```
plotA273 = Plot3D[Evaluate[vordAB] /. kfet → .3 /. kIA → .001 /. kB → .01,
    {a, .00001, .005}, {b, .00001, .05}, PlotRange → {0, .3},
    ViewPoint → {-2, -2, 1}, AxesLabel → {"[A]", "[B]", ""},
    Lighting -> {{"Ambient", GrayLevel[1]}}, PlotLabel → "v273"];
```

```
GraphicsArray[{{plotA298, plotA273}}]
```

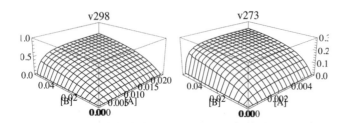

Fig. 3.12 (a) Velocity of ordered A + B → products at 298.15 K. (b) Velocity of ordered A + B → products at 273.15 K.

The determination of the activation energy is exactly the same as in Appendix 3 in Chapter 2, and so it is not repeated here. The following two sections deal with the two Michaelis constants.

■ **3.6.1 Effects of temperature on EA = E + A where K_{IA} = [E][A]/[EA]**

The following equations were used in Appendix 3 of Chapter 2:

$$\Delta_r G'^\circ = -RT\ln K' \tag{A3.3}$$

$$\Delta_r H'^\circ = \ln(K_2'/K_1')RT_1T_2/(T_2 - T_1) \tag{A3.5}$$

$$\Delta_r S'^\circ = (\Delta_r H'^\circ - \Delta_r G'^\circ)/T \tag{A3.7}$$

Calculation of the standard transformed Gibbs energy of reaction for EA = E + A

Using equation A3.3, the standard transformed Gibbs energy for this reaction at 298.15 K in J mol^{-1} is obtained from

```
-8.3145 * 298.15 * Log[.005]
```

```
13 134.4
```

The standard transformed Gibbs energy for this reaction at 273.15 K in J mol^{-1} is obtained from

```
-8.3145 * 273.15 * Log[.001]
```

15 688.2

Calculation of the standard transformed enthalpy of reaction for EA = E + A

Using equation A3.5, the standard transformed enthalpy of reaction in J mol^{-1} is given by

```
Log[.001 / .005] * 8.3145 * 298.15 * 273.15 / (-25)
```

43 592.

It is assumed that the standard transformed enthalpy of reaction is independent of temperature. Since this is an endothermic reaction, the dissociation increases when the temperature is raised.

Calculation of the standard transformed entropy of reaction for EA = E + A

Equation A3.7 is used to calculate the standard transformed entropy of reaction.

```
(43 592 - 13 134) / 298.15
```

102.157

This value is in J K^{-1} mol^{-1}, and it is independent of temperature. The fact that the standard transformed entropy of reaction is positive shows that E + A is more random than EA.

Table 3.9 Standard transformed thermodynamic properties for EA = E + A calculated using the assumed kinetic parameters at two temperatures

```
TableForm[{{13 134, 43 592, 102.16, .005}, {15 688, 43 592, 102.16, .001}},
  TableHeadings → {{"298.15 K", "273.15 K"}, {G, H, S, K'}}]
```

	G	H	S	K'
298.15 K	13 134	43 592	102.16	0.005
273.15 K	15 688	43 592	102.16	0.001

G is $\Delta_r G'^\circ$, H is $\Delta_r H'^\circ$, and S is $\Delta_r S'^\circ$.

■ 3.6.2 Effects of temperature on EAB = EA + B where K_B = [EA][B]/[EAB]

Calculation of the standard transformed Gibbs energy of reaction for EAB = EA + B

Using equation A3.3, the standard transformed Gibbs energy for this reaction at 298.15 K in J mol^{-1} is obtained from

```
-8.3145 * 298.15 * Log[.02]
```

9697.78

The standard transformed Gibbs energy for this reaction at 273.15 K in J mol^{-1} is obtained from

```
-8.3145 * 273.15 * Log[.01]
```

10 458.8

Calculation of the standard transformed enthalpy of reaction for EAB = EA + B

Using equation A3.5, the standard transformed enthalpy of reaction in J mol^{-1} is given by

Log[.01 / .02] * 8.3145 * 298.15 * 273.15 / (-25)

18 774.

This standard transformed enthalpy of reaction is independent of temperature. Since EAB = EA + B is an endothermic reaction, the dissociation increases when the temperature is raised.

Calculation of the standard transformed entropy of reaction for EAB = EA + B

Using equation A3.7, yields

(18 774 - 9697) / 298.15

30.4444

This is the standard transformed entropy of reaction in J K^{-1} mol^{-1}. It is independent of temperature. The fact that the standard transformed entropy of reaction is positive shows that EA + B is more random than EAB.

Table 3.10 Standard transformed thermodynamic properties for EAB = EA + B calculated from the assumed kinetic properties at two temperatures.

TableForm[{{9698, 18 774, 30.44, 0.02}, {10 458, 18 774, 30.44, 0.01}},
** TableHeadings → {{"298.15 K", "273.15 K"}, {G, H, S, K'}}]**

	G	H	S	K'
298.15 K	9698	18 774	30.44	0.02
273.15 K	10 458	18 774	30.44	0.01

where G is $\Delta_r G'^{\circ}$, H is $\Delta_r H'^{\circ}$, and S is $\Delta_r S'^{\circ}$.

Chapter 2 discusses the relative importance of $\Delta_r H'^{\circ}$ and $\Delta_r S'^{\circ}$ in determining K' and the use of the terms weak binding and strong binding, so that discussion is not repeated here.

These calculations are based on the determination of three kinetic parameters at two temperatures. This can be done with six velocity determinations. The two limiting velocities make it possible to calculate the activation energy for kfet. The values of the two Michaelis constants make it possible to calculate the standard transformed Gibbs energies for the two reactions in the mechanism. The temperature dependencies of the Michaelis constants make it possible to calculate the standard transformed enthalpies of the two reactions. Knowledge of the standard transformed Gibbs energies of reaction and standard transformed enthalpies of reaction make it possible to calculate the standard transformed entropies of reaction. It is of interest to see the relative contributions of enthalpy and entropy to Michaelis constants.

3.7 Discussion

This chapter provides further examples of the use of Solve to derive rapid-equilibrium rate equations. The rate equation vordB can be used to calculate velocities when arbitrary values of the kinetic parameters are used and substrate concentrations are specified. These velocities can be treated like experimental data and can be used to estimate kinetic parameters using the minimum number of velocity measurements. Figure 3.6 shows that the best pairs of velocities to use to obtain the three kinetic parameters are {high [A],high[B]}, {low [A],high [B]}, and {low [A],low [B]}. The actual concentrations used in the laboratory are very dependent on the analytical instrumentation.

The effects of pH are discussed first when the enzymatic site and the enzyme-substrate complexes each have two pKs. Two ways to derive the pH-dependent rate equation are demonstrated: (1) Section 3.3.1 uses the factor method. (2) Section 3.3.2 uses Solve. The use of Solve has the advantage that the nine kinetic parameters can be estimated from nine velocity measurements. Section 3.3 demonstrated that when a reaction is studied for the first time, there is an advantage in using the most general mechanism for that type of reaction [32-35].

The program calckinparsordABsimp calculates nine kinetic parameters for ordered A + B → products with 6 pKs from 9 experimental velocities. It has been shown that when this program is applied to data for a mechanism with 5 pKs, the output clearly shows which pK is missing and yields correct pKs for the other 5 acid dissociations. The program calckinparsordAB-simph estimates 9 kinetic parameters for ordered A + B → products with 6 pKs when a single hydrogen ion is consumed in the rate-determining reaction. Both of these programs were tested by applying 5% errors to 9 velocities, one at a time, to show the effects on the estimation of kinetic parameters. These programs are easy to write in the sense that they each contain 9 versions of a rate equation. This provides an organized way to put sets of velocities, hydrogen ion concentrations, and substrate equations into Solve in *Mathematica*. It is important to use more than one set of substrate concentrations to be sure that the calculated values of kinetic parameters are not changing with the choice of substrate concentrations.

An integer number of hydrogen ions can be consumed in the rate-determining reaction, and this produces effects on the velocity of an entirely different type than pKs. When n hydrogen ions are consumed in the rate-determining reaction this adds a factor $[H^+]^n$ in the rate equation. The effect of this term is so large that n can be determined by plotting the velocity at high concentrations of substrates as a function of pH. It is important to distinguish between the change in binding of hydrogen ions $\Delta_r N_H$ in the reaction that is catalyzed (Section 1.7) and n. $\Delta_r N_H$ can have any value, but n is an integer [27]. $\Delta_r N_H$ can be obtained from thermodynamic studies, but n can only be determined from kinetic measurements,

The calculation of 5% errors in the velocities and putting these results in tables takes a lot of steps, but these operations can be done by copying and pasteing with a few critical changes. These calculations emphasize the importance of using velocity measurements at the lowest possible concentrations. Replicate measurements can be made to improve the accuracy of velocity measurements.

These calculations can be extended in various ways.
1. Different sets of parameters can be used.
2. Different substrate concentrations can be used.
3. Different numbers of hydrogen ions can be consumed in the rate-determining reaction.
4. Different rapid-equilibrium mechanisms can be used.

Eight rate equations for ordered A + B → products have been used:

velordAB	Rate equation derived with Solve (Section 3.1)
simpvelordAB	Simplified rate equation derived with Solve (Section 3.1)
vtestordAB2	Rate equation with arbitrary kinetic parameters inserted (Section 3.1)
velordABea	Rate equation with HA = $EA^- + H^+$ missing (Section 3.4)
simpvelordABea	Simplified version of preceding rate equation (Section 3.4)
vtestordAB2ea	simpvelordABea with arbitrary kinetic parameters inserted (Section 3.4)
simpvelordABh	Simplified rate equation when a single hydrogen ion is consumed (Section 3.5)
	when *n* hydrogen ions are consumed
simpvtestordABh	Rate equation with arbitrary kinetic parameters (Section 3.5)

The following programs have been written to calculate kinetic parameters from the mnimum number of velocities:

calckinparsordAB (Section 3.2.1)

calckinparsordABsimp (Section 3.3.5)

calckinpaarsordABsimph (Section 3.4.1)

Chapter 4 Random A + B → Products

4.1 Two rapid-equilibrium rate equations for random A + B → products

- 4.1.1 Comparison of two mechanisms and derivations of two rate equations

- 4.1.2 Use of the two mechanisms to calculate velocities

4.2 Use of Solve to estimate the pH-dependent kinetic parameters from the minimum number of velocity measurements for random A + B → products

- 4.2.1 Estimation of kinetic parameters using {[A],[B]} = {100,100}, {1,100}, {100,1}, and {5,5}

- 4.2.2 Estimation of kinetic parameters using {[A],[B]} = {150,150}, {1,150}, {150,1}, and {1,1}

- 4.2.3 Estimation of kinetic parameters using {[A],[B]} = {70,70}, {5,70}, {70,5}, and {7,7}

4.3 Application of calckinparsrandAB to velocities for the ordered mechanism

---

4.4 Effects of pH on the kinetics of random A + B → products when no hydrogen ions are consumed in the rate-determining reaction

- 4.4.1 Mechanism and rate equation for random A + B → products when the enzymatic site and enzyme-substrate complexes each have two pKs and the factor method is used to derive the rate equation

- 4.4.2 Estimation of pH-independent kinetic parameters using the rate equation derived using the factor method

- 4.4.3 Mechanism and rate equation for random A + B → products when the enzymatic site and enzyme-substrate complexes each have two pKs and Solve is used to derive the rate equation

- 4.4.4 Estimation of pH-independent kinetic parameters using the rate equation derived by use of Solve

---

4.5 Effects of pH on the kinetics of random A + B → products when hydrogen ions are consumed in the rate-determining reaction

- 4.5.1 When one hydrogen ion is consumed

- 4.5.2 When n hydrogen ions are consumed

4.6 Discussion

4.1 Two rapid-equilibrium rate equations for random A + B → products

■ **4.1.1 Comparison of two mechanisms and derivations of two rate equations**

The preceding chapter has emphasized that rapid-equilibrium rate equations can be derived by use of Solve, but this can only be done with an independent set of reactions. Each random mechanism must include A, B, E, EA, EB, and EAB, and so N' = 6. There are 3 components (A, B, and E), and since $R' = N' - C'$, possible mechanisms must each involve 6 - 3 = 3 independent equilibrium reactions [23].

There are two equivalent rapid-equilibrium mechanisms for random A + B → products. These mechanisms involve E, EA, EB, and EAB that are related by the following cycle:

$$
\begin{array}{ccc}
& K_{IA} & \\
E + A & = & EA \\
+ & & + \\
B & & B \\
\| \; K_{IB} & & \| \; K_B \\
EB + A & = & EAB \\
& K_A &
\end{array}
\tag{1}
$$

In this thermodynamic cycle, K_{IA} = [E][A]/[EA], K_{IB} = [E][B]/[EB], K_A = [EB][A]/[EAB], and K_B = [EA][B]/[EAB]. These are pH-dependent kinetic parameters. In Section 4.4, the estimation of pH-independent kinetic parameters will be discussed. This thermodynamic cycle shows that $K_{IA}K_B = K_{IB}K_A$ or, in *Mathematica*, kIA kB = kIB kA. This relation has been known for a long time.

Mechanism I is

E + A = EA	kIA = e*a/ea	K_{IA} = [E][A]/[EA]	(2)
E + B = EB	kIB = e*b/eb	K_{IB} = [E][B]/[EB]	(3)
EA + B = EAB	kB = ea*b/eab	K_B = [EA][B]/[EAB]	(4)
EAB → products	v = kf*eab = kf $*$ et $*$ (eab/et) = kfet*(eab/et) and $V_{fexp} = k_f[E]_t$		(5)

K_A is not involved, but note $K_A = K_{IA}K_B/K_{IB}$ at any pH.

Mechanism II is

E + A = EA	kIA = e*a/ea	K_{IA} = [E][A]/[EA]	(6)
E + B = EB	kIB = e*b/eb	K_{IB} = [E][B]/[EB]	(7)
EB + A = EAB	kA = eb*a/eab	K_A = [EB][A]/[EAB]	(8)
EAB → products	v = kf*eab	$V_{fexp} = k_f[E]_t$	(9)

K_B is not involved, but note $K_B = K_A K_{IB}/K_{IA}$. It is possible for one investigator to use mechanism I and report values for the kinetic parameters K_{IA}, K_{IB}, K_B and V_{fexp}. Another investigator could use mechanism II and report values for the kinetic parameters for K_{IA}, K_{IB}, K_A and V_{fexp}. This chapter makes the point that both investigators should report values for K_{IA}, K_{IB}, K_A, K_A, and V_{fexp}. This is true at each pH, and so the pH dependencies of the five kinetic parameters can be determined using either mechanism.

Solve[vars,eqs,elims] in *Mathematica* can be used to derive rapid-equilibrium rate equations. To calculate the equilibrium concentration eab, there are three equilibrium equations and one conservation equation for each mechanism. Note that e, ea, and eb have to be eliminated in using Solve.

The rapid-equilibrium rate equation for mechanism I is derived as follows:

$$\text{Solve}[\{kIA == e * a / ea,\ kIB == e * b / eb,\ kB == ea * b / eab,\ et == e + ea + eb + eab\},\ \{eab\},\ \{e,\ ea,\ eb\}]$$

$$\left\{\left\{eab \to \frac{a\ b\ et\ kIB}{b\ kB\ kIA + a\ b\ kIB + a\ kB\ kIB + kB\ kIA\ kIB}\right\}\right\}$$

Since v = kf*eab, it is necessary to replace et with vfexp = kf*et = kfet to obtain the rate equation. At a specified pH, vfexp = kfet.

$$\text{vrandAB1} = \frac{a\ b\ vfexp\ kIB}{b\ kB\ kIA + a\ b\ kIB + a\ kB\ kIB + kB\ kIA\ kIB}$$

$$\frac{a\ b\ kIB\ vfexp}{b\ kB\ kIA + a\ b\ kIB + a\ kB\ kIB + kB\ kIA\ kIB}$$

The expression for vrandAB1 can be converted to the usual form by dividing the numerator and denominator by a b kIB.

Divide the denominator by a*b*kIB, one term at a time.

$$\{b\ kB\ kIA,\ a\ b\ kIB,\ a\ kB\ kIB,\ kB\ kIA\ kIB\}\ /\ (a * b * kIB)$$

$$\left\{\frac{kB\ kIA}{a\ kIB},\ 1,\ \frac{kB}{b},\ \frac{kB\ kIA}{a\ b}\right\}$$

This yields the following rate equation for mechanism I:

$$\text{vfexp}\Big/\left(1 + \frac{kB\ kIA}{a\ kIB} + \frac{kB}{b} + \frac{kB\ kIA}{a\ b}\right)$$

$$\frac{vfexp}{1 + \frac{kB}{b} + \frac{kB\ kIA}{a\ b} + \frac{kB\ kIA}{a\ kIB}}$$

The rapid-equilibrium rate equation for mechanism II is derived as follows:

$$\text{Solve}[\{kIA == e * a / ea,\ kIB == e * b / eb,\ kA == eb * a / eab,\ et == e + ea + eb + eab\},\ \{eab\},\ \{e,\ ea,\ eb\}]$$

$$\left\{\left\{eab \to \frac{a\ b\ et\ kIA}{a\ b\ kIA + b\ kA\ kIA + a\ kA\ kIB + kA\ kIA\ kIB}\right\}\right\}$$

Since v = kf*eab, it is necessary to replace et with vfexp to obtain the rate equation.

$$\text{vrandAB2} = \frac{a\ b\ vfexp\ kIA}{a\ b\ kIA + b\ kA\ kIA + a\ kA\ kIB + kA\ kIA\ kIB}$$

$$\frac{a\ b\ kIA\ vfexp}{a\ b\ kIA + b\ kA\ kIA + a\ kA\ kIB + kA\ kIA\ kIB}$$

The expression for vrandAB2 can be converted to the usual form as follows:

Divide the denominator by a*b*kIA one term at a time.

{a b kIA, b kA kIA, a kA kIB, kA kIA kIB} / (a * b * kIA)

$$\left\{ 1, \frac{kA}{a}, \frac{kA\ kIB}{b\ kIA}, \frac{kA\ kIB}{a\ b} \right\}$$

This yields the following rate equation for mechanism II:

$$\mathbf{vfexp} \Big/ \left(1 + \frac{\mathbf{kA}}{\mathbf{a}} + \frac{\mathbf{kA\ kIB}}{\mathbf{b\ kIA}} + \frac{\mathbf{kA\ kIB}}{\mathbf{a\ b}} \right)$$

$$\frac{vfexp}{1 + \frac{kA}{a} + \frac{kA\ kIB}{a\ b} + \frac{kA\ kIB}{b\ kIA}}$$

These two rate equations can be interconverted using kIA kB = kIB kA.

■ **4.1.2 Use of the two mechanisms to calculate velocities**

In order to make numerical calculations, the values of the four kinetic parameters in mechanism I are arbitraily taken to be vfexp = 1, kIA = 5, kIB = 40, and kB = 20. kA is given by

kA = kIA kB/kIB = 5*20/40 = 2.5   (10)

The two rate equations can be tested by arbitrarily choosing [A] = 10 and [B] = 20.

vrandAB1 /. vfexp → 1 /. kIA → 5 /. kIB → 40 /. kB → 20 /. a → 10 /. b → 20 // N

0.363636

vrandAB2 /. vfexp → 1 /. kIA → 5 /. kIB → 40 /. kA → 2.5 /. a → 10 /. b → 20 // N

0.363636

When mechanism I is used to interpret velocities for random A + B → products, the values of vfexp, kIA, kIB, and kB are obtained directly, and they can be used to calculate kA = kIA kB/kIB. This calculation should be made, and the value of kA should be reported because it is the equilibrium constant for EB + A = EAB. When more complicated enzyme-catalyzed reactions are studied, more equilibrium constants can be obtained in this way. For example, when A + B + C → products has a completely random mechanism, seven equilibrium constants can be obtained using the rapid-equilibrium rate equation, but five more equilibrium constants can be obtained using equations like kA kIB = kIA kB (see Chapter 6 and [34]).

There are two specialized forms of the rate equations that are useful; one is obtained by specifying the kinetic parameters, and the other is obtained by specifying substrate concentrations. When the values of the kinetic parameters are specified, the velocity is a function of the substrate concentrations. To demonstrate this using mechanism I, the following values are arbitrarily assigned to the four kinetic parameters: vfexp = 1, kIA = 5, kB = 20, and kIB = 40. The rate equation is given by

vrandAB1 /. vfexp → 1 /. kIA → 5 /. kB → 20 /. kIB → 40

$$\frac{40\ a\ b}{4000 + 800\ a + 100\ b + 40\ a\ b}$$

This form of the rate equation will be used in this section to make plots of velocities for various [A] and [B].

The other form of the rate equation is obtained by substitution of specified values of a and b.

```
vrandAB1 /. a → 100 /. b → 80
```

$$\frac{8000 \text{ kIB vfexp}}{80 \text{ kB kIA} + 8000 \text{ kIB} + 100 \text{ kB kIB} + \text{kB kIA kIB}}$$

Since vrandAB1/. vfexp → 1 /. kIA → 5 /. kB → 20 /. kIB → 40 is a function of [A] and [B], the velocity can be plotted versus [A] and [B]. More information about the shape of this surface can be obtained by taking partial derivatives with respect to [A] and [B], and the mixed second derivative with respect to [A] and [B]. Plots of these derivatives are useful in choosing pairs of substrate concentrations {[A],[B]} to use in estimating the kinetic parameters with the minimum number of velocity measurements.

```
plotI = Plot3D[Evaluate[vrandAB1 /. vfexp → 1 /. kIA → 5 /. kB → 20 /. kIB → 40],
    {a, .0001, 15}, {b, .0001, 60}, PlotRange → {0, 1}, ViewPoint → {-2, -2, 1},
    AxesLabel → {"[A]", "[B]", ""}, Lighting -> {{"Ambient", GrayLevel[1]}}, PlotLabel -> "v"];

plotJ = Plot3D[Evaluate[10 * D[vrandAB1 /. vfexp → 1 /. kIA → 5 /. kB → 20 /. kIB → 40, a]],
    {a, .0001, 15}, {b, .0001, 60}, PlotRange → {0, 1.5},
    ViewPoint → {-2, -2, 1}, AxesLabel → {"[A]", "[B]", ""},
    Lighting -> {{"Ambient", GrayLevel[1]}}, PlotLabel -> "10dv/d[A]"];

plotK = Plot3D[Evaluate[30 * D[vrandAB1 /. vfexp → 1 /. kIA → 5 /. kB → 20 /. kIB → 40, b]],
    {a, .0001, 15}, {b, .0001, 60}, PlotRange → {0, 1.3},
    ViewPoint → {-2, -2, 1}, AxesLabel → {"[A]", "[B]", ""},
    Lighting -> {{"Ambient", GrayLevel[1]}}, PlotLabel -> "30dv/d[B]"];

plotL = Plot3D[Evaluate[750 * D[vrandAB1 /. vfexp → 1 /. kIA → 5 /. kB → 20 /. kIB → 40, a, b]],
    {a, .0001, 15}, {b, .0001, 60}, PlotRange → {-1, 7},
    ViewPoint → {-2, -2, 1}, AxesLabel → {"[A]", "[B]", ""},
    Lighting -> {{"Ambient", GrayLevel[1]}}, PlotLabel -> "750d(dv/d[A]/db"];
```

```
fig22rand = GraphicsArray[{{plotI, plotJ}, {plotK, plotL}}]
```

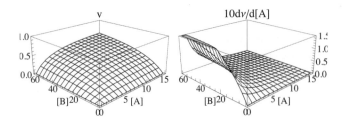

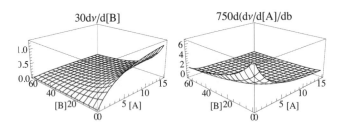

Fig. 4.1 (a) v versus [A] and [B] for random mechanism I with V_{fexp} = 1.00, K_{IA} = 5.00, K_B = 20.00, and K_{IB} = 40.00. (b) 10dv/d[A]. (c) 30dv/d[B]. (d) 750d(dv/d[A]/db.

The plot of 10dv/d[A]shows that the velocity is most sensitive to [A] at high [B] and low [A]. The plot of 10dv/d[B] shows that the velocity is most sensitive to [B] at high [A] and low [B]. The plot of 1000d(dv/d[A]/d[B] shows that the velocity is most sensitive to [A] and [B] together at low [A] and low [B]. d(dv/d[A]/db shows a distinguishing feature of the random mechanism, as can be seen by comparing this figure with Fig. 3.6 in the preceeding chapter.

Since the two rate equations yield the same velocities, these plots apply to both mechanisms.

4.2 Use of Solve to estimate the pH-dependent kinetic parameters from the minimum number of velocity measurements for random A + B → products

■ 4.2.1 Estimation of kinetic parameters using {[A],[B]} = {100,100}, {1,100}, {100,1}, and {5,5}

As Duggleby pointed out in 1979 [9], when a rate equation has N kinetic parameters, it is only necessary to determine N velocities. The application of this method to ordered A + B → products and random A + B → products has been discussed [32]. As a test, arbitrary values of the 4 kinetic parameters in mechanism I are used to calculate velocities at {[A],[B]} equal to {100,100}, {1,100}, {100,1}, and {5,5}. These velocities are then treated as experimental data and a computer program calckinparsrandABI has been written to calculate the 4 kinetic parameters from the 4 velocities. This program yields exact values of the kinetic parameters when there are no experimental errors, but it is necessary to take experimental errors in measuring the velocities into account. This is accomplished by introducing 5% errors in the 4 velocities, one at a time. This process is repeated with {[A],[B]} equal to {150,150}, {1,150}, {150,1}, and {1,1} to show how to obtain more accurate values of the 4 parameters. The process is repeated again with {70,70}, {5,70}, {70,5}, and {7,7} to show that less accurate values of the 4 parameters are obtained when a narrower range of substrate concentrations is used.

This program yields the values of vfexp, kIA, kB, and kIB from four measured velocities at four pairs of substrate concentrations for mechanism I. This program is based on vrandAB1.

vrandAB1

$$\frac{a\ b\ kIB\ vfexp}{b\ kB\ kIA + a\ b\ kIB + a\ kB\ kIB + kB\ kIA\ kIB}$$

```
calckinparsrandAB1[v1_, a1_, b1_, v2_, a2_, b2_, v3_, a3_, b3_, v4_, a4_, b4_] :=
  Module[{}, (*This program calculates vfexp, kIA,
   kIB and kB from four experimental velocities for A + B →
    products at four pairs of substrate concentrations on the assumption that
     the mechanism is random.  The first velocity is at high [A] and high [B],
    the second velocity is at low [A] and high [B], the third velocity is at
     high [A] and low [B], and the fourth velocity is at low [A] and low [B].*)
   Solve[{v1 == a1 * b1 * kIB * vfexp / (b1 * kB * kIA + a1 * b1 * kIB + a1 * kB * kIB + kB * kIA * kIB),
      v2 == a2 * b2 * kIB * vfexp / (b2 * kB * kIA + a2 * b2 * kIB + a2 * kB * kIB + kB * kIA * kIB),
      v3 == a3 * b3 * kIB * vfexp / (b3 * kB * kIA + a3 * b3 * kIB + a3 * kB * kIB + kB * kIA * kIB),
      v4 == a4 * b4 * kIB * vfexp / (b4 * kB * kIA + a4 * b4 * kIB + a4 * kB * kIB + kB * kIA * kIB)},
    {vfexp, kIA, kIB, kB}]]
```

The velocities at specified temperature, pH, and ionic strength are readily calculated at $\{[A],[B]\} = \{100,100\},\{1,100\},\{100,1\}$, and $\{5,5\}$.

vrandAB1 /. kIA → 5 /. kIB → 40 /. kB → 20 /. vfexp → 1 /. a → 100 /. b → 100 // N

0.809717

vrandAB1 /. kIA → 5 /. kIB → 40 /. kB → 20 /. vfexp → 1 /. a → 1 /. b → 100 // N

0.212766

vrandAB1 /. kIA → 5 /. kIB → 40 /. kB → 20 /. vfexp → 1 /. a → 100 /. b → 1 // N

0.045403

vrandAB1 /. kIA → 5 /. kIB → 40 /. kB → 20 /. vfexp → 1 /. a → 5 /. b → 5 // N

0.105263

These are now considered to be experimental velocities. The 4 velocities are expressed by the following four rate equations:

vrandAB1 /. a -> 100 /. b → 100

$$\frac{10\,000\ kIB\ vfexp}{100\ kB\ kIA + 10\,000\ kIB + 100\ kB\ kIB + kB\ kIA\ kIB}$$

vrandAB1 /. a -> 1 /. b → 100

$$\frac{100\ kIB\ vfexp}{100\ kB\ kIA + 100\ kIB + kB\ kIB + kB\ kIA\ kIB}$$

vrandAB1 /. a -> 100 /. b → 1

$$\frac{100\ kIB\ vfexp}{kB\ kIA + 100\ kIB + 100\ kB\ kIB + kB\ kIA\ kIB}$$

vrandAB1 /. a → 5 /. b → 5

$$\frac{25\ kIB\ vfexp}{5\ kB\ kIA + 25\ kIB + 5\ kB\ kIB + kB\ kIA\ kIB}$$

The program calckinparsrandAB1 solves these 4 simultaneous equations for vfexp, kIA, kB, and kIB. The velocities are rounded to five digits.

The 4 kinetic parameters for random A + B → products are given by

> **calckinparsrandAB1[.80972, 100, 100, .21277, 1, 100, .045403, 100, 1, .10526, 5, 5]**
>
> {{vfexp → 1., kIA → 5.00046, kB → 20., kIB → 40.0062}}

line11 is needed to obtain the calculated values of the 4 parameters in the form to make a table.

> **line11 = {vfexp, kIA, kB, kIB} /.**
>   **calckinparsrandAB1[.80972, 100, 100, .21277, 1, 100, .045403, 100, 1, .10526, 5, 5]**
>
> {{1., 5.00046, 20., 40.0062}}

These values are correct, but experimental errors in velocities have to be taken into account. 5% errors in the velocities are introduced, one at a time.

> **calckinparsrandAB1[1.05 * .80972, 100, 100, .21277, 1, 100, .045403, 100, 1, .10526, 5, 5]**
>
> {{vfexp → 1.06372, kIA → 5.03821, kB → 21.3267, kIB → 39.6252}}

> **line12 = {vfexp, kIA, kB, kIB} /.**
>   **calckinparsrandAB1[1.05 * .80972, 100, 100, .21277, 1, 100, .045403, 100, 1, .10526, 5, 5]**
>
> {{1.06372, 5.03821, 21.3267, 39.6252}}

> **calckinparsrandAB1[.80972, 100, 100, 1.05 * .21277, 1, 100, .045403, 100, 1, .10526, 5, 5]**
>
> {{vfexp → 0.997629, kIA → 5.06296, kB → 19.9406, kIB → 44.7411}}

> **line13 = {vfexp, kIA, kB, kIB} /.**
>   **calckinparsrandAB1[.80972, 100, 100, 1.05 * .21277, 1, 100, .045403, 100, 1, .10526, 5, 5]**
>
> {{0.997629, 5.06296, 19.9406, 44.7411}}

> **calckinparsrandAB1[.80972, 100, 100, .21277, 1, 100, 1.05 * .045403, 100, 1, .10526, 5, 5]**
>
> {{vfexp → 0.988975, kIA → 5.59099, kB → 18.6765, kIB → 43.2002}}

> **line14 = {vfexp, kIA, kB, kIB} /.**
>   **calckinparsrandAB1[.80972, 100, 100, .21277, 1, 100, 1.05 * .045403, 100, 1, .10526, 5, 5]**
>
> {{0.988975, 5.59099, 18.6765, 43.2002}}

> **calckinparsrandAB1[.80972, 100, 100, .21277, 1, 100, .045403, 100, 1, 1.05 * .10526, 5, 5]**
>
> {{vfexp → 1.00126, kIA → 4.34663, kB → 20.1505, kIB → 33.3227}}

> **line15 = {vfexp, kIA, kB, kIB} /.**
>   **calckinparsrandAB1[.80972, 100, 100, .21277, 1, 100, .045403, 100, 1, 1.05 * .10526, 5, 5]**
>
> {{1.00126, 4.34663, 20.1505, 33.3227}}

Table 4.1 Values of kinetic parameters obtained using {100,100}, {1,100}, {100,1}, and {5,5} and testing the effects of 5% errors in the measured velocities, one at a time.

```
TableForm[Round[{line11[[1]], line12[[1]], line13[[1]], line14[[1]], line15[[1]]}, 0.01],
  TableHeadings →
   {{"No errors", "1.05*v1", "1.05*v2", "1.05*v3", "1.05*v4"}, {"vfet", "kIA", "kB", "kIB"}}]
```

	vfet	kIA	kB	kIB
No errors	1.	5.	20.	40.01
1.05*v1	1.06	5.04	21.33	39.63
1.05*v2	1.	5.06	19.94	44.74
1.05*v3	0.99	5.59	18.68	43.2
1.05*v4	1.	4.35	20.15	33.32

These values of dissociation constants can be used to calculate values of kA using kA = kIA kB/kIB.

Fortunately, the columns in Table 4.1 can be copied and used to calculate the corresponding values of kA.

$$\text{Round}\left[\frac{\begin{matrix}5.\grave{} \\ 5.04\grave{} \\ 5.0600000000000005\grave{} \\ 5.59\grave{} \\ 4.350000000000005\grave{}\end{matrix}}{} * \frac{\begin{matrix}20.\grave{} \\ 21.330000000000002\grave{} \\ 19.94\grave{} \\ 18.68\grave{} \\ 20.150000000000002\grave{}\end{matrix}}{} \Big/ \frac{\begin{matrix}40.01\grave{} \\ 39.63\grave{} \\ 44.74\grave{} \\ 43.2\grave{} \\ 33.32\grave{}\end{matrix}}{}, 0.01\right]$$

{{2.5}, {2.71}, {2.26}, {2.42}, {2.63}}

These values can be inserted at the end of each line in Table 1. line11 becomes augmented line11a.

```
line11a =
  {{1.000003292201511`, 5.000455337744712`, 19.999961325866373`, 40.006166902473936`, 2.50}}

{{1., 5.00046, 20., 40.0062, 2.5}}


line12a =
  {{1.0637169182861397`, 5.03820821056791`, 21.326738245762204`, 39.62519882525381`, 2.71}}

{{1.06372, 5.03821, 21.3267, 39.6252, 2.71}}


line13a =
  {{0.9976292844615865`, 5.062958705312499`, 19.940611419572804`, 44.74113269985607`, 2.26}}

{{0.997629, 5.06296, 19.9406, 44.7411, 2.26}}


line14a =
  {{0.9889746023395195`, 5.590991142356963`, 18.676523325976987`, 43.20020929800704`, 2.42}}

{{0.988975, 5.59099, 18.6765, 43.2002, 2.42}}


line15a =
  {{1.0012580433022464`, 4.346631357956853`, 20.150530913722786`, 33.32266922630096`, 2.63}}

{{1.00126, 4.34663, 20.1505, 33.3227, 2.63}}
```

Table 4.2 Values of kinetic parameters including kA obtained using {100,100}, {1,100}, {100,1}, and {5,5} and testing the effects of 5% errors in the measured velocities, one at a time.

```
TableForm[Round[{line11a[[1]], line12a[[1]], line13a[[1]], line14a[[1]], line15a[[1]]}, 0.01],
  TableHeadings → {{"No errors", "1.05*v1", "1.05*v2", "1.05*v3", "1.05*v4"},
    {"vfet", "kIA", "kB", "kIB", "kA"}}]
```

	vfet	kIA	kB	kIB	kA
No errors	1.	5.	20.	40.01	2.5
1.05*v1	1.06	5.04	21.33	39.63	2.71
1.05*v2	1.	5.06	19.94	44.74	2.26
1.05*v3	0.99	5.59	18.68	43.2	2.42
1.05*v4	1.	4.35	20.15	33.32	2.63

Thus measuring 4 velocities yields 5 kinetic parameters, four directly and a fifth using kA = kIA kB/kIB.

- ### 4.2.2 Estimation of kinetic parameters using {[A],[B]} = {150,150}, {1,150}, {150,1}, and {1,1}

    Increasing the range of substrate concentrations improves the estimation of the kinetic parameters.

    vrandAB1 /. kIA → 5 /. kIB → 40 /. kB → 20 /. vfexp → 1 /. a → 150 /. b → 150 // N

    0.866218

    vrandAB1 /. kIA → 5 /. kIB → 40 /. kB → 20 /. vfexp → 1 /. a → 1 /. b → 150 // N

    0.232558

    vrandAB1 /. kIA → 5 /. kIB → 40 /. kB → 20 /. vfexp → 1 /. a → 150 /. b → 1 // N

    0.0461184

    vrandAB1 /. kIA → 5 /. kIB → 40 /. kB → 20 /. vfexp → 1 /. a → 1 /. b → 1 // N

    0.00809717

The enzyme concentration can be increased to obtain the fourth velocity.

The four kinetic parameters are calculated as follows:

    calckinparsrandAB1[.86622, 150, 150, .23256, 1, 150, .046118, 150, 1, .0080972, 1, 1]

    {{vfexp → 1., kIA → 4.99992, kB → 20.0003, kIB → 40.0002}}

    line21 = {vfexp, kIA, kB, kIB} /.
      calckinparsrandAB1[.86622, 150, 150, .23256, 1, 150, .046118, 150, 1, .0080972, 1, 1]
    {{1., 4.99992, 20.0003, 40.0002}}

5% errors in the velocities are introduced, one at a time.

    calckinparsrandAB1[1.05 * .86622, 150, 150, .23256, 1, 150, .046118, 150, 1, .0080972, 1, 1]

    {{vfexp → 1.05901, kIA → 4.98325, kB → 21.2393, kIB → 39.1064}}

    line22 = {vfexp, kIA, kB, kIB} /.
      calckinparsrandAB1[1.05 * .86622, 150, 150, .23256, 1, 150, .046118, 150, 1, .0080972, 1, 1]
    {{1.05901, 4.98325, 21.2393, 39.1064}}

    calckinparsrandAB1[.86622, 150, 150, 1.05 * .23256, 1, 150, .046118, 150, 1, .0080972, 1, 1]

    {{vfexp → 0.998623, kIA → 5.01064, kB → 19.9713, kIB → 43.7116}}

    line23 = {vfexp, kIA, kB, kIB} /.
      calckinparsrandAB1[.86622, 150, 150, 1.05 * .23256, 1, 150, .046118, 150, 1, .0080972, 1, 1]
    {{0.998623, 5.01064, 19.9713, 43.7116}}

    calckinparsrandAB1[.86622, 150, 150, .23256, 1, 150, 1.05 * .046118, 150, 1, .0080972, 1, 1]

    {{vfexp → 0.993076, kIA → 5.33118, kB → 18.8225, kIB → 40.5319}}

```
line24 = {vfexp, kIA, kB, kIB} /.
   calckinparsrandAB1[.86622, 150, 150, .23256, 1, 150, 1.05 * .046118, 150, 1, .0080972, 1, 1]
```

{{0.993076, 5.33118, 18.8225, 40.5319}}

```
calckinparsrandAB1[.86622, 150, 150, .23256, 1, 150, .046118, 150, 1, 1.05 * .0080972, 1, 1]
```

{{vfexp → 1.00027, kIA → 4.69259, kB → 20.0453, kIB → 37.0276}}

```
line25 = {vfexp, kIA, kB, kIB} /.
   calckinparsrandAB1[.86622, 150, 150, .23256, 1, 150, .046118, 150, 1, 1.05 * .0080972, 1, 1]
```

{{1.00027, 4.69259, 20.0453, 37.0276}}

Table 4.3   Calculations of effects of 5% errors in velocities on kinetic parameters for random A + B → products using {150,150},{1,150},{150,1},{1,1}

```
TableForm[Round[{line21[[1]], line22[[1]], line23[[1]], line24[[1]], line25[[1]]}, 0.01],
  TableHeadings →
    {{"No errors", "1.05*v1", "1.05*v2", "1.05*v3", "1.05*v4"}, {"vfexp", "kIA", "kB", "kIB"}}]
```

	vfexp	kIA	kB	kIB
No errors	1.	5.	20.	40.
1.05*v1	1.06	4.98	21.24	39.11
1.05*v2	1.	5.01	19.97	43.71
1.05*v3	0.99	5.33	18.82	40.53
1.05*v4	1.	4.69	20.05	37.03

This shows that when the substrate concentrations cover a wider range (with respect to the unknown Michaelis constants), more accurate values of kinetic parameters are obtained. Thus the objective of the velocity measurements is to cover as wide a range of substrate concentrations as practical.

These values of dissociation constants can be used to calculate values of kA using kA = kIA kB/kIB.

```
5 * 20 / 40 // N
```

2.5

```
4.983 * 21.24 / 39.11
```

2.70619

The columns in Table 4.3 can be copied and used to calculate the corresponding values of kA.

$$
\text{Round}\left[\begin{array}{c} 5.\grave{} \\ \hline 4.98\grave{} \\ 5.01\grave{} \\ 5.33\grave{} \\ 4.69\grave{} \end{array} * \begin{array}{c} 20.\grave{} \\ \hline 21.240000000000002\grave{} \\ 19.97\grave{} \\ 18.82\grave{} \\ 20.05\grave{} \end{array} \Bigg/ \begin{array}{c} 40.\grave{} \\ \hline 39.11\grave{} \\ 43.71\grave{} \\ 40.53\grave{} \\ 37.03\grave{} \end{array}, 0.01 \right]
$$

{{2.5}, {2.7}, {2.29}, {2.47}, {2.54}}

This column can be added to Table 4.3.

```
line21a =
  {{1.0000043218428698`, 4.999920674427261`, 20.000269600576466`, 40.000160958433504`, 2.50}}
```

{{1., 4.99992, 20.0003, 40.0002, 2.5}}

```
line22a =
  {{1.059005771687429`, 4.983253349602989`, 21.239310599196592`, 39.10636087929682`, 2.70}}
```

```
{{1.05901, 4.98325, 21.2393, 39.1064, 2.7}}
```

```
line23a =
  {{0.9986227682097853`, 5.010643073171724`, 19.971256727203116`, 43.71158321572731`, 2.29}}
```

```
{{0.998623, 5.01064, 19.9713, 43.7116, 2.29}}
```

```
line24a =
  {{0.9930762333264954`, 5.331182250633359`, 18.822497775120926`, 40.53185428377968`, 2.47}}
```

```
{{0.993076, 5.33118, 18.8225, 40.5319, 2.47}}
```

```
line25a =
  {{1.000269289041362`, 4.69259410333882`, 20.045313901081048`, 37.02757207803543`, 2.54}}
```

```
{{1.00027, 4.69259, 20.0453, 37.0276, 2.54}}
```

Table 4.4   Calculations of effects of 5% errors in velocities on kinetic parameters including kA for random A + B → products using {150,150},{1,150},{150,1},{1,1}

```
TableForm[Round[{line21a[[1]], line22a[[1]], line23a[[1]], line24a[[1]], line25a[[1]]}, 0.01],
  TableHeadings → {{"No errors", "1.05*v1", "1.05*v2", "1.05*v3", "1.05*v4"},
    {"vfexp", "kIA", "kB", "kIB", "kA"}}]
```

	vfexp	kIA	kB	kIB	kA
No errors	1.	5.	20.	40.	2.5
1.05*v1	1.06	4.98	21.24	39.11	2.7
1.05*v2	1.	5.01	19.97	43.71	2.29
1.05*v3	0.99	5.33	18.82	40.53	2.47
1.05*v4	1.	4.69	20.05	37.03	2.54

This shows that more accurate values are obtained with a broader range of substrate concentrations.

■ 4.2.3 Estimation of kinetic parameters using {[A],[B]} = {70,70}, {5,70}, {70,5}, and {7,7}

The velocities at specified temperature, pH, and ionic strength are readily calculated at {70,70},{5,70},{70,5},{7,7}.

```
vrandAB1 /. kIA → 5 /. kIB → 40 /. kB → 20 /. vfexp → 1 /. a → 70 /. b → 70 // N
```

```
0.745247
```

```
vrandAB1 /. kIA → 5 /. kIB → 40 /. kB → 20 /. vfexp → 1 /. a → 5 /. b → 70 // N
```

```
0.482759
```

```
vrandAB1 /. kIA → 5 /. kIB → 40 /. kB → 20 /. vfexp → 1 /. a → 70 /. b → 5 // N
```

```
0.187919
```

```
vrandAB1 /. kIA → 5 /. kIB → 40 /. kB → 20 /. vfexp → 1 /. a → 7 /. b → 7 // N
```

```
0.159869
```

These velocities are rounded to five digits to calculate the values of kinetic parameters.

```
calckinparsrandAB1[.74525, 70, 70, .48276, 5, 70, .18792, 70, 5, .15987, 7, 7]
```

```
{{vfexp → 1., kIA → 4.99999, kB → 20., kIB → 39.9998}}
```

```
line31 = {vfexp, kIA, kB, kIB} /.
   calckinparsrandAB1[.74525, 70, 70, .48276, 5, 70, .18792, 70, 5, .15987, 7, 7]
```

{{1., 4.99999, 20., 39.9998}}

5% errors in the velocities are introduced, one at a time.

```
line32 = {vfexp, kIA, kB, kIB} /.
   calckinparsrandAB1[1.05 * .74525, 70, 70, .48276, 5, 70, .18792, 70, 5, .15987, 7, 7]
```

{{1.07995, 4.8373, 21.9935, 34.3814}}

```
line33 = {vfexp, kIA, kB, kIB} /.
   calckinparsrandAB1[.74525, 70, 70, 1.05 * .48276, 5, 70, .18792, 70, 5, .15987, 7, 7]
```

{{0.991644, 5.22195, 19.7743, 54.5228}}

```
line34 = {vfexp, kIA, kB, kIB} /.
   calckinparsrandAB1[.74525, 70, 70, .48276, 5, 70, 1.05 * .18792, 70, 5, .15987, 7, 7]
```

{{0.978805, 5.98425, 18.0921, 47.1011}}

```
line35 = {vfexp, kIA, kB, kIB} /.
   calckinparsrandAB1[.74525, 70, 70, .48276, 5, 70, .18792, 70, 5, 1.05 * .15987, 7, 7]
```

{{1.0037, 4.04696, 20.3322, 29.7311}}

Table 4.5 Calculations of effects of 5% errors in velocities on kinetic parameters for random A + B → products using {70,70},{5,70},{70,5},{7,7}

```
TableForm[Round[{line31[[1]], line32[[1]], line33[[1]], line34[[1]], line35[[1]]}, 0.01],
  TableHeadings →
   {{"No errors", "1.05*v1", "1.05*v2", "1.05*v3", "1.05*v4"}, {"vfexp", "kIA", "kB", "kIB"}}]
```

	vfexp	kIA	kB	kIB
No errors	1.	5.	20.	40.
1.05*v1	1.08	4.84	21.99	34.38
1.05*v2	0.99	5.22	19.77	54.52
1.05*v3	0.98	5.98	18.09	47.1
1.05*v4	1.	4.05	20.33	29.73

The corrresponding values of kA can now be calculated.

$$
\text{Round}\left[
\begin{matrix}
\dfrac{5.\grave{}}{4.84\grave{}} & \dfrac{20.\grave{}}{21.990000000000002\grave{}} & \dfrac{40.\grave{}}{34.38\grave{}} \\
5.22\grave{} & * & 19.77\grave{} \\
5.98\grave{} & & 18.09\grave{} \\
4.05\grave{} & & 20.330000000000002\grave{}
\end{matrix}
\Bigg/
\begin{matrix}
\\
54.52\grave{} \\
47.1\grave{} \\
29.73\grave{}
\end{matrix}
, 0.01\right]
$$

{{2.5}, {3.1}, {1.89}, {2.3}, {2.77}}

This column can be added to Table 4.5 as follows:

```
line31a =
  {{1.0000043878370606`, 4.999990422237514`, 20.000039705009982`, 39.99979610181423`, 2.5}}
```
{{1., 4.99999, 20., 39.9998, 2.5}}

```
line32a =
  {{1.0799541951724063`, 4.837303442380841`, 21.993536705344436`, 34.38135857844797`, 3.10}}
```
{{1.07995, 4.8373, 21.9935, 34.3814, 3.1}}

```
line33a =
  {{0.9916440986112875`, 5.221951717334113`, 19.7743125544227`, 54.52283924412567`, 1.89}}
```
{{0.991644, 5.22195, 19.7743, 54.5228, 1.89}}

```
line34a =
  {{0.9788051622407055`, 5.98424545554048`, 18.09211693127682`, 47.10114192930412`, 2.30}}
```
{{0.978805, 5.98425, 18.0921, 47.1011, 2.3}}

```
line35a =
  {{1.0036952897437839`, 4.046963751958782`, 20.332219565611272`, 29.731063294412817`, 2.77}}
```
{{1.0037, 4.04696, 20.3322, 29.7311, 2.77}}

Table 4.6 Calculations of effects of 5% errors in velocities on kinetic parameters including kA for random A + B → products using {70,70},{5,70},{70,5},{7,7}

```
TableForm[Round[{line31a[[1]], line32a[[1]], line33a[[1]], line34a[[1]], line35a[[1]]}, 0.01],
  TableHeadings → {{"No errors", "1.05*v1", "1.05*v2", "1.05*v3", "1.05*v4"},
    {"vfexp", "kIA", "kB", "kIB", "kA"}}]
```

	vfexp	kIA	kB	kIB	kA
No errors	1.	5.	20.	40.	2.5
1.05*v1	1.08	4.84	21.99	34.38	3.1
1.05*v2	0.99	5.22	19.77	54.52	1.89
1.05*v3	0.98	5.98	18.09	47.1	2.3
1.05*v4	1.	4.05	20.33	29.73	2.77

Using a narrower range of substrate concentrations has resulted in larger errors in the five kinetic parameters.

4.3 Application of calckinparsrandAB1 to velocities for the ordered mechanism

Of course, when calckinparsrandAB1 is applied to experimental data, it is not known whether the mechanism is ordered or random. It of interest to see what this program does when it is applied to data for the ordered mechanism. Velocities were calculated in Section 3.2.1 for the ordered mechanism at {100,100}, {1,100}, and {100,1}. In order to use calckinparsrandAB, it is necessary to have one more experimental velocity. That is calculated at {5,5}.

$$\text{vordAB} = \frac{a\,b\,kfet}{a\,b + a\,kB + kB\,kIA};$$

```
vordAB /. kfet → 1 /. kIA → 5 /. kB → 20 /. a → 100 /. b → 100 // N
```

0.826446

```
vordAB /. kfet → 1 /. kIA → 5 /. kB → 20 /. a → 1 /. b → 100 // N
```

0.454545

```
vordAB /. kfet → 1 /. kIA → 5 /. kB → 20 /. a → 100 /. b → 1 // N
```

0.0454545

```
vordAB /. kfet → 1 /. kIA → 5 /. kB → 20 /. a → 5 /. b → 5 // N
```

0.111111

Now calckinparsrandAB1 is used to calculate the kinetic parameters assuming the mechanism is random.

```
calckinparsrandAB1[.826446, 100, 100, .454545, 1, 100, .0454545, 100, 1, 0.111111, 5, 5]
```

$$\{\{\text{vfexp} → 1., \text{kIA} → 5., \text{kB} → 20., \text{kIB} → 1.49174 \times 10^8\}\}$$

This yields the correct values for kIA and kB, and it indicates that kIB is infinite. This is in accord with the rate equation for the random mechanism that is

vrandAB1

$$\frac{a\,b\,kIB\,vfexp}{b\,kB\,kIA + a\,b\,kIB + a\,kB\,kIB + kB\,kIA\,kIB}$$

This rate equation can also be written as

$$\frac{a\,b\,vfexp}{b\,kB\,kIA\,/\,kIB + a\,b + a\,kB + kB\,kIA}$$

$$\frac{a\,b\,vfexp}{a\,b + a\,kB + kB\,kIA + \frac{b\,kB\,kIA}{kIB}}$$

Thus when kIB = ∞, the rate equation for the random mechanism reduces to the rate equation for the ordered mechanism.

A more familiar form of the rate equation for ordered A + B → products is

v = vfexp/(1 + kB/b + kIAkB/a b) = vfexp/(1 + (kB/b)(1 + kIA/a))

Thus when velocity data is for an ordered mechanism, the use of calckinparsrandAB1 shows that the mechanism is not random by giving a very large value for kIB. It gives the correct values of vfexp, kIA, and kB for the ordered mechanism.

This shows that a computer program for a more general mechanism can be used to calculate kinetic parameters for a less general mechanism [32].

4.4 Effects of pH on the kinetics of random A + B → products when no hydrogen ions are consumed in the rate-determining reaction

■ **4.4.1 Mechanism and rate equation for random A + B → products when the enzymatic site and enzyme-substrate complexes each have two pKs and the factor method is used to derive the rate equation**

It is assumed that the enzymatic site and enzyme-substrate complexes each have two pKs and that the substrates A and B do not have pKs in the pH range of interest.

$$
\begin{array}{ll}
\mathrm{E^-} & \mathrm{EA^-} \\
\| \, \mathrm{pK1e} & \| \, \mathrm{pK1ea} \\
\mathrm{HE \; + \; A \; = \; HEA} \\
\| \, \mathrm{pK2e} & \| \, \mathrm{pK2ea} \\
\mathrm{H_2\,E^+} & \mathrm{H_2\,EA^+}
\end{array}
\qquad \mathrm{K_{HEA} = [HE][A]/[HEA]} \qquad \mathrm{kcHEA = he*a/hea} \qquad\qquad (11)
$$

$$
\begin{array}{ll}
\mathrm{E^-} & \mathrm{EB^-} \\
\| \, \mathrm{pK1e} & \| \, \mathrm{pK1eb} \\
\mathrm{HE \; + \; B \; = \; HEB} \\
\| \, \mathrm{pK2e} & \| \, \mathrm{pK2eb} \\
\mathrm{H_2\,E^+} & \mathrm{H_2\,EB^+}
\end{array}
\qquad \mathrm{K_{HEB} = [HE][B]/[HEB]} \qquad \mathrm{kcHEB = he*b/heb} \qquad\qquad (12)
$$

$$
\begin{array}{ll}
\mathrm{EA^-} & \mathrm{EAB^-} \\
\| & \| \, \mathrm{pK1eab} \\
\mathrm{HEA \; + \; B \; = \; HEAB} \\
\| & \| \, \mathrm{pK2eab} \\
\mathrm{H_2\,EA^+} & \mathrm{H_2\,EAB^+}
\end{array}
\qquad \mathrm{K_{HEAB} = [HEA][B]/[HEAB]} \qquad \mathrm{kcHEAB = hea*b/heab} \qquad (13)
$$

$$
\begin{array}{l}
\mathrm{EAB^-} \\
\| \quad \mathrm{k_f} \\
\mathrm{HEAB \; \rightarrow \; products} \\
\| \\
\mathrm{H_2\,EAB^+}
\end{array}
\qquad v = k_f\mathrm{[HEAB]} = \mathrm{kf*et*eab/et} = \mathrm{kfet*eab/et} \qquad\qquad (14)
$$

The chemical equilibrium constants $\mathrm{K_{HEA}}$, $\mathrm{K_{HEB}}$, and $\mathrm{K_{HEAB}}$ are always obeyed. kfet is used for $k_f\mathrm{[E]_t}$ because the relative contributions of k_f and $\mathrm{[E]_t}$ are not discussed here. This mechanism is written in terms of species.

The dissociation constants involved in this mechanism are defined as follows:

kHE = h*e/he = 10^{-pK1e}

kH2E = h*he/h2e = 10^{-pK2e}

kHEA = h*ea/hea = 10^{-pK1ea}

kH2EA = h*hea/h2ea = 10^{-pK2ea}

kHEB = h*eb/heb = 10^{-pK1eb}

kH2EB = h*heb/h2eb = 10^{-pK2eb}

kHEAB = h*eab/heab = $10^{-pK1eab}$

kH2EAB = h*heab/h2eab = $10^{-pK2eab}$

kcHEA = he*a/hea

kcHEB = he*b/heb

kcHEAB = hea*b/heab

The specified concentration of hydrogen ions is represented by h. The total concentration of enzymatic sites $[E]_t$ = et is given by

$$et = e + he + h2e + ea + hea + h2ea + eb + heb + h2eb + eab + heab + h2eab \qquad (15)$$

There are two ways to derive the rate equation for random A + B → products when there are pKs: (1) the factor method and (2) the use of Solve.

In the factor method, the expressions for the pH-dependent kinetic parameters are typed into *Mathematica*, and then these expressions are inserted into the expression of the velocity in terms of the pH-dependent kinetic parameters V_{fexp}, K_{IA}, K_{IB}, and K_B. See vrandAB1 in Section 4.1.1. This equation can be rearranged into the following familiar form:

$$\mathbf{vfexp} \Big/ \left(1 + \frac{\mathbf{kB\ kIA}}{\mathbf{a\ kIB}} + \frac{\mathbf{kB}}{\mathbf{b}} + \frac{\mathbf{kB\ kIA}}{\mathbf{a\ b}}\right)$$

$$\frac{vfexp}{1 + \frac{kB}{b} + \frac{kB\ kIA}{a\ b} + \frac{kB\ kIA}{a\ kIB}}$$

The expression for vfexp as a function of the hydrogen ion concentration is [33]

$$\mathbf{vfexp = kfet\ /\ (1 + kHEAB\ /\ h + h\ /\ kH2EAB)}$$

$$\frac{kfet}{1 + \frac{h}{kH2EAB} + \frac{kHEAB}{h}}$$

The expression for K_{IA} as a function of the hydrogen ion concentration is

$$\mathbf{kIA = kcHEA * (1 + kHE\ /\ h + h\ /\ kH2E)\ /\ (1 + kHEA\ /\ h + h\ /\ kH2EA)}$$

$$\frac{kcHEA\ \left(1 + \frac{h}{kH2E} + \frac{kHE}{h}\right)}{1 + \frac{h}{kH2EA} + \frac{kHEA}{h}}$$

The expression for K_{IB} as a function of the hydrogen ion concentration is

$$\mathbf{kIB = kcHEB * (1 + kHE\ /\ h + h\ /\ kH2E)\ /\ (1 + kHEB\ /\ h + h\ /\ kH2EB)}$$

$$\frac{kcHEB\ \left(1 + \frac{h}{kH2E} + \frac{kHE}{h}\right)}{1 + \frac{h}{kH2EB} + \frac{kHEB}{h}}$$

The expression for K_B as a function of the hydrogen ion concentration is

```
kB = kcHEAB * (1 + kHEA / h + h / kH2EA) / (1 + kHEAB / h + h / kH2EAB)
```

$$\frac{kcHEAB \left(1 + \frac{h}{kH2EA} + \frac{kHEA}{h}\right)}{1 + \frac{h}{kH2EAB} + \frac{kHEAB}{h}}$$

Since $K_{IA}K_B = K_{IB}K_A$ or, in *Mathematica*, kIA kB = kIB kA,

```
kA = kIA * kB / kIB
```

$$\frac{kcHEA \; kcHEAB \left(1 + \frac{h}{kH2EB} + \frac{kHEB}{h}\right)}{kcHEB \left(1 + \frac{h}{kH2EAB} + \frac{kHEAB}{h}\right)}$$

As shown in the first equation in this chapter, when the reactions E + A = EA, E + B = B, and EA + B = EAB are at equilibrium, the reaction EB + A = EAB is also at equilibrium. The pH dependence of this reaction is represented by

EB⁻   EAB⁻
 ‖ pK1eb   ‖ pK1eab
HEB + A = HEAB K_{HEBA} = [HEA][B]/[HEAB]   kcHEBA = heb*a/heab   (16)
 ‖ pK2eb   ‖ pK2eab
$H_2 EB^+$ $H_2 EAB^+$

Notice that the chemical equilibrium constant for this reaction is represented by kcHEBA because kcHEAB has already been used. This chemical equilibrium constant is related to the other three by kcHEBA = kcHEA kcHEAB/kcHEB. These 4 chemical equilibrium constants are not independent. Thus kA is given by

$$\frac{kcHEBA \left(1 + \frac{h}{kH2EB} + \frac{kHEB}{h}\right)}{\left(1 + \frac{h}{kH2EAB} + \frac{kHEAB}{h}\right)}$$

Derivation of the general form of the rate equation.

These four functions of h can be put into the expression for the velocity by use of /.x->. The general expression for the velocity vgen = is obtained by inserting the expressions for the 4 kinetic parameters.

```
vgen = (vfexp / (1 + kB kIA / (a kIB) + kB / b + kB kIA / (a b))) /. vfexp -> kfet / (1 + kHEAB / h + h / kH2EAB) /.
    kIA -> kcHEA * (1 + kHE / h + h / kH2E) / (1 + kHEA / h + h / kH2EA) /.
    kIB -> kcHEB * (1 + kHE / h + h / kH2E) / (1 + kHEB / h + h / kH2EB) /.
    kB -> kcHEAB * (1 + kHEA / h + h / kH2EA) / (1 + kHEAB / h + h / kH2EAB)
```

$$\frac{kfet}{\left(1 + \frac{h}{kH2EAB} + \frac{kHEAB}{h}\right) \left(1 + \frac{kcHEA \; kcHEAB \left(1+\frac{h}{kH2E}+\frac{kHE}{h}\right)}{a b \left(1+\frac{h}{kH2EAB}+\frac{kHEAB}{h}\right)} + \frac{kcHEAB \left(1+\frac{h}{kH2EA}+\frac{kHEA}{h}\right)}{b \left(1+\frac{h}{kH2EAB}+\frac{kHEAB}{h}\right)} + \frac{kcHEA \; kcHEAB \left(1+\frac{h}{kH2EB}+\frac{kHEB}{h}\right)}{a \; kcHEB \left(1+\frac{h}{kH2EAB}+\frac{kHEAB}{h}\right)}\right)}$$

```
Simplify[vgen]
```

$$(a\, b\, h\, kcHEB\, kfet\, kH2E\, kH2EA\, kH2EAB\, kH2EB) \Big/$$
$$\Big(kcHEAB\, kcHEB\, kH2EAB\, kH2EB \left(kcHEA\, kH2EA \left(h^2 + h\, kH2E + kH2E\, kHE\right) + a\, kH2E \left(h^2 + h\, kH2EA + kH2EA\, kHEA\right)\right) +$$
$$b\, kH2E\, kH2EA \left(h \left(kcHEA\, kcHEAB + a\, kcHEB\right) kH2EAB\, kH2EB +\right.$$
$$\left. h^2 \left(kcHEA\, kcHEAB\, kH2EAB + a\, kcHEB\, kH2EB\right) + kH2EAB\, kH2EB \left(a\, kcHEB\, kHEAB + kcHEA\, kcHEAB\, kHEB\right)\right)\Big)$$

This is the same simplified rate equation as found with Solve later in Section 4.4.2. vgen can be used to calculate the velocity for any set of the 12 pH-independent parameters and any set of h, a, and b.

The following values for the kinetic parameters are arbitrarily chosen for test calculations:

```
Grid[{{pK1e, pK2e, pK1ea, pK2ea, pK1eb, pK2eb, pK1eab, pK2eab, kfet, kcHEA, kcHEB, kcHEAB},
   {7, 6, 7.5, 6.5, 7.8, 6.2, 8, 6, 2, 5, 10, 20}}]
```

pK1e	pK2e	pK1ea	pK2ea	pK1eb	pK2eb	pK1eab	pK2eab	kfet	kcHEA	kcHEB	kcHEAB
7	6	7.5	6.5	7.8	6.2	8	6	2	5	10	20

```
Grid[{{kHE, kH2E, kHEA, kH2EA, kHEB, kH2EB, kHEAB, kH2EAB, kfet, kcHEA, kcHEB, kcHEAB},
   {10^-7, 10^-6, 10^-7.5, 10^-6.5, 10^-7.8, 10^-6.2, 10^-8, 10^-6, 2, 5, 10, 20}}] // N
```

kHE	kH2E	kHEA	kH2EA	kHEB	kH2EB	kHEAB	kH2EAB	kfet	kcHEA	kcHEB	kcHEAB
$1. \times 10^{-7}$	$1. \times 10^{-6}$	3.16228×10^{-8}	3.16228×10^{-7}	1.58489×10^{-8}	6.30957×10^{-7}	$1. \times 10^{-8}$	$1. \times 10^{-6}$	2.	5.	10.	20.

These values of the 12 pH-independent kinetic parameters are inserted into the general expression for the velocity vgen to obtain the expression for vfhab, the velocity as a function of h, a, and b.

```
vfhab =
  vgen /. kHE → 10^-7 /. kH2E → 10^-6 /. kHEA → 10^-7.5 /. kH2EA → 10^-6.5 /. kHEB → 10^-7.8 /.
      kH2EB → 10^-6.2 /. kHEAB → 10^-8 /. kH2EAB → 10^-6 /.
    kfet → 2 /. kcHEA → 5 /. kcHEB → 10 /. kcHEAB → 20
```

$$2 \bigg/ \left(\left(1 + \frac{1}{100\,000\,000\,h} + 1\,000\,000\,h \right) \left(1 + \frac{100 \left(1 + \frac{1}{10\,000\,000\,h} + 1\,000\,000\,h \right)}{a\,b \left(1 + \frac{1}{100\,000\,000\,h} + 1\,000\,000\,h \right)} + \right. \right.$$
$$\left. \left. \frac{10 \left(1 + \frac{1.58489 \times 10^{-8}}{h} + 1.58489 \times 10^{6}\,h \right)}{a \left(1 + \frac{1}{100\,000\,000\,h} + 1\,000\,000\,h \right)} + \frac{20 \left(1 + \frac{3.16228 \times 10^{-8}}{h} + 3.16228 \times 10^{6}\,h \right)}{b \left(1 + \frac{1}{100\,000\,000\,h} + 1\,000\,000\,h \right)} \right) \right)$$

vfhab can be used to calculate the velocity for any set of h, a, and b.

Calculate the expressions for vfexp, kIA, kIB, and kB as functions of h and pH with the test kinetic parameters

V$_{\text{fexp}}$

```
fnvfexp =
  kfet / (1 + kHEAB / h + h / kH2EAB) /. kHE → 10^-7 /. kH2E → 10^-6 /. kHEA → 10^-7.5 /. kH2EA →
        10^-6.5 /. kHEB → 10^-7.8 /. kH2EB → 10^-6.2 /. kHEAB → 10^-8 /.
    kH2EAB → 10^-6 /. kfet → 2 /. kcHEA → 5 /. kcHEB → 10 /. kcHEAB → 20
```

$$\frac{2}{1 + \frac{1}{100\,000\,000\,h} + 1\,000\,000\,h}$$

It is convenient to put in all the test parameters, even though only 3 are needed. To plot the pH-dependent kinetic parameters versus pH, it is necessary to use h->10^-pH.

```
fnvfexppH = fnvfexp /. h → 10^-pH
```

$$\frac{2}{1 + 10^{6-pH} + 10^{-8+pH}}$$

```
plot1 = Plot[fnvfexppH, {pH, 5, 9}, AxesLabel → {"pH", "Vfexp"}];
```

K$_{\text{IA}}$

```
fnkia =
 kcHEA * (1 + kHE / h + h / kH2E) / (1 + kHEA / h + h / kH2EA) /. kHE → 10^-7 /. kH2E → 10^-6 /. kHEA →
        10^-7.5 /. kH2EA → 10^-6.5 /. kHEB → 10^-7.8 /. kH2EB → 10^-6.2 /.
      kHEAB → 10^-8 /. kH2EAB → 10^-6 /. kfet → 2 /. kcHEA → 5 /. kcHEB → 10 /. kcHEAB → 20
```

$$\frac{5\left(1+\frac{1}{10\,000\,000\,h}+1\,000\,000\,h\right)}{1+\frac{3.16228\times10^{-8}}{h}+3.16228\times10^{6}\,h}$$

```
fnkiapH = fnkia /. h → 10^-pH
```

$$\frac{5\left(1+10^{6-pH}+10^{-7+pH}\right)}{1+3.16228\times10^{6}\,10^{-pH}+3.16228\times10^{-8}\,10^{pH}}$$

```
plot2 = Plot[fnkiapH, {pH, 5, 9}, AxesLabel → {"pH", "K_IA"}, PlotRange → {0, 20}];
```

K_{IB}

```
fnkib =
 kcHEB * (1 + kHE / h + h / kH2E) / (1 + kHEB / h + h / kH2EB) /. kHE → 10^-7 /. kH2E → 10^-6 /. kHEA →
        10^-7.5 /. kH2EA → 10^-6.5 /. kHEB → 10^-7.8 /. kH2EB → 10^-6.2 /.
      kHEAB → 10^-8 /. kH2EAB → 10^-6 /. kfet → 2 /. kcHEA → 5 /. kcHEB → 10 /. kcHEAB → 20
```

$$\frac{10\left(1+\frac{1}{10\,000\,000\,h}+1\,000\,000\,h\right)}{1+\frac{1.58489\times10^{-8}}{h}+1.58489\times10^{6}\,h}$$

```
fnkibpH = fnkib /. h → 10^-pH
```

$$\frac{10\left(1+10^{6-pH}+10^{-7+pH}\right)}{1+1.58489\times10^{6}\,10^{-pH}+1.58489\times10^{-8}\,10^{pH}}$$

```
plot3 = Plot[fnkibpH, {pH, 5, 9}, AxesLabel → {"pH", "K_IB"}, PlotRange → {0, 70}];
```

K_B

```
fnkb = kcHEAB * (1 + kHEA / h + h / kH2EA) / (1 + kHEAB / h + h / kH2EAB) /. kHE → 10^-7 /. kH2E → 10^-6 /.
         kHEA → 10^-7.5 /. kH2EA → 10^-6.5 /. kHEB → 10^-7.8 /. kH2EB → 10^-6.2 /.
      kHEAB → 10^-8 /. kH2EAB → 10^-6 /. kfet → 2 /. kcHEA → 5 /. kcHEB → 10 /. kcHEAB → 20
```

$$\frac{20\left(1+\frac{3.16228\times10^{-8}}{h}+3.16228\times10^{6}\,h\right)}{1+\frac{1}{100\,000\,000\,h}+1\,000\,000\,h}$$

```
fnkbpH = fnkb /. h → 10^-pH
```

$$\frac{20\left(1+3.16228\times10^{6}\,10^{-pH}+3.16228\times10^{-8}\,10^{pH}\right)}{1+10^{6-pH}+10^{-8+pH}}$$

```
plot4 = Plot[fnkbpH, {pH, 5, 9}, AxesLabel → {"pH", "K_B"}, PlotRange → {0, 70}];
```

K_A

K_A is not in the mechanism for mechanism I (equations 2-5), but it can be calculated using $K_B\,K_{IA}/K_{IB}$.

```
fnkapH = fnkibpH * fnkiapH / fnkibpH
```

$$\frac{5\left(1+10^{6-pH}+10^{-7+pH}\right)}{1+3.16228\times10^{6}\,10^{-pH}+3.16228\times10^{-8}\,10^{pH}}$$

```
plot5 = Plot[fnkapH, {pH, 5, 9}, AxesLabel → {"pH", "K_A"}, PlotRange → {0, 20}];

GraphicsArray[{{plot1, plot2}, {plot3, plot4}, {plot5}}]
```

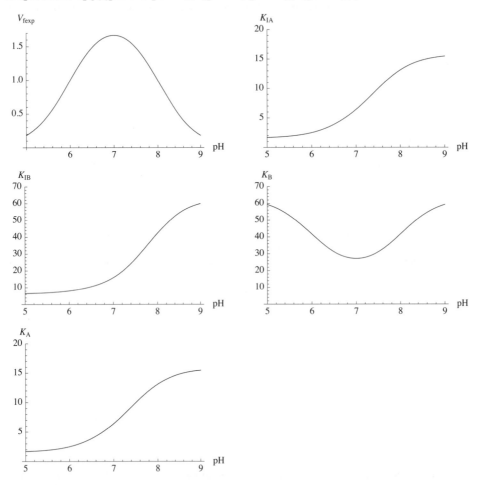

Fig. 4.2 Plots of the pH-dependent kinetic parameters as functions of pH for random A + B → products for an arbitrary set of kinetic parameters.

■ 4.4.2 Estimation of pH-dependent kinetic parameters using the rate equation derived using the factor method

Since there are 12 kinetic parameters in vgen, it is necessary to measure 12 velocities.

Calculate 12 velocities using the chosen kinetic parameters (see Section 4.4.1).

vfhab

$$
2 \left/ \left(\left(1 + \frac{1}{100\,000\,000\,\mathrm{h}} + 1\,000\,000\,\mathrm{h} \right) \left(1 + \frac{100 \left(1 + \frac{1}{10\,000\,000\,\mathrm{h}} + 1\,000\,000\,\mathrm{h} \right)}{\mathrm{a}\,\mathrm{b} \left(1 + \frac{1}{100\,000\,000\,\mathrm{h}} + 1\,000\,000\,\mathrm{h} \right)} + \right.\right.\right.
$$

$$
\left.\left.\left. \frac{10 \left(1 + \frac{1.58489 \times 10^{-8}}{\mathrm{h}} + 1.58489 \times 10^6\,\mathrm{h} \right)}{\mathrm{a} \left(1 + \frac{1}{100\,000\,000\,\mathrm{h}} + 1\,000\,000\,\mathrm{h} \right)} + \frac{20 \left(1 + \frac{3.16228 \times 10^{-8}}{\mathrm{h}} + 3.16228 \times 10^6\,\mathrm{h} \right)}{\mathrm{b} \left(1 + \frac{1}{100\,000\,000\,\mathrm{h}} + 1\,000\,000\,\mathrm{h} \right)} \right) \right) \right)
$$

The strategy in choosing triplets of concentrations {h,a,b} is to set {a,b} to be {500,500},{5,500},{500,5},{5,5} at pH 6, 7, and 8. The {5,5} is replaced with {10,10} when the velocity is so low that a more concentrated enzyme would have to be used.

vfhab /. h → 10^-6 /. a → 500 /. b → 500

0.896615

vfhab /. h → 10^-6 /. a → 5 /. b → 500

0.26798

vfhab /. h → 10^-6 /. a → 500 /. b → 5

0.105699

vfhab /. h → 10^-6 /. a → 10 /. b → 10

0.132463

vfhab /. h → 10^-7 /. a → 500 /. b → 500

1.54742

vfhab /. h → 10^-7 /. a → 5 /. b → 500

0.502102

vfhab /. h → 10^-7 /. a → 500 /. b → 5

0.255097

vfhab /. h → 10^-7 /. a → 5 /. b → 5

0.106588

vfhab /. h → 10^-8 /. a → 500 /. b → 500

0.895185

vfhab /. h → 10^-8 /. a → 5 /. b → 500

0.255766

vfhab /. h → 10^-8 /. a → 500 /. b → 5

0.103745

vfhab /. h → 10^-8 /. a → 10 /. b → 10

0.0833037

These velocities and concentrations are put into the program calckinparsrandAB to calculate the 12 kinetic parameters.
The program to calculate the 12 kinetic parameters from 12 velocity measurements is based on vgen. The following 12 forms of the rate equation are needed for the program to calculate the 12 kinetic parameters from 12 measured velocities.

$$v == \left(vfexp \middle/ \left(1 + \frac{kB\ kIA}{a\ kIB} + \frac{kB}{b} + \frac{kB\ kIA}{a\ b}\right)\right) \, / . \, vfexp \rightarrow kfet \, / \, (1 + kHEAB\,/\,h + h\,/\,kH2EAB) \, / .$$
$$kIA \rightarrow kcHEA * (1 + kHE\,/\,h + h\,/\,kH2E)\,/\,(1 + kHEA\,/\,h + h\,/\,kH2EA)\,/.$$
$$kIB \rightarrow kcHEB * (1 + kHE\,/\,h + h\,/\,kH2E)\,/\,(1 + kHEB\,/\,h + h\,/\,kH2EB)\,/.$$
$$kB \rightarrow kcHEAB * (1 + kHEA\,/\,h + h\,/\,kH2EA)\,/\,(1 + kHEAB\,/\,h + h\,/\,kH2EAB)$$

$$v == \frac{kfet}{\left(1 + \frac{h}{kH2EAB} + \frac{kHEAB}{h}\right)\left(1 + \frac{kcHEA\ kcHEAB\left(1 + \frac{h}{kH2E} + \frac{kHE}{h}\right)}{a\ b\left(1 + \frac{h}{kH2EAB} + \frac{kHEAB}{h}\right)} + \frac{kcHEAB\left(1 + \frac{h}{kH2EA} + \frac{kHEA}{h}\right)}{b\left(1 + \frac{h}{kH2EAB} + \frac{kHEAB}{h}\right)} + \frac{kcHEA\ kcHEAB\left(1 + \frac{h}{kH2EB} + \frac{kHEB}{h}\right)}{a\ kcHEB\left(1 + \frac{h}{kH2EAB} + \frac{kHEAB}{h}\right)}\right)}$$

This double equal sign has a different significance than a single equal sign. The double equal sign indicates a delayed calculation

$$v == \left(vfexp \middle/ \left(1 + \frac{kB\ kIA}{a\ kIB} + \frac{kB}{b} + \frac{kB\ kIA}{a\ b}\right)\right) \, / . \, vfexp \rightarrow kfet \, / \, (1 + kHEAB\,/\,h + h\,/\,kH2EAB) \, / .$$
$$kIA \rightarrow kcHEA * (1 + kHE\,/\,h + h\,/\,kH2E)\,/\,(1 + kHEA\,/\,h + h\,/\,kH2EA)\,/.$$
$$kIB \rightarrow kcHEB * (1 + kHE\,/\,h + h\,/\,kH2E)\,/\,(1 + kHEB\,/\,h + h\,/\,kH2EB)\,/.$$
$$kB \rightarrow kcHEAB * (1 + kHEA\,/\,h + h\,/\,kH2EA)\,/\,(1 + kHEAB\,/\,h + h\,/\,kH2EAB)\,/.$$
$$v \rightarrow v1 \, / . \, h \rightarrow h1 \, / . \, a \rightarrow a1 \, / . \, b \rightarrow b1$$

$$v1 == \frac{kfet}{\left(1 + \frac{h1}{kH2EAB} + \frac{kHEAB}{h1}\right)\left(1 + \frac{kcHEA\ kcHEAB\left(1 + \frac{h1}{kH2E} + \frac{kHE}{h1}\right)}{a1\ b1\left(1 + \frac{h1}{kH2EAB} + \frac{kHEAB}{h1}\right)} + \frac{kcHEAB\left(1 + \frac{h1}{kH2EA} + \frac{kHEA}{h1}\right)}{b1\left(1 + \frac{h1}{kH2EAB} + \frac{kHEAB}{h1}\right)} + \frac{kcHEA\ kcHEAB\left(1 + \frac{h1}{kH2EB} + \frac{kHEB}{h1}\right)}{a1\ kcHEB\left(1 + \frac{h1}{kH2EAB} + \frac{kHEAB}{h1}\right)}\right)}$$

This is the first line of the program, and here are the next 11 lines.

$$v == \left(vfexp \middle/ \left(1 + \frac{kB\ kIA}{a\ kIB} + \frac{kB}{b} + \frac{kB\ kIA}{a\ b}\right)\right) \, / . \, vfexp \rightarrow kfet \, / \, (1 + kHEAB\,/\,h + h\,/\,kH2EAB) \, / .$$
$$kIA \rightarrow kcHEA * (1 + kHE\,/\,h + h\,/\,kH2E)\,/\,(1 + kHEA\,/\,h + h\,/\,kH2EA)\,/.$$
$$kIB \rightarrow kcHEB * (1 + kHE\,/\,h + h\,/\,kH2E)\,/\,(1 + kHEB\,/\,h + h\,/\,kH2EB)\,/.$$
$$kB \rightarrow kcHEAB * (1 + kHEA\,/\,h + h\,/\,kH2EA)\,/\,(1 + kHEAB\,/\,h + h\,/\,kH2EAB)\,/.$$
$$v \rightarrow v2 \, / . \, h \rightarrow h2 \, / . \, a \rightarrow a2 \, / . \, b \rightarrow b2$$

$$v2 == \frac{kfet}{\left(1 + \frac{h2}{kH2EAB} + \frac{kHEAB}{h2}\right)\left(1 + \frac{kcHEA\ kcHEAB\left(1 + \frac{h2}{kH2E} + \frac{kHE}{h2}\right)}{a2\ b2\left(1 + \frac{h2}{kH2EAB} + \frac{kHEAB}{h2}\right)} + \frac{kcHEAB\left(1 + \frac{h2}{kH2EA} + \frac{kHEA}{h2}\right)}{b2\left(1 + \frac{h2}{kH2EAB} + \frac{kHEAB}{h2}\right)} + \frac{kcHEA\ kcHEAB\left(1 + \frac{h2}{kH2EB} + \frac{kHEB}{h2}\right)}{a2\ kcHEB\left(1 + \frac{h2}{kH2EAB} + \frac{kHEAB}{h2}\right)}\right)}$$

$$v == \left(vfexp \middle/ \left(1 + \frac{kB\ kIA}{a\ kIB} + \frac{kB}{b} + \frac{kB\ kIA}{a\ b}\right)\right) \, / . \, vfexp \rightarrow kfet \, / \, (1 + kHEAB\,/\,h + h\,/\,kH2EAB) \, / .$$
$$kIA \rightarrow kcHEA * (1 + kHE\,/\,h + h\,/\,kH2E)\,/\,(1 + kHEA\,/\,h + h\,/\,kH2EA)\,/.$$
$$kIB \rightarrow kcHEB * (1 + kHE\,/\,h + h\,/\,kH2E)\,/\,(1 + kHEB\,/\,h + h\,/\,kH2EB)\,/.$$
$$kB \rightarrow kcHEAB * (1 + kHEA\,/\,h + h\,/\,kH2EA)\,/\,(1 + kHEAB\,/\,h + h\,/\,kH2EAB)\,/.$$
$$v \rightarrow v3 \, / . \, h \rightarrow h3 \, / . \, a \rightarrow a3 \, / . \, b \rightarrow b3$$

$$v3 == \frac{kfet}{\left(1 + \frac{h3}{kH2EAB} + \frac{kHEAB}{h3}\right)\left(1 + \frac{kcHEA\ kcHEAB\left(1 + \frac{h3}{kH2E} + \frac{kHE}{h3}\right)}{a3\ b3\left(1 + \frac{h3}{kH2EAB} + \frac{kHEAB}{h3}\right)} + \frac{kcHEAB\left(1 + \frac{h3}{kH2EA} + \frac{kHEA}{h3}\right)}{b3\left(1 + \frac{h3}{kH2EAB} + \frac{kHEAB}{h3}\right)} + \frac{kcHEA\ kcHEAB\left(1 + \frac{h3}{kH2EB} + \frac{kHEB}{h3}\right)}{a3\ kcHEB\left(1 + \frac{h3}{kH2EAB} + \frac{kHEAB}{h3}\right)}\right)}$$

$$v == \left(vfexp \Big/ \left(1 + \frac{kB\,kIA}{a\,kIB} + \frac{kB}{b} + \frac{kB\,kIA}{a\,b}\right)\right) \,/.\; vfexp \to kfet\,/\,(1 + kHEAB\,/\,h + h\,/\,kH2EAB)\;/.$$
$$kIA \to kcHEA * (1 + kHE\,/\,h + h\,/\,kH2E)\,/\,(1 + kHEA\,/\,h + h\,/\,kH2EA)\;/.$$
$$kIB \to kcHEB * (1 + kHE\,/\,h + h\,/\,kH2E)\,/\,(1 + kHEB\,/\,h + h\,/\,kH2EB)\;/.$$
$$kB \to kcHEAB * (1 + kHEA\,/\,h + h\,/\,kH2EA)\,/\,(1 + kHEAB\,/\,h + h\,/\,kH2EAB)\;/.$$
$$v \to v4\;/.\; h \to h4\;/.\; a \to a4\;/.\; b \to b4$$

$$v4 == \frac{kfet}{\left(1 + \frac{h4}{kH2EAB} + \frac{kHEAB}{h4}\right)\left(1 + \frac{kcHEA\,kcHEAB\left(1 + \frac{h4}{kH2E} + \frac{kHE}{h4}\right)}{a4\,b4\left(1 + \frac{h4}{kH2EAB} + \frac{kHEAB}{h4}\right)} + \frac{kcHEAB\left(1 + \frac{h4}{kH2EA} + \frac{kHEA}{h4}\right)}{b4\left(1 + \frac{h4}{kH2EAB} + \frac{kHEAB}{h4}\right)} + \frac{kcHEA\,kcHEAB\left(1 + \frac{h4}{kH2EB} + \frac{kHEB}{h4}\right)}{a4\,kcHEB\left(1 + \frac{h4}{kH2EAB} + \frac{kHEAB}{h4}\right)}\right)}$$

$$v == \left(vfexp \Big/ \left(1 + \frac{kB\,kIA}{a\,kIB} + \frac{kB}{b} + \frac{kB\,kIA}{a\,b}\right)\right) \,/.\; vfexp \to kfet\,/\,(1 + kHEAB\,/\,h + h\,/\,kH2EAB)\;/.$$
$$kIA \to kcHEA * (1 + kHE\,/\,h + h\,/\,kH2E)\,/\,(1 + kHEA\,/\,h + h\,/\,kH2EA)\;/.$$
$$kIB \to kcHEB * (1 + kHE\,/\,h + h\,/\,kH2E)\,/\,(1 + kHEB\,/\,h + h\,/\,kH2EB)\;/.$$
$$kB \to kcHEAB * (1 + kHEA\,/\,h + h\,/\,kH2EA)\,/\,(1 + kHEAB\,/\,h + h\,/\,kH2EAB)\;/.$$
$$v \to v5\;/.\; h \to h5\;/.\; a \to a5\;/.\; b \to b5$$

$$v5 == \frac{kfet}{\left(1 + \frac{h5}{kH2EAB} + \frac{kHEAB}{h5}\right)\left(1 + \frac{kcHEA\,kcHEAB\left(1 + \frac{h5}{kH2E} + \frac{kHE}{h5}\right)}{a5\,b5\left(1 + \frac{h5}{kH2EAB} + \frac{kHEAB}{h5}\right)} + \frac{kcHEAB\left(1 + \frac{h5}{kH2EA} + \frac{kHEA}{h5}\right)}{b5\left(1 + \frac{h5}{kH2EAB} + \frac{kHEAB}{h5}\right)} + \frac{kcHEA\,kcHEAB\left(1 + \frac{h5}{kH2EB} + \frac{kHEB}{h5}\right)}{a5\,kcHEB\left(1 + \frac{h5}{kH2EAB} + \frac{kHEAB}{h5}\right)}\right)}$$

$$v == \left(vfexp \Big/ \left(1 + \frac{kB\,kIA}{a\,kIB} + \frac{kB}{b} + \frac{kB\,kIA}{a\,b}\right)\right) \,/.\; vfexp \to kfet\,/\,(1 + kHEAB\,/\,h + h\,/\,kH2EAB)\;/.$$
$$kIA \to kcHEA * (1 + kHE\,/\,h + h\,/\,kH2E)\,/\,(1 + kHEA\,/\,h + h\,/\,kH2EA)\;/.$$
$$kIB \to kcHEB * (1 + kHE\,/\,h + h\,/\,kH2E)\,/\,(1 + kHEB\,/\,h + h\,/\,kH2EB)\;/.$$
$$kB \to kcHEAB * (1 + kHEA\,/\,h + h\,/\,kH2EA)\,/\,(1 + kHEAB\,/\,h + h\,/\,kH2EAB)\;/.$$
$$v \to v6\;/.\; h \to h6\;/.\; a \to a6\;/.\; b \to b6$$

$$v6 == \frac{kfet}{\left(1 + \frac{h6}{kH2EAB} + \frac{kHEAB}{h6}\right)\left(1 + \frac{kcHEA\,kcHEAB\left(1 + \frac{h6}{kH2E} + \frac{kHE}{h6}\right)}{a6\,b6\left(1 + \frac{h6}{kH2EAB} + \frac{kHEAB}{h6}\right)} + \frac{kcHEAB\left(1 + \frac{h6}{kH2EA} + \frac{kHEA}{h6}\right)}{b6\left(1 + \frac{h6}{kH2EAB} + \frac{kHEAB}{h6}\right)} + \frac{kcHEA\,kcHEAB\left(1 + \frac{h6}{kH2EB} + \frac{kHEB}{h6}\right)}{a6\,kcHEB\left(1 + \frac{h6}{kH2EAB} + \frac{kHEAB}{h6}\right)}\right)}$$

$$v == \left(vfexp \Big/ \left(1 + \frac{kB\,kIA}{a\,kIB} + \frac{kB}{b} + \frac{kB\,kIA}{a\,b}\right)\right) \,/.\; vfexp \to kfet\,/\,(1 + kHEAB\,/\,h + h\,/\,kH2EAB)\;/.$$
$$kIA \to kcHEA * (1 + kHE\,/\,h + h\,/\,kH2E)\,/\,(1 + kHEA\,/\,h + h\,/\,kH2EA)\;/.$$
$$kIB \to kcHEB * (1 + kHE\,/\,h + h\,/\,kH2E)\,/\,(1 + kHEB\,/\,h + h\,/\,kH2EB)\;/.$$
$$kB \to kcHEAB * (1 + kHEA\,/\,h + h\,/\,kH2EA)\,/\,(1 + kHEAB\,/\,h + h\,/\,kH2EAB)\;/.$$
$$v \to v7\;/.\; h \to h7\;/.\; a \to a7\;/.\; b \to b7$$

$$v7 == \frac{kfet}{\left(1 + \frac{h7}{kH2EAB} + \frac{kHEAB}{h7}\right)\left(1 + \frac{kcHEA\,kcHEAB\left(1 + \frac{h7}{kH2E} + \frac{kHE}{h7}\right)}{a7\,b7\left(1 + \frac{h7}{kH2EAB} + \frac{kHEAB}{h7}\right)} + \frac{kcHEAB\left(1 + \frac{h7}{kH2EA} + \frac{kHEA}{h7}\right)}{b7\left(1 + \frac{h7}{kH2EAB} + \frac{kHEAB}{h7}\right)} + \frac{kcHEA\,kcHEAB\left(1 + \frac{h7}{kH2EB} + \frac{kHEB}{h7}\right)}{a7\,kcHEB\left(1 + \frac{h7}{kH2EAB} + \frac{kHEAB}{h7}\right)}\right)}$$

$$v == \left(vfexp \Big/ \left(1 + \frac{kB\,kIA}{a\,kIB} + \frac{kB}{b} + \frac{kB\,kIA}{a\,b}\right)\right) \,/.\; vfexp \to kfet\,/\,(1 + kHEAB\,/\,h + h\,/\,kH2EAB)\;/.$$
$$kIA \to kcHEA * (1 + kHE\,/\,h + h\,/\,kH2E)\,/\,(1 + kHEA\,/\,h + h\,/\,kH2EA)\;/.$$
$$kIB \to kcHEB * (1 + kHE\,/\,h + h\,/\,kH2E)\,/\,(1 + kHEB\,/\,h + h\,/\,kH2EB)\;/.$$
$$kB \to kcHEAB * (1 + kHEA\,/\,h + h\,/\,kH2EA)\,/\,(1 + kHEAB\,/\,h + h\,/\,kH2EAB)\;/.$$
$$v \to v8\;/.\; h \to h8\;/.\; a \to a8\;/.\; b \to b8$$

$$v8 == \frac{kfet}{\left(1 + \frac{h8}{kH2EAB} + \frac{kHEAB}{h8}\right)\left(1 + \frac{kcHEA\,kcHEAB\left(1 + \frac{h8}{kH2E} + \frac{kHE}{h8}\right)}{a8\,b8\left(1 + \frac{h8}{kH2EAB} + \frac{kHEAB}{h8}\right)} + \frac{kcHEAB\left(1 + \frac{h8}{kH2EA} + \frac{kHEA}{h8}\right)}{b0\left(1 + \frac{h8}{kH2EAB} + \frac{kHEAB}{h8}\right)} + \frac{kcHEA\,kcHEAB\left(1 + \frac{h8}{kH2EB} + \frac{kHEB}{h8}\right)}{a8\,kcHEB\left(1 + \frac{h8}{kH2EAB} + \frac{kHEAB}{h8}\right)}\right)}$$

$$v == \left(vfexp \middle/ \left(1 + \frac{kB\ kIA}{a\ kIB} + \frac{kB}{b} + \frac{kB\ kIA}{a\ b}\right)\right) /.\ vfexp \rightarrow kfet\ /\ (1 + kHEAB\ /\ h + h\ /\ kH2EAB)\ /.$$

$$kIA \rightarrow kcHEA * (1 + kHE\ /\ h + h\ /\ kH2E)\ /\ (1 + kHEA\ /\ h + h\ /\ kH2EA)\ /.$$
$$kIB \rightarrow kcHEB * (1 + kHE\ /\ h + h\ /\ kH2E)\ /\ (1 + kHEB\ /\ h + h\ /\ kH2EB)\ /.$$
$$kB \rightarrow kcHEAB * (1 + kHEA\ /\ h + h\ /\ kH2EA)\ /\ (1 + kHEAB\ /\ h + h\ /\ kH2EAB)\ /.$$
$$v \rightarrow v9\ /.\ h \rightarrow h9\ /.\ a \rightarrow a9\ /.\ b \rightarrow b9$$

$$v9 == \frac{kfet}{\left(1 + \frac{h9}{kH2EAB} + \frac{kHEAB}{h9}\right)\left(1 + \frac{kcHEA\ kcHEAB\left(1 + \frac{h9}{kH2E} + \frac{kHE}{h9}\right)}{a9\ b9\left(1 + \frac{h9}{kH2EAB} + \frac{kHEAB}{h9}\right)} + \frac{kcHEAB\left(1 + \frac{h9}{kH2EA} + \frac{kHEA}{h9}\right)}{b9\left(1 + \frac{h9}{kH2EAB} + \frac{kHEAB}{h9}\right)} + \frac{kcHEA\ kcHEAB\left(1 + \frac{h9}{kH2EB} + \frac{kHEB}{h9}\right)}{a9\ kcHEB\left(1 + \frac{h9}{kH2EAB} + \frac{kHEAB}{h9}\right)}\right)}$$

$$v == \left(vfexp \middle/ \left(1 + \frac{kB\ kIA}{a\ kIB} + \frac{kB}{b} + \frac{kB\ kIA}{a\ b}\right)\right) /.\ vfexp \rightarrow kfet\ /\ (1 + kHEAB\ /\ h + h\ /\ kH2EAB)\ /.$$

$$kIA \rightarrow kcHEA * (1 + kHE\ /\ h + h\ /\ kH2E)\ /\ (1 + kHEA\ /\ h + h\ /\ kH2EA)\ /.$$
$$kIB \rightarrow kcHEB * (1 + kHE\ /\ h + h\ /\ kH2E)\ /\ (1 + kHEB\ /\ h + h\ /\ kH2EB)\ /.$$
$$kB \rightarrow kcHEAB * (1 + kHEA\ /\ h + h\ /\ kH2EA)\ /\ (1 + kHEAB\ /\ h + h\ /\ kH2EAB)\ /.$$
$$v \rightarrow v10\ /.\ h \rightarrow h10\ /.\ a \rightarrow a10\ /.\ b \rightarrow b10$$

$$v10 == \frac{kfet}{\left(1 + \frac{h10}{kH2EAB} + \frac{kHEAB}{h10}\right)\left(1 + \frac{kcHEA\ kcHEAB\left(1 + \frac{h10}{kH2E} + \frac{kHE}{h10}\right)}{a10\ b10\left(1 + \frac{h10}{kH2EAB} + \frac{kHEAB}{h10}\right)} + \frac{kcHEAB\left(1 + \frac{h10}{kH2EA} + \frac{kHEA}{h10}\right)}{b10\left(1 + \frac{h10}{kH2EAB} + \frac{kHEAB}{h10}\right)} + \frac{kcHEA\ kcHEAB\left(1 + \frac{h10}{kH2EB} + \frac{kHEB}{h10}\right)}{a10\ kcHEB\left(1 + \frac{h10}{kH2EAB} + \frac{kHEAB}{h10}\right)}\right)}$$

$$v == \left(vfexp \middle/ \left(1 + \frac{kB\ kIA}{a\ kIB} + \frac{kB}{b} + \frac{kB\ kIA}{a\ b}\right)\right) /.\ vfexp \rightarrow kfet\ /\ (1 + kHEAB\ /\ h + h\ /\ kH2EAB)\ /.$$

$$kIA \rightarrow kcHEA * (1 + kHE\ /\ h + h\ /\ kH2E)\ /\ (1 + kHEA\ /\ h + h\ /\ kH2EA)\ /.$$
$$kIB \rightarrow kcHEB * (1 + kHE\ /\ h + h\ /\ kH2E)\ /\ (1 + kHEB\ /\ h + h\ /\ kH2EB)\ /.$$
$$kB \rightarrow kcHEAB * (1 + kHEA\ /\ h + h\ /\ kH2EA)\ /\ (1 + kHEAB\ /\ h + h\ /\ kH2EAB)\ /.$$
$$v \rightarrow v11\ /.\ h \rightarrow h11\ /.\ a \rightarrow a11\ /.\ b \rightarrow b11$$

$$v11 == \frac{kfet}{\left(1 + \frac{h11}{kH2EAB} + \frac{kHEAB}{h11}\right)\left(1 + \frac{kcHEA\ kcHEAB\left(1 + \frac{h11}{kH2E} + \frac{kHE}{h11}\right)}{a11\ b11\left(1 + \frac{h11}{kH2EAB} + \frac{kHEAB}{h11}\right)} + \frac{kcHEAB\left(1 + \frac{h11}{kH2EA} + \frac{kHEA}{h11}\right)}{b11\left(1 + \frac{h11}{kH2EAB} + \frac{kHEAB}{h11}\right)} + \frac{kcHEA\ kcHEAB\left(1 + \frac{h11}{kH2EB} + \frac{kHEB}{h11}\right)}{a11\ kcHEB\left(1 + \frac{h11}{kH2EAB} + \frac{kHEAB}{h11}\right)}\right)}$$

$$v == \left(vfexp \middle/ \left(1 + \frac{kB\ kIA}{a\ kIB} + \frac{kB}{b} + \frac{kB\ kIA}{a\ b}\right)\right) /.\ vfexp \rightarrow kfet\ /\ (1 + kHEAB\ /\ h + h\ /\ kH2EAB)\ /.$$

$$kIA \rightarrow kcHEA * (1 + kHE\ /\ h + h\ /\ kH2E)\ /\ (1 + kHEA\ /\ h + h\ /\ kH2EA)\ /.$$
$$kIB \rightarrow kcHEB * (1 + kHE\ /\ h + h\ /\ kH2E)\ /\ (1 + kHEB\ /\ h + h\ /\ kH2EB)\ /.$$
$$kB \rightarrow kcHEAB * (1 + kHEA\ /\ h + h\ /\ kH2EA)\ /\ (1 + kHEAB\ /\ h + h\ /\ kH2EAB)\ /.$$
$$v \rightarrow v12\ /.\ h \rightarrow h12\ /.\ a \rightarrow a12\ /.\ b \rightarrow b12$$

$$v12 == \frac{kfet}{\left(1 + \frac{h12}{kH2EAB} + \frac{kHEAB}{h12}\right)\left(1 + \frac{kcHEA\ kcHEAB\left(1 + \frac{h12}{kH2E} + \frac{kHE}{h12}\right)}{a12\ b12\left(1 + \frac{h12}{kH2EAB} + \frac{kHEAB}{h12}\right)} + \frac{kcHEAB\left(1 + \frac{h12}{kH2EA} + \frac{kHEA}{h12}\right)}{b12\left(1 + \frac{h12}{kH2EAB} + \frac{kHEAB}{h12}\right)} + \frac{kcHEA\ kcHEAB\left(1 + \frac{h12}{kH2EB} + \frac{kHEB}{h12}\right)}{a12\ kcHEB\left(1 + \frac{h12}{kH2EAB} + \frac{kHEAB}{h12}\right)}\right)}$$

The first and the last parts of the program are added to the list of 12 rate equations

```
calckinparsrandAB[v1_, h1_, a1_, b1_, v2_, h2_, a2_, b2_, v3_, h3_, a3_, b3_, v4_, h4_, a4_, b4_,
    v5_, h5_, a5_, b5_, v6_, h6_, a6_, b6_, v7_, h7_, a7_, b7_, v8_, h8_, a8_, b8_, v9_, h9_,
    a9_, b9_, v10_, h10_, a10_, b10_, v11_, h11_, a11_, b11_, v12_, h12_, a12_, b12_] :=
 Module[{}, (*This program calculates 12 kinetic parameters from 12 experimental
```

velocities for A + B → products at 12 triplets of $[H^+]$, $[A]$,
and $[B]$ on the assumption that the mechanism is random. Velocities are
measured at {high [A],high [B]}, {low [A], high[B]}, {high [A], low [B]}
and {low [A], low [B]} at high $[H^+]$, intermediate $[H^+]$, and low $[H^+]$.*)

$$\text{Solve}\Big[\Big\{v1 = kfet \Big/ \Big(\Big(1 + \frac{h1}{kH2EAB} + \frac{kHEAB}{h1}\Big)\Big(1 + \frac{kcHEA\ kcHEAB\ \big(1 + \frac{h1}{kH2E} + \frac{kHE}{h1}\big)}{a1\ b1\ \big(1 + \frac{h1}{kH2EAB} + \frac{kHEAB}{h1}\big)} +$$

$$\frac{kcHEAB\ \big(1 + \frac{h1}{kH2EA} + \frac{kHEA}{h1}\big)}{b1\ \big(1 + \frac{h1}{kH2EAB} + \frac{kHEAB}{h1}\big)} + \frac{kcHEA\ kcHEAB\ \big(1 + \frac{h1}{kH2EB} + \frac{kHEB}{h1}\big)}{a1\ kcHEB\ \big(1 + \frac{h1}{kH2EAB} + \frac{kHEAB}{h1}\big)}\Big)\Big),$$

$$v2 = kfet \Big/ \Big(\Big(1 + \frac{h2}{kH2EAB} + \frac{kHEAB}{h2}\Big)\Big(1 + \frac{kcHEA\ kcHEAB\ \big(1 + \frac{h2}{kH2E} + \frac{kHE}{h2}\big)}{a2\ b2\ \big(1 + \frac{h2}{kH2EAB} + \frac{kHEAB}{h2}\big)} +$$

$$\frac{kcHEAB\ \big(1 + \frac{h2}{kH2EA} + \frac{kHEA}{h2}\big)}{b2\ \big(1 + \frac{h2}{kH2EAB} + \frac{kHEAB}{h2}\big)} + \frac{kcHEA\ kcHEAB\ \big(1 + \frac{h2}{kH2EB} + \frac{kHEB}{h2}\big)}{a2\ kcHEB\ \big(1 + \frac{h2}{kH2EAB} + \frac{kHEAB}{h2}\big)}\Big)\Big),$$

$$v3 = kfet \Big/ \Big(\Big(1 + \frac{h3}{kH2EAB} + \frac{kHEAB}{h3}\Big)\Big(1 + \frac{kcHEA\ kcHEAB\ \big(1 + \frac{h3}{kH2E} + \frac{kHE}{h3}\big)}{a3\ b3\ \big(1 + \frac{h3}{kH2EAB} + \frac{kHEAB}{h3}\big)} +$$

$$\frac{kcHEAB\ \big(1 + \frac{h3}{kH2EA} + \frac{kHEA}{h3}\big)}{b3\ \big(1 + \frac{h3}{kH2EAB} + \frac{kHEAB}{h3}\big)} + \frac{kcHEA\ kcHEAB\ \big(1 + \frac{h3}{kH2EB} + \frac{kHEB}{h3}\big)}{a3\ kcHEB\ \big(1 + \frac{h3}{kH2EAB} + \frac{kHEAB}{h3}\big)}\Big)\Big),$$

$$v4 = kfet \Big/ \Big(\Big(1 + \frac{h4}{kH2EAB} + \frac{kHEAB}{h4}\Big)\Big(1 + \frac{kcHEA\ kcHEAB\ \big(1 + \frac{h4}{kH2E} + \frac{kHE}{h4}\big)}{a4\ b4\ \big(1 + \frac{h4}{kH2EAB} + \frac{kHEAB}{h4}\big)} +$$

$$\frac{kcHEAB\ \big(1 + \frac{h4}{kH2EA} + \frac{kHEA}{h4}\big)}{b4\ \big(1 + \frac{h4}{kH2EAB} + \frac{kHEAB}{h4}\big)} + \frac{kcHEA\ kcHEAB\ \big(1 + \frac{h4}{kH2EB} + \frac{kHEB}{h4}\big)}{a4\ kcHEB\ \big(1 + \frac{h4}{kH2EAB} + \frac{kHEAB}{h4}\big)}\Big)\Big),$$

$$v5 = kfet \Big/ \Big(\Big(1 + \frac{h5}{kH2EAB} + \frac{kHEAB}{h5}\Big)\Big(1 + \frac{kcHEA\ kcHEAB\ \big(1 + \frac{h5}{kH2E} + \frac{kHE}{h5}\big)}{a5\ b5\ \big(1 + \frac{h5}{kH2EAB} + \frac{kHEAB}{h5}\big)} +$$

$$\left. \frac{\text{kcHEAB}\left(1 + \frac{h5}{\text{kH2EA}} + \frac{\text{kHEA}}{h5}\right)}{b5\left(1 + \frac{h5}{\text{kH2EAB}} + \frac{\text{kHEAB}}{h5}\right)} + \frac{\text{kcHEA kcHEAB}\left(1 + \frac{h5}{\text{kH2EB}} + \frac{\text{kHEB}}{h5}\right)}{a5\,\text{kcHEB}\left(1 + \frac{h5}{\text{kH2EAB}} + \frac{\text{kHEAB}}{h5}\right)}\right)\right),$$

$$v6 == \text{kfet} \Bigg/ \left(\left(1 + \frac{h6}{\text{kH2EAB}} + \frac{\text{kHEAB}}{h6}\right)\left(1 + \frac{\text{kcHEA kcHEAB}\left(1 + \frac{h6}{\text{kH2E}} + \frac{\text{kHE}}{h6}\right)}{a6\,b6\left(1 + \frac{h6}{\text{kH2EAB}} + \frac{\text{kHEAB}}{h6}\right)} + \right.\right.$$

$$\left.\left. \frac{\text{kcHEAB}\left(1 + \frac{h6}{\text{kH2EA}} + \frac{\text{kHEA}}{h6}\right)}{b6\left(1 + \frac{h6}{\text{kH2EAB}} + \frac{\text{kHEAB}}{h6}\right)} + \frac{\text{kcHEA kcHEAB}\left(1 + \frac{h6}{\text{kH2EB}} + \frac{\text{kHEB}}{h6}\right)}{a6\,\text{kcHEB}\left(1 + \frac{h6}{\text{kH2EAB}} + \frac{\text{kHEAB}}{h6}\right)}\right)\right),$$

$$v7 == \text{kfet} \Bigg/ \left(\left(1 + \frac{h7}{\text{kH2EAB}} + \frac{\text{kHEAB}}{h7}\right)\left(1 + \frac{\text{kcHEA kcHEAB}\left(1 + \frac{h7}{\text{kH2E}} + \frac{\text{kHE}}{h7}\right)}{a7\,b7\left(1 + \frac{h7}{\text{kH2EAB}} + \frac{\text{kHEAB}}{h7}\right)} + \right.\right.$$

$$\left.\left. \frac{\text{kcHEAB}\left(1 + \frac{h7}{\text{kH2EA}} + \frac{\text{kHEA}}{h7}\right)}{b7\left(1 + \frac{h7}{\text{kH2EAB}} + \frac{\text{kHEAB}}{h7}\right)} + \frac{\text{kcHEA kcHEAB}\left(1 + \frac{h7}{\text{kH2EB}} + \frac{\text{kHEB}}{h7}\right)}{a7\,\text{kcHEB}\left(1 + \frac{h7}{\text{kH2EAB}} + \frac{\text{kHEAB}}{h7}\right)}\right)\right),$$

$$v8 == \text{kfet} \Bigg/ \left(\left(1 + \frac{h8}{\text{kH2EAB}} + \frac{\text{kHEAB}}{h8}\right)\left(1 + \frac{\text{kcHEA kcHEAB}\left(1 + \frac{h8}{\text{kH2E}} + \frac{\text{kHE}}{h8}\right)}{a8\,b8\left(1 + \frac{h8}{\text{kH2EAB}} + \frac{\text{kHEAB}}{h8}\right)} + \right.\right.$$

$$\left.\left. \frac{\text{kcHEAB}\left(1 + \frac{h8}{\text{kH2EA}} + \frac{\text{kHEA}}{h8}\right)}{b8\left(1 + \frac{h8}{\text{kH2EAB}} + \frac{\text{kHEAB}}{h8}\right)} + \frac{\text{kcHEA kcHEAB}\left(1 + \frac{h8}{\text{kH2EB}} + \frac{\text{kHEB}}{h8}\right)}{a8\,\text{kcHEB}\left(1 + \frac{h8}{\text{kH2EAB}} + \frac{\text{kHEAB}}{h8}\right)}\right)\right),$$

$$v9 == \text{kfet} \Bigg/ \left(\left(1 + \frac{h9}{\text{kH2EAB}} + \frac{\text{kHEAB}}{h9}\right)\left(1 + \frac{\text{kcHEA kcHEAB}\left(1 + \frac{h9}{\text{kH2E}} + \frac{\text{kHE}}{h9}\right)}{a9\,b9\left(1 + \frac{h9}{\text{kH2EAB}} + \frac{\text{kHEAB}}{h9}\right)} + \right.\right.$$

$$\left.\left. \frac{\text{kcHEAB}\left(1 + \frac{h9}{\text{kH2EA}} + \frac{\text{kHEA}}{h9}\right)}{b9\left(1 + \frac{h9}{\text{kH2EAB}} + \frac{\text{kHEAB}}{h9}\right)} + \frac{\text{kcHEA kcHEAB}\left(1 + \frac{h9}{\text{kH2EB}} + \frac{\text{kHEB}}{h9}\right)}{a9\,\text{kcHEB}\left(1 + \frac{h9}{\text{kH2EAB}} + \frac{\text{kHEAB}}{h9}\right)}\right)\right),$$

$$v10 == \text{kfet} \Bigg/ \left(\left(1 + \frac{h10}{\text{kH2EAB}} + \frac{\text{kHEAB}}{h10}\right)\left(1 + \frac{\text{kcHEA kcHEAB}\left(1 + \frac{h10}{\text{kH2E}} + \frac{\text{kHE}}{h10}\right)}{a10\,b10\left(1 + \frac{h10}{\text{kH2EAB}} + \frac{\text{kHEAB}}{h10}\right)} + \right.\right.$$

$$\left.\left. \frac{\text{kcHEAB}\left(1 + \frac{h10}{\text{kH2EA}} + \frac{\text{kHEA}}{h10}\right)}{b10\left(1 + \frac{h10}{\text{kH2EAB}} + \frac{\text{kHEAB}}{h10}\right)} + \frac{\text{kcHEA kcHEAB}\left(1 + \frac{h10}{\text{kH2EB}} + \frac{\text{kHEB}}{h10}\right)}{a10\,\text{kcHEB}\left(1 + \frac{h10}{\text{kH2EAB}} + \frac{\text{kHEAB}}{h10}\right)}\right)\right),$$

$$v11 == kfet \Big/ \left(\left(1 + \frac{h11}{kH2EAB} + \frac{kHEAB}{h11}\right)\left(1 + \frac{kcHEA\; kcHEAB\left(1 + \frac{h11}{kH2E} + \frac{kHE}{h11}\right)}{a11\; b11\left(1 + \frac{h11}{kH2EAB} + \frac{kHEAB}{h11}\right)} + \right.\right.$$

$$\left.\left.\frac{kcHEAB\left(1 + \frac{h11}{kH2EA} + \frac{kHEA}{h11}\right)}{b11\left(1 + \frac{h11}{kH2EAB} + \frac{kHEAB}{h11}\right)} + \frac{kcHEA\; kcHEAB\left(1 + \frac{h11}{kH2EB} + \frac{kHEB}{h11}\right)}{a11\; kcHEB\left(1 + \frac{h11}{kH2EAB} + \frac{kHEAB}{h11}\right)}\right)\right),$$

$$v12 == kfet \Big/ \left(\left(1 + \frac{h12}{kH2EAB} + \frac{kHEAB}{h12}\right)\left(1 + \frac{kcHEA\; kcHEAB\left(1 + \frac{h12}{kH2E} + \frac{kHE}{h12}\right)}{a12\; b12\left(1 + \frac{h12}{kH2EAB} + \frac{kHEAB}{h12}\right)} + \right.\right.$$

$$\left.\left.\frac{kcHEAB\left(1 + \frac{h12}{kH2EA} + \frac{kHEA}{h12}\right)}{b12\left(1 + \frac{h12}{kH2EAB} + \frac{kHEAB}{h12}\right)} + \frac{kcHEA\; kcHEAB\left(1 + \frac{h12}{kH2EB} + \frac{kHEB}{h12}\right)}{a12\; kcHEB\left(1 + \frac{h12}{kH2EAB} + \frac{kHEAB}{h12}\right)}\right)\right)\Big\},$$

$$\{kHE, kH2E, kHEA, kH2EA, kHEB, kH2EB, kHEAB, kH2EAB, kfet, kcHEA, kcHEAB, kcHEB\}\Big]\Big]$$

In calculating the 12 kinetic parameters, the velocities are rounded to five digits.

```
calckinparsrandAB[0.8966, 10^-6, 500, 500, 0.2680, 10^-6, 5, 500, 0.1057,
  10^-6, 500, 5, 0.1325, 10^-6, 10, 10, 1.547, 10^-7, 500, 500, 0.5021, 10^-7,
  5, 500, 0.2551, 10^-7, 500, 5, 0.1066, 10^-7, 5, 5, 0.8952, 10^-8, 500,
  500, 0.2558, 10^-8, 5, 500, 0.1037, 10^-8, 500, 5, 0.08330, 10^-8, 10, 10]
```

$$\{\{kfet \to 1.9991, kHEAB \to 9.98989 \times 10^{-9}, kHEA \to 3.1661 \times 10^{-8}, kHEB \to 1.58475 \times 10^{-8},$$
$$kHE \to 9.99393 \times 10^{-8}, kcHEA \to 5.00356, kcHEAB \to 19.983, kcHEB \to 10.0043,$$
$$kH2E \to 1.00463 \times 10^{-6}, kH2EB \to 6.30923 \times 10^{-7}, kH2EA \to 3.16064 \times 10^{-7}, kH2EAB \to 1.00085 \times 10^{-6}\}\}$$

The following values are expected:

kHE	kH2E	kHEA	kH2EA	kHEB	kH2EB	kHEAB	kH2EAB	kfet	kcHEA	kcHEB	kcHEAB
$1. \times 10^{-7}$	$1. \times 10^{-6}$	3.16228×10^{-8}	3.16228×10^{-7}	1.58489×10^{-8}	6.30957×10^{-7}	$1. \times 10^{-8}$	$1. \times 10^{-6}$	$2.$	$5.$	$10.$	$20.$

The calculated velocities yield the correct kinetic parameters, but it is important to check on the sensitivity of these parameters to errors in the measured velocities. Construct a table of kinetic parameters with 13 lines.

```
line1 = {kfet, kHEAB, kHEA, kHEB, kHE, kcHEA, kcHEAB, kcHEB, kH2E, kH2EB, kH2EA, kH2EAB} /.
  calckinparsrandAB[0.8966, 10^-6, 500, 500, 0.2680, 10^-6, 5, 500, 0.1057,
    10^-6, 500, 5, 0.1325, 10^-6, 10, 10, 1.547, 10^-7, 500, 500, 0.5021, 10^-7,
    5, 500, 0.2551, 10^-7, 500, 5, 0.1066, 10^-7, 5, 5, 0.8952, 10^-8, 500,
    500, 0.2558, 10^-8, 5, 500, 0.1037, 10^-8, 500, 5, 0.08330, 10^-8, 10, 10]
```

$$\{\{1.9991, 9.98989 \times 10^{-9}, 3.1661 \times 10^{-8}, 1.58475 \times 10^{-8}, 9.99393 \times 10^{-8}, 5.00356,$$
$$19.983, 10.0043, 1.00463 \times 10^{-6}, 6.30923 \times 10^{-7}, 3.16064 \times 10^{-7}, 1.00085 \times 10^{-6}\}\}$$

```
calckinparsrandAB[1.05 * 0.8966, 10^-6, 500, 500, 0.2680, 10^-6, 5, 500,
  0.1057, 10^-6, 500, 5, 0.1325, 10^-6, 10, 10, 1.547, 10^-7, 500, 500, 0.5021,
  10^-7, 5, 500, 0.2551, 10^-7, 500, 5, 0.1066, 10^-7, 5, 5, 0.8952, 10^-8, 500,
  500, 0.2558, 10^-8, 5, 500, 0.1037, 10^-8, 500, 5, 0.08330, 10^-8, 10, 10]
```

$$\{\{kfet \to 1.97272, kHEAB \to 9.7381 \times 10^{-9}, kHEA \to 3.17964 \times 10^{-8}, kHEB \to 1.60137 \times 10^{-8},$$
$$kHE \to 9.99543 \times 10^{-8}, kcHEA \to 5.0195, kcHEAB \to 19.654, kcHEB \to 10.0696,$$
$$kH2E \to 1.00323 \times 10^{-6}, kH2EB \to 6.03815 \times 10^{-7}, kH2EA \to 3.12048 \times 10^{-7}, kH2EAB \to 1.15474 \times 10^{-6}\}\}$$

```
line2 = {kfet, kHEAB, kHEA, kHEB, kHE, kcHEA, kcHEAB, kcHEB, kH2E, kH2EB, kH2EA, kH2EAB} /.
    calckinparsrandAB[1.05 * 0.8966, 10^-6, 500, 500, 0.2680, 10^-6, 5, 500,
      0.1057, 10^-6, 500, 5, 0.1325, 10^-6, 10, 10, 1.547, 10^-7, 500, 500, 0.5021,
      10^-7, 5, 500, 0.2551, 10^-7, 500, 5, 0.1066, 10^-7, 5, 5, 0.8952, 10^-8, 500,
      500, 0.2558, 10^-8, 5, 500, 0.1037, 10^-8, 500, 5, 0.08330, 10^-8, 10, 10]
```

$$\left\{\left\{1.97272,\ 9.7381 \times 10^{-9},\ 3.17964 \times 10^{-8},\ 1.60137 \times 10^{-8},\ 9.99543 \times 10^{-8},\ 5.0195,\right.\right.$$
$$\left.\left.19.654,\ 10.0696,\ 1.00323 \times 10^{-6},\ 6.03815 \times 10^{-7},\ 3.12048 \times 10^{-7},\ 1.15474 \times 10^{-6}\right\}\right\}$$

```
calckinparsrandAB[0.8966, 10^-6, 500, 500, 1.05 * 0.2680, 10^-6, 5, 500,
  0.1057, 10^-6, 500, 5, 0.1325, 10^-6, 10, 10, 1.547, 10^-7, 500, 500, 0.5021,
  10^-7, 5, 500, 0.2551, 10^-7, 500, 5, 0.1066, 10^-7, 5, 5, 0.8952, 10^-8, 500,
  500, 0.2558, 10^-8, 5, 500, 0.1037, 10^-8, 500, 5, 0.08330, 10^-8, 10, 10]
```

$$\left\{\left\{\text{kfet} \to 2.,\ \text{kHEAB} \to 9.99852 \times 10^{-9},\ \text{kHEA} \to 3.16518 \times 10^{-8},\ \text{kHEB} \to 1.5296 \times 10^{-8},\right.\right.$$
$$\text{kHE} \to 1.02461 \times 10^{-7},\ \text{kcHEA} \to 4.88936,\ \text{kcHEAB} \to 19.9965,\ \text{kcHEB} \to 9.56194,$$
$$\left.\left.\text{kH2E} \to 8.13908 \times 10^{-7},\ \text{kH2EB} \to 7.4134 \times 10^{-7},\ \text{kH2EA} \to 3.16341 \times 10^{-7},\ \text{kH2EAB} \to 9.96302 \times 10^{-7}\right\}\right\}$$

```
line3 = {kfet, kHEAB, kHEA, kHEB, kHE, kcHEA, kcHEAB, kcHEB, kH2E, kH2EB, kH2EA, kH2EAB} /.
    calckinparsrandAB[0.8966, 10^-6, 500, 500, 1.05 * 0.2680, 10^-6, 5, 500,
      0.1057, 10^-6, 500, 5, 0.1325, 10^-6, 10, 10, 1.547, 10^-7, 500, 500, 0.5021,
      10^-7, 5, 500, 0.2551, 10^-7, 500, 5, 0.1066, 10^-7, 5, 5, 0.8952, 10^-8, 500,
      500, 0.2558, 10^-8, 5, 500, 0.1037, 10^-8, 500, 5, 0.08330, 10^-8, 10, 10]
```

$$\left\{\left\{2.,\ 9.99852 \times 10^{-9},\ 3.16518 \times 10^{-8},\ 1.5296 \times 10^{-8},\ 1.02461 \times 10^{-7},\ 4.88936,\right.\right.$$
$$\left.\left.19.9965,\ 9.56194,\ 8.13908 \times 10^{-7},\ 7.4134 \times 10^{-7},\ 3.16341 \times 10^{-7},\ 9.96302 \times 10^{-7}\right\}\right\}$$

```
calckinparsrandAB[0.8966, 10^-6, 500, 500, 0.2680, 10^-6, 5, 500, 1.05 * 0.1057,
  10^-6, 500, 5, 0.1325, 10^-6, 10, 10, 1.547, 10^-7, 500, 500, 0.5021, 10^-7,
  5, 500, 0.2551, 10^-7, 500, 5, 0.1066, 10^-7, 5, 5, 0.8952, 10^-8, 500,
  500, 0.2558, 10^-8, 5, 500, 0.1037, 10^-8, 500, 5, 0.08330, 10^-8, 10, 10]
```

$$\left\{\left\{\text{kfet} \to 2.00139,\ \text{kHEAB} \to 1.00118 \times 10^{-8},\ \text{kHEA} \to 3.0525 \times 10^{-8},\ \text{kHEB} \to 1.58189 \times 10^{-8},\right.\right.$$
$$\text{kHE} \to 1.06568 \times 10^{-7},\ \text{kcHEA} \to 4.58533,\ \text{kcHEAB} \to 20.5796,\ \text{kcHEB} \to 9.42014,$$
$$\left.\left.\text{kH2E} \to 6.21672 \times 10^{-7},\ \text{kH2EB} \to 6.35828 \times 10^{-7},\ \text{kH2EA} \to 3.5432 \times 10^{-7},\ \text{kH2EAB} \to 9.89388 \times 10^{-7}\right\}\right\}$$

```
line4 = {kfet, kHEAB, kHEA, kHEB, kHE, kcHEA, kcHEAB, kcHEB, kH2E, kH2EB, kH2EA, kH2EAB} /.
    calckinparsrandAB[0.8966, 10^-6, 500, 500, 0.2680, 10^-6, 5, 500, 1.05 * 0.1057,
      10^-6, 500, 5, 0.1325, 10^-6, 10, 10, 1.547, 10^-7, 500, 500, 0.5021, 10^-7,
      5, 500, 0.2551, 10^-7, 500, 5, 0.1066, 10^-7, 5, 5, 0.8952, 10^-8, 500,
      500, 0.2558, 10^-8, 5, 500, 0.1037, 10^-8, 500, 5, 0.08330, 10^-8, 10, 10]
```

$$\left\{\left\{2.00139,\ 1.00118 \times 10^{-8},\ 3.0525 \times 10^{-8},\ 1.58189 \times 10^{-8},\ 1.06568 \times 10^{-7},\ 4.58533,\right.\right.$$
$$\left.\left.20.5796,\ 9.42014,\ 6.21672 \times 10^{-7},\ 6.35828 \times 10^{-7},\ 3.5432 \times 10^{-7},\ 9.89388 \times 10^{-7}\right\}\right\}$$

```
calckinparsrandAB[0.8966, 10^-6, 500, 500, 0.2680, 10^-6, 5, 500, 0.1057,
  10^-6, 500, 5, 1.05 * 0.1325, 10^-6, 10, 10, 1.547, 10^-7, 500, 500, 0.5021,
  10^-7, 5, 500, 0.2551, 10^-7, 500, 5, 0.1066, 10^-7, 5, 5, 0.8952, 10^-8, 500,
  500, 0.2558, 10^-8, 5, 500, 0.1037, 10^-8, 500, 5, 0.08330, 10^-8, 10, 10]
```

$$\left\{\left\{\text{kfet} \to 1.99902,\ \text{kHEAB} \to 9.98919 \times 10^{-9},\ \text{kHEA} \to 3.16987 \times 10^{-8},\ \text{kHEB} \to 1.58936 \times 10^{-8},\right.\right.$$
$$\text{kHE} \to 9.072 \times 10^{-8},\ \text{kcHEA} \to 5.47079,\ \text{kcHEAB} \to 19.9638,\ \text{kcHEB} \to 10.9486,$$
$$\left.\left.\text{kH2E} \to 7.01686 \times 10^{-6},\ \text{kH2EB} \to 6.23151 \times 10^{-7},\ \text{kH2EA} \to 3.14936 \times 10^{-7},\ \text{kH2EAB} \to 1.00123 \times 10^{-6}\right\}\right\}$$

```
line5 = {kfet, kHEAB, kHEA, kHEB, kHE, kcHEA, kcHEAB, kcHEB, kH2E, kH2EB, kH2EA, kH2EAB} /.
  calckinparsrandAB[0.8966, 10^-6, 500, 500, 0.2680, 10^-6, 5, 500, 0.1057,
    10^-6, 500, 5, 1.05 * 0.1325, 10^-6, 10, 10, 1.547, 10^-7, 500, 500, 0.5021,
    10^-7, 5, 500, 0.2551, 10^-7, 500, 5, 0.1066, 10^-7, 5, 5, 0.8952, 10^-8, 500,
    500, 0.2558, 10^-8, 5, 500, 0.1037, 10^-8, 500, 5, 0.08330, 10^-8, 10, 10]
```

$\{\{1.99902, 9.98919 \times 10^{-9}, 3.16987 \times 10^{-8}, 1.58936 \times 10^{-8}, 9.072 \times 10^{-8}, 5.47079,$

$\quad 19.9638, 10.9486, 7.01686 \times 10^{-6}, 6.23151 \times 10^{-7}, 3.14936 \times 10^{-7}, 1.00123 \times 10^{-6}\}\}$

```
calckinparsrandAB[0.8966, 10^-6, 500, 500, 0.2680, 10^-6, 5, 500, 0.1057,
  10^-6, 500, 5, 0.1325, 10^-6, 10, 10, 1.05 * 1.547, 10^-7, 500, 500, 0.5021,
  10^-7, 5, 500, 0.2551, 10^-7, 500, 5, 0.1066, 10^-7, 5, 5, 0.8952, 10^-8, 500,
  500, 0.2558, 10^-8, 5, 500, 0.1037, 10^-8, 500, 5, 0.08330, 10^-8, 10, 10]
```

$\{\{kfet \to 2.1689, kHEAB \to 1.16794 \times 10^{-8}, kHEA \to 3.08625 \times 10^{-8}, kHEB \to 1.4877 \times 10^{-8},$

$\quad kHE \to 1.02132 \times 10^{-7}, kcHEA \to 4.81137, kcHEAB \to 22.1049, kcHEB \to 9.43878,$

$\quad kH2E \to 9.66156 \times 10^{-7}, kH2EB \to 6.72077 \times 10^{-7}, kH2EA \to 3.24243 \times 10^{-7}, kH2EAB \to 8.56085 \times 10^{-7}\}\}$

```
line6 = {kfet, kHEAB, kHEA, kHEB, kHE, kcHEA, kcHEAB, kcHEB, kH2E, kH2EB, kH2EA, kH2EAB} /.
  calckinparsrandAB[0.8966, 10^-6, 500, 500, 0.2680, 10^-6, 5, 500, 0.1057,
    10^-6, 500, 5, 0.1325, 10^-6, 10, 10, 1.05 * 1.547, 10^-7, 500, 500, 0.5021,
    10^-7, 5, 500, 0.2551, 10^-7, 500, 5, 0.1066, 10^-7, 5, 5, 0.8952, 10^-8, 500,
    500, 0.2558, 10^-8, 5, 500, 0.1037, 10^-8, 500, 5, 0.08330, 10^-8, 10, 10]
```

$\{\{2.1689, 1.16794 \times 10^{-8}, 3.08625 \times 10^{-8}, 1.4877 \times 10^{-8}, 1.02132 \times 10^{-7}, 4.81137,$

$\quad 22.1049, 9.43878, 9.66156 \times 10^{-7}, 6.72077 \times 10^{-7}, 3.24243 \times 10^{-7}, 8.56085 \times 10^{-7}\}\}$

```
calckinparsrandAB[0.8966, 10^-6, 500, 500, 0.2680, 10^-6, 5, 500, 0.1057,
  10^-6, 500, 5, 0.1325, 10^-6, 10, 10, 1.547, 10^-7, 500, 500, 1.05 * 0.5021,
  10^-7, 5, 500, 0.2551, 10^-7, 500, 5, 0.1066, 10^-7, 5, 5, 0.8952, 10^-8, 500,
  500, 0.2558, 10^-8, 5, 500, 0.1037, 10^-8, 500, 5, 0.08330, 10^-8, 10, 10]
```

$\{\{kfet \to 1.99429, kHEAB \to 9.94203 \times 10^{-9}, kHEA \to 3.16861 \times 10^{-8}, kHEB \to 1.9381 \times 10^{-8},$

$\quad kHE \to 9.36916 \times 10^{-8}, kcHEA \to 5.30853, kcHEAB \to 19.9229, kcHEB \to 12.0634,$

$\quad kH2E \to 1.1332 \times 10^{-6}, kH2EB \to 5.15899 \times 10^{-7}, kH2EA \to 3.15814 \times 10^{-7}, kH2EAB \to 1.00567 \times 10^{-6}\}\}$

```
line7 = {kfet, kHEAB, kHEA, kHEB, kHE, kcHEA, kcHEAB, kcHEB, kH2E, kH2EB, kH2EA, kH2EAB} /.
  calckinparsrandAB[0.8966, 10^-6, 500, 500, 0.2680, 10^-6, 5, 500, 0.1057,
    10^-6, 500, 5, 0.1325, 10^-6, 10, 10, 1.547, 10^-7, 500, 500, 1.05 * 0.5021,
    10^-7, 5, 500, 0.2551, 10^-7, 500, 5, 0.1066, 10^-7, 5, 5, 0.8952, 10^-8, 500,
    500, 0.2558, 10^-8, 5, 500, 0.1037, 10^-8, 500, 5, 0.08330, 10^-8, 10, 10]
```

$\{\{1.99429, 9.94203 \times 10^{-9}, 3.16861 \times 10^{-8}, 1.9381 \times 10^{-8}, 9.36916 \times 10^{-8}, 5.30853,$

$\quad 19.9229, 12.0634, 1.1332 \times 10^{-6}, 5.15899 \times 10^{-7}, 3.15814 \times 10^{-7}, 1.00567 \times 10^{-6}\}\}$

```
calckinparsrandAB[0.8966, 10^-6, 500, 500, 0.2680, 10^-6, 5, 500, 0.1057,
  10^-6, 500, 5, 0.1325, 10^-6, 10, 10, 1.547, 10^-7, 500, 500, 0.5021, 10^-7,
  5, 500, 1.05 * 0.2551, 10^-7, 500, 5, 0.1066, 10^-7, 5, 5, 0.8952, 10^-8, 500,
  500, 0.2558, 10^-8, 5, 500, 0.1037, 10^-8, 500, 5, 0.08330, 10^-8, 10, 10]
```

$\{\{kfet \to 1.98965, kHEAB \to 9.89591 \times 10^{-9}, kHEA \to 3.72637 \times 10^{-8}, kHEB \to 1.59088 \times 10^{-8},$

$\quad kHE \to 8.82842 \times 10^{-8}, kcHEA \to 6.35208, kcHEAB \to 17.526, kcHEB \to 11.2185,$

$\quad kH2E \to 1.27436 \times 10^{-6}, kH2EB \to 6.28492 \times 10^{-7}, kH2EA \to 2.68536 \times 10^{-7}, kH2EAB \to 1.01036 \times 10^{-6}\}\}$

```
line8 = {kfet, kHEAB, kHEA, kHEB, kHE, kcHEA, kcHEAB, kcHEB, kH2E, kH2EB, kH2EA, kH2EAB} /.
   calckinparsrandAB[0.8966, 10^-6, 500, 500, 0.2680, 10^-6, 5, 500, 0.1057,
     10^-6, 500, 5, 0.1325, 10^-6, 10, 10, 1.547, 10^-7, 500, 500, 0.5021, 10^-7,
     5, 500, 1.05 * 0.2551, 10^-7, 500, 5, 0.1066, 10^-7, 5, 5, 0.8952, 10^-8, 500,
     500, 0.2558, 10^-8, 5, 500, 0.1037, 10^-8, 500, 5, 0.08330, 10^-8, 10, 10]
```

$$\left\{\left\{1.98965,\ 9.89591 \times 10^{-9},\ 3.72637 \times 10^{-8},\ 1.59088 \times 10^{-8},\ 8.82842 \times 10^{-8},\ 6.35208,\right.\right.$$
$$\left.\left.17.526,\ 11.2185,\ 1.27436 \times 10^{-6},\ 6.28492 \times 10^{-7},\ 2.68536 \times 10^{-7},\ 1.01036 \times 10^{-6}\right\}\right\}$$

```
calckinparsrandAB[0.8966, 10^-6, 500, 500, 0.2680, 10^-6, 5, 500, 0.1057,
   10^-6, 500, 5, 0.1325, 10^-6, 10, 10, 1.547, 10^-7, 500, 500, 0.5021, 10^-7,
   5, 500, 0.2551, 10^-7, 500, 5, 1.05 * 0.1066, 10^-7, 5, 5, 0.8952, 10^-8, 500,
   500, 0.2558, 10^-8, 5, 500, 0.1037, 10^-8, 500, 5, 0.08330, 10^-8, 10, 10]
```

$$\left\{\left\{\text{kfet} \to 1.99933,\ \text{kHEAB} \to 9.99215 \times 10^{-9},\ \text{kHEA} \to 3.15432 \times 10^{-8},\ \text{kHEB} \to 1.57019 \times 10^{-8},\right.\right.$$
$$\text{kHE} \to 1.43522 \times 10^{-7},\ \text{kcHEA} \to 3.57205,\ \text{kcHEAB} \to 20.0421,\ \text{kcHEB} \to 7.1219,$$
$$\left.\left.\text{kH2E} \to 5.60791 \times 10^{-7},\ \text{kH2EB} \to 6.3677 \times 10^{-7},\ \text{kH2EA} \to 3.17245 \times 10^{-7},\ \text{kH2EAB} \to 1.00063 \times 10^{-6}\right\}\right\}$$

```
line9 = {kfet, kHEAB, kHEA, kHEB, kHE, kcHEA, kcHEAB, kcHEB, kH2E, kH2EB, kH2EA, kH2EAB} /.
   calckinparsrandAB[0.8966, 10^-6, 500, 500, 0.2680, 10^-6, 5, 500, 0.1057,
     10^-6, 500, 5, 0.1325, 10^-6, 10, 10, 1.547, 10^-7, 500, 500, 0.5021, 10^-7,
     5, 500, 0.2551, 10^-7, 500, 5, 1.05 * 0.1066, 10^-7, 5, 5, 0.8952, 10^-8, 500,
     500, 0.2558, 10^-8, 5, 500, 0.1037, 10^-8, 500, 5, 0.08330, 10^-8, 10, 10]
```

$$\left\{\left\{1.99933,\ 9.99215 \times 10^{-9},\ 3.15432 \times 10^{-8},\ 1.57019 \times 10^{-8},\ 1.43522 \times 10^{-7},\ 3.57205,\right.\right.$$
$$\left.\left.20.0421,\ 7.1219,\ 5.60791 \times 10^{-7},\ 6.3677 \times 10^{-7},\ 3.17245 \times 10^{-7},\ 1.00063 \times 10^{-6}\right\}\right\}$$

```
calckinparsrandAB[0.8966, 10^-6, 500, 500, 0.2680, 10^-6, 5, 500, 0.1057,
   10^-6, 500, 5, 0.1325, 10^-6, 10, 10, 1.547, 10^-7, 500, 500, 0.5021, 10^-7,
   5, 500, 0.2551, 10^-7, 500, 5, 0.1066, 10^-7, 5, 5, 1.05 * 0.8952, 10^-8, 500,
   500, 0.2558, 10^-8, 5, 500, 0.1037, 10^-8, 500, 5, 0.08330, 10^-8, 10, 10]
```

$$\left\{\left\{\text{kfet} \to 1.97268,\ \text{kHEAB} \to 8.65636 \times 10^{-9},\ \text{kHEA} \to 3.20688 \times 10^{-8},\ \text{kHEB} \to 1.65601 \times 10^{-8},\right.\right.$$
$$\text{kHE} \to 9.99657 \times 10^{-8},\ \text{kcHEA} \to 5.01952,\ \text{kcHEAB} \to 19.6535,\ \text{kcHEB} \to 10.0697,$$
$$\left.\left.\text{kH2E} \to 1.00437 \times 10^{-6},\ \text{kH2EB} \to 6.24364 \times 10^{-7},\ \text{kH2EA} \to 3.14716 \times 10^{-7},\ \text{kH2EAB} \to 1.02677 \times 10^{-6}\right\}\right\}$$

```
line10 = {kfet, kHEAB, kHEA, kHEB, kHE, kcHEA, kcHEAB, kcHEB, kH2E, kH2EB, kH2EA, kH2EAB} /.
   calckinparsrandAB[0.8966, 10^-6, 500, 500, 0.2680, 10^-6, 5, 500, 0.1057,
     10^-6, 500, 5, 0.1325, 10^-6, 10, 10, 1.547, 10^-7, 500, 500, 0.5021, 10^-7,
     5, 500, 0.2551, 10^-7, 500, 5, 0.1066, 10^-7, 5, 5, 1.05 * 0.8952, 10^-8, 500,
     500, 0.2558, 10^-8, 5, 500, 0.1037, 10^-8, 500, 5, 0.08330, 10^-8, 10, 10]
```

$$\left\{\left\{1.97268,\ 8.65636 \times 10^{-9},\ 3.20688 \times 10^{-8},\ 1.65601 \times 10^{-8},\ 9.99657 \times 10^{-8},\ 5.01952,\right.\right.$$
$$\left.\left.19.6535,\ 10.0697,\ 1.00437 \times 10^{-6},\ 6.24364 \times 10^{-7},\ 3.14716 \times 10^{-7},\ 1.02677 \times 10^{-6}\right\}\right\}$$

```
calckinparsrandAB[0.8966, 10^-6, 500, 500, 0.2680, 10^-6, 5, 500, 0.1057,
   10^-6, 500, 5, 0.1325, 10^-6, 10, 10, 1.547, 10^-7, 500, 500, 0.5021, 10^-7,
   5, 500, 0.2551, 10^-7, 500, 5, 0.1066, 10^-7, 5, 5, 0.8952, 10^-8, 500, 500,
   1.05 * 0.2558, 10^-8, 5, 500, 0.1037, 10^-8, 500, 5, 0.08330, 10^-8, 10, 10]
```

$$\left\{\left\{\text{kfet} \to 2.00005,\ \text{kHEAB} \to 1.00377 \times 10^{-8},\ \text{kHEA} \to 3.16319 \times 10^{-8},\ \text{kHEB} \to 1.33768 \times 10^{-8},\right.\right.$$
$$\text{kHE} \to 1.04568 \times 10^{-7},\ \text{kcHEA} \to 4.88392,\ \text{kcHEAB} \to 19.9972,\ \text{kcHEB} \to 9.54133,$$
$$\left.\left.\text{kH2E} \to 9.60078 \times 10^{-7},\ \text{kH2EB} \to 6.54768 \times 10^{-7},\ \text{kH2EA} \to 3.16161 \times 10^{-7},\ \text{kH2EAB} \to 9.99947 \times 10^{-7}\right\}\right\}$$

```
line11 = {kfet, kHEAB, kHEA, kHEB, kHE, kcHEA, kcHEAB, kcHEB, kH2E, kH2EB, kH2EA, kH2EAB} /.
    calckinparsrandAB[0.8966, 10^-6, 500, 500, 0.2680, 10^-6, 5, 500, 0.1057,
      10^-6, 500, 5, 0.1325, 10^-6, 10, 10, 1.547, 10^-7, 500, 500, 0.5021, 10^-7,
      5, 500, 0.2551, 10^-7, 500, 5, 0.1066, 10^-7, 5, 5, 0.8952, 10^-8, 500, 500,
      1.05 * 0.2558, 10^-8, 5, 500, 0.1037, 10^-8, 500, 5, 0.08330, 10^-8, 10, 10]
```

$\{\{2.00005,\ 1.00377\times10^{-8},\ 3.16319\times10^{-8},\ 1.33768\times10^{-8},\ 1.04568\times10^{-7},\ 4.88392,$
$\quad 19.9972,\ 9.54133,\ 9.60078\times10^{-7},\ 6.54768\times10^{-7},\ 3.16161\times10^{-7},\ 9.99947\times10^{-7}\}\}$

```
calckinparsrandAB[0.8966, 10^-6, 500, 500, 0.2680, 10^-6, 5, 500, 0.1057,
  10^-6, 500, 5, 0.1325, 10^-6, 10, 10, 1.547, 10^-7, 500, 500, 0.5021, 10^-7,
  5, 500, 0.2551, 10^-7, 500, 5, 0.1066, 10^-7, 5, 5, 0.8952, 10^-8, 500, 500,
  0.2558, 10^-8, 5, 500, 1.05 * 0.1037, 10^-8, 500, 5, 0.08330, 10^-8, 10, 10]
```

$\{\{\text{kfet}\to2.00144,\ \text{kHEAB}\to1.01079\times10^{-8},\ \text{kHEA}\to2.81802\times10^{-8},\ \text{kHEB}\to1.57228\times10^{-8},$
$\quad \text{kHE}\to1.11779\times10^{-7},\ \text{kcHEA}\to4.57749,\ \text{kcHEAB}\to20.5911,\ \text{kcHEB}\to9.40889,$
$\quad \text{kH2E}\to8.98038\times10^{-7},\ \text{kH2EB}\to6.32084\times10^{-7},\ \text{kH2EA}\to3.28057\times10^{-7},\ \text{kH2EAB}\to9.98621\times10^{-7}\}\}$

```
line12 = {kfet, kHEAB, kHEA, kHEB, kHE, kcHEA, kcHEAB, kcHEB, kH2E, kH2EB, kH2EA, kH2EAB} /.
    calckinparsrandAB[0.8966, 10^-6, 500, 500, 0.2680, 10^-6, 5, 500, 0.1057,
      10^-6, 500, 5, 0.1325, 10^-6, 10, 10, 1.547, 10^-7, 500, 500, 0.5021, 10^-7,
      5, 500, 0.2551, 10^-7, 500, 5, 0.1066, 10^-7, 5, 5, 0.8952, 10^-8, 500, 500,
      0.2558, 10^-8, 5, 500, 1.05 * 0.1037, 10^-8, 500, 5, 0.08330, 10^-8, 10, 10]
```

$\{\{2.00144,\ 1.01079\times10^{-8},\ 2.81802\times10^{-8},\ 1.57228\times10^{-8},\ 1.11779\times10^{-7},\ 4.57749,$
$\quad 20.5911,\ 9.40889,\ 8.98038\times10^{-7},\ 6.32084\times10^{-7},\ 3.28057\times10^{-7},\ 9.98621\times10^{-7}\}\}$

```
calckinparsrandAB[0.8966, 10^-6, 500, 500, 0.2680, 10^-6, 5, 500, 0.1057,
  10^-6, 500, 5, 0.1325, 10^-6, 10, 10, 1.547, 10^-7, 500, 500, 0.5021, 10^-7,
  5, 500, 0.2551, 10^-7, 500, 5, 0.1066, 10^-7, 5, 5, 0.8952, 10^-8, 500, 500,
  0.2558, 10^-8, 5, 500, 0.1037, 10^-8, 500, 5, 1.05 * 0.08330, 10^-8, 10, 10]
```

$\{\{\text{kfet}\to1.99898,\ \text{kHEAB}\to9.98397\times10^{-9},\ \text{kHEA}\to3.18414\times10^{-8},\ \text{kHEB}\to1.61622\times10^{-8},$
$\quad \text{kHE}\to7.54911\times10^{-8},\ \text{kcHEA}\to5.74716,\ \text{kcHEAB}\to19.9524,\ \text{kcHEB}\to11.508,$
$\quad \text{kH2E}\to1.33082\times10^{-6},\ \text{kH2EB}\to6.28009\times10^{-7},\ \text{kH2EA}\to3.15466\times10^{-7},\ \text{kH2EAB}\to1.00096\times10^{-6}\}\}$

```
line13 = {kfet, kHEAB, kHEA, kHEB, kHE, kcHEA, kcHEAB, kcHEB, kH2E, kH2EB, kH2EA, kH2EAB} /.
    calckinparsrandAB[0.8966, 10^-6, 500, 500, 0.2680, 10^-6, 5, 500, 0.1057,
      10^-6, 500, 5, 0.1325, 10^-6, 10, 10, 1.547, 10^-7, 500, 500, 0.5021, 10^-7,
      5, 500, 0.2551, 10^-7, 500, 5, 0.1066, 10^-7, 5, 5, 0.8952, 10^-8, 500, 500,
      0.2558, 10^-8, 5, 500, 0.1037, 10^-8, 500, 5, 1.05 * 0.08330, 10^-8, 10, 10]
```

$\{\{1.99898,\ 9.98397\times10^{-9},\ 3.18414\times10^{-8},\ 1.61622\times10^{-8},\ 7.54911\times10^{-8},\ 5.74716,$
$\quad 19.9524,\ 11.508,\ 1.33082\times10^{-6},\ 6.28009\times10^{-7},\ 3.15466\times10^{-7},\ 1.00096\times10^{-6}\}\}$

Table 4.7 Calculations of effects of 5% errors in the velocities on kinetic parameters for random A + B → products using the factor method. (The negative base 10 logarithms of the kinetic parameters are plotted because it easier to see the errors on the pKs, rather than the dissociation constants.)

```
TableForm[
  Round[-Log[10, {line1[[1]], line2[[1]], line3[[1]], line4[[1]], line5[[1]], line6[[1]],
      line7[[1]], line8[[1]], line9[[1]], line10[[1]], line11[[1]], line12[[1]], line13[[1]]}],
    0.01], TableHeadings → {{"No errors", "1.05*v1", "1.05*v2", "1.05*v3", "1.05*v4", "1.05*v5",
      "1.05*v6", "1.05*v7", "1.05*v8", "1.05*v9", "1.05*v10", "1.05*v10", "1.05*v12"},
    {"kfet", "pK1eab", "pK1ea", "pK1ea", "pK1e", "kcHEA", "kcHEAB",
      "kcHEB", "pK2e", "pK2eb", "pK2ea", "pK2eab"}}]
```

	kfet	pK1eab	pK1ea	pK1ea	pK1e	kcHEA	kcHEAB	kcHEB	pK2e	pK2eb	pK2ea	pK2eab
No errors	-0.3	8.	7.5	7.8	7.	-0.7	-1.3	-1.	6.	6.2	6.5	6.
1.05*v1	-0.3	8.01	7.5	7.8	7.	-0.7	-1.29	-1.	6.	6.22	6.51	5.94
1.05*v2	-0.3	8.	7.5	7.82	6.99	-0.69	-1.3	-0.98	6.09	6.13	6.5	6.
1.05*v3	-0.3	8.	7.52	7.8	6.97	-0.66	-1.31	-0.97	6.21	6.2	6.45	6.
1.05*v4	-0.3	8.	7.5	7.8	7.04	-0.74	-1.3	-1.04	5.15	6.21	6.5	6.
1.05*v5	-0.34	7.93	7.51	7.83	6.99	-0.68	-1.34	-0.97	6.01	6.17	6.49	6.07
1.05*v6	-0.3	8.	7.5	7.71	7.03	-0.72	-1.3	-1.08	5.95	6.29	6.5	6.
1.05*v7	-0.3	8.	7.43	7.8	7.05	-0.8	-1.24	-1.05	5.89	6.2	6.57	6.
1.05*v8	-0.3	8.	7.5	7.8	6.84	-0.55	-1.3	-0.85	6.25	6.2	6.5	6.
1.05*v9	-0.3	8.06	7.49	7.78	7.	-0.7	-1.29	-1.	6.	6.2	6.5	5.99
1.05*v10	-0.3	8.	7.5	7.87	6.98	-0.69	-1.3	-0.98	6.02	6.18	6.5	6.
1.05*v10	-0.3	8.	7.55	7.8	6.95	-0.66	-1.31	-0.97	6.05	6.2	6.48	6.
1.05*v12	-0.3	8.	7.5	7.79	7.12	-0.76	-1.3	-1.06	5.88	6.2	6.5	6.

In making this table, 4 velocities were at pH 6, 4 velocities were at pH 7, and 4 velocities were at pH 8. Note that the columns for kfet, kcHEA, kcHEAB, and kcHEB are for the negative base 10 logarithms. Note that kcHEBA is not determined directly, but it can be calculated using kcHEBA = kcHEA kcHEAB/kcHEB (Section 4.1.1).

```
-Log[10, {2, 5, 10, 20}] // N
```

$$\{-0.30103, -0.69897, -1., -1.30103\}$$

◾ 4.4.3 Mechanism and rate equation for random A + B → products when the enzymatic site and enzyme-substrate complexes each have two pKs and Solve is used to derive the rate equation

Since the reactions up to the rate-determining reaction are at equilibrium, the expression for the equilibrium concentration of HEAB can be derived using Solve[eqns,vars,elims].

```
Solve[{kHE == h * e / he, kH2E == h * he / h2e, kHEA == h * ea / hea,
  kH2EA == h * hea / h2ea, kHEB == h * eb / heb, kH2EB == h * heb / h2eb,
  kHEAB == h * eab / heab, kH2EAB == h * heab / h2eab, kcHEA == he * a / hea,
  kcHEB == he * b / heb, kcHEAB == hea * b / heab,
  et == e + he + h2e + ea + hea + h2ea + eb + heb + h2eb + eab + heab + h2eab},
  {heab}, {e, he, h2e, ea, hea, h2ea, eb, heb, h2eb, eab, h2eab}]
```

$$\{\{heab \rightarrow (a\ b\ et\ h\ kcHEB\ kH2E\ kH2EA\ kH2EAB\ kH2EB)\ /$$
$$\quad (b\ h^2\ kcHEA\ kcHEAB\ kH2E\ kH2EA\ kH2EAB + a\ b\ h^2\ kcHEB\ kH2E\ kH2EA\ kH2EB +$$
$$\quad a\ h^2\ kcHEAB\ kcHEB\ kH2E\ kH2EAB\ kH2EB + h^2\ kcHEA\ kcHEAB\ kcHEB\ kH2EA\ kH2EAB\ kH2EB +$$
$$\quad b\ h\ kcHEA\ kcHEAB\ kH2E\ kH2EA\ kH2EAB\ kH2EB + a\ b\ h\ kcHEB\ kH2E\ kH2EA\ kH2EAB\ kH2EB +$$
$$\quad a\ h\ kcHEAB\ kcHEB\ kH2E\ kH2EA\ kH2EAB\ kH2EB + h\ kcHEA\ kcHEAB\ kcHEB\ kH2E\ kH2EA\ kH2EAB\ kH2EB +$$
$$\quad kcHEA\ kcHEAB\ kcHEB\ kH2E\ kH2EA\ kH2EAB\ kH2EB\ kHE + a\ kcHEAB\ kcHEB\ kH2E\ kH2EA\ kH2EAB\ kH2EB\ kHEA +$$
$$\quad a\ b\ kcHEB\ kH2E\ kH2EA\ kH2EAB\ kH2EB\ kHEAB + b\ kcHEA\ kcHEAB\ kH2E\ kH2EA\ kH2EAB\ kH2EB\ kHEB)\}\}$$

It takes my laptop computer 35 seconds to derive the equilibrium expression for heab.

This rate equation has 12 terms in the denominator because there are 12 kinetic parameters. To obtain the expression for the rapid-equilibrium velocity velrandAB in terms of a and b, et is replaced with kfet.

$$
\begin{aligned}
\text{velrandAB} = &(a\,b\,\text{kfet}\,h\,\text{kcHEB}\,\text{kH2E}\,\text{kH2EA}\,\text{kH2EAB}\,\text{kH2EB}) \Big/ \\
&\Big(b\,h^2\,\text{kcHEA}\,\text{kcHEAB}\,\text{kH2E}\,\text{kH2EA}\,\text{kH2EAB} + a\,b\,h^2\,\text{kcHEB}\,\text{kH2E}\,\text{kH2EA}\,\text{kH2EB} + \\
&a\,h^2\,\text{kcHEAB}\,\text{kcHEB}\,\text{kH2E}\,\text{kH2EAB}\,\text{kH2EB} + h^2\,\text{kcHEA}\,\text{kcHEAB}\,\text{kcHEB}\,\text{kH2EA}\,\text{kH2EAB}\,\text{kH2EB} + \\
&b\,h\,\text{kcHEA}\,\text{kcHEAB}\,\text{kH2E}\,\text{kH2EA}\,\text{kH2EAB}\,\text{kH2EB} + a\,b\,h\,\text{kcHEB}\,\text{kH2E}\,\text{kH2EA}\,\text{kH2EAB}\,\text{kH2EB} + \\
&a\,h\,\text{kcHEAB}\,\text{kcHEB}\,\text{kH2E}\,\text{kH2EA}\,\text{kH2EAB}\,\text{kH2EB} + h\,\text{kcHEA}\,\text{kcHEAB}\,\text{kcHEB}\,\text{kH2E}\,\text{kH2EA}\,\text{kH2EAB}\,\text{kH2EB} + \\
&\text{kcHEA}\,\text{kcHEAB}\,\text{kcHEB}\,\text{kH2E}\,\text{kH2EA}\,\text{kH2EAB}\,\text{kH2EB}\,\text{kHE} + a\,\text{kcHEA}\,\text{kcHEB}\,\text{kH2E}\,\text{kH2EA}\,\text{kH2EAB}\,\text{kH2EB}\,\text{kHEA} + \\
&a\,b\,\text{kcHEB}\,\text{kH2E}\,\text{kH2EA}\,\text{kH2EAB}\,\text{kH2EB}\,\text{kHEAB} + b\,\text{kcHEA}\,\text{kcHEAB}\,\text{kH2E}\,\text{kH2EA}\,\text{kH2EAB}\,\text{kH2EB}\,\text{kHEB}\Big);
\end{aligned}
$$

This rate equation can be simplified by using Simplify.

simpvelrandAB = Simplify[velrandAB]

$$
\begin{aligned}
&(a\,b\,h\,\text{kcHEB}\,\text{kfet}\,\text{kH2E}\,\text{kH2EA}\,\text{kH2EAB}\,\text{kH2EB}) \Big/ \\
&\Big(\text{kcHEAB}\,\text{kcHEB}\,\text{kH2EAB}\,\text{kH2EB}\,\big(\text{kcHEA}\,\text{kH2EA}\,(h^2 + h\,\text{kH2E} + \text{kH2E}\,\text{kHE}) + a\,\text{kH2E}\,(h^2 + h\,\text{kH2EA} + \text{kH2EA}\,\text{kHEA})\big) + \\
&\quad b\,\text{kH2E}\,\text{kH2EA}\,\big(h\,(\text{kcHEA}\,\text{kcHEAB} + a\,\text{kcHEB})\,\text{kH2EAB}\,\text{kH2EB} + \\
&\qquad h^2\,(\text{kcHEA}\,\text{kcHEAB}\,\text{kH2EAB} + a\,\text{kcHEB}\,\text{kH2EB}) + \text{kH2EAB}\,\text{kH2EB}\,(a\,\text{kcHEB}\,\text{kHEAB} + \text{kcHEA}\,\text{kcHEAB}\,\text{kHEB})\big)\Big)
\end{aligned}
$$

The following values for the kinetic parameters are arbitrarily chosen for test calculations:

**Grid[{{pK1e, pK2e, pK1ea, pK2ea, pK1eb, pK2eb, pK1eab, pK2eab, kfet, kcHEA, kcHEB, kcHEAB},
   {7, 6, 7.5, 6.5, 7.8, 6.2, 8, 6, 2, 5, 10, 20}}]**

pK1e	pK2e	pK1ea	pK2ea	pK1eb	pK2eb	pK1eab	pK2eab	kfet	kcHEA	kcHEB	kcHEAB
7	6	7.5	6.5	7.8	6.2	8	6	2	5	10	20

**Grid[{{kHE, kH2E, kHEA, kH2EA, kHEB, kH2EB, kHEAB, kH2EAB, kfet, kcHEA, kcHEB, kcHEAB},
   {10^-7, 10^-6, 10^-7.5, 10^-6.5, 10^-7.8, 10^-6.2, 10^-8, 10^-6, 2, 5, 10, 20}}] // N**

kHE	kH2E	kHEA	kH2EA	kHEB	kH2EB	kHEAB	kH2EAB	kfet	kcHEA	kcHEB	kcHEAB
$1. \times 10^{-7}$	$1. \times 10^{-6}$	3.16228×10^{-8}	3.16228×10^{-7}	1.58489×10^{-8}	6.30957×10^{-7}	$1. \times 10^{-8}$	$1. \times 10^{-6}$	2.	5.	10.	20.

This is the same Grid that was used earlier in Section 4.4.1. These arbitrary values are substituted in the rate equation to obtain the equation vtestrandAB2 that can be used to calculate velocities for chosen {h,a,b}.

**vtestrandAB2 =
   simpvelrandAB /. kHE → 10^-7 /. kH2E → 10^-6 /. kHEA → 10^-7.5 /. kH2EA → 10^-6.5 /.
      kHEB → 10^-7.8 /. kH2EB → 10^-6.2 /. kHEAB → 10^-8 /.
   kH2EAB → 10^-6 /. kfet → 2 /. kcHEA → 5 /. kcHEB → 10 /. kcHEAB → 20**

$$
\begin{aligned}
&\left(3.99052 \times 10^{-24}\,a\,b\,h\right) \Big/ \\
&\left(3.16228 \times 10^{-13}\,b\left(6.30957 \times 10^{-13}\left(1.58489 \times 10^{-6} + \frac{a}{10\,000\,000}\right) + 6.30957 \times 10^{-13}\,(100 + 10\,a)\,h + \right.\right. \\
&\quad \left(\frac{1}{10\,000} + 6.30957 \times 10^{-6}\,a\right)h^2\Bigg) + 1.26191 \times 10^{-10} \\
&\quad \left.\left.\left(\frac{a\left(1. \times 10^{-14} + 3.16228 \times 10^{-7}\,h + h^2\right)}{1\,000\,000} + 1.58114 \times 10^{-6}\left(\frac{1}{10\,000\,000\,000\,000} + \frac{h}{1\,000\,000} + h^2\right)\right)\right)\right)
\end{aligned}
$$

The rate equation at pH 7 is given by

```
simpvelrandAB /. kHE → 10^-7 /. kH2E → 10^-6 /. kHEA → 10^-7.5 /. kH2EA → 10^-6.5 /.
     kHEB → 10^-7.8 /. kH2EB → 10^-6.2 /. kHEAB → 10^-8 /. kH2EAB → 10^-6 /.
     kfet → 2 /. kcHEA → 5 /. kcHEB → 10 /. kcHEAB → 20 /. h → 10^-7
```

$$\left(3.99052 \times 10^{-31}\, a\, b\right) \Big/$$

$$\left(1.26191 \times 10^{-10} \left(3.32039 \times 10^{-19} + 5.16228 \times 10^{-20}\, a\right) + 3.16228 \times 10^{-13} \left(6.30957 \times 10^{-13}\right.\right.$$

$$\left.\left. \left(1.58489 \times 10^{-6} + \frac{a}{10\,000\,000}\right) + \frac{\frac{1}{10\,000} + 6.30957 \times 10^{-6}\, a}{100\,000\,000\,000\,000} + 6.30957 \times 10^{-20}\, (100 + 10\, a)\right) b\right)$$

■ 4.4.4 Estimation of pH-independent kinetic parameters using the rate equation derived by use of Solve

Rate equation simpvelrandAB can be used to write a program calckinparsrandABsimp that can calculate the 12 kinetic parameters from 12 velocity measurements. This program has to contain 12 copies of simpvelrandAB. These different versions of the rate equation can be produced by use of ReplaceAll.

```
simpvelrandAB
```

$$(a\, b\, h\, kcHEB\, kfet\, kH2E\, kH2EA\, kH2EAB\, kH2EB) \Big/$$
$$\left(kcHEAB\, kcHEB\, kH2EAB\, kH2EB \left(kcHEA\, kH2EA \left(h^2 + h\, kH2E + kH2E\, kHE\right) + a\, kH2E \left(h^2 + h\, kH2EA + kH2EA\, kHEA\right)\right) + \right.$$
$$b\, kH2E\, kH2EA \left(h\, (kcHEA\, kcHEAB + a\, kcHEB)\, kH2EAB\, kH2EB + \right.$$
$$\left.\left. h^2 \left(kcHEA\, kcHEAB\, kH2EAB + a\, kcHEB\, kH2EB\right) + kH2EAB\, kH2EB \left(a\, kcHEB\, kHEAB + kcHEA\, kcHEAB\, kHEB\right)\right)\right)$$

To calculate 12 kinetic parameters, 12 versions of the rate equation are needed. The first line of the program is

```
v == simpvelrandAB /. h → h1 /. a → a1 /. b → b1 /. v → v1
```

$$v1 == (a1\, b1\, h1\, kcHEB\, kfet\, kH2E\, kH2EA\, kH2EAB\, kH2EB) \Big/ \left(kcHEAB\, kcHEB\, kH2EAB\, kH2EB\right.$$
$$\left(kcHEA\, kH2EA \left(h1^2 + h1\, kH2E + kH2E\, kHE\right) + a1\, kH2E \left(h1^2 + h1\, kH2EA + kH2EA\, kHEA\right)\right) + b1\, kH2E\, kH2EA$$
$$\left(h1\, (kcHEA\, kcHEAB + a1\, kcHEB)\, kH2EAB\, kH2EB + h1^2 \left(kcHEA\, kcHEAB\, kH2EAB + a1\, kcHEB\, kH2EB\right) + \right.$$
$$\left.\left. kH2EAB\, kH2EB\, (a1\, kcHEB\, kHEAB + kcHEA\, kcHEAB\, kHEB)\right)\right)$$

The full list of rate equations is obtained from

```
{v == simpvelrandAB /. h → h1 /. a → a1 /. b → b1 /. v → v1,
 v == simpvelrandAB /. h → h2 /. a → a2 /. b → b2 /. v → v2,
 v == simpvelrandAB /. h → h3 /. a → a3 /. b → b3 /. v → v3,
 v == simpvelrandAB /. h → h4 /. a → a4 /. b → 41 /. v → v4,
 v == simpvelrandAB /. h → h5 /. a → a5 /. b → b5 /. v → v5,
 v == simpvelrandAB /. h → h6 /. a → a6 /. b → b6 /. v → v6,
 v == simpvelrandAB /. h → h7 /. a → a7 /. b → b7 /. v → v7,
 v == simpvelrandAB /. h → h8 /. a → a8 /. b → b8 /. v → v8,
 v == simpvelrandAB /. h → h9 /. a → a9 /. b → b9 /. v → v9,
 v == simpvelrandAB /. h → h10 /. a → a10 /. b → b10 /. v → v10,
 v == simpvelrandAB /. h → h11 /. a → a11 /. b → b11 /. v → v11,
 v == simpvelrandAB /. h → h12 /. a → a12 /. b → b12 /. v → v12}
```

$\{$ v1 == (a1 b1 h1 kcHEB kfet kH2E kH2EA kH2EAB kH2EB) / (kcHEAB kcHEB kH2EAB kH2EB
$\quad$ (kcHEA kH2EA ($h1^2$ + h1 kH2E + kH2E kHE) + a1 kH2E ($h1^2$ + h1 kH2EA + kH2EA kHEA)) + b1 kH2E kH2EA
$\quad$ (h1 (kcHEA kcHEAB + a1 kcHEB) kH2EAB kH2EB + $h1^2$ (kcHEA kcHEAB kH2EAB + a1 kcHEB kH2EB) +
$\quad$ kH2EAB kH2EB (a1 kcHEB kHEAB + kcHEA kcHEAB kHEB))),

v2 == (a2 b2 h2 kcHEB kfet kH2E kH2EA kH2EAB kH2EB) / (kcHEAB kcHEB kH2EAB kH2EB
$\quad$ (kcHEA kH2EA ($h2^2$ + h2 kH2E + kH2E kHE) + a2 kH2E ($h2^2$ + h2 kH2EA + kH2EA kHEA)) + b2 kH2E kH2EA
$\quad$ (h2 (kcHEA kcHEAB + a2 kcHEB) kH2EAB kH2EB + $h2^2$ (kcHEA kcHEAB kH2EAB + a2 kcHEB kH2EB) +
$\quad$ kH2EAB kH2EB (a2 kcHEB kHEAB + kcHEA kcHEAB kHEB))),

v3 == (a3 b3 h3 kcHEB kfet kH2E kH2EA kH2EAB kH2EB) / (kcHEAB kcHEB kH2EAB kH2EB
$\quad$ (kcHEA kH2EA ($h3^2$ + h3 kH2E + kH2E kHE) + a3 kH2E ($h3^2$ + h3 kH2EA + kH2EA kHEA)) + b3 kH2E kH2EA
$\quad$ (h3 (kcHEA kcHEAB + a3 kcHEB) kH2EAB kH2EB + $h3^2$ (kcHEA kcHEAB kH2EAB + a3 kcHEB kH2EB) +
$\quad$ kH2EAB kH2EB (a3 kcHEB kHEAB + kcHEA kcHEAB kHEB))),

v4 == (41 a4 h4 kcHEB kfet kH2E kH2EA kH2EAB kH2EB) / (kcHEAB kcHEB kH2EAB kH2EB
$\quad$ (kcHEA kH2EA ($h4^2$ + h4 kH2E + kH2E kHE) + a4 kH2E ($h4^2$ + h4 kH2EA + kH2EA kHEA)) + 41 kH2E kH2EA
$\quad$ (h4 (kcHEA kcHEAB + a4 kcHEB) kH2EAB kH2EB + $h4^2$ (kcHEA kcHEAB kH2EAB + a4 kcHEB kH2EB) +
$\quad$ kH2EAB kH2EB (a4 kcHEB kHEAB + kcHEA kcHEAB kHEB))),

v5 == (a5 b5 h5 kcHEB kfet kH2E kH2EA kH2EAB kH2EB) / (kcHEAB kcHEB kH2EAB kH2EB
$\quad$ (kcHEA kH2EA ($h5^2$ + h5 kH2E + kH2E kHE) + a5 kH2E ($h5^2$ + h5 kH2EA + kH2EA kHEA)) + b5 kH2E kH2EA
$\quad$ (h5 (kcHEA kcHEAB + a5 kcHEB) kH2EAB kH2EB + $h5^2$ (kcHEA kcHEAB kH2EAB + a5 kcHEB kH2EB) +
$\quad$ kH2EAB kH2EB (a5 kcHEB kHEAB + kcHEA kcHEAB kHEB))),

v6 == (a6 b6 h6 kcHEB kfet kH2E kH2EA kH2EAB kH2EB) / (kcHEAB kcHEB kH2EAB kH2EB
$\quad$ (kcHEA kH2EA ($h6^2$ + h6 kH2E + kH2E kHE) + a6 kH2E ($h6^2$ + h6 kH2EA + kH2EA kHEA)) + b6 kH2E kH2EA
$\quad$ (h6 (kcHEA kcHEAB + a6 kcHEB) kH2EAB kH2EB + $h6^2$ (kcHEA kcHEAB kH2EAB + a6 kcHEB kH2EB) +
$\quad$ kH2EAB kH2EB (a6 kcHEB kHEAB + kcHEA kcHEAB kHEB))),

v7 == (a7 b7 h7 kcHEB kfet kH2E kH2EA kH2EAB kH2EB) / (kcHEAB kcHEB kH2EAB kH2EB
$\quad$ (kcHEA kH2EA ($h7^2$ + h7 kH2E + kH2E kHE) + a7 kH2E ($h7^2$ + h7 kH2EA + kH2EA kHEA)) + b7 kH2E kH2EA
$\quad$ (h7 (kcHEA kcHEAB + a7 kcHEB) kH2EAB kH2EB + $h7^2$ (kcHEA kcHEAB kH2EAB + a7 kcHEB kH2EB) +
$\quad$ kH2EAB kH2EB (a7 kcHEB kHEAB + kcHEA kcHEAB kHEB))),

v8 == (a8 b8 h8 kcHEB kfet kH2E kH2EA kH2EAB kH2EB) / (kcHEAB kcHEB kH2EAB kH2EB
$\quad$ (kcHEA kH2EA ($h8^2$ + h8 kH2E + kH2E kHE) + a8 kH2E ($h8^2$ + h8 kH2EA + kH2EA kHEA)) + b8 kH2E kH2EA
$\quad$ (h8 (kcHEA kcHEAB + a8 kcHEB) kH2EAB kH2EB + $h8^2$ (kcHEA kcHEAB kH2EAB + a8 kcHEB kH2EB) +
$\quad$ kH2EAB kH2EB (a8 kcHEB kHEAB + kcHEA kcHEAB kHEB))),

v9 == (a9 b9 h9 kcHEB kfet kH2E kH2EA kH2EAB kH2EB) / (kcHEAB kcHEB kH2EAB kH2EB
$\quad$ (kcHEA kH2EA ($h9^2$ + h9 kH2E + kH2E kHE) + a9 kH2E ($h9^2$ + h9 kH2EA + kH2EA kHEA)) + b9 kH2E kH2EA
$\quad$ (h9 (kcHEA kcHEAB + a9 kcHEB) kH2EAB kH2EB + $h9^2$ (kcHEA kcHEAB kH2EAB + a9 kcHEB kH2EB) +
$\quad$ kH2EAB kH2EB (a9 kcHEB kHEAB + kcHEA kcHEAB kHEB))),

v10 == (a10 b10 h10 kcHEB kfet kH2E kH2EA kH2EAB kH2EB) / (kcHEAB kcHEB kH2EAB kH2EB

$$\left(\text{kcHEA kH2EA}\left(\text{h10}^2+\text{h10 kH2E}+\text{kH2E kHE}\right)+\text{a10 kH2E}\left(\text{h10}^2+\text{h10 kH2EA}+\text{kH2EA kHEA}\right)\right)+$$
$$\text{b10 kH2E kH2EA}\left(\text{h10 (kcHEA kcHEAB}+\text{a10 kcHEB) kH2EAB kH2EB}+\text{h10}^2\text{ (kcHEA kcHEAB kH2EAB}+\right.$$
$$\left.\left.\text{a10 kcHEB kH2EB)}+\text{kH2EAB kH2EB (a10 kcHEB kHEAB}+\text{kcHEA kcHEAB kHEB)}\right)\right),$$

$$\text{v11} = (\text{a11 b11 h11 kcHEB kfet kH2E kH2EA kH2EAB kH2EB}) \Big/ \left(\text{kcHEAB kcHEB kH2EAB kH2EB}\right.$$
$$\left(\text{kcHEA kH2EA}\left(\text{h11}^2+\text{h11 kH2E}+\text{kH2E kHE}\right)+\text{a11 kH2E}\left(\text{h11}^2+\text{h11 kH2EA}+\text{kH2EA kHEA}\right)\right)+$$
$$\text{b11 kH2E kH2EA}\left(\text{h11 (kcHEA kcHEAB}+\text{a11 kcHEB) kH2EAB kH2EB}+\text{h11}^2\text{ (kcHEA kcHEAB kH2EAB}+\right.$$
$$\left.\left.\text{a11 kcHEB kH2EB)}+\text{kH2EAB kH2EB (a11 kcHEB kHEAB}+\text{kcHEA kcHEAB kHEB)}\right)\right),$$

$$\text{v12} = (\text{a12 b12 h12 kcHEB kfet kH2E kH2EA kH2EAB kH2EB}) \Big/ \left(\text{kcHEAB kcHEB kH2EAB kH2EB}\right.$$
$$\left(\text{kcHEA kH2EA}\left(\text{h12}^2+\text{h12 kH2E}+\text{kH2E kHE}\right)+\text{a12 kH2E}\left(\text{h12}^2+\text{h12 kH2EA}+\text{kH2EA kHEA}\right)\right)+$$
$$\text{b12 kH2E kH2EA}\left(\text{h12 (kcHEA kcHEAB}+\text{a12 kcHEB) kH2EAB kH2EB}+\text{h12}^2\text{ (kcHEA kcHEAB kH2EAB}+\right.$$
$$\left.\left.\left.\text{a12 kcHEB kH2EB)}+\text{kH2EAB kH2EB (a12 kcHEB kHEAB}+\text{kcHEA kcHEAB kHEB)}\right)\right)\right\}$$

The first part and the last part of the program calcknparsrandABsimp are added to the list of 12 equations.

```
calckinparsrandABsimp[v1_, h1_, a1_, b1_, v2_, h2_, a2_, b2_, v3_, h3_, a3_, b3_, v4_, h4_, a4_,
    b4_, v5_, h5_, a5_, b5_, v6_, h6_, a6_, b6_, v7_, h7_, a7_, b7_, v8_, h8_, a8_, b8_, v9_,
    h9_, a9_, b9_, v10_, h10_, a10_, b10_, v11_, h11_, a11_, b11_, v12_, h12_, a12_, b12_] :=
Module[{}, (*This program calculates 12 kinetic parameters from 12 experimental
        velocities for A + B → products at 12 triplets of [H⁺], [A],
        and [B] on the assumption that the mechanism is random.  Velocities are
        measured at {high [A],high [B]}, {low [A], high[B]}, {high [A], low [B]}
    and {low [A], low [B]} at high [H⁺], intermediate [H⁺], and low [H⁺].*)
Solve[{v1 == (a1 b1 h1 kcHEB kfet kH2E kH2EA kH2EAB kH2EB) /
        (kcHEAB kcHEB kH2EAB kH2EB (kcHEA kH2EA (h1² + h1 kH2E + kH2E kHE) +
            a1 kH2E (h1² + h1 kH2EA + kH2EA kHEA)) + b1 kH2E kH2EA
            (h1 (kcHEA kcHEAB + a1 kcHEB) kH2EAB kH2EB + h1² (kcHEA kcHEAB kH2EAB + a1 kcHEB kH2EB) +
            kH2EAB kH2EB (a1 kcHEB kHEAB + kcHEA kcHEAB kHEB))),
        v2 == (a2 b2 h2 kcHEB kfet kH2E kH2EA kH2EAB kH2EB) / (kcHEAB kcHEB kH2EAB kH2EB
            (kcHEA kH2EA (h2² + h2 kH2E + kH2E kHE) + a2 kH2E (h2² + h2 kH2EA + kH2EA kHEA)) +
            b2 kH2E kH2EA (h2 (kcHEA kcHEAB + a2 kcHEB) kH2EAB kH2EB + h2² (kcHEA kcHEAB kH2EAB +
            a2 kcHEB kH2EB) + kH2EAB kH2EB (a2 kcHEB kHEAB + kcHEA kcHEAB kHEB))),
        v3 == (a3 b3 h3 kcHEB kfet kH2E kH2EA kH2EAB kH2EB) / (kcHEAB kcHEB kH2EAB kH2EB
            (kcHEA kH2EA (h3² + h3 kH2E + kH2E kHE) + a3 kH2E (h3² + h3 kH2EA + kH2EA kHEA)) +
            b3 kH2E kH2EA (h3 (kcHEA kcHEAB + a3 kcHEB) kH2EAB kH2EB + h3² (kcHEA kcHEAB kH2EAB +
            a3 kcHEB kH2EB) + kH2EAB kH2EB (a3 kcHEB kHEAB + kcHEA kcHEAB kHEB))),
        v4 == (a4 b4 h4 kcHEB kfet kH2E kH2EA kH2EAB kH2EB) / (kcHEAB kcHEB kH2EAB kH2EB
            (kcHEA kH2EA (h4² + h4 kH2E + kH2E kHE) + a4 kH2E (h4² + h4 kH2EA + kH2EA kHEA)) +
            b4 kH2E kH2EA (h4 (kcHEA kcHEAB + a4 kcHEB) kH2EAB kH2EB + h4² (kcHEA kcHEAB kH2EAB +
            a4 kcHEB kH2EB) + kH2EAB kH2EB (a4 kcHEB kHEAB + kcHEA kcHEAB kHEB))),
        v5 == (a5 b5 h5 kcHEB kfet kH2E kH2EA kH2EAB kH2EB) / (kcHEAB kcHEB kH2EAB kH2EB
```

$$\left(kcHEA\ kH2EA\ \left(h5^2 + h5\ kH2E + kH2E\ kHE \right) + a5\ kH2E\ \left(h5^2 + h5\ kH2EA + kH2EA\ kHEA \right) \right) +$$
$$b5\ kH2E\ kH2EA\ \left(h5\ \left(kcHEA\ kcHEAB + a5\ kcHEB \right) kH2EAB\ kH2EB + h5^2\ \left(kcHEA\ kcHEAB\ kH2EAB + \right.\right.$$
$$\left.\left.\left. a5\ kcHEB\ kH2EB \right) + kH2EAB\ kH2EB\ \left(a5\ kcHEB\ kHEAB + kcHEA\ kcHEAB\ kHEB \right) \right) \right),$$

$$v6 == \left(a6\ b6\ h6\ kcHEB\ kfet\ kH2E\ kH2EA\ kH2EAB\ kH2EB \right) \Big/ \left(kcHEAB\ kcHEB\ kH2EAB\ kH2EB \right.$$
$$\left(kcHEA\ kH2EA\ \left(h6^2 + h6\ kH2E + kH2E\ kHE \right) + a6\ kH2E\ \left(h6^2 + h6\ kH2EA + kH2EA\ kHEA \right) \right) +$$
$$b6\ kH2E\ kH2EA\ \left(h6\ \left(kcHEA\ kcHEAB + a6\ kcHEB \right) kH2EAB\ kH2EB + h6^2\ \left(kcHEA\ kcHEAB\ kH2EAB + \right.\right.$$
$$\left.\left.\left. a6\ kcHEB\ kH2EB \right) + kH2EAB\ kH2EB\ \left(a6\ kcHEB\ kHEAB + kcHEA\ kcHEAB\ kHEB \right) \right) \right),$$

$$v7 == \left(a7\ b7\ h7\ kcHEB\ kfet\ kH2E\ kH2EA\ kH2EAB\ kH2EB \right) \Big/ \left(kcHEAB\ kcHEB\ kH2EAB\ kH2EB \right.$$
$$\left(kcHEA\ kH2EA\ \left(h7^2 + h7\ kH2E + kH2E\ kHE \right) + a7\ kH2E\ \left(h7^2 + h7\ kH2EA + kH2EA\ kHEA \right) \right) +$$
$$b7\ kH2E\ kH2EA\ \left(h7\ \left(kcHEA\ kcHEAB + a7\ kcHEB \right) kH2EAB\ kH2EB + h7^2\ \left(kcHEA\ kcHEAB\ kH2EAB + \right.\right.$$
$$\left.\left.\left. a7\ kcHEB\ kH2EB \right) + kH2EAB\ kH2EB\ \left(a7\ kcHEB\ kHEAB + kcHEA\ kcHEAB\ kHEB \right) \right) \right),$$

$$v8 == \left(a8\ b8\ h8\ kcHEB\ kfet\ kH2E\ kH2EA\ kH2EAB\ kH2EB \right) \Big/ \left(kcHEAB\ kcHEB\ kH2EAB\ kH2EB \right.$$
$$\left(kcHEA\ kH2EA\ \left(h8^2 + h8\ kH2E + kH2E\ kHE \right) + a8\ kH2E\ \left(h8^2 + h8\ kH2EA + kH2EA\ kHEA \right) \right) +$$
$$b8\ kH2E\ kH2EA\ \left(h8\ \left(kcHEA\ kcHEAB + a8\ kcHEB \right) kH2EAB\ kH2EB + h8^2\ \left(kcHEA\ kcHEAB\ kH2EAB + \right.\right.$$
$$\left.\left.\left. a8\ kcHEB\ kH2EB \right) + kH2EAB\ kH2EB\ \left(a8\ kcHEB\ kHEAB + kcHEA\ kcHEAB\ kHEB \right) \right) \right),$$

$$v9 == \left(a9\ b9\ h9\ kcHEB\ kfet\ kH2E\ kH2EA\ kH2EAB\ kH2EB \right) \Big/ \left(kcHEAB\ kcHEB\ kH2EAB\ kH2EB \right.$$
$$\left(kcHEA\ kH2EA\ \left(h9^2 + h9\ kH2E + kH2E\ kHE \right) + a9\ kH2E\ \left(h9^2 + h9\ kH2EA + kH2EA\ kHEA \right) \right) +$$
$$b9\ kH2E\ kH2EA\ \left(h9\ \left(kcHEA\ kcHEAB + a9\ kcHEB \right) kH2EAB\ kH2EB + h9^2\ \left(kcHEA\ kcHEAB\ kH2EAB + \right.\right.$$
$$\left.\left.\left. a9\ kcHEB\ kH2EB \right) + kH2EAB\ kH2EB\ \left(a9\ kcHEB\ kHEAB + kcHEA\ kcHEAB\ kHEB \right) \right) \right),$$

$$v10 == \left(a10\ b10\ h10\ kcHEB\ kfet\ kH2E\ kH2EA\ kH2EAB\ kH2EB \right) \Big/ \left(kcHEAB\ kcHEB\ kH2EAB\ kH2EB \right.$$
$$\left(kcHEA\ kH2EA\ \left(h10^2 + h10\ kH2E + kH2E\ kHE \right) + a10\ kH2E\ \left(h10^2 + h10\ kH2EA + kH2EA\ kHEA \right) \right) +$$
$$b10\ kH2E\ kH2EA\ \left(h10\ \left(kcHEA\ kcHEAB + a10\ kcHEB \right) kH2EAB\ kH2EB + h10^2\ \left(kcHEA\ kcHEAB\ kH2EAB + \right.\right.$$
$$\left.\left.\left. a10\ kcHEB\ kH2EB \right) + kH2EAB\ kH2EB\ \left(a10\ kcHEB\ kHEAB + kcHEA\ kcHEAB\ kHEB \right) \right) \right),$$

$$v11 == \left(a11\ b11\ h11\ kcHEB\ kfet\ kH2E\ kH2EA\ kH2EAB\ kH2EB \right) \Big/ \left(kcHEAB\ kcHEB\ kH2EAB\ kH2EB \right.$$
$$\left(kcHEA\ kH2EA\ \left(h11^2 + h11\ kH2E + kH2E\ kHE \right) + a11\ kH2E\ \left(h11^2 + h11\ kH2EA + kH2EA\ kHEA \right) \right) +$$
$$b11\ kH2E\ kH2EA\ \left(h11\ \left(kcHEA\ kcHEAB + a11\ kcHEB \right) kH2EAB\ kH2EB + h11^2\ \left(kcHEA\ kcHEAB\ kH2EAB + \right.\right.$$
$$\left.\left.\left. a11\ kcHEB\ kH2EB \right) + kH2EAB\ kH2EB\ \left(a11\ kcHEB\ kHEAB + kcHEA\ kcHEAB\ kHEB \right) \right) \right),$$

$$v12 == \left(a12\ b12\ h12\ kcHEB\ kfet\ kH2E\ kH2EA\ kH2EAB\ kH2EB \right) \Big/ \left(kcHEAB\ kcHEB\ kH2EAB\ kH2EB \right.$$
$$\left(kcHEA\ kH2EA\ \left(h12^2 + h12\ kH2E + kH2E\ kHE \right) + a12\ kH2E\ \left(h12^2 + h12\ kH2EA + kH2EA\ kHEA \right) \right) +$$
$$b12\ kH2E\ kH2EA\ \left(h12\ \left(kcHEA\ kcHEAB + a12\ kcHEB \right) kH2EAB\ kH2EB + h12^2\ \left(kcHEA\ kcHEAB\ kH2EAB + \right.\right.$$
$$\left.\left.\left. a12\ kcHEB\ kH2EB \right) + kH2EAB\ kH2EB\ \left(a12\ kcHEB\ kHEAB + kcHEA\ kcHEAB\ kHEB \right) \right) \right) \right\},$$
$$\left\{ kHE,\ kH2E,\ kHEA,\ kH2EA,\ kHEB,\ kH2EB,\ kHEAB,\ kH2EAB,\ kfet,\ kcHEA,\ kcHEAB,\ kcHEB \right\} \Big] \Big]$$

Calculate 12 velocities using vtestrandAB2 that was made in Section 4.4.3.

```
vtestrandAB2 /. h → 10^-6 /. a → 500 /. b → 500
```

```
0.896615
```

```
vtestrandAB2 /. h → 10^-6 /. a → 1 /. b → 500
```

```
0.0699174
```

```
vtestrandAB2 /. h → 10^-6 /. a → 500 /. b → 1
```

```
0.0231589
```

```
vtestrandAB2 /. h → 10^-6 /. a → 5 /. b → 5
```

```
0.061753
```

```
vtestrandAB2 /. h → 10^-7 /. a → 500 /. b → 500
```

1.54742

```
vtestrandAB2 /. h → 10^-7 /. a → 1 /. b → 500
```

0.134634

```
vtestrandAB2 /. h → 10^-7 /. a → 500 /. b → 1
```

0.0583168

```
vtestrandAB2 /. h → 10^-7 /. a → 5 /. b → 5
```

0.106588

```
vtestrandAB2 /. h → 10^-8 /. a → 500 /. b → 500
```

0.895185

```
vtestrandAB2 /. h → 10^-8 /. a → 1 /. b → 500
```

0.0658172

```
vtestrandAB2 /. h → 10^-8 /. a → 500 /. b → 1
```

0.0226907

```
vtestrandAB2 /. h → 10^-8 /. a → 5 /. b → 5
```

0.0294001

These velocities are now considered to be experimental data to be used to estimate the 12 kinetic parameters.

```
calckinparsrandABsimp[.8966, 10^-6, 500, 500, 0.06992, 10^-6, 1, 500,
  0.02316, 10^-6, 500, 1, 0.06175, 10^-6, 5, 5, 1.547, 10^-7, 500, 500, 0.1346,
  10^-7, 1, 500, 0.05832, 10^-7, 500, 1, 0.1066, 10^-7, 5, 5, 0.8952, 10^-8,
  500, 500, 0.06582, 10^-8, 1, 500, .02269, 10^-8, 500, 1, .02940, 10^-8, 5, 5]
```

$\{\{kfet \to 1.99914, kHEAB \to 9.99108 \times 10^{-9}, kHEA \to 3.16291 \times 10^{-8}, kHEB \to 1.58355 \times 10^{-8},$
$\quad kHE \to 1.001 \times 10^{-7}, kcHEA \to 4.99602, kcHEAB \to 19.989, kcHEB \to 9.98612,$
$\quad kH2E \to 9.97516 \times 10^{-7}, kH2EB \to 6.31491 \times 10^{-7}, kH2EA \to 3.162 \times 10^{-7}, kH2EAB \to 1.0008 \times 10^{-6}\}\}$

These values are correct, but it is important to test for errors in the velocity measurements.

```
Grid[{{pK1e, pK2e, pK1ea, pK2ea, pK1eb, pK2eb, pK1eab, pK2eab, kfet, kcHEA, kcHEB, kcHEAB},
  {7, 6, 7.5, 6.5, 7.8, 6.2, 8, 6, 2, 5, 10, 20}}]
```

pK1e	pK2e	pK1ea	pK2ea	pK1eb	pK2eb	pK1eab	pK2eab	kfet	kcHEA	kcHEB	kcHEAB
7	6	7.5	6.5	7.8	6.2	8	6	2	5	10	20

```
Grid[{{kHE, kH2E, kHEA, kH2EA, kHEB, kH2EB, kHEAB, kH2EAB, kfet, kcHEA, kcHEB, kcHEAB},
  {10^-7, 10^-6, 10^-7.5, 10^-6.5, 10^-7.8, 10^-6.2, 10^-8, 10^-6, 2, 5, 10, 20}}] // N
```

kHE	kH2E	kHEA	kH2EA	kHEB	kH2EB	kHEAB	kH2EAB	kfet	kcHEA	kcHEB	kcHEAB
$1. \times 10^{-7}$	$1. \times 10^{-6}$	3.16228×10^{-8}	3.16228×10^{-7}	1.58489×10^{-8}	6.30957×10^{-7}	$1. \times 10^{-8}$	$1. \times 10^{-6}$	2.	5.	10.	20.

Construct a table with 13 lines to show the effects of 5% errors in the measured velocities, one at a time.

```
lin1 = {kfet, kHEAB, kHEA, kHEB, kHE, kcHEA, kcHEAB, kcHEB, kH2E, kH2EB, kH2EA, kH2EAB} /.
   calckinparsrandABsimp[.8966, 10^-6, 500, 500, 0.06992, 10^-6, 1, 500,
    0.02316, 10^-6, 500, 1, 0.06175, 10^-6, 5, 5, 1.547, 10^-7, 500, 500, 0.1346,
    10^-7, 1, 500, 0.05832, 10^-7, 500, 1, 0.1066, 10^-7, 5, 5, 0.8952, 10^-8,
    500, 500, 0.06582, 10^-8, 1, 500, .02269, 10^-8, 500, 1, .02940, 10^-8, 5, 5]
```

$$\{\{1.99914, 9.99108 \times 10^{-9}, 3.16291 \times 10^{-8}, 1.58355 \times 10^{-8}, 1.001 \times 10^{-7}, 4.99602,$$
$$19.989, 9.98612, 9.97516 \times 10^{-7}, 6.31491 \times 10^{-7}, 3.162 \times 10^{-7}, 1.0008 \times 10^{-6}\}\}$$

```
calckinparsrandABsimp[1.05 * .8966, 10^-6, 500, 500, 0.06992, 10^-6, 1, 500,
 0.02316, 10^-6, 500, 1, 0.06175, 10^-6, 5, 5, 1.547, 10^-7, 500, 500, 0.1346,
 10^-7, 1, 500, 0.05832, 10^-7, 500, 1, 0.1066, 10^-7, 5, 5, 0.8952, 10^-8,
 500, 500, 0.06582, 10^-8, 1, 500, .02269, 10^-8, 500, 1, .02940, 10^-8, 5, 5]
```

$$\{\{kfet \to 1.97318, kHEAB \to 9.74322 \times 10^{-9}, kHEA \to 3.1655 \times 10^{-8}, kHEB \to 1.58673 \times 10^{-8},$$
$$kHE \to 1.00321 \times 10^{-7}, kcHEA \to 4.98911, kcHEAB \to 19.7168, kcHEB \to 9.97865,$$
$$kH2E \to 9.77598 \times 10^{-7}, kH2EB \to 6.26111 \times 10^{-7}, kH2EA \to 3.15421 \times 10^{-7}, kH2EAB \to 1.15189 \times 10^{-6}\}\}$$

```
lin2 = {kfet, kHEAB, kHEA, kHEB, kHE, kcHEA, kcHEAB, kcHEB, kH2E, kH2EB, kH2EA, kH2EAB} /.
   calckinparsrandABsimp[1.05 * .8966, 10^-6, 500, 500, 0.06992, 10^-6, 1, 500,
    0.02316, 10^-6, 500, 1, 0.06175, 10^-6, 5, 5, 1.547, 10^-7, 500, 500, 0.1346,
    10^-7, 1, 500, 0.05832, 10^-7, 500, 1, 0.1066, 10^-7, 5, 5, 0.8952, 10^-8,
    500, 500, 0.06582, 10^-8, 1, 500, .02269, 10^-8, 500, 1, .02940, 10^-8, 5, 5]
```

$$\{\{1.97318, 9.74322 \times 10^{-9}, 3.1655 \times 10^{-8}, 1.58673 \times 10^{-8}, 1.00321 \times 10^{-7}, 4.98911,$$
$$19.7168, 9.97865, 9.77598 \times 10^{-7}, 6.26111 \times 10^{-7}, 3.15421 \times 10^{-7}, 1.15189 \times 10^{-6}\}\}$$

```
calckinparsrandABsimp[.8966, 10^-6, 500, 500, 1.05 * 0.06992, 10^-6, 1, 500,
 0.02316, 10^-6, 500, 1, 0.06175, 10^-6, 5, 5, 1.547, 10^-7, 500, 500, 0.1346,
 10^-7, 1, 500, 0.05832, 10^-7, 500, 1, 0.1066, 10^-7, 5, 5, 0.8952, 10^-8,
 500, 500, 0.06582, 10^-8, 1, 500, .02269, 10^-8, 500, 1, .02940, 10^-8, 5, 5]
```

$$\{\{kfet \to 1.99982, kHEAB \to 9.99757 \times 10^{-9}, kHEA \to 3.16256 \times 10^{-8}, kHEB \to 1.54186 \times 10^{-8},$$
$$kHE \to 1.01038 \times 10^{-7}, kcHEA \to 4.95305, kcHEAB \to 19.9975, kcHEB \to 9.73544,$$
$$kH2E \to 9.18104 \times 10^{-7}, kH2EB \to 7.11745 \times 10^{-7}, kH2EA \to 3.16304 \times 10^{-7}, kH2EAB \to 9.97375 \times 10^{-7}\}\}$$

```
lin3 = {kfet, kHEAB, kHEA, kHEB, kHE, kcHEA, kcHEAB, kcHEB, kH2E, kH2EB, kH2EA, kH2EAB} /.
   calckinparsrandABsimp[.8966, 10^-6, 500, 500, 1.05 * 0.06992, 10^-6, 1, 500,
    0.02316, 10^-6, 500, 1, 0.06175, 10^-6, 5, 5, 1.547, 10^-7, 500, 500, 0.1346,
    10^-7, 1, 500, 0.05832, 10^-7, 500, 1, 0.1066, 10^-7, 5, 5, 0.8952, 10^-8,
    500, 500, 0.06582, 10^-8, 1, 500, .02269, 10^-8, 500, 1, .02940, 10^-8, 5, 5]
```

$$\{\{1.99982, 9.99757 \times 10^{-9}, 3.16256 \times 10^{-8}, 1.54186 \times 10^{-8}, 1.01038 \times 10^{-7}, 4.95305,$$
$$19.9975, 9.73544, 9.18104 \times 10^{-7}, 7.11745 \times 10^{-7}, 3.16304 \times 10^{-7}, 9.97375 \times 10^{-7}\}\}$$

```
calckinparsrandABsimp[.8966, 10^-6, 500, 500, 0.06992, 10^-6, 1, 500,
 1.05 * 0.02316, 10^-6, 500, 1, 0.06175, 10^-6, 5, 5, 1.547, 10^-7, 500, 500,
 0.1346, 10^-7, 1, 500, 0.05832, 10^-7, 500, 1, 0.1066, 10^-7, 5, 5, 0.8952, 10^-8,
 500, 500, 0.06582, 10^-8, 1, 500, .02269, 10^-8, 500, 1, .02940, 10^-8, 5, 5]
```

$$\{\{kfet \to 2.0012, kHEAB \to 1.00107 \times 10^{-8}, kHEA \to 3.0609 \times 10^{-8}, kHEB \to 1.58227 \times 10^{-8},$$
$$kHE \to 1.02982 \times 10^{-7}, kcHEA \to 4.74561, kcHEAB \to 20.5237, kcHEB \to 9.72433,$$
$$kH2E \to 7.88088 \times 10^{-7}, kH2EB \to 6.33684 \times 10^{-7}, kH2EA \to 3.5019 \times 10^{-7}, kH2EAB \to 9.90518 \times 10^{-7}\}\}$$

```
lin4 = {kfet, kHEAB, kHEA, kHEB, kHE, kcHEA, kcHEAB, kcHEB, kH2E, kH2EB, kH2EA, kH2EAB} /.
   calckinparsrandABsimp[.8966, 10^-6, 500, 500, 0.06992, 10^-6, 1, 500,
      1.05 * 0.02316, 10^-6, 500, 1, 0.06175, 10^-6, 5, 5, 1.547, 10^-7, 500, 500,
      0.1346, 10^-7, 1, 500, 0.05832, 10^-7, 500, 1, 0.1066, 10^-7, 5, 5, 0.8952, 10^-8,
      500, 500, 0.06582, 10^-8, 1, 500, .02269, 10^-8, 500, 1, .02940, 10^-8, 5, 5]
```

$\{\{2.0012, 1.00107 \times 10^{-8}, 3.0609 \times 10^{-8}, 1.58227 \times 10^{-8}, 1.02982 \times 10^{-7}, 4.74561,$
$\quad 20.5237, 9.72433, 7.88088 \times 10^{-7}, 6.33684 \times 10^{-7}, 3.5019 \times 10^{-7}, 9.90518 \times 10^{-7}\}\}$

```
calckinparsrandABsimp[.8966, 10^-6, 500, 500, 0.06992, 10^-6, 1, 500, 0.02316,
   10^-6, 500, 1, 1.05 * 0.06175, 10^-6, 5, 5, 1.547, 10^-7, 500, 500, 0.1346,
   10^-7, 1, 500, 0.05832, 10^-7, 500, 1, 0.1066, 10^-7, 5, 5, 0.8952, 10^-8,
   500, 500, 0.06582, 10^-8, 1, 500, .02269, 10^-8, 500, 1, .02940, 10^-8, 5, 5]
```

$\{\{\text{kfet} \to 1.9991, \text{kHEAB} \to 9.99071 \times 10^{-9}, \text{kHEA} \to 3.16489 \times 10^{-8}, \text{kHEB} \to 1.58598 \times 10^{-8},$
$\quad \text{kHE} \to 9.50383 \times 10^{-8}, \text{kcHEA} \to 5.24144, \text{kcHEAB} \to 19.9789, \text{kcHEB} \to 10.4818,$
$\quad \text{kH2E} \to 1.87069 \times 10^{-6}, \text{kH2EB} \to 6.27381 \times 10^{-7}, \text{kH2EA} \to 3.15606 \times 10^{-7}, \text{kH2EAB} \to 1.001 \times 10^{-6}\}\}$

```
lin5 = {kfet, kHEAB, kHEA, kHEB, kHE, kcHEA, kcHEAB, kcHEB, kH2E, kH2EB, kH2EA, kH2EAB} /.
   calckinparsrandABsimp[.8966, 10^-6, 500, 500, 0.06992, 10^-6, 1, 500, 0.02316,
      10^-6, 500, 1, 1.05 * 0.06175, 10^-6, 5, 5, 1.547, 10^-7, 500, 500, 0.1346,
      10^-7, 1, 500, 0.05832, 10^-7, 500, 1, 0.1066, 10^-7, 5, 5, 0.8952, 10^-8,
      500, 500, 0.06582, 10^-8, 1, 500, .02269, 10^-8, 500, 1, .02940, 10^-8, 5, 5]
```

$\{\{1.9991, 9.99071 \times 10^{-9}, 3.16489 \times 10^{-8}, 1.58598 \times 10^{-8}, 9.50383 \times 10^{-8}, 5.24144,$
$\quad 19.9789, 10.4818, 1.87069 \times 10^{-6}, 6.27381 \times 10^{-7}, 3.15606 \times 10^{-7}, 1.001 \times 10^{-6}\}\}$

```
calckinparsrandABsimp[.8966, 10^-6, 500, 500, 0.06992, 10^-6, 1, 500, 0.02316,
   10^-6, 500, 1, 0.06175, 10^-6, 5, 5, 1.05 * 1.547, 10^-7, 500, 500, 0.1346,
   10^-7, 1, 500, 0.05832, 10^-7, 500, 1, 0.1066, 10^-7, 5, 5, 0.8952, 10^-8,
   500, 500, 0.06582, 10^-8, 1, 500, .02269, 10^-8, 500, 1, .02940, 10^-8, 5, 5]
```

$\{\{\text{kfet} \to 2.166, \text{kHEAB} \to 1.16513 \times 10^{-8}, \text{kHEA} \to 3.14748 \times 10^{-8}, \text{kHEB} \to 1.56452 \times 10^{-8},$
$\quad \text{kHE} \to 9.88152 \times 10^{-8}, \text{kcHEA} \to 5.03631, \text{kcHEAB} \to 21.7381, \text{kcHEB} \to 10.0294,$
$\quad \text{kH2E} \to 1.02123 \times 10^{-6}, \text{kH2EB} \to 6.39175 \times 10^{-7}, \text{kH2EA} \to 3.1775 \times 10^{-7}, \text{kH2EAB} \to 8.58199 \times 10^{-7}\}\}$

```
lin6 = {kfet, kHEAB, kHEA, kHEB, kHE, kcHEA, kcHEAB, kcHEB, kH2E, kH2EB, kH2EA, kH2EAB} /.
   calckinparsrandABsimp[.8966, 10^-6, 500, 500, 0.06992, 10^-6, 1, 500, 0.02316,
      10^-6, 500, 1, 0.06175, 10^-6, 5, 5, 1.05 * 1.547, 10^-7, 500, 500, 0.1346,
      10^-7, 1, 500, 0.05832, 10^-7, 500, 1, 0.1066, 10^-7, 5, 5, 0.8952, 10^-8,
      500, 500, 0.06582, 10^-8, 1, 500, .02269, 10^-8, 500, 1, .02940, 10^-8, 5, 5]
```

$\{\{2.166, 1.16513 \times 10^{-8}, 3.14748 \times 10^{-8}, 1.56452 \times 10^{-8}, 9.88152 \times 10^{-8}, 5.03631,$
$\quad 21.7381, 10.0294, 1.02123 \times 10^{-6}, 6.39175 \times 10^{-7}, 3.1775 \times 10^{-7}, 8.58199 \times 10^{-7}\}\}$

```
calckinparsrandABsimp[.8966, 10^-6, 500, 500, 0.06992, 10^-6, 1, 500, 0.02316,
   10^-6, 500, 1, 0.06175, 10^-6, 5, 5, 1.547, 10^-7, 500, 500, 1.05 * 0.1346,
   10^-7, 1, 500, 0.05832, 10^-7, 500, 1, 0.1066, 10^-7, 5, 5, 0.8952, 10^-8,
   500, 500, 0.06582, 10^-8, 1, 500, .02269, 10^-8, 500, 1, .02940, 10^-8, 5, 5]
```

$\{\{\text{kfet} \to 1.99558, \text{kHEAB} \to 9.95563 \times 10^{-9}, \text{kHEA} \to 3.16476 \times 10^{-8}, \text{kHEB} \to 1.83578 \times 10^{-8},$
$\quad \text{kHE} \to 9.53948 \times 10^{-8}, \text{kcHEA} \to 5.22162, \text{kcHEAB} \to 19.9445, \text{kcHEB} \to 11.4548,$
$\quad \text{kH2E} \to 1.09021 \times 10^{-6}, \text{kH2EB} \to 5.44729 \times 10^{-7}, \text{kH2EA} \to 3.16015 \times 10^{-7}, \text{kH2EAB} \to 1.00437 \times 10^{-6}\}\}$

```
lin7 = {kfet, kHEAB, kHEA, kHEB, kHE, kcHEA, kcHEAB, kcHEB, kH2E, kH2EB, kH2EA, kH2EAB} /.
   calckinparsrandABsimp[.8966, 10^-6, 500, 500, 0.06992, 10^-6, 1, 500, 0.02316,
      10^-6, 500, 1, 0.06175, 10^-6, 5, 5, 1.547, 10^-7, 500, 500, 1.05 * 0.1346,
      10^-7, 1, 500, 0.05832, 10^-7, 500, 1, 0.1066, 10^-7, 5, 5, 0.8952, 10^-8,
      500, 500, 0.06582, 10^-8, 1, 500, .02269, 10^-8, 500, 1, .02940, 10^-8, 5, 5]
```

$$\{\{1.99558, 9.95563 \times 10^{-9}, 3.16476 \times 10^{-8}, 1.83578 \times 10^{-8}, 9.53948 \times 10^{-8}, 5.22162,$$
$$19.9445, 11.4548, 1.09021 \times 10^{-6}, 5.44729 \times 10^{-7}, 3.16015 \times 10^{-7}, 1.00437 \times 10^{-6}\}\}$$

```
calckinparsrandABsimp[.8966, 10^-6, 500, 500, 0.06992, 10^-6, 1, 500,
   0.02316, 10^-6, 500, 1, 0.06175, 10^-6, 5, 5, 1.547, 10^-7, 500, 500, 0.1346,
   10^-7, 1, 500, 1.05 * 0.05832, 10^-7, 500, 1, 0.1066, 10^-7, 5, 5, 0.8952, 10^-8,
   500, 500, 0.06582, 10^-8, 1, 500, .02269, 10^-8, 500, 1, .02940, 10^-8, 5, 5]
```

$$\{\{kfet \to 1.99094, kHEAB \to 9.90945 \times 10^{-9}, kHEA \to 3.64009 \times 10^{-8}, kHEB \to 1.58887 \times 10^{-8},$$
$$kHE \to 8.98155 \times 10^{-8}, kcHEA \to 6.14457, kcHEAB \to 17.8554, kcHEB \to 11.0388,$$
$$kH2E \to 1.22521 \times 10^{-6}, kH2EB \to 6.2938 \times 10^{-7}, kH2EA \to 2.74748 \times 10^{-7}, kH2EAB \to 1.00905 \times 10^{-6}\}\}$$

```
lin8 = {kfet, kHEAB, kHEA, kHEB, kHE, kcHEA, kcHEAB, kcHEB, kH2E, kH2EB, kH2EA, kH2EAB} /.
   calckinparsrandABsimp[.8966, 10^-6, 500, 500, 0.06992, 10^-6, 1, 500, 0.02316,
      10^-6, 500, 1, 0.06175, 10^-6, 5, 5, 1.547, 10^-7, 500, 500, 0.1346, 10^-7,
      1, 500, 1.05 * 0.05832, 10^-7, 500, 1, 0.1066, 10^-7, 5, 5, 0.8952, 10^-8,
      500, 500, 0.06582, 10^-8, 1, 500, .02269, 10^-8, 500, 1, .02940, 10^-8, 5, 5]
```

$$\{\{1.99094, 9.90945 \times 10^{-9}, 3.64009 \times 10^{-8}, 1.58887 \times 10^{-8}, 8.98155 \times 10^{-8}, 6.14457,$$
$$17.8554, 11.0388, 1.22521 \times 10^{-6}, 6.2938 \times 10^{-7}, 2.74748 \times 10^{-7}, 1.00905 \times 10^{-6}\}\}$$

```
calckinparsrandABsimp[.8966, 10^-6, 500, 500, 0.06992, 10^-6, 1, 500,
   0.02316, 10^-6, 500, 1, 0.06175, 10^-6, 5, 5, 1.547, 10^-7, 500, 500, 0.1346,
   10^-7, 1, 500, 0.05832, 10^-7, 500, 1, 1.05 * 0.1066, 10^-7, 5, 5, 0.8952, 10^-8,
   500, 500, 0.06582, 10^-8, 1, 500, .02269, 10^-8, 500, 1, .02940, 10^-8, 5, 5]
```

$$\{\{kfet \to 1.99937, kHEAB \to 9.99334 \times 10^{-9}, kHEA \to 3.15114 \times 10^{-8}, kHEB \to 1.56902 \times 10^{-8},$$
$$kHE \to 1.43822 \times 10^{-7}, kcHEA \to 3.56493, kcHEAB \to 20.0481, kcHEB \to 7.10552,$$
$$kH2E \to 5.57267 \times 10^{-7}, kH2EB \to 6.37342 \times 10^{-7}, kH2EA \to 3.17381 \times 10^{-7}, kH2EAB \to 1.00058 \times 10^{-6}\}\}$$

```
lin9 = {kfet, kHEAB, kHEA, kHEB, kHE, kcHEA, kcHEAB, kcHEB, kH2E, kH2EB, kH2EA, kH2EAB} /.
   calckinparsrandABsimp[.8966, 10^-6, 500, 500, 0.06992, 10^-6, 1, 500,
      0.02316, 10^-6, 500, 1, 0.06175, 10^-6, 5, 5, 1.547, 10^-7, 500, 500, 0.1346,
      10^-7, 1, 500, 0.05832, 10^-7, 500, 1, 1.05 * 0.1066, 10^-7, 5, 5, 0.8952, 10^-8,
      500, 500, 0.06582, 10^-8, 1, 500, .02269, 10^-8, 500, 1, .02940, 10^-8, 5, 5]
```

$$\{\{1.99937, 9.99334 \times 10^{-9}, 3.15114 \times 10^{-8}, 1.56902 \times 10^{-8}, 1.43822 \times 10^{-7}, 3.56493,$$
$$20.0481, 7.10552, 5.57267 \times 10^{-7}, 6.37342 \times 10^{-7}, 3.17381 \times 10^{-7}, 1.00058 \times 10^{-6}\}\}$$

```
calckinparsrandABsimp[.8966, 10^-6, 500, 500, 0.06992, 10^-6, 1, 500,
   0.02316, 10^-6, 500, 1, 0.06175, 10^-6, 5, 5, 1.547, 10^-7, 500, 500, 0.1346,
   10^-7, 1, 500, 0.05832, 10^-7, 500, 1, 0.1066, 10^-7, 5, 5, 1.05 * 0.8952, 10^-8,
   500, 500, 0.06582, 10^-8, 1, 500, .02269, 10^-8, 500, 1, .02940, 10^-8, 5, 5]
```

$$\{\{kfet \to 1.97314, kHEAB \to 8.67847 \times 10^{-9}, kHEA \to 3.17073 \times 10^{-8}, kHEB \to 1.59718 \times 10^{-8},$$
$$kHE \to 1.00488 \times 10^{-7}, kcHEA \to 4.9891, kcHEAB \to 19.7164, kcHEB \to 9.97864,$$
$$kH2E \to 9.93676 \times 10^{-7}, kH2EB \to 6.30225 \times 10^{-7}, kH2EA \to 3.1594 \times 10^{-7}, kH2EAB \to 1.0263 \times 10^{-6}\}\}$$

```
lin10 = {kfet, kHEAB, kHEA, kHEB, kHE, kcHEA, kcHEAB, kcHEB, kH2E, kH2EB, kH2EA, kH2EAB} /.
   calckinparsrandABsimp[.8966, 10^-6, 500, 500, 0.06992, 10^-6, 1, 500, 0.02316,
    10^-6, 500, 1, 0.06175, 10^-6, 5, 5, 1.547, 10^-7, 500, 500, 0.1346, 10^-7,
    1, 500, 0.05832, 10^-7, 500, 1, 0.1066, 10^-7, 5, 5, 1.05 * 0.8952, 10^-8,
    500, 500, 0.06582, 10^-8, 1, 500, .02269, 10^-8, 500, 1, .02940, 10^-8, 5, 5]
```

$$\{\{1.97314, 8.67847 \times 10^{-9}, 3.17073 \times 10^{-8}, 1.59718 \times 10^{-8}, 1.00488 \times 10^{-7}, 4.9891,$$
$$19.7164, 9.97864, 9.93676 \times 10^{-7}, 6.30225 \times 10^{-7}, 3.1594 \times 10^{-7}, 1.0263 \times 10^{-6}\}\}$$

```
calckinparsrandABsimp[.8966, 10^-6, 500, 500, 0.06992, 10^-6, 1, 500,
  0.02316, 10^-6, 500, 1, 0.06175, 10^-6, 5, 5, 1.547, 10^-7, 500, 500, 0.1346,
  10^-7, 1, 500, 0.05832, 10^-7, 500, 1, 0.1066, 10^-7, 5, 5, 0.8952, 10^-8, 500,
  500, 1.05 * 0.06582, 10^-8, 1, 500, .02269, 10^-8, 500, 1, .02940, 10^-8, 5, 5]
```

$$\{\{kfet \rightarrow 1.99987, kHEAB \rightarrow 1.00276 \times 10^{-8}, kHEA \rightarrow 3.1618 \times 10^{-8}, kHEB \rightarrow 1.39407 \times 10^{-8},$$
$$kHE \rightarrow 1.01845 \times 10^{-7}, kcHEA \rightarrow 4.95037, kcHEAB \rightarrow 19.9981, kcHEB \rightarrow 9.7201,$$
$$kH2E \rightarrow 9.80446 \times 10^{-7}, kH2EB \rightarrow 6.49643 \times 10^{-7}, kH2EA \rightarrow 3.16237 \times 10^{-7}, kH2EAB \rightarrow 1.00011 \times 10^{-6}\}\}$$

```
lin11 = {kfet, kHEAB, kHEA, kHEB, kHE, kcHEA, kcHEAB, kcHEB, kH2E, kH2EB, kH2EA, kH2EAB} /.
   calckinparsrandABsimp[.8966, 10^-6, 500, 500, 0.06992, 10^-6, 1, 500, 0.02316,
    10^-6, 500, 1, 0.06175, 10^-6, 5, 5, 1.547, 10^-7, 500, 500, 0.1346, 10^-7,
    1, 500, 0.05832, 10^-7, 500, 1, 0.1066, 10^-7, 5, 5, 0.8952, 10^-8, 500, 500,
    1.05 * 0.06582, 10^-8, 1, 500, .02269, 10^-8, 500, 1, .02940, 10^-8, 5, 5]
```

$$\{\{1.99987, 1.00276 \times 10^{-8}, 3.1618 \times 10^{-8}, 1.39407 \times 10^{-8}, 1.01845 \times 10^{-7}, 4.95037,$$
$$19.9981, 9.7201, 9.80446 \times 10^{-7}, 6.49643 \times 10^{-7}, 3.16237 \times 10^{-7}, 1.00011 \times 10^{-6}\}\}$$

```
calckinparsrandABsimp[.8966, 10^-6, 500, 500, 0.06992, 10^-6, 1, 500,
  0.02316, 10^-6, 500, 1, 0.06175, 10^-6, 5, 5, 1.547, 10^-7, 500, 500, 0.1346,
  10^-7, 1, 500, 0.05832, 10^-7, 500, 1, 0.1066, 10^-7, 5, 5, 0.8952, 10^-8, 500,
  500, 0.06582, 10^-8, 1, 500, 1.05 * .02269, 10^-8, 500, 1, .02940, 10^-8, 5, 5]
```

$$\{\{kfet \rightarrow 2.00124, kHEAB \rightarrow 1.0097 \times 10^{-8}, kHEA \rightarrow 2.84974 \times 10^{-8}, kHEB \rightarrow 1.57796 \times 10^{-8},$$
$$kHE \rightarrow 1.05249 \times 10^{-7}, kcHEA \rightarrow 4.74056, kcHEAB \rightarrow 20.5348, kcHEB \rightarrow 9.71891,$$
$$kH2E \rightarrow 9.48751 \times 10^{-7}, kH2EB \rightarrow 6.32012 \times 10^{-7}, kH2EA \rightarrow 3.26958 \times 10^{-7}, kH2EAB \rightarrow 9.98801 \times 10^{-7}\}\}$$

```
lin12 = {kfet, kHEAB, kHEA, kHEB, kHE, kcHEA, kcHEAB, kcHEB, kH2E, kH2EB, kH2EA, kH2EAB} /.
   calckinparsrandABsimp[.8966, 10^-6, 500, 500, 0.06992, 10^-6, 1, 500, 0.02316,
    10^-6, 500, 1, 0.06175, 10^-6, 5, 5, 1.547, 10^-7, 500, 500, 0.1346, 10^-7,
    1, 500, 0.05832, 10^-7, 500, 1, 0.1066, 10^-7, 5, 5, 0.8952, 10^-8, 500, 500,
    0.06582, 10^-8, 1, 500, 1.05 * .02269, 10^-8, 500, 1, .02940, 10^-8, 5, 5]
```

$$\{\{2.00124, 1.0097 \times 10^{-8}, 2.84974 \times 10^{-8}, 1.57796 \times 10^{-8}, 1.05249 \times 10^{-7}, 4.74056,$$
$$20.5348, 9.71891, 9.48751 \times 10^{-7}, 6.32012 \times 10^{-7}, 3.26958 \times 10^{-7}, 9.98801 \times 10^{-7}\}\}$$

```
calckinparsrandABsimp[.8966, 10^-6, 500, 500, 0.06992, 10^-6, 1, 500,
  0.02316, 10^-6, 500, 1, 0.06175, 10^-6, 5, 5, 1.547, 10^-7, 500, 500, 0.1346,
  10^-7, 1, 500, 0.05832, 10^-7, 500, 1, 0.1066, 10^-7, 5, 5, 0.8952, 10^-8, 500,
  500, 0.06582, 10^-8, 1, 500, .02269, 10^-8, 500, 1, 1.05 * .02940, 10^-8, 5, 5]
```

$$\{\{kfet \rightarrow 1.99906, kHEAB \rightarrow 9.98696 \times 10^{-9}, kHEA \rightarrow 3.17542 \times 10^{-8}, kHEB \rightarrow 1.60537 \times 10^{-8},$$
$$kHE \rightarrow 8.24043 \times 10^{-8}, kcHEA \rightarrow 5.51176, kcHEAB \rightarrow 19.9678, kcHEB \rightarrow 11.0282,$$
$$kH2E \rightarrow 1.21155 \times 10^{-6}, kH2EB \rightarrow 6.29466 \times 10^{-7}, kH2EA \rightarrow 3.15785 \times 10^{-7}, kH2EAB \rightarrow 1.00088 \times 10^{-6}\}\}$$

```
lin13 = {kfet, kHEAB, kHEA, kHEB, kHE, kcHEA, kcHEAB, kcHEB, kH2E, kH2EB, kH2EA, kH2EAB} /.
   calckinparsrandABsimp[.8966, 10^-6, 500, 500, 0.06992, 10^-6, 1, 500, 0.02316,
     10^-6, 500, 1, 0.06175, 10^-6, 5, 5, 1.547, 10^-7, 500, 500, 0.1346, 10^-7,
     1, 500, 0.05832, 10^-7, 500, 1, 0.1066, 10^-7, 5, 5, 0.8952, 10^-8, 500, 500,
     0.06582, 10^-8, 1, 500, .02269, 10^-8, 500, 1, 1.05 * .02940, 10^-8, 5, 5]
```

$$\{\{1.99906, 9.98696 \times 10^{-9}, 3.17542 \times 10^{-8}, 1.60537 \times 10^{-8}, 8.24043 \times 10^{-8}, 5.51176,$$
$$19.9678, 11.0282, 1.21155 \times 10^{-6}, 6.29466 \times 10^{-7}, 3.15785 \times 10^{-7}, 1.00088 \times 10^{-6}\}\}$$

Table 4.8 Estimation of 12 kinetic parameters from 12 velocities when Solve is used to derive the rate equation and to write the program calckinparsrandABsimp. The velocities were at slightly different velocities than for Table 4.7.

```
TableForm[
  Round[-Log[10, {lin1[[1]], lin2[[1]], lin3[[1]], lin4[[1]], lin5[[1]], lin6[[1]], lin7[[1]],
       lin8[[1]], lin9[[1]], lin10[[1]], lin11[[1]], lin12[[1]], lin13[[1]]}], 0.01],
  TableHeadings → {{"No errors", "1.05*v1", "1.05*v2", "1.05*v3", "1.05*v4", "1.05*v5",
     "1.05*v6", "1.05*v7", "1.05*v8", "1.05*v9", "1.05*v10", "1.05*v11", "1.05*v12"},
    {"kfet", "pK1eab", "pK1ea", "pK1ea", "pK1e", "kcHEA", "kcHEAB",
     "kcHEB", "pK2e", "pK2eb", "pK2ea", "pK2eab"}}]
```

	kfet	pK1eab	pK1ea	pK1ea	pK1e	kcHEA	kcHEAB	kcHEB	pK2e	pK2eb	pK2ea	pK2eab
No errors	-0.3	8.	7.5	7.8	7.	-0.7	-1.3	-1.	6.	6.2	6.5	6.
1.05*v1	-0.3	8.01	7.5	7.8	7.	-0.7	-1.29	-1.	6.01	6.2	6.5	5.94
1.05*v2	-0.3	8.	7.5	7.81	7.	-0.69	-1.3	-0.99	6.04	6.15	6.5	6.
1.05*v3	-0.3	8.	7.51	7.8	6.99	-0.68	-1.31	-0.99	6.1	6.2	6.46	6.
1.05*v4	-0.3	8.	7.5	7.8	7.02	-0.72	-1.3	-1.02	5.73	6.2	6.5	6.
1.05*v5	-0.34	7.93	7.5	7.81	7.01	-0.7	-1.34	-1.	5.99	6.19	6.5	6.07
1.05*v6	-0.3	8.	7.5	7.74	7.02	-0.72	-1.3	-1.06	5.96	6.26	6.5	6.
1.05*v7	-0.3	8.	7.44	7.8	7.05	-0.79	-1.25	-1.04	5.91	6.2	6.56	6.
1.05*v8	-0.3	8.	7.5	7.8	6.84	-0.55	-1.3	-0.85	6.25	6.2	6.5	6.
1.05*v9	-0.3	8.06	7.5	7.8	7.	-0.7	-1.29	-1.	6.	6.2	6.5	5.99
1.05*v10	-0.3	8.	7.5	7.86	6.99	-0.69	-1.3	-0.99	6.01	6.19	6.5	6.
1.05*v11	-0.3	8.	7.55	7.8	6.98	-0.68	-1.31	-0.99	6.02	6.2	6.49	6.
1.05*v12	-0.3	8.	7.5	7.79	7.08	-0.74	-1.3	-1.04	5.92	6.2	6.5	6.

Note that the columns for kfet, kcHEA, kcHEAB, and kcHEB are for the negative base 10 logarithms. These values are the same as those in Table 4.7. Notice that four columns are not pKs.

```
-Log[10, {2, 5, 10, 20}] // N
```

$$\{-0.30103, -0.69897, -1., -1.30103\}$$

4.5 Effects of pH on the kinetics of random A + B → products when hydrogen ions are consumed in the rate-determining reaction

■ **4.5.1 When one hydrogen ion is consumed**

There are two different ways that hydrogen ions are involved in the thermodynamics and rapid-equilibrium kinetics of reactions at a specified pH. In addition to the effects of pKs, an integer number of hydrogen ions may be consumed in the rate-determining reaction. This is often the case for reactions involving NAD. Up to 8 hydrogen ions may be consumed in reductase reactions [29]. When a single hydrogen ion is consumed, the last reaction in the mechanism in Section 4.4.1 is replaced with

$$EAB^-$$
$$\| \quad k_f$$
$$H^+ + HEAB \; \rightarrow \; products \qquad\qquad v = k_f[H^+][HEAB] \qquad v = kfet*h*eab/et$$
$$\|$$
$$H_2\,EAB^+$$

There is a simple but powerful change in the expression for the velocity. If one hydrogen ion is consumed, kfet is replaced with h*kfet. If the rate equation is expressed in terms of the pH, kfet is replaced with 10^{-pH}*kfet. Since this introduces large numbers, it is convenient to use 10^{-pH+7}*kfet = $10^{-(pH-7)}$*kfet. This is equivalent to using $h/10^{-7}$*kfet. When a single hydrogen ion is consumed, the rate equation is given by simpvelordABh = h 10^7 simpvelordAB. The 10^7 has been omitted because it is arbitrary.

simpvelrandABh = h * simpvelrandAB

$$\left(a\, b\, h^2\, kcHEB\, kfet\, kH2E\, kH2EA\, kH2EAB\, kH2EB\right) \Big/$$
$$\Big(kcHEAB\, kcHEB\, kH2EAB\, kH2EB\, \big(kcHEA\, kH2EA\, (h^2 + h\, kH2E + kH2E\, kHE) + a\, kH2E\, (h^2 + h\, kH2EA + kH2EA\, kHEA)\big) +$$
$$b\, kH2E\, kH2EA\, \big(h\, (kcHEA\, kcHEAB + a\, kcHEB)\, kH2EAB\, kH2EB +$$
$$h^2\, (kcHEA\, kcHEAB\, kH2EAB + a\, kcHEB\, kH2EB) + kH2EAB\, kH2EB\, (a\, kcHEB\, kHEAB + kcHEA\, kcHEAB\, kHEB)\big)\Big)$$

This rate equation is used to write a program to estimate the 12 kinetic parameters from 12 measured velocities. The 12 simultaneous rate equations can be used to write the following computer program.

```
calckinparsrandABsimph[v1_, h1_, a1_, b1_, v2_, h2_, a2_, b2_, v3_, h3_, a3_, b3_, v4_, h4_, a4_,
    b4_, v5_, h5_, a5_, b5_, v6_, h6_, a6_, b6_, v7_, h7_, a7_, b7_, v8_, h8_, a8_, b8_, v9_,
    h9_, a9_, b9_, v10_, h10_, a10_, b10_, v11_, h11_, a11_, b11_, v12_, h12_, a12_, b12_] :=
Module[{}, (*This program calculates 12 kinetic parameters from 12 experimental
        velocities for A + B → products at 12 triplets of [H+], [A],
        and [B] on the assumption that the mechanism is random.  Velocities are
        measured at {high [A],high [B]}, {low [A], high[B]}, {high [A], low [B]}
    and {low [A], low [B]} at high [H+], intermediate [H+], and low [H+].*)
    Solve[{v1 == (a1 b1 h1² kcHEB kfet kH2E kH2EA kH2EAB kH2EB) /
```

$$\Big(\text{kcHEAB kcHEB kH2EAB kH2EB} \big(\text{kcHEA kH2EA} \left(h1^2 + h1\ kH2E + kH2E\ kHE\right) +$$
$$a1\ kH2E \left(h1^2 + h1\ kH2EA + kH2EA\ kHEA\right)\big) + b1\ kH2E\ kH2EA$$
$$\big(h1\ (\text{kcHEA kcHEAB} + a1\ \text{kcHEB})\ \text{kH2EAB kH2EB} + h1^2\ (\text{kcHEA kcHEAB kH2EAB} + a1\ \text{kcHEB kH2EB}) +$$
$$\text{kH2EAB kH2EB}\ (a1\ \text{kcHEB kHEAB} + \text{kcHEA kcHEAB kHEB})\big)\Big),$$

$$v2 == \left(a2\ b2\ h2^2\ \text{kcHEB kfet kH2E kH2EA kH2EAB kH2EB}\right) \Big/ \big(\text{kcHEAB kcHEB kH2EAB kH2EB}$$
$$\big(\text{kcHEA kH2EA} \left(h2^2 + h2\ kH2E + kH2E\ kHE\right) + a2\ kH2E \left(h2^2 + h2\ kH2EA + kH2EA\ kHEA\right)\big) +$$
$$b2\ kH2E\ kH2EA \big(h2\ (\text{kcHEA kcHEAB} + a2\ \text{kcHEB})\ \text{kH2EAB kH2EB} + h2^2\ (\text{kcHEA kcHEAB kH2EAB} +$$
$$a2\ \text{kcHEB kH2EB}) + \text{kH2EAB kH2EB}\ (a2\ \text{kcHEB kHEAB} + \text{kcHEA kcHEAB kHEB})\big)\big),$$

$$v3 == \left(a3\ b3\ h3^2\ \text{kcHEB kfet kH2E kH2EA kH2EAB kH2EB}\right) \Big/ \big(\text{kcHEAB kcHEB kH2EAB kH2EB}$$
$$\big(\text{kcHEA kH2EA} \left(h3^2 + h3\ kH2E + kH2E\ kHE\right) + a3\ kH2E \left(h3^2 + h3\ kH2EA + kH2EA\ kHEA\right)\big) +$$
$$b3\ kH2E\ kH2EA \big(h3\ (\text{kcHEA kcHEAB} + a3\ \text{kcHEB})\ \text{kH2EAB kH2EB} + h3^2\ (\text{kcHEA kcHEAB kH2EAB} +$$
$$a3\ \text{kcHEB kH2EB}) + \text{kH2EAB kH2EB}\ (a3\ \text{kcHEB kHEAB} + \text{kcHEA kcHEAB kHEB})\big)\big),$$

$$v4 == \left(a4\ b4\ h4^2\ \text{kcHEB kfet kH2E kH2EA kH2EAB kH2EB}\right) \Big/ \big(\text{kcHEAB kcHEB kH2EAB kH2EB}$$
$$\big(\text{kcHEA kH2EA} \left(h4^2 + h4\ kH2E + kH2E\ kHE\right) + a4\ kH2E \left(h4^2 + h4\ kH2EA + kH2EA\ kHEA\right)\big) +$$
$$b4\ kH2E\ kH2EA \big(h4\ (\text{kcHEA kcHEAB} + a4\ \text{kcHEB})\ \text{kH2EAB kH2EB} + h4^2\ (\text{kcHEA kcHEAB kH2EAB} +$$
$$a4\ \text{kcHEB kH2EB}) + \text{kH2EAB kH2EB}\ (a4\ \text{kcHEB kHEAB} + \text{kcHEA kcHEAB kHEB})\big)\big),$$

$$v5 == \left(a5\ b5\ h5^2\ \text{kcHEB kfet kH2E kH2EA kH2EAB kH2EB}\right) \Big/ \big(\text{kcHEAB kcHEB kH2EAB kH2EB}$$
$$\big(\text{kcHEA kH2EA} \left(h5^2 + h5\ kH2E + kH2E\ kHE\right) + a5\ kH2E \left(h5^2 + h5\ kH2EA + kH2EA\ kHEA\right)\big) +$$
$$b5\ kH2E\ kH2EA \big(h5\ (\text{kcHEA kcHEAB} + a5\ \text{kcHEB})\ \text{kH2EAB kH2EB} + h5^2\ (\text{kcHEA kcHEAB kH2EAB} +$$
$$a5\ \text{kcHEB kH2EB}) + \text{kH2EAB kH2EB}\ (a5\ \text{kcHEB kHEAB} + \text{kcHEA kcHEAB kHEB})\big)\big),$$

$$v6 == \left(a6\ b6\ h6^2\ \text{kcHEB kfet kH2E kH2EA kH2EAB kH2EB}\right) \Big/ \big(\text{kcHEAB kcHEB kH2EAB kH2EB}$$
$$\big(\text{kcHEA kH2EA} \left(h6^2 + h6\ kH2E + kH2E\ kHE\right) + a6\ kH2E \left(h6^2 + h6\ kH2EA + kH2EA\ kHEA\right)\big) +$$
$$b6\ kH2E\ kH2EA \big(h6\ (\text{kcHEA kcHEAB} + a6\ \text{kcHEB})\ \text{kH2EAB kH2EB} + h6^2\ (\text{kcHEA kcHEAB kH2EAB} +$$
$$a6\ \text{kcHEB kH2EB}) + \text{kH2EAB kH2EB}\ (a6\ \text{kcHEB kHEAB} + \text{kcHEA kcHEAB kHEB})\big)\big),$$

$$v7 == \left(a7\ b7\ h7^2\ \text{kcHEB kfet kH2E kH2EA kH2EAB kH2EB}\right) \Big/ \big(\text{kcHEAB kcHEB kH2EAB kH2EB}$$
$$\big(\text{kcHEA kH2EA} \left(h7^2 + h7\ kH2E + kH2E\ kHE\right) + a7\ kH2E \left(h7^2 + h7\ kH2EA + kH2EA\ kHEA\right)\big) +$$
$$b7\ kH2E\ kH2EA \big(h7\ (\text{kcHEA kcHEAB} + a7\ \text{kcHEB})\ \text{kH2EAB kH2EB} + h7^2\ (\text{kcHEA kcHEAB kH2EAB} +$$
$$a7\ \text{kcHEB kH2EB}) + \text{kH2EAB kH2EB}\ (a7\ \text{kcHEB kHEAB} + \text{kcHEA kcHEAB kHEB})\big)\big),$$

$$v8 == \left(a8\ b8\ h8^2\ \text{kcHEB kfet kH2E kH2EA kH2EAB kH2EB}\right) \Big/ \big(\text{kcHEAB kcHEB kH2EAB kH2EB}$$
$$\big(\text{kcHEA kH2EA} \left(h8^2 + h8\ kH2E + kH2E\ kHE\right) + a8\ kH2E \left(h8^2 + h8\ kH2EA + kH2EA\ kHEA\right)\big) +$$
$$b8\ kH2E\ kH2EA \big(h8\ (\text{kcHEA kcHEAB} + a8\ \text{kcHEB})\ \text{kH2EAB kH2EB} + h8^2\ (\text{kcHEA kcHEAB kH2EAB} +$$
$$a8\ \text{kcHEB kH2EB}) + \text{kH2EAB kH2EB}\ (a8\ \text{kcHEB kHEAB} + \text{kcHEA kcHEAB kHEB})\big)\big),$$

$$v9 == \left(a9\ b9\ h9^2\ \text{kcHEB kfet kH2E kH2EA kH2EAB kH2EB}\right) \Big/ \big(\text{kcHEAB kcHEB kH2EAB kH2EB}$$
$$\big(\text{kcHEA kH2EA} \left(h9^2 + h9\ kH2E + kH2E\ kHE\right) + a9\ kH2E \left(h9^2 + h9\ kH2EA + kH2EA\ kHEA\right)\big) +$$
$$b9\ kH2E\ kH2EA \big(h9\ (\text{kcHEA kcHEAB} + a9\ \text{kcHEB})\ \text{kH2EAB kH2EB} + h9^2\ (\text{kcHEA kcHEAB kH2EAB} +$$
$$a9\ \text{kcHEB kH2EB}) + \text{kH2EAB kH2EB}\ (a9\ \text{kcHEB kHEAB} + \text{kcHEA kcHEAB kHEB})\big)\big),$$

$$v10 == \left(a10\ b10\ h10^2\ \text{kcHEB kfet kH2E kH2EA kH2EAB kH2EB}\right) \Big/ \big(\text{kcHEAB kcHEB kH2EAB kH2EB}$$

$$\left(\text{kcHEA kH2EA}\left(\text{h10}^2 + \text{h10 kH2E} + \text{kH2E kHE}\right) + \text{a10 kH2E}\left(\text{h10}^2 + \text{h10 kH2EA} + \text{kH2EA kHEA}\right)\right) +$$
$$\text{b10 kH2E kH2EA}\left(\text{h10}\left(\text{kcHEA kcHEAB} + \text{a10 kcHEB}\right)\text{kH2EAB kH2EB} + \text{h10}^2\left(\text{kcHEA kcHEAB kH2EAB} + \right.\right.$$
$$\left.\left.\text{a10 kcHEB kH2EB}\right) + \text{kH2EAB kH2EB}\left(\text{a10 kcHEB kHEAB} + \text{kcHEA kcHEAB kHEB}\right)\right)\right),$$
$$\text{v11} = \left(\text{a11 b11 h11}^2\,\text{kcHEB kfet kH2E kH2EA kH2EAB kH2EB}\right) \Big/ \Big(\text{kcHEAB kcHEB kH2EAB kH2EB}$$
$$\left(\text{kcHEA kH2EA}\left(\text{h11}^2 + \text{h11 kH2E} + \text{kH2E kHE}\right) + \text{a11 kH2E}\left(\text{h11}^2 + \text{h11 kH2EA} + \text{kH2EA kHEA}\right)\right) +$$
$$\text{b11 kH2E kH2EA}\left(\text{h11}\left(\text{kcHEA kcHEAB} + \text{a11 kcHEB}\right)\text{kH2EAB kH2EB} + \text{h11}^2\left(\text{kcHEA kcHEAB kH2EAB} + \right.\right.$$
$$\left.\left.\text{a11 kcHEB kH2EB}\right) + \text{kH2EAB kH2EB}\left(\text{a11 kcHEB kHEAB} + \text{kcHEA kcHEAB kHEB}\right)\right)\right),$$
$$\text{v12} = \left(\text{a12 b12 h12}^2\,\text{kcHEB kfet kH2E kH2EA kH2EAB kH2EB}\right) \Big/ \Big(\text{kcHEAB kcHEB kH2EAB kH2EB}$$
$$\left(\text{kcHEA kH2EA}\left(\text{h12}^2 + \text{h12 kH2E} + \text{kH2E kHE}\right) + \text{a12 kH2E}\left(\text{h12}^2 + \text{h12 kH2EA} + \text{kH2EA kHEA}\right)\right) +$$
$$\text{b12 kH2E kH2EA}\left(\text{h12}\left(\text{kcHEA kcHEAB} + \text{a12 kcHEB}\right)\text{kH2EAB kH2EB} + \text{h12}^2\left(\text{kcHEA kcHEAB kH2EAB} + \right.\right.$$
$$\left.\left.\text{a12 kcHEB kH2EB}\right) + \text{kH2EAB kH2EB}\left(\text{a12 kcHEB kHEAB} + \text{kcHEA kcHEAB kHEB}\right)\right)\right)\Big\},$$

```
{kHE, kH2E, kHEA, kH2EA, kHEB, kH2EB, kHEAB, kH2EAB, kfet, kcHEA,
 kcHEAB,
 kcHEB}]]
```

This program yields the same values of the 12 kinetic parameters as when no hydrogen ions are consumed in the rate-determining reaction (see Section 4.4.4), but V_{fexp} is proportional to $10^{-\text{pH}}$.

■ 4.5.2 When *n* hydrogen ions are consumed

$V_{\text{fexp}} = 10^{n(\text{pH}-7)}\text{kfet} \times \text{eab/et}$, where $n = 0, -1, -2,....$ The values of *n* are negative because that is the convention for reactants in chemical reactions. The discussion in Section 3.5 applies here. The number of hydrogen ions consumed in the rate-determining reaction can be determined by studying V_{fexp} as described in Section 3.5.3. The effects of the consumption of hydrogen ions is powerful.

4.6 Discussion

An important distinction between the rapid-equilibrium rate equations for the random mechanism and the ordered mechanism is that the random mechanism involves a cycle. This makes it possible to calculate an additional Michaelis constant. Publication of rapid-equilbrium kinetic data for the random mechanism should include this additional Michaelis constant. Another source of programs on random A + B → products is *MathSource* [*MathSource*5].

In using the minimum number of velocity determinations at a specified pH, Figure 4.1 makes it clear what pairs {[A],[B]} to use for velocity measurements to obtain the most accurate values of the kinetic parameters. The major determinant of the accuracy is the determination of velocities at low substrate concentrations. Replicate measurements of velocities can be used to improve the accuracy. Section 4.3 shows again that when a reaction A + B → products is studied for the first time, the more general computer program (that is the program for the random mechanism) should be used first.

The inclusion of pKs for the enzymatic site and enzyme-substrate complexes leads to a large increase in the number of kinetic parameters. Rapid-equilibrium rate equations can be derived using [H^+] or pH, but rate equations in [H^+], h in *Mathematica*, have to be used to determine the kinetic parameters using the minimum number of velocity measurements because Solve was developed to solve sets of simultaneous polynomial equations. For test calculations, vfhab is used to calculate twelve velocities, and vgen is used to write the program calckinparsrandAB. The calculations of kinetic parameters and effects of 5% errors in velocity measurements look long, but it can be accomplished quickly by copying and pasteing.

This chapter provides derivations of five rate equations and demonstrates their applications.

vrandAB1 is the rapid-equilibrium rate equation for mechanism I. The velocity is a function of [A], [B], and four kinetic parameters.

vrandAB2 is the rapid-equilibrium rate equation for mechanism II. The velocity is a function of [A], [B], and four kinetic parameters.

vgen is the rapid-equilibrium rate equation for mechanism I when the enzymatic site and enzyme-substrate complexes each have two pKs. The velocity is a function of [A], [B], [H^+], and 12 pH-independent kinetic parameters. The velocity is derived using the factor method. **vgen** was used to calculate 12 velocities and then use these velocities to calculate kinetic parameters that are given in Table 4.7.

simpvelrandAB is like vgen, except that it is derived using Solve. **simpvelrandAB** was used to calculate 12 velocities and then use these velocities to calculate kinetic parameters that are given in Table 4.8. These tables are really equivalent even though different choices of {h,a,b} were used.

simpvelrandABh was derived by multiplying **simpvelrandAB** by h. The consumption of H^+ in the rate determining reaction causes large effects, but this does not affect the estimation of pKs.

Chapter 5 A + B = P + Q

5.1 Ordered A + B = ordered P + Q

- 5.1.1 Derivation of the rapid-equilibrium rate equation for ordered A + B = ordered P + Q

- 5.1.2 Estimation of pH-dependent kinetic parameters with the minimum number of velocity measurements

- 5.1.3 Derivation of the rapid-equilibrium rate equation when K' is very large and A, B, and Q are present initially

- 5.1.4 Estimation of pH-dependent kinetic parameters when K' is very large and A, B, P, and Q are present initially

- 5.1.5 pH effects and the estimation of pH-independent kinetic parameters

- 5.1.6 Dead-end enzyme-substrate complexes

5.2 Ordered A + B = random P + Q

- 5.2.1 Derivation of the rapid-equilibrium rate equation

- 5.2.2 Estimation of pH-dependent kinetic parameters with the minimum number of velocity measurements

---

5.3 Random A + B = random P + Q

- 5.3.1 Derivation of the rapid-equilibrium rate equation

- 5.3.2 Estimation of pH-dependent kinetic parameters with the minimum number of measured velocities

- 5.3.3 Application of calckinparsrandABPQfr to data for ordered A + B = ordered P + Q

- 5.3.4 Application of calckinparsrandABPQfr to data for ordered A + B = random P + Q

- 5.3.5 Table of velocities when calckinparsrandABPQfr is applied to velocity data for 3 mechanisms

---

5.4 Appendix

- Alternate rate equation for ordered A + B = random P + Q

5.5 Discussion

5.1 Ordered A + B = ordered P + Q

■ 5.1.1 Derivation of the rapid-equilibrium rate equation for ordered A + B = ordered P + Q

The rapid-equilibrium kinetics of ordered A + B → products was discussed in Chapter 3, but there are advantages in discussing the forward and reverse reactions at the same time. The rate-determining reaction for ordered A + B = P + Q is EAB ⇌ EPQ. The assumption is that other reactions in the mechanism are at equilibrium. For ordered A + B = ordered P + Q, these other reactions are

E + A = EA   kIA = e*a/ea
EA + B = EAB   kB = ea*b/eab
E + Q = EQ   kIQ = e*q/eq
EQ + P = EPQ   kP = eq*p/epq

The thermodynamic constraints are the expressions for the equilibrium constants of these four reactions and the conservation equation for the total concentration of enzymatic sites et:

et == e + ea + eab + eq + epq

There are 9 reactants, four independent equilibrium equations, and 5 components (A, B, P, Q, and E). Linear algebra [23] shows that the number of reactants N' must be equal to the number of components C' plus the number of independent reactions R' so that $N' = C' + R'$ is 9 = 5 + 4. It is convenient to solve the 5 equations that have to be satisfied using Solve[eqs,vars,elims].

First, derive the expression for the equilibrium concentration eab.

```
Solve[{kIA == e * a / ea, kB == ea * b / eab, kIQ == e * q / eq,
  kP == eq * p / epq, et == e + ea + eab + eq + epq}, {eab}, {e, ea, eq, epq}]
```

$$\left\{\left\{eab \rightarrow \frac{a\,b\,et\,kIQ\,kP}{a\,b\,kIQ\,kP + a\,kB\,kIQ\,kP + kB\,kIA\,kIQ\,kP + kB\,kIA\,kP\,q + kB\,kIA\,p\,q}\right\}\right\}$$

The equilibrium concentration eab is proportional to et, as expected.

Second, derive the expression for the equilibrium concentration epq.

```
Solve[{kIA == e * a / ea, kB == ea * b / eab, kIQ == e * q / eq,
  kP == eq * p / epq, et == e + ea + eab + eq + epq}, {epq}, {e, ea, ep, eq, eab}]
```

$$\left\{\left\{epq \rightarrow \frac{et\,kB\,kIA\,p\,q}{a\,b\,kIQ\,kP + a\,kB\,kIQ\,kP + kB\,kIA\,kIQ\,kP + kB\,kIA\,kP\,q + kB\,kIA\,p\,q}\right\}\right\}$$

The whole mechanism is

E + A = EA   kIA = e*a/ea
EA + B = EAB   kB = ea*b/eab
EAB ⇌ EPQ   v = kf*eab - kr*epq
E + Q = EQ   kIQ = e*q/eq
EQ + P = EPQ   kP = eq*p/epq

To calculate the velocity in the forward direction, the expression for eab is multiplied by kf. When this is done, kf*et can then be replaced with the symbol vfexp = kf*et for the limiting velocity in the forward direction to obtain the expression for the velocity vf of the forward reaction. To obtain the rate of the reverse reaction, epq is multiplied by kr so that vrexp = kr*et. The factor kr*et in the rate equation for the reverse reaction is replaced with the symbol vrexp for the limiting velocity in the reverse direction. Thus the reaction velocity is given by

v = vf − vr = kf*eab - kr*epq

Note that when v = 0, kf*eab = kr*epq. Substituting the expressions for eab and epq in this equation and changing et in the expression eab to vfexp and et in the expression for epq to vrexp yields the rapid-equilibrium velocity.

$$\frac{\text{a b vfexp kIQ kP}}{\text{a b kIQ kP + a kB kIQ kP + kB kIA kIQ kP + kB kIA kP q + kB kIA p q}} -$$

$$\frac{\text{vrexp kB kIA p q}}{\text{a b kIQ kP + a kB kIQ kP + kB kIA kIQ kP + kB kIA kP q + kB kIA p q}};$$

Since the denominators are the same, the velocity vordABordPQ of A + B = P + Q in the forward direction is given by

$$\text{vordABordPQ} = \frac{\text{(a b vfexp kIQ kP – vrexp kB kIA p q)}}{\text{a b kIQ kP + a kB kIQ kP + kB kIA kIQ kP + kB kIA kP q + kB kIA p q}};$$

There are other ways to write this equation [21,25], but this is the way *Mathematica* writes it. An advantage in using Solve to derive the rate equation is that the rate equation is obtained in computer-readable form. Another advantage is that more complicated mechanisms can be studied more easily because no hand calculations are required.

Since there are 4 Michaelis constants and 2 limiting velocities, there are 6 kinetic parameters, and so 6 velocities have to be measured to estimate the values of the 6 kinetic parameters. When the reverse reaction is included, the number of kinetic parameters is one more than the number of terms in the denominator of the rate equation.

When vordABordPQ = 0, the Haldane relation, which expresses the apparent equilibrium constant K' in terms of kinetic parameters is

$$K' = \frac{\text{vfexp kP kIQ}}{\text{vrexp kIA kB}} = \frac{\text{kf kP kIQ}}{\text{kr kIA kB}}$$

Rate equation vordABordPQ can be used to calculate velocities for any set of {a,b,p,q}.

For these velocity calculations, the 6 kinetic parameters are arbitrarily set at vfexp = 1, vrexp = 0.5 , kIA = 5, kB = 20, kIQ = 10, and kP = 15. Since there are 6 kinetic parameters, it is necessary to measure 6 velocities to calculate the kinetic parameters with the minimum number of velocities. To illustrate this method, 6 velocities at the following quadruplets of substrate concentrations {a,b,p,q} are used: {100,100,0,0}, {1,100,0,0}, {100,1,0,0}, {0,0,100,100}, {0,0,1,100}, {0,0,100,1}. This follows the recomendations used in Chapter 3. The 6 test velocities are calculated by use of ReplaceAll (/.x->).

```
vordABordPQ /. vfexp → 1 /. vrexp → .5 /. kIA → 5 /. kB → 20 /. kIQ → 10 /. kP → 15 /. a → 100 /.
    b → 100 /. p → 0 /. q → 0 // N

0.826446
```

```
vordABordPQ /. vfexp → 1 /. vrexp → .5 /. kIA → 5 /. kB → 20 /. kIQ → 10 /. kP → 15 /. a → 1 /.
    b → 100 /. p → 0 /. q → 0 // N

0.454545
```

```
vordABordPQ /. vfexp → 1 /. vrexp → .5 /. kIA → 5 /. kB → 20 /. kIQ → 10 /. kP → 15 /. a → 100 /.
    b → 1 /. p → 0 /. q → 0 // N

0.0454545
```

```
vordABordPQ /. vfexp → 1 /. vrexp → .5 /. kIA → 5 /. kB → 20 /. kIQ → 10 /. kP → 15 /. a → 0 /. b → 0 /.
   p → 100 /. q → 100 // N
```

-0.429185

```
vordABordPQ /. vfexp → 1 /. vrexp → .5 /. kIA → 5 /. kB → 20 /. kIQ → 10 /. kP → 15 /. a -> 0 /.
   b → 0 /. p → 100 /. q → 1 // N
```

-0.188679

```
vordABordPQ /. vfexp → 1 /. vrexp → .5 /. kIA → 5 /. kB → 20 /. kIQ → 10 /. kP → 15 /. a → 0 /. b → 0 /.
   p → 1 /. q → 100 // N
```

-0.0285714

The following two velocities will be used later in Section 5.3.3, where a more general program to determine kinetic parameters is used.

```
vordABordPQ /. vfexp → 1 /. vrexp → .5 /. kIA → 5 /. kB → 20 /. kIQ → 10 /. kP → 15 /. a → 5 /. b → 5 /.
   p → 0 /. q → 0 // N
```

0.111111

```
vordABordPQ /. vfexp → 1 /. vrexp → .5 /. kIA → 5 /. kB → 20 /. kIQ → 10 /. kP → 15 /. a → 0 /. b → 0 /.
   p → 5 /. q → 5 // N
```

-0.05

In the next section these velocities will be treated as experimental measurements.

■ 5.1.2 Estimation of pH-dependent kinetic parameters with the minimum number of velocity measurements

The program calckinparsordABordPQfr was written to use Solve to estimate the 6 kinetic parameters from 6 measured velocities.

```
calckinparsordABordPQfr[v1_, a1_, b1_, p1_, q1_, v2_, a2_, b2_, p2_, q2_, v3_, a3_, b3_,
  p3_, q3_, v4_, a4_, b4_, p4_, q4_, v5_, a5_, b5_, p5_, q5_, v6_, a6_, b6_, p6_, q6_] :=
 Module[{}, (*This program calculates vfexp, vrexp, kIA, kB,
   kIQ and kP from 6 experimental velocities for ordered A + B = ordered P +
     Q. As an example, the following quadruplets of velocities can be used:{100,100,0,0},
   {1,100,0,0},{100,1,0,0},{0,0,100,100},{0,0,1,100}, and {0,0,100,1}.*)
  Solve[{v1 == (a1 b1 kIQ kP vfexp - kB kIA p1 q1 vrexp) /
       (a1 b1 kIQ kP + a1 kB kIQ kP + kB kIA kIQ kP + kB kIA kP q1 + kB kIA p1 q1),
    v2 == (a2 b2 kIQ kP vfexp - kB kIA p2 q2 vrexp) /
       (a2 b2 kIQ kP + a2 kB kIQ kP + kB kIA kIQ kP + kB kIA kP q2 + kB kIA p2 q2),
    v3 == (a3 b3 kIQ kP vfexp - kB kIA p3 q3 vrexp) / (a3 b3 kIQ kP + a3 kB kIQ kP + kB kIA kIQ kP +
         kB kIA kP q3 + kB kIA p3 q3), v4 == (a4 b4 kIQ kP vfexp - kB kIA p4 q4 vrexp) /
       (a4 b4 kIQ kP + a4 kB kIQ kP + kB kIA kIQ kP + kB kIA kP q4 + kB kIA p4 q4),
    v5 == (a5 b5 kIQ kP vfexp - kB kIA p5 q5 vrexp) /
       (a5 b5 kIQ kP + a5 kB kIQ kP + kB kIA kIQ kP + kB kIA kP q5 + kB kIA p5 q5),
    v6 == (a6 b6 kIQ kP vfexp - kB kIA p6 q6 vrexp) / (a6 b6 kIQ kP + a6 kB kIQ kP +
         kB kIA kIQ kP + kB kIA kP q6 + kB kIA p6 q6)}, {vfexp, vrexp, kIA, kB, kIQ, kP}]]
```

The 6 lines for v1 to v6 in this program are different versions of the rate equation vordABordPQ. Programs like this can also be written with steady-state rate equations or empirical rate equations. The formation of dead-end complexes can be included. This program can be used to estimate the kinetic parameters using the 6 velocities calculated in the previous section.

```
calckinparsordABordPQfr[.82645, 100, 100, 0, 0, .454545, 1, 100, 0, 0, .0454545,
  100, 1, 0, 0, -.429185, 0, 0, 100, 100, -.18868, 0, 0, 100, 1, -.028571, 0, 0, 1, 100]
```

{{vfexp → 1.00001, vrexp → 0.500002, kIA → 5.00003, kIQ → 9.99975, kP → 15.0003, kB → 20.0001}}

These values of the kinetic parameters are correct, and other sets of substrate concentrations that follow the general recommendations yield the same exact values. But there are experimental errors in the measurements of velocities and in substrate concentrations. Errors in substrate concentrations are not presented in tables because they are easier to calculate; a 5% error in a substrate concentration causes a 5% error in the corresponding Michaelis constant. The next line yields the form of a line in Table 5.1 of kinetic parameters.

```
line13 = {vfexp, vrexp, kIA, kIQ, kP, kB} /.
   calckinparsordABordPQfr[.82645, 100, 100, 0, 0, .454545, 1, 100, 0, 0, .0454545, 100,
     1, 0, 0, -.429185, 0, 0, 100, 100, -.18868, 0, 0, 100, 1, -.028571, 0, 0, 1, 100]
```

{{1.00001, 0.500002, 5.00003, 9.99975, 15.0003, 20.0001}}

The effects of 5% errors in the measured velocities are calculated as follows:

```
calckinparsordABordPQfr[1.05 * .82645, 100, 100, 0, 0, .454545, 1, 100, 0, 0, .0454545,
   100, 1, 0, 0, -.429185, 0, 0, 100, 100, -.18868, 0, 0, 100, 1, -.028571, 0, 0, 1, 100]
```

{{vfexp → 1.0618, vrexp → 0.500002, kIA → 5.29104, kIQ → 9.99975, kP → 15.0003, kB → 21.2361}}

```
line23 = {vfexp, vrexp, kIA, kIQ, kP, kB} /.
   calckinparsordABordPQfr[1.05 * .82645, 100, 100, 0, 0, .454545, 1, 100, 0, 0, .0454545,
     100, 1, 0, 0, -.429185, 0, 0, 100, 100, -.18868, 0, 0, 100, 1, -.028571, 0, 0, 1, 100]
```

{{1.0618, 0.500002, 5.29104, 9.99975, 15.0003, 21.2361}}

```
calckinparsordABordPQfr[.82645, 100, 100, 0, 0, 1.05 * .454545, 1, 100, 0, 0, .0454545,
   100, 1, 0, 0, -.429185, 0, 0, 100, 100, -.18868, 0, 0, 100, 1, -.028571, 0, 0, 1, 100]
```

{{vfexp → 1.00001, vrexp → 0.500002, kIA → 4.4474, kIQ → 9.99975, kP → 15.0003, kB → 20.106}}

```
line33 = {vfexp, vrexp, kIA, kIQ, kP, kB} /.
   calckinparsordABordPQfr[.82645, 100, 100, 0, 0, 1.05 * .454545, 1, 100, 0, 0, .0454545,
     100, 1, 0, 0, -.429185, 0, 0, 100, 100, -.18868, 0, 0, 100, 1, -.028571, 0, 0, 1, 100]
```

{{1.00001, 0.500002, 4.4474, 9.99975, 15.0003, 20.106}}

```
calckinparsordABordPQfr[.82645, 100, 100, 0, 0, .454545, 1, 100, 0, 0, 1.05 * .0454545,
   100, 1, 0, 0, -.429185, 0, 0, 100, 100, -.18868, 0, 0, 100, 1, -.028571, 0, 0, 1, 100]
```

{{vfexp → 0.989534, vrexp → 0.500002, kIA → 5.27937, kIQ → 9.99975, kP → 15.0003, kB → 18.7436}}

```
line43 = {vfexp, vrexp, kIA, kIQ, kP, kB} /.
   calckinparsordABordPQfr[.82645, 100, 100, 0, 0, .454545, 1, 100, 0, 0, 1.05 * .0454545,
     100, 1, 0, 0, -.429185, 0, 0, 100, 100, -.18868, 0, 0, 100, 1, -.028571, 0, 0, 1, 100]
```

{{0.989534, 0.500002, 5.27937, 9.99975, 15.0003, 18.7436}}

```
calckinparsordABordPQfr[.82645, 100, 100, 0, 0, .454545, 1, 100, 0, 0, .0454545, 100, 1,
   0, 0, 1.05 * -.429185, 0, 0, 100, 100, -.18868, 0, 0, 100, 1, -.028571, 0, 0, 1, 100]
```

{{vfexp → 1.00001, vrexp → 0.529684, kIA → 5.00003, kIQ → 10.3733, kP → 15.8908, kB → 20.0001}}

```
line53 = {vfexp, vrexp, kIA, kIQ, kP, kB} /.
   calckinparsordABordPQfr[.82645, 100, 100, 0, 0, .454545, 1, 100, 0, 0, .0454545, 100,
     1, 0, 0, 1.05 * -.429185, 0, 0, 100, 100, -.18868, 0, 0, 100, 1, -.028571, 0, 0, 1, 100]
```

{{1.00001, 0.529684, 5.00003, 10.3733, 15.8908, 20.0001}}

```
calckinparsordABordPQfr[.82645, 100, 100, 0, 0, .454545, 1, 100, 0, 0, .0454545, 100, 1,
   0, 0, -.429185, 0, 0, 100, 100, 1.05 * -.18868, 0, 0, 100, 1, -.028571, 0, 0, 1, 100]
```

{{vfexp → 1.00001, vrexp → 0.500002, kIA → 5.00003, kIQ → 9.07291, kP → 15.1278, kB → 20.0001}}

```
line63 = {vfexp, vrexp, kIA, kIQ, kP, kB} /.
   calckinparsordABordPQfr[.82645, 100, 100, 0, 0, .454545, 1, 100, 0, 0, .0454545, 100,
      1, 0, 0, -.429185, 0, 0, 100, 100, 1.05 * -.18868, 0, 0, 100, 1, -.028571, 0, 0, 1, 100]
```

{{1.00001, 0.500002, 5.00003, 9.07291, 15.1278, 20.0001}}

```
calckinparsordABordPQfr[.82645, 100, 100, 0, 0, .454545, 1, 100, 0, 0, .0454545, 100, 1,
   0, 0, -.429185, 0, 0, 100, 100, -.18868, 0, 0, 100, 1, 1.05 * -.028571, 0, 0, 1, 100]
```

{{vfexp → 1.00001, vrexp → 0.495828, kIA → 5.00003, kIQ → 10.5943, kP → 14.0404, kB → 20.0001}}

```
line73 = {vfexp, vrexp, kIA, kIQ, kP, kB} /.
   calckinparsordABordPQfr[.82645, 100, 100, 0, 0, .454545, 1, 100, 0, 0, .0454545, 100,
      1, 0, 0, -.429185, 0, 0, 100, 100, -.18868, 0, 0, 100, 1, 1.05 * -.028571, 0, 0, 1, 100]
```

{{1.00001, 0.495828, 5.00003, 10.5943, 14.0404, 20.0001}}

The estimated values of kinetic parameters are summarized in Table 5.1.

Table 5.1 Kinetic parameters for ordered A + B = ordered P + Q calculated from 6 velocities at {100,100,0,0}, {1,100,0,0}, {100,1,0,0}, {0,0,100,100}, {0,0,1,100}, {0,0,100,1}

```
TableForm[Round[{line13[[1]], line23[[1]], line33[[1]],
   line43[[1]], line53[[1]], line63[[1]], line73[[1]]}, 0.01], TableHeadings →
   {{"No errors", "1.05*v1", "1.05*v2", "1.05*v3", "1.05*v4", "1.05*v5", "1.05*v6"},
      {"vfexp", "vrexp", "kIA", "kIQ", "kP", "kB"}}]
```

	vfexp	vrexp	kIA	kIQ	kP	kB
No errors	1.	0.5	5.	10.	15.	20.
1.05*v1	1.06	0.5	5.29	10.	15.	21.24
1.05*v2	1.	0.5	4.45	10.	15.	20.11
1.05*v3	0.99	0.5	5.28	10.	15.	18.74
1.05*v4	1.	0.53	5.	10.37	15.89	20.
1.05*v5	1.	0.5	5.	9.07	15.13	20.
1.05*v6	1.	0.5	5.	10.59	14.04	20.

The values for the parameters in Table 5.1 can be used to calculate K' using the Haldane relation.

$$K' = \frac{\text{vfexp kP kIQ}}{\text{vrexp kIA kB}}$$

Seven values of K' can be obtained by copying columns from Table 5.1.

$$\begin{pmatrix} 1.` \\ 1.06` \\ 1.` \\ 0.99` \\ 1.` \\ 1.` \\ 1.` \end{pmatrix} * \begin{pmatrix} 15.` \\ 15.` \\ 15.` \\ 15.` \\ 15.89` \\ 15.13` \\ 14.040000000000001` \end{pmatrix} * \begin{pmatrix} 10.` \\ 10.` \\ 10.` \\ 10.` \\ 10.370000000000001` \\ 9.07` \\ 10.59` \end{pmatrix} \Bigg/ \begin{pmatrix} 0.5` & 5.` & 20.` \\ 0.5` & 5.29` & 21.240000000000002` \\ 0.5` & 4.45` & 20.11` \\ 0.5` & 5.28` & 18.740000000000002` \\ 0.53` & 5.` & 20.` \\ 0.5` & 5.` & 20.` \\ 0.5` & 5.` & 20.` \end{pmatrix}$$

{{3.}, {2.8302}, {3.35235}, {3.0016}, {3.10904}, {2.74458}, {2.97367}}

These values could be added to Table 5.1 to show how 5% errors in the 6 velocities affect the apparent equilibrium constant calculated using the Haldane relation.

As shown by the figures in reference [32], each of the velocity determinations is targeted on a particular kinetic parameter. The first velocity determination {100,100,0,0} is targetted on vfexp, and the 5% error in v1 yields vfexp = 1.06. The second velocity determination {1,100,0,0} is targetted on kIA, and the 5% error in v2 yields kIA = 4.45. The third velocity determination {100,1,0,0} is targetted on kB, and the 5% error in v3 yields kB = 18.74. The fourth velocity determination {0,0,100,100} is targetted on vrexp, and the 5% error in v4 yields vrexp = 0.53. The fifth velocity determination {0,0,1,100} is targetted on kIQ, and the 5% error in v5 yields kIQ = 9.07. The sixth velocity determination {0,0,100,1} is targetted on kP, and the 5% error in v6 yields kP = 14.04.

More accurate values of kinetic parameters can be obtained by increasing the range of substrate concentrations. For example, for ordered A + B = ordered P + Q, the following choices of substrate concentrations are used: {200,200,0,0}, {0.1,200,0,0}, {200,0.1,0,0}, {0,0,200,200}, {0,0,0.1,200}, {0,0,200,0.1} can be used. Of course, when a rate equation is being explored for the first time, the limiting velocities and Michaelis constants are not known. However, the recommendations for choices of substrate concentrations are very general; high a, b, p, and q means as high as practical, and low a, b, p, and q means as low as practical, considering the analytical method. When low substrate concentrations are used, rates are low and more difficult to determine, but the enzyme concentration can be increased. When the enyzme concentration is increased 10 fold for a measurement, the velocity has to be divided by 10 before comparing it with velocities at lower enzyme concentrations.

■ 5.1.3 Derivation of the rapid-equilibrium rate equation when K' is very large and A, B, and Q are present initially

When the apparent equilibrium constant for the enzyme-catalyzed reaction is very large, it is not possible to measure velocities of the reverse reation, but all the kinetic parameters, except vrexp, can be determined by studying the forward reaction.

When A, B, and Q are present initially, the mechanism is

E + A = EA   kIA = e*a/ea
EA + B = EAB   kB = ea*b/eab
EAB → products   vfexp = kf*eab
E + Q = EQ   kIQ = e*q/eq

$N' = 7$, $C' = 4$, and $R' = 3$, so $N' = C' + R'$ is 7 = 4 + 3. There are 4 kinetic parameters, and so 4 velocities have to be measured. The velocity of the forward reaction derived in the Section 5.1.1 is given by

vordABordPQ

$$\frac{a\ b\ kIQ\ kP\ vfexp - kB\ kIA\ p\ q\ vrexp}{a\ b\ kIQ\ kP + a\ kB\ kIQ\ kP + kB\ kIA\ kIQ\ kP + kB\ kIA\ kP\ q + kB\ kIA\ p\ q}$$

Deleting the second term in the numerator, yields the following velocity:

$$\frac{a\ b\ vfexp\ kIQ\ kP}{a\ b\ kIQ\ kP + a\ kB\ kIQ\ kP + kB\ kIA\ kIQ\ kP + kB\ kIA\ kP\ q + kB\ kIA\ p\ q};$$

Dividing numerator and denominator by kP and setting p = 0 yields the velocity, which is vordABordQirr when Q is present initially, and P is not present.

$$\mathbf{vordABordQirr} = \frac{a\ b\ kIQ\ vfexp}{a\ b\ kIQ + a\ kB\ kIQ + kB\ kIA\ kIQ + kB\ kIA\ q};$$

This equation is used to write a program to estimate four kinetic parameters from four velocity measurements.

Estimation of pH-dependent parameters when K' is very large and A, B, and Q are present initally

Since there are 4 kinetic parameters, 4 velocities have to be measured to estimate the kinetic parameters.

Calculate velocities at {[A],[B],[Q]} = {100, 100, 0}, {1, 100, 0}, {100, 1, 0}, {5, 5, 100}.

```
vordABordQirr /. vfexp → 1 /. kIA → 5 /. kB → 20 /. kIQ → 10 /. a → 100 /. b → 100 /. q → 0 // N
```

0.826446

```
vordABordQirr /. vfexp → 1 /. kIA → 5 /. kB → 20 /. kIQ → 10 /. a → 1 /. b → 100 /. q → 0 // N
```

0.454545

```
vordABordQirr /. vfexp → 1 /. kIA → 5 /. kB → 20 /. kIQ → 10 /. a → 100 /. b → 1 /. q → 0 // N
```

0.0454545

```
vordABordQirr /. vfexp → 1 /. kIA → 5 /. kB → 20 /. kIQ → 10 /. a → 5 /. b → 5 /. q → 100 // N
```

0.0204082

```
calckinparsordABordQirr[v1_, a1_, b1_, q1_, v2_, a2_, b2_, q2_, v3_, a3_, b3_, q3_,
   v4_, a4_, b4_, q4_] := Module[{}, (*This program calculates vfexp, kIA, kB,
   and kIQ from 4 experimental velocities for ordered A + B = ordered P + Q with a very
      large apparent equilibrium constant and Q is present initially. As an example,
   the following triplets of substrate concentrations can be used:{100,100,0},
   {1,100,0},{100,1,0},{5,5,100}.*)
```

$$
\text{Solve}\Bigg[\Bigg\{ \frac{a1\ b1\ kIQ\ vfexp}{a1\ b1\ kIQ + a1\ kB\ kIQ + kB\ kIA\ kIQ + kB\ kIA\ q1} == v1,
$$

$$
\frac{a2\ b2\ kIQ\ vfexp}{a2\ b2\ kIQ + a2\ kB\ kIQ + kB\ kIA\ kIQ + kB\ kIA\ q2} == v2,
$$

$$
\frac{a3\ b3\ kIQ\ vfexp}{a3\ b3\ kIQ + a3\ kB\ kIQ + kB\ kIA\ kIQ + kB\ kIA\ q3} == v3,
$$

$$
\frac{a4\ b4\ kIQ\ vfexp}{a4\ b4\ kIQ + a4\ kB\ kIQ + kB\ kIA\ kIQ + kB\ kIA\ q4} == v4 \Bigg\}, \{vfexp, kIA, kB, kIQ\}\Bigg]\Bigg]
$$

```
calckinparsordABordQirr[.826446, 100, 100, 0,
 .454545, 1, 100, 0, .0454545, 100, 1, 0, .020408, 5, 5, 100]
```

{{vfexp → 1., kIQ → 9.99992, kIA → 5., kB → 20.}}

These values are correct, but it is important to explore the sensitivity of these values to errors in measured velocities.

```
line16 = {vfexp, kIQ, kIA, kB} /. calckinparsordABordQirr[.826446,
   100, 100, 0, .454545, 1, 100, 0, .0454545, 100, 1, 0, .020408, 5, 5, 100]
```

{{1., 9.99992, 5., 20.}}

Calculate effects of 5% errors in the measurements of velocities, one at a time.

```
calckinparsordABordQirr[1.05 * .826446, 100, 100,
 0, .454545, 1, 100, 0, .0454545, 100, 1, 0, .020408, 5, 5, 100]
```

{{vfexp → 1.0618, kIQ → 10.6283, kIA → 5.29101, kB → 21.236}}

```
line26 = {vfexp, kIQ, kIA, kB} /. calckinparsordABordQirr[1.05 * .826446,
   100, 100, 0, .454545, 1, 100, 0, .0454545, 100, 1, 0, .020408, 5, 5, 100]
```

{{1.0618, 10.6283, 5.29101, 21.236}}

```
calckinparsordABordQirr[.826446, 100, 100, 0,
    1.05 * .454545, 1, 100, 0, .0454545, 100, 1, 0, .020408, 5, 5, 100]
```

$\{\{vfexp \to 1., kIQ \to 8.85274, kIA \to 4.44737, kB \to 20.1058\}\}$

```
line36 = {vfexp, kIQ, kIA, kB} /. calckinparsordABordQirr[.826446,
    100, 100, 0, 1.05 * .454545, 1, 100, 0, .0454545, 100, 1, 0, .020408, 5, 5, 100]
```

$\{\{1., 8.85274, 4.44737, 20.1058\}\}$

```
calckinparsordABordQirr[.826446, 100, 100, 0,
    .454545, 1, 100, 0, 1.05 * .0454545, 100, 1, 0, .020408, 5, 5, 100]
```

$\{\{vfexp \to 0.989529, kIQ \to 9.94991, kIA \to 5.27933, kB \to 18.7435\}\}$

```
line46 = {vfexp, kIQ, kIA, kB} /. calckinparsordABordQirr[.826446,
    100, 100, 0, .454545, 1, 100, 0, 1.05 * .0454545, 100, 1, 0, .020408, 5, 5, 100]
```

$\{\{0.989529, 9.94991, 5.27933, 18.7435\}\}$

```
calckinparsordABordQirr[.826446, 100, 100, 0,
    .454545, 1, 100, 0, .0454545, 100, 1, 0, 1.05 * .020408, 5, 5, 100]
```

$\{\{vfexp \to 1., kIQ \to 10.6194, kIA \to 5., kB \to 20.\}\}$

```
line56 = {vfexp, kIQ, kIA, kB} /. calckinparsordABordQirr[.826446,
    100, 100, 0, .454545, 1, 100, 0, .0454545, 100, 1, 0, 1.05 * .020408, 5, 5, 100]
```

$\{\{1., 10.6194, 5., 20.\}\}$

Table 5.2 Kinetic parameters for ordered A + B = ordered P + Q when K' is very large and Q is present initially. The following concentrations are used: $\{100, 100, 0\}, \{1, 100, 0\}, \{100, 1, 0\}, \{5, 5, 100\}$.

```
TableForm[Round[{line16[[1]], line26[[1]], line36[[1]], line46[[1]], line56[[1]]}, 0.01],
    TableHeadings →
    {{"No errors", "1.05*v1", "1.05*v2", "1.05*v3", "1.05*v4"}, {"vfexp", "kIQ", "kIA", "kB"}}]
```

	vfexp	kIQ	kIA	kB
No errors	1.	10.	5.	20.
1.05*v1	1.06	10.63	5.29	21.24
1.05*v2	1.	8.85	4.45	20.11
1.05*v3	0.99	9.95	5.28	18.74
1.05*v4	1.	10.62	5.	20.

These velocity determinations yield kIQ, which is the Michaelis constant for Q. It is not necessary to discuss the case that P is present, but no Q, because P is only involved in EQ + P = EPQ.

■ 5.1.4 Estimation of kinetic parameters when K' is very large and A, B, P, and Q are present initially

The mechanism is

E + A = EA	kIA = e*a/ea
EA + B = EAB	kB = ea*b/eab
EAB → products	vfexp = kf*eab
E + Q = EQ	kIQ = e*q/eq
EQ + P = EPQ	kP = eq*p/epq

There are nine reactants, five components, and four independent equilibrium, and so $N' = C' + R'$ is 9 = 5 + 4. There are 5 kinetic parameters, and so 5 velocities have to be measured. The velocity of the forward reaction derived in Section 5.1.1 is

vordABordPQ

$$\frac{a \ b \ kIQ \ kP \ vfexp - kB \ kIA \ p \ q \ vrexp}{a \ b \ kIQ \ kP + a \ kB \ kIQ \ kP + kB \ kIA \ kIQ \ kP + kB \ kIA \ kP \ q + kB \ kIA \ p \ q}$$

Deleting the second term in the numerator, the velocity is given by

$$\frac{a \ b \ vfexp \ kIQ \ kP}{a \ b \ kIQ \ kP + a \ kB \ kIQ \ kP + kB \ kIA \ kIQ \ kP + kB \ kIA \ kP \ q + kB \ kIA \ p \ q};$$

This velocity is named vordABordPQirr.

$$\textbf{vordABordPQirr} = \frac{a \ b \ kIQ \ kP \ vfexp}{a \ b \ kIQ \ kP + a \ kB \ kIQ \ kP + kB \ kIA \ kIQ \ kP + kB \ kIA \ kP \ q + kB \ kIA \ p \ q};$$

This rate equation is used to write a program to calculate five kinetic parameters from five velocity measurements.

Calculate velocities at {[A],[B],[P],[Q]} = {100, 100, 0, 0}, {1, 100, 0, 0}, {100, 1, 0, 0}, {5, 5, 100, 5}, and {5, 5, 5, 100}. Notice that the last 2 quadruplets have the predominant effect in determining kIQ and kP, respectively. All 4 substrates must be present in the last two quadruplets.

```
vordABordPQirr /. vfexp → 1 /. kIA → 5 /. kB → 20 /. kIQ → 10 /. kP → 15 /. a → 100 /. b → 100 /.
   p → 0 /. q → 0 // N
```

0.826446

```
vordABordPQirr /. vfexp → 1 /. kIA → 5 /. kB → 20 /. kIQ → 10 /. kP → 15 /. a → 1 /. b → 100 /. p → 0 /.
   q → 0 // N
```

0.454545

```
vordABordPQirr /. vfexp → 1 /. kIA → 5 /. kB → 20 /. kIQ → 10 /. kP → 15 /. a → 100 /. b → 1 /. p → 0 /.
   q → 0 // N
```

0.0454545

```
vordABordPQirr /. vfexp → 1 /. kIA → 5 /. kB → 20 /. kIQ → 10 /. kP → 15 /. a → 5 /. b → 5 /. p → 100 /.
   q → 5 // N
```

0.0410959

```
vordABordPQirr /. vfexp → 1 /. kIA → 5 /. kB → 20 /. kIQ → 10 /. kP → 15 /. a → 5 /. b → 5 /. p → 5 /.
   q → 100 // N
```

0.0160428

These velocities are treated as experimental data to estimate the 5 kinetic parameters involved by using the following program.

```
calckinparsordABordPQirr[v1_, a1_, b1_, p1_, q1_, v2_, a2_, b2_, p2_, q2_,
  v3_, a3_, b3_, p3_, q3_, v4_, a4_, b4_, p4_, q4_, v5_, a5_, b5_, p5_, q5_] :=
 Module[{}, (*This program calculates vfexp, kIA, kB,
  kIQ and kP from 5 experimental velocities for ordered A + B =
   ordered P + Q with a very large apparent equilibrium constant. As an example,
  the following quadruplets of velocities can be used:{100,100,0,0},
  {1,100,0,0},{100,1,0,0},{5,5,100,5}, and {5,5,5,100}.*)
```

$$\text{Solve}\left[\left\{\frac{a1\ b1\ kIQ\ kP\ vfexp}{a1\ b1\ kIQ\ kP + a1\ kB\ kIQ\ kP + kB\ kIA\ kIQ\ kP + kB\ kIA\ kP\ q1 + kB\ kIA\ p1\ q1} == v1,\right.\right.$$

$$\frac{a2\ b2\ kIQ\ kP\ vfexp}{a2\ b2\ kIQ\ kP + a2\ kB\ kIQ\ kP + kB\ kIA\ kIQ\ kP + kB\ kIA\ kP\ q2 + kB\ kIA\ p2\ q2} == v2,$$

$$\frac{a3\ b3\ kIQ\ kP\ vfexp}{a3\ b3\ kIQ\ kP + a3\ kB\ kIQ\ kP + kB\ kIA\ kIQ\ kP + kB\ kIA\ kP\ q3 + kB\ kIA\ p3\ q3} == v3,$$

$$\frac{a4\ b4\ kIQ\ kP\ vfexp}{a4\ b4\ kIQ\ kP + a4\ kB\ kIQ\ kP + kB\ kIA\ kIQ\ kP + kB\ kIA\ kP\ q4 + kB\ kIA\ p4\ q4} == v4,$$

$$\left.\frac{a5\ b5\ kIQ\ kP\ vfexp}{a5\ b5\ kIQ\ kP + a5\ kB\ kIQ\ kP + kB\ kIA\ kIQ\ kP + kB\ kIA\ kP\ q5 + kB\ kIA\ p5\ q5} == v5\right\},$$

$$\left.\left.\{vfexp, kIA, kB, kIQ, kP\}\right]\right]$$

```
calckinparsordABordPQirr[.826446, 100, 100, 0, 0, .454545, 1, 100,
  0, 0, .0454545, 100, 1, 0, 0, .041096, 5, 5, 100, 5, .016043, 5, 5, 5, 100]
```

{{vfexp → 1., kIQ → 10.0002, kIA → 5., kP → 14.9997, kB → 20.}}

These values are correct, but it is important to test the sensitivity of these parameters to experimental errors in the measured velocities.

```
line15 = {vfexp, kIQ, kIA, kP, kB} /. calckinparsordABordPQirr[.826446, 100, 100, 0, 0, .454545,
   1, 100, 0, 0, .0454545, 100, 1, 0, 0, .041096, 5, 5, 100, 5, .016043, 5, 5, 5, 100]
```

{{1., 10.0002, 5., 14.9997, 20.}}

Test 5% errors in velocities.

```
calckinparsordABordPQirr[1.05 * .826446, 100, 100, 0, 0, .454545, 1,
  100, 0, 0, .0454545, 100, 1, 0, 0, .041096, 5, 5, 100, 5, .016043, 5, 5, 5, 100]
```

{{vfexp → 1.0618, kIQ → 10.5822, kIA → 5.29101, kP → 15.1988, kB → 21.236}}

```
line25 = {vfexp, kIQ, kIA, kP, kB} /. calckinparsordABordPQirr[1.05 * .826446, 100, 100, 0, 0,
   .454545, 1, 100, 0, 0, .0454545, 100, 1, 0, 0, .041096, 5, 5, 100, 5, .016043, 5, 5, 5, 100]
```

{{1.0618, 10.5822, 5.29101, 15.1988, 21.236}}

```
calckinparsordABordPQirr[.826446, 100, 100, 0, 0, 1.05 * .454545, 1,
  100, 0, 0, .0454545, 100, 1, 0, 0, .041096, 5, 5, 100, 5, .016043, 5, 5, 5, 100]
```

{{vfexp → 1., kIQ → 8.942, kIA → 4.44737, kP → 14.5606, kB → 20.1058}}

```
line35 = {vfexp, kIQ, kIA, kP, kB} /.
  calckinparsordABordPQirr[.826446, 100, 100, 0, 0, 1.05 * .454545, 1, 100,
   0, 0, .0454545, 100, 1, 0, 0, .041096, 5, 5, 100, 5, .016043, 5, 5, 5, 100]
```

{{1., 8.942, 4.44737, 14.5606, 20.1058}}

```
calckinparsordABordPQirr[.826446, 100, 100, 0, 0, .454545, 1, 100, 0, 0,
  1.05 * .0454545, 100, 1, 0, 0, 1.05 * .041096, 5, 5, 100, 5, .016043, 5, 5, 5, 100]
```

{{vfexp → 0.989529, kIQ → 9.70431, kIA → 5.27933, kP → 16.7357, kB → 18.7435}}

```
line45 = {vfexp, kIQ, kIA, kP, kB} /. calckinparsordABordPQirr[.826446, 100, 100, 0, 0, .454545,
    1, 100, 0, 0, 1.05 * .0454545, 100, 1, 0, 0, 1.05 * .041096, 5, 5, 100, 5, .016043, 5, 5, 5, 100]
```

{{0.989529, 9.70431, 5.27933, 16.7357, 18.7435}}

```
calckinparsordABordPQirr[.826446, 100, 100, 0, 0, .454545, 1, 100, 0,
    0, .0454545, 100, 1, 0, 0, 1.05 * .041096, 5, 5, 100, 5, .016043, 5, 5, 5, 100]
```

{{vfexp → 1., kIQ → 9.70431, kIA → 5., kP → 17.0135, kB → 20.}}

```
line55 = {vfexp, kIQ, kIA, kP, kB} /. calckinparsordABordPQirr[.826446, 100, 100, 0, 0, .454545,
    1, 100, 0, 0, .0454545, 100, 1, 0, 0, 1.05 * .041096, 5, 5, 100, 5, .016043, 5, 5, 5, 100]
```

{{1., 9.70431, 5., 17.0135, 20.}}

```
calckinparsordABordPQirr[.826446, 100, 100, 0, 0, .454545, 1, 100, 0,
    0, .0454545, 100, 1, 0, 0, .041096, 5, 5, 100, 5, 1.05 * .016043, 5, 5, 5, 100]
```

{{vfexp → 1., kIQ → 10.8476, kIA → 5., kP → 13.6679, kB → 20.}}

```
line65 = {vfexp, kIQ, kIA, kP, kB} /. calckinparsordABordPQirr[.826446, 100, 100, 0, 0, .454545,
    1, 100, 0, 0, .0454545, 100, 1, 0, 0, .041096, 5, 5, 100, 5, 1.05 * .016043, 5, 5, 5, 100]
```

{{1., 10.8476, 5., 13.6679, 20.}}

Table 5.3 Kinetic parameters for ordered A + B = ordered P + Q when K' is very large and P and Q are present. The following substrate concentrations are used: $\{100, 100, 0, 0\}$, $\{1, 100, 0, 0\}$, $\{100, 1, 0, 0\}$, $\{5, 5, 100, 5\}$, and $\{5, 5, 5, 100\}$

```
TableForm[
  Round[{line15[[1]], line25[[1]], line35[[1]], line45[[1]], line55[[1]], line65[[1]]}, 0.01],
  TableHeadings → {{"No errors", "1.05*v1", "1.05*v2", "1.05*v3", "1.05*v4", "1.05*v5"},
    {"vfexp", "kIQ", "kIA", "kP", "kB"}}]
```

	vfexp	kIQ	kIA	kP	kB
No errors	1.	10.	5.	15.	20.
1.05*v1	1.06	10.58	5.29	15.2	21.24
1.05*v2	1.	8.94	4.45	14.56	20.11
1.05*v3	0.99	9.7	5.28	16.74	18.74
1.05*v4	1.	9.7	5.	17.01	20.
1.05*v5	1.	10.85	5.	13.67	20.

These velocity determinations yield kIQ and kP, which are the equilibrium constant constants for Q and P in the rapid-equilibrium rate equation.

If the apparent equilibrium constant is very large, the kinetic parameters cannot be used to verify the Haldane equation. But if the apparent equilibrium constant for A + B = P + Q can be calculated from standard transformed Gibbs energies of formation of the reactants at the temperature, pH and ionic strength for the rate measurements, this information can be used to calculate K' and the Haldane equation can be used to calculate vfexp. It is not necessary that equilibrium measurements have been made for the reaction studied kinetically because tables [23] have been constructed for standard transformed thermodynamic properties of reactants in enzyme-catalyzed reactions for which apparent equilibrium constants can be determined. When the standard transformed Gibbs energies of formation are known for all the reactants in the reaction studied kinetically, the apparent equilibrium constant can be calculated.

▪ 5.1.5 pH effects and the estimation of pH-independent kinetic parameters

Acid dissociations can be included in the mechanism in Section 5.1.1, but this may significantly increase the minimum number of velocities that have to be measured. If 2 acid dissociations are included for E, EA, and EB, 8 more velocities have to be measured to determine the kinetic parameters in the forward reaction. Another approach to determine the pKs and chemical equilibrium constants involved is to determine the 6 pH-dependent kinetic parameters at several pHs. These parameters and certain ratios can be plotted and 3 pH-independent kinetic parameters can be calculated from each bell-shaped curve [33]. Alternatively, the mechanism can be expanded to include the acid dissociations of E, EA, and EB and Solve can be used to derive the rate equation.

The pKs of substrates affect the kinetic parmeters and K'. Thus K' may vary with pH, even when H$^+$ is not consumed in the rate-determining reaction. The effects of the consumption of H$^+$ in the rate determining reaction are much larger than the effects of pKs because they introduce a factor of 10^{-npH}, where n = -1, -2, -3, ... into K'. This is discussed in more detail in Chapters 7 and 8 because large values of n are found in some oxidoreductase reactions [29].

▪ 5.1.6 Dead - end enzyme - substrate complexes

The following dead-end complexes can be added to any mechanism for A + B = P + Q.

EA + P = EAP   kAP = ea*p/eap
EB + Q = EPQ   kBQ = eb*q/ebq

Equilibrium constants that are found empirically can also be added to any rate equation.

5.2 Ordered A + B = random P + Q

▪ 5.2.1 Derivation of the rapid-equilibrium rate equation

The equilibrium concentrations eab and epq are calculated using the following equilibria :

E + A = EA   kIA = e*a/ea
EA + B = EAB   kB = ea*b/eab
E + P = EP   kIP = e*p/ep   This is the new reaction.
E + Q = EQ   kIQ = e*q/eq
EQ + P = EPQ   kP = eq*p/epq

First, derive the expression for the equilibrium concentration of eab.

```
Solve[{kIA == e * a / ea, kB == ea * b / eab, kIP == e * p / ep, kIQ == e * q / eq,
   kP == eq * p / epq, et == e + ea + eab + ep + eq + epq}, {eab}, {e, ea, ep, eq, epq}]
```

```
{{eab → (a b et kIP kIQ kP) / (a b kIP kIQ kP + a kB kIP kIQ kP +
    kB kIA kIP kIQ kP + kB kIA kIQ kP p + kB kIA kIP kP q + kB kIA kIP p q)}}
```

Second, derive the expression for epq.

```
Solve[{kIA == e * a / ea, kB == ea * b / eab, kIP == e * p / ep, kIQ == e * q / eq,
   kP == eq * p / epq, et == e + ea + eab + ep + eq + epq}, {epq}, {e, ea, ep, eq, eab}]
```

```
{{epq → (et kB kIA kIP p q) / (a b kIP kIQ kP + a kB kIP kIQ kP +
    kB kIA kIP kIQ kP + kB kIA kIQ kP p + kB kIA kIP kP q + kB kIA kIP p q)}}
```

The complete mechanism including the rate − determining reaction is

E + A = EA	kIA = e∗a/ea
EA + B = EAB	kB = ea∗b/eab
EAB ⇌ EPQ	v = kf∗eab − kr∗epq
E + P = EP	kIP = e∗p/ep
E + Q = EQ	kIQ = e∗q/eq
EQ + P = EPQ	kP = eq∗p/epq

The number of kinetic parameters is 7, and the number of denominator terms of the rate equation is 6. The number of reactants is 10. The number of components is 5. The number of independent equilibrium expressions is 5, and so $N' = C' + R'$ is $10 = 5 + 5$.

The rapid-equilibrium velocity is given by kf*eab-kr*epq. To obtain the expression for the velocity, copy eab and subtract epq, and then change et in the expression for eab to vfexp and change et in the expression for epq to vrexp.

```
(a b vfexp kIP kIQ kP) / (a b kIP kIQ kP + a kB kIP kIQ kP + kB kIA kIP kIQ kP +
    kB kIA kIQ kP p + kB kIA kIP kP q + kB kIA kIP p q) - (vrexp kB kIA kIP p q) /
  (a b kIP kIQ kP + a kB kIP kIQ kP + kB kIA kIP kIQ kP + kB kIA kIQ kP p + kB kIA kIP kP q + kB kIA kIP p q);
```

Since the denominators are the same, the velocity of ordered A + B = random P + Q can be written as

```
vfordABrandPQ = (a b vfexp kIP kIQ kP - vrexp kB kIA kIP p q) /
  (a b kIP kIQ kP + a kB kIP kIQ kP + kB kIA kIP kIQ kP + kB kIA kIQ kP p + kB kIA kIP kP q + kB kIA kIP p q);
```

The same expression for the Haldane equation is obtained as in Section 5.1.1.

The rate equation vfordABrandPQ can be used to calculate velocities for seven quadruplets of {a,b,p,q}.

vfordABrandPQ is used to calculate the velocity for various quadruplets of substrate concentrations. Since there are 7 kinetic parameters, 7 velocity measurements are required to estimate the 7 kinetic parameters. Since kIP is new, there is one more quadruplet. This last quadruplet should be like {a,b,p,q} = {0,0,1,1}, but {0,0,5,5} is used to obtain a higher velocity: {100,100,0,0}, {1,100,0,0), {100,1,0,0}, {0,0,100,100}, {0,0,1,100}, {0,0,100,1}, and {0,0,5,5}.

```
vfordABrandPQ /. vfexp → 1 /. vrexp → .5 /. kIA → 5 /. kB → 20 /. kIP → 25 /. kIQ → 10 /. kP → 15 /.
    a → 100 /. b → 100 /. p → 0 /. q → 0 // N
```

```
0.826446
```

```
vfordABrandPQ /. vfexp → 1 /. vrexp → .5 /. kIA → 5 /. kB → 20 /. kIP → 25 /. kIQ → 10 /. kP → 15 /.
    a → 1 /. b → 100 /. p → 0 /. q → 0 // N
```

```
0.454545
```

```
vfordABrandPQ /. vfexp → 1 /. vrexp → .5 /. kIA → 5 /. kB → 20 /. kIP → 25 /. kIQ → 10 /. kP → 15 /.
    a → 100 /. b → 1 /. p → 0 /. q → 0 // N
```

```
0.0454545
```

```
vfordABrandPQ /. vfexp → 1 /. vrexp → .5 /. kIA → 5 /. kB → 20 /. kIP → 25 /. kIQ → 10 /. kP → 15 /.
    a → 0 /. b → 0 /. p → 100 /. q → 100 // N
```

```
-0.408163
```

```
vfordABrandPQ /. vfexp → 1 /. vrexp → .5 /. kIA → 5 /. kB → 20 /. kIP → 25 /. kIQ → 10 /. kP → 15 /.
    a → 0 /. b → 0 /. p → 1 /. q → 100 // N
```

```
-0.0284738
```

```
vfordABrandPQ /. vfexp → 1 /. vrexp → .5 /. kIA → 5 /. kB → 20 /. kIP → 25 /. kIQ → 10 /. kP → 15 /.
    a → 0 /. b → 0 /. p → 100 /. q → 1 // N
```

```
-0.0578035
```

```
vfordABrandPQ /. vfexp → 1 /. vrexp → .5 /. kIA → 5 /. kB → 20 /. kIP → 25 /. kIQ → 10 /. kP → 15 /.
   a → 0 /. b → 0 /. p → 5 /. q → 5 // N
```

```
- 0.0446429
```

The following velocity will be used later in Section 5.3.4.

```
vfordABrandPQ /. vfexp → 1 /. vrexp → .5 /. kIA → 5 /. kB → 20 /. kIP → 25 /. kIQ → 10 /. kP → 15 /.
   a → 5 /. b → 5 /. p → 0 /. q → 0 // N
```

```
0.111111
```

■ 5.2.2 Estimation of pH-dependent kinetic parameters with the minimum number of velocity measurements

The 7 equations in the following program are given by vfordABrandPQ with different subscripts on v, a, b, p, and q.

```
calckinparsordABrandPQfr[v1_, a1_, b1_, p1_, q1_, v2_, a2_,
   b2_, p2_, q2_, v3_, a3_, b3_, p3_, q3_, v4_, a4_, b4_, p4_, q4_, v5_,
   a5_, b5_, p5_, q5_, v6_, a6_, b6_, p6_, q6_, v7_, a7_, b7_, p7_, q7_] :=
 Module[{}, (*This program calculates vfexp, vrexp, kIA, kB,kIP,
  kIQ and kP from 7 experimental velocities for ordered A + B = random P +
    Q. As an example, the following quadruplets of velocities can be used:{100,100,0,0},
   {0,0,100,100},{1,100,0,0},{100,1,0,0},{0,0,1,100},{0,0,100,1},{0,0,5,5}.*)
  Solve[{v1 == (a1 b1 vfexp kIP kIQ kP - vrexp kB kIA kIP p1 q1) / (a1 b1 kIP kIQ kP +
       a1 kB kIP kIQ kP + kB kIA kIP kIQ kP + kB kIA kIQ kP p1 + kB kIA kIP kP q1 + kB kIA kIP p1 q1),
    v2 == (a2 b2 vfexp kIP kIQ kP - vrexp kB kIA kIP p2 q2) / (a2 b2 kIP kIQ kP + a2 kB kIP kIQ kP +
       kB kIA kIP kIQ kP + kB kIA kIQ kP p2 + kB kIA kIP kP q2 + kB kIA kIP p2 q2),
    v3 == (a3 b3 vfexp kIP kIQ kP - vrexp kB kIA kIP p3 q3) / (a3 b3 kIP kIQ kP + a3 kB kIP kIQ kP +
       kB kIA kIP kIQ kP + kB kIA kIQ kP p3 + kB kIA kIP kP q3 + kB kIA kIP p3 q3),
    v4 == (a4 b4 vfexp kIP kIQ kP - vrexp kB kIA kIP p4 q4) / (a4 b4 kIP kIQ kP + a4 kB kIP kIQ kP +
       kB kIA kIP kIQ kP + kB kIA kIQ kP p4 + kB kIA kIP kP q4 + kB kIA kIP p4 q4),
    v5 == (a5 b5 vfexp kIP kIQ kP - vrexp kB kIA kIP p5 q5) / (a5 b5 kIP kIQ kP + a5 kB kIP kIQ kP +
       kB kIA kIP kIQ kP + kB kIA kIQ kP p5 + kB kIA kIP kP q5 + kB kIA kIP p5 q5),
    v6 == (a6 b6 vfexp kIP kIQ kP - vrexp kB kIA kIP p6 q6) / (a6 b6 kIP kIQ kP + a6 kB kIP kIQ kP +
       kB kIA kIP kIQ kP + kB kIA kIQ kP p6 + kB kIA kIP kP q6 + kB kIA kIP p6 q6),
    v7 == (a7 b7 vfexp kIP kIQ kP - vrexp kB kIA kIP p7 q7) / (a7 b7 kIP kIQ kP + a7 kB kIP kIQ kP +
       kB kIA kIP kIQ kP + kB kIA kIQ kP p7 + kB kIA kIP kP q7 + kB kIA kIP p7 q7)},
   {vfexp, vrexp, kIA, kB, kIP, kIQ, kP}]]
```

Values of the seven kinetic parameters are obtained as follows:

```
calckinparsordABrandPQfr[.82645, 100, 100, 0, 0, .454545, 1, 100, 0, 0, .0454545, 100, 1, 0, 0,
   -.40816, 0, 0, 100, 100, -.028474, 0, 0, 1, 100, -.057804, 0, 0, 100, 1, -.044643, 0, 0, 5, 5]
```

```
{{vfexp → 1.00001, vrexp → 0.499994, kIA → 5.00003,
   kB → 20.0001, kIQ → 10.0001, kP → 14.9997, kIP → 25.0004}}
```

```
line12 = {vfexp, vrexp, kIA, kB, kIQ, kP, kIP} /.
   calckinparsordABrandPQfr[.82645, 100, 100, 0, 0, .454545, 1, 100, 0, 0, .0454545, 100, 1, 0, 0,
     -.40816, 0, 0, 100, 100, -.028474, 0, 0, 1, 100, -.057804, 0, 0, 100, 1, -.044643, 0, 0, 5, 5]
```

```
{{1.00001, 0.499994, 5.00003, 20.0001, 10.0001, 14.9997, 25.0004}}
```

These are the expected values for the 7 kinetic parameters. Now put in 5% errors in the measured velocities, one at a time.

```
calckinparsordABrandPQfr[1.05 * .82645, 100, 100, 0, 0,
   .454545, 1, 100, 0, 0, .0454545, 100, 1, 0, 0, -.40816, 0, 0, 100, 100,
   -.028474, 0, 0, 1, 100, -.057804, 0, 0, 100, 1, -.044643, 0, 0, 5, 5]
```

```
{{vfexp → 1.0618, vrexp → 0.499994, kIA → 5.29104,
   kB → 21.2361, kIQ → 10.0001, kP → 14.9997, kIP → 25.0004}}
```

```
line22 = {vfexp, vrexp, kIA, kB, kIQ, kP, kIP} /. calckinparsordABrandPQfr[1.05 * .82645,
    100, 100, 0, 0, .454545, 1, 100, 0, 0, .0454545, 100, 1, 0, 0, -.40816, 0, 0,
    100, 100, -.028474, 0, 0, 1, 100, -.057804, 0, 0, 100, 1, -.044643, 0, 0, 5, 5]
```

{{1.0618, 0.499994, 5.29104, 21.2361, 10.0001, 14.9997, 25.0004}}

```
calckinparsordABrandPQfr[.82645, 100, 100, 0, 0, 1.05 * .454545,
    1, 100, 0, 0, .0454545, 100, 1, 0, 0, -.40816, 0, 0, 100, 100,
    -.028474, 0, 0, 1, 100, -.057804, 0, 0, 100, 1, -.044643, 0, 0, 5, 5]
```

{{vfexp → 1.00001, vrexp → 0.499994, kIA → 4.4474,
  kB → 20.106, kIQ → 10.0001, kP → 14.9997, kIP → 25.0004}}

```
line32 = {vfexp, vrexp, kIA, kB, kIQ, kP, kIP} /. calckinparsordABrandPQfr[.82645,
    100, 100, 0, 0, 1.05 * .454545, 1, 100, 0, 0, .0454545, 100, 1, 0, 0, -.40816, 0,
    0, 100, 100, -.028474, 0, 0, 1, 100, -.057804, 0, 0, 100, 1, -.044643, 0, 0, 5, 5]
```

{{1.00001, 0.499994, 4.4474, 20.106, 10.0001, 14.9997, 25.0004}}

```
calckinparsordABrandPQfr[.82645, 100, 100, 0, 0, .454545,
    1, 100, 0, 0, 1.05 * .0454545, 100, 1, 0, 0, -.40816, 0, 0, 100, 100,
    -.028474, 0, 0, 1, 100, -.057804, 0, 0, 100, 1, -.044643, 0, 0, 5, 5]
```

{{vfexp → 0.989534, vrexp → 0.499994, kIA → 5.27937,
  kB → 18.7436, kIQ → 10.0001, kP → 14.9997, kIP → 25.0004}}

```
line42 = {vfexp, vrexp, kIA, kB, kIQ, kP, kIP} /. calckinparsordABrandPQfr[.82645,
    100, 100, 0, 0, .454545, 1, 100, 0, 0, 1.05 * .0454545, 100, 1, 0, 0, -.40816, 0,
    0, 100, 100, -.028474, 0, 0, 1, 100, -.057804, 0, 0, 100, 1, -.044643, 0, 0, 5, 5]
```

{{0.989534, 0.499994, 5.27937, 18.7436, 10.0001, 14.9997, 25.0004}}

```
calckinparsordABrandPQfr[.82645, 100, 100, 0, 0, .454545,
    1, 100, 0, 0, .0454545, 100, 1, 0, 0, 1.05 * -.40816, 0, 0, 100, 100,
    -.028474, 0, 0, 1, 100, -.057804, 0, 0, 100, 1, -.044643, 0, 0, 5, 5]
```

{{vfexp → 1.00001, vrexp → 0.531576, kIA → 5.00003,
  kB → 20.0001, kIQ → 10.0337, kP → 15.9992, kIP → 24.9626}}

```
line52 = {vfexp, vrexp, kIA, kB, kIQ, kP, kIP} /. calckinparsordABrandPQfr[.82645,
    100, 100, 0, 0, .454545, 1, 100, 0, 0, .0454545, 100, 1, 0, 0, 1.05 * -.40816, 0,
    0, 100, 100, -.028474, 0, 0, 1, 100, -.057804, 0, 0, 100, 1, -.044643, 0, 0, 5, 5]
```

{{1.00001, 0.531576, 5.00003, 20.0001, 10.0337, 15.9992, 24.9626}}

```
calckinparsordABrandPQfr[.82645, 100, 100, 0, 0, .454545,
    1, 100, 0, 0, .0454545, 100, 1, 0, 0, -.40816, 0, 0, 100, 100,
    1.05 * -.028474, 0, 0, 1, 100, -.057804, 0, 0, 100, 1, -.044643, 0, 0, 5, 5]
```

{{vfexp → 1.00001, vrexp → 0.495588, kIA → 5.00003,
  kB → 20.0001, kIQ → 10.9452, kP → 13.9863, kIP → 25.9334}}

```
line62 = {vfexp, vrexp, kIA, kB, kIQ, kP, kIP} /. calckinparsordABrandPQfr[.82645,
    100, 100, 0, 0, .454545, 1, 100, 0, 0, .0454545, 100, 1, 0, 0, -.40816, 0, 0, 100,
    100, 1.05 * -.028474, 0, 0, 1, 100, -.057804, 0, 0, 100, 1, -.044643, 0, 0, 5, 5]
```

{{1.00001, 0.495588, 5.00003, 20.0001, 10.9452, 13.9863, 25.9334}}

```
calckinparsordABrandPQfr[.82645, 100, 100, 0, 0, .454545, 1,
    100, 0, 0, .0454545, 100, 1, 0, 0, -.40816, 0, 0, 100, 100, -.028474,
    0, 0, 1, 100, 1.05 * -.057804, 0, 0, 100, 1, -.044643, 0, 0, 5, 5]
```

{{vfexp → 1.00001, vrexp → 0.497814, kIA → 5.00003,
  kB → 20.0001, kIQ → 10.1609, kP → 14.9125, kIP → 27.3626}}

```
line72 = {vfexp, vrexp, kIA, kB, kIQ, kP, kIP} /. calckinparsordABrandPQfr[.82645,
    100, 100, 0, 0, .454545, 1, 100, 0, 0, .0454545, 100, 1, 0, 0, -.40816, 0, 0, 100,
    100, -.028474, 0, 0, 1, 100, 1.05 * -.057804, 0, 0, 100, 1, -.044643, 0, 0, 5, 5]
```

{{1.00001, 0.497814, 5.00003, 20.0001, 10.1609, 14.9125, 27.3626}}

```
calckinparsordABrandPQfr[.82645, 100, 100, 0, 0, .454545, 1,
  100, 0, 0, .0454545, 100, 1, 0, 0, -.40816, 0, 0, 100, 100, -.028474,
  0, 0, 1, 100, -.057804, 0, 0, 100, 1, 1.05 * -.044643, 0, 0, 5, 5]
```

{{vfexp → 1.00001, vrexp → 0.500734, kIA → 5.00003,
  kB → 20.0001, kIQ → 8.92723, kP → 15.1698, kIP → 21.9964}}

```
line82 = {vfexp, vrexp, kIA, kB, kIQ, kP, kIP} /. calckinparsordABrandPQfr[.82645,
    100, 100, 0, 0, .454545, 1, 100, 0, 0, .0454545, 100, 1, 0, 0, -.40816, 0, 0, 100,
    100, -.028474, 0, 0, 1, 100, -.057804, 0, 0, 100, 1, 1.05 * -.044643, 0, 0, 5, 5]
```

{{1.00001, 0.500734, 5.00003, 20.0001, 8.92723, 15.1698, 21.9964}}

The estimated values of kinetic parameters are summarized in Table 5.4.

Table 5.4 Kinetic parameters for ordered A + B = random P + Q calculated from velocities at $\{100, 100, 0, 0\}$, $\{1, 100, 0, 0)$, $\{100, 1, 0, 0\}$, $\{0, 0, 100, 100\}$, $\{0, 0, 1, 100\}$, $\{0, 0, 100, 1\}$, and $\{0, 0, 5, 5\}$

```
TableForm[Round[{line12[[1]], line22[[1]], line32[[1]],
    line42[[1]], line52[[1]], line62[[1]], line72[[1]], line82[[1]]}, 0.01],
  TableHeadings → {{"No errors", "1.05*v1", "1.05*v2", "1.05*v3", "1.05*v4", "1.05*v5",
    "1.05*v6", "1.05*v7"}, {"vfexp", "vrexp", "kIA", "kB", "kIQ", "kP", "kIP"}}]
```

	vfexp	vrexp	kIA	kB	kIQ	kP	kIP
No errors	1.	0.5	5.	20.	10.	15.	25.
1.05*v1	1.06	0.5	5.29	21.24	10.	15.	25.
1.05*v2	1.	0.5	4.45	20.11	10.	15.	25.
1.05*v3	0.99	0.5	5.28	18.74	10.	15.	25.
1.05*v4	1.	0.53	5.	20.	10.03	16.	24.96
1.05*v5	1.	0.5	5.	20.	10.95	13.99	25.93
1.05*v6	1.	0.5	5.	20.	10.16	14.91	27.36
1.05*v7	1.	0.5	5.	20.	8.93	15.17	22.

The random mechanism for the reverse reaction brings in 3 sides of the following thermodynamic cycle:

```
        kIQ = 10
E + Q =  EQ
+          +
P          P
|| kIP=25 || kP = 15
EP + Q = EPQ
        kQ
```

Since kIQ kP = kIP kQ, it is possible to calculate kQ from the information in Table 5.4 when there are no experimental errors:

kQ = kIQ kP/kIP = 10*15/25 = 6

The effects of experimental errors on the values of kQ can be calculated by copying columns from Table 5.4 and repeating the calculation of kQ:

$$\text{Round}\left[\frac{\begin{matrix}10.^` \\ 10.^` \\ 10.^` \\ 10.^` \\ 10.03^` \\ 10.950000000000001^` \\ 10.16^` \\ 8.93^`\end{matrix}}{} * \frac{\begin{matrix}15.^` \\ 15.^` \\ 15.^` \\ 15.^` \\ 16.^` \\ 13.99^` \\ 14.91^` \\ 15.17^`\end{matrix}}{} \Big/ \frac{\begin{matrix}25.^` \\ 25.^` \\ 25.^` \\ 25.^` \\ 24.96^` \\ 25.93^` \\ 27.36^` \\ 22.^`\end{matrix}}{}, 0.01\right]$$

{{6.}, {6.}, {6.}, {6.}, {6.43}, {5.91}, {5.54}, {6.16}}

These values can be added to Table 5.4. They should be reported in an experimental article because they are equilibrium constants.

The values for the parameters in Table 5.4 can be used to calculate K' using the Haldane relation.

$$K' = \frac{\text{vfexp kP kIQ}}{\text{vrexp kIA kB}}$$

Eight values of K' can be obtained by copying columns from Table 5.4.

$$\frac{\begin{matrix}1.^` \\ 1.06^` \\ 1.^` \\ 0.99^` \\ 1.^` \\ 1.^` \\ 1.^`\end{matrix}}{} * \frac{\begin{matrix}15.^` \\ 15.^` \\ 15.^` \\ 15.^` \\ 16.^` \\ 13.99^` \\ 14.91^` \\ 15.17^`\end{matrix}}{} * \frac{\begin{matrix}10.^` \\ 10.^` \\ 10.^` \\ 10.^` \\ 10.03^` \\ 10.950000000000001^` \\ 10.16^` \\ 8.93^`\end{matrix}}{} \Big/ \left(\frac{\begin{matrix}0.5^` \\ 0.5^` \\ 0.5^` \\ 0.5^` \\ 0.53^` \\ 0.5^` \\ 0.5^` \\ 0.5^`\end{matrix}}{} * \frac{\begin{matrix}5.^` \\ 5.29^` \\ 4.45^` \\ 5.28^` \\ 5.^` \\ 5.^` \\ 5.^`\end{matrix}}{} * \frac{\begin{matrix}20.^` \\ 21.240000000000002^` \\ 20.11^` \\ 18.740000000000002^` \\ 20.^` \\ 20.^` \\ 20.^` \\ 20.^`\end{matrix}}{}\right)$$

{{3.}, {2.8302}, {3.35235}, {3.0016}, {3.02792}, {3.06381}, {3.02971}, {2.70936}}

These values could be added to Table 5.4 to show how 5% errors in the 6 velocities (one at a time) affect the apparent equilibrium constant calculated using the Haldane relation.

There is an alternate rate equation for ordered A + B = random P + Q because EP + Q = EPQ can be used rather than EQ + P = EPQ. However, this rate equation yields the same results as the mechanism in this section, even though there are different parameters in the rate equation. Therefore, the derivation of the alternate rate equation has been put in the Appendix of this chapter.

The effects of adding P and Q when the mechanism is ordered A + B = random P + Q can be discussed in the way they were in Sections 5.1.3 and 5.1.4. The estimates of pH-independent kinetic parameters can also be discussed (see preceding chapter). Dead-end enzyme-substrate complexes like EAP and EBQ can also be added to the mechanism.

5.3 Random A + B = random P + Q

■ 5.3.1 Derivation of the rapid-equilibrium rate equation

The expressions for the equilibrium concentrations eab and epq can be calculated using the following reactions:

E + A = EA	kIA = e*a/ea	
E + B = EB	kIB = e*b/eb	This is the new reaction.
EA + B = EAB	kB = ea*b/eab	
E + P = EP	kIP = e*p/ep	
E + Q = EQ	kIQ = e*q/eq	
EQ + P = EPQ	kP = eq*p/epq	

As stated earlier, reactions that can be obtained by adding and subtracting these 6 reactions will also be at equilibrium; examples are EB + A = EAB and EP + Q = EPQ. The rapid-equilibrium rate equation for this mechanism can be derived by using Solve. The thermodynamic constraints are 6 equilibrium expressions and one conservation equation:

et = e + ea + eb + eab + ep + eq + epq

The total concentration of enzymatic sites is et. There are 11 reactants, 5 components (a, b, p, q, and e), and 6 independent reactions, and so $N' = C' + R'$ is 11 = 5 + 6. The 7 unknown concentrations on the right hand side of the equation for et can be calculated by applying the 7 constraints and specifying et. Solve[eqs,vars,elims] can be used to calculate the equilibrium expression for any of the 7 reactants that make up et. This is illustrated by deriving equilibrium expressions for eab and epq.

```
Solve[{kIA == e * a / ea, kIB == e * b / eb, kB == ea * b / eab, kIP == e * p / ep, kIQ == e * q / eq,
   kP == eq * p / epq, et == e + ea + eb + eab + ep + eq + epq}, {eab}, {e, ea, eb, ep, eq, epq}]
```

{{eab → (a b et kIB kIP kIQ kP) / (b kB kIA kIP kIQ kP + a b kIB kIP kIQ kP + a kB kIB kIP kIQ kP +
   kB kIA kIB kIP kIQ kP + kB kIA kIB kIQ kP p + kB kIA kIB kIP kP q + kB kIA kIB kIP p q)}}

```
Solve[{kIA == e * a / ea, kIB == e * b / eb, kB == ea * b / eab, kIP == e * p / ep, kIQ == e * q / eq,
   kP == eq * p / epq, et == e + ea + eb + eab + ep + eq + epq}, {epq}, {e, ea, eb, ep, eq, eab}]
```

{{epq → (et kB kIA kIB kIP p q) / (b kB kIA kIP kIQ kP + a b kIB kIP kIQ kP + a kB kIB kIP kIQ kP +
   kB kIA kIB kIP kIQ kP + kB kIA kIB kIQ kP p + kB kIA kIB kIP kP q + kB kIA kIB kIP p q)}}

Note that the 7 denominator terms are the same in these two expressions.

The discussion of the rapid-equilibrium kinetics of random A + B = random P + Q is based on the following mechanism:

E + A = EA	kIA = e*a/ea
E + B = EB	kIB = e*b/eb
EA + B = EAB	kB = ea*b/eab
EAB ⇌ EPQ	v = kf*eab - kr*epq
E + P = EP	kIP = e*p/ep
E + Q = EQ	kIQ = e*q/eq
EQ + P = EPQ	kP = eq*p/epq

There are 8 kinetic parameters. This is one more than the number of denominator terms in the rate equation.

These expressions for eab and epq can be used to derive the rapid-equilibrium rate equation for random A + B = random P + Q. The velocity of the forward reaction is proportional to eab.

The rapid-equilibrium velocity of the forward reaction is given by

```
(a b vfexp kIB kIP kIQ kP) / (b kB kIA kIP kIQ kP + a b kIB kIP kIQ kP + a kB kIB kIP kIQ kP +
   kB kIA kIB kIP kIQ kP + kB kIA kIB kIQ kP p + kB kIA kIB kIP kP q + kB kIA kIB kIP p q)
```

```
(a b kIB kIP kIQ kP vfexp) /
  (b kB kIA kIP kIQ kP + a b kIB kIP kIQ kP + a kB kIB kIP kIQ kP + kB kIA kIB kIP kIQ kP +
    kB kIA kIB kIQ kP p + kB kIA kIB kIP kP q + kB kIA kIB kIP p q)
```

The rapid-equilibrium velocity of the reverse reaction is proportional to epq. The velocity of the reverse reaction is given by

```
(vrexp kB kIA kIB kIP p q) / (b kB kIA kIP kIQ kP + a b kIB kIP kIQ kP + a kB kIB kIP kIQ kP +
   kB kIA kIB kIP kIQ kP + kB kIA kIB kIQ kP p + kB kIA kIB kIP kP q + kB kIA kIB kIP p q)
```

```
(kB kIA kIB kIP p q vrexp) /
  (b kB kIA kIP kIQ kP + a b kIB kIP kIQ kP + a kB kIB kIP kIQ kP + kB kIA kIB kIP kIQ kP +
    kB kIA kIB kIQ kP p + kB kIA kIB kIP kP q + kB kIA kIB kIP p q)
```

The rapid-equilibrium velocity of the reaction A + B = P + Q is given by v = vf - vr = kf*eab - vr*epq. Since the denominators for eab and epq are the same, the velocity v in the forward direction is given by

```
vfrandABPQ2 = (a b kIB kIP kIQ kP vfexp – kB kIA kIB kIP p q vrexp) /
  (b kB kIA kIP kIQ kP + a b kIB kIP kIQ kP + a kB kIB kIP kIQ kP +
    kB kIA kIB kIP kIQ kP + kB kIA kIB kIQ kP p + kB kIA kIB kIP kP q + kB kIA kIB kIP p q)
```

```
(a b kIB kIP kIQ kP vfexp – kB kIA kIB kIP p q vrexp) /
  (b kB kIA kIP kIQ kP + a b kIB kIP kIQ kP + a kB kIB kIP kIQ kP +
    kB kIA kIB kIP kIQ kP + kB kIA kIB kIQ kP p + kB kIA kIB kIP kP q + kB kIA kIB kIP p q)
```

This rate equation is used to calculate the velocity of random A + B = random P + Q for any 8 arbitrarily specified kinetic parameters and any set of 4 substrate concentrations.

■ 5.3.2 Estimation of pH-dependent kinetic parameters with the minimum number of measured velocities

The function vfrandABPQ2 of a, b, p, and q can be used to calculate the velocity for any quadruplet of substrate equations. The choice of 8 quadruplets of substrate concentration follows from the recommendations for the forward reaction A + B → products, in the absence of P and Q [35]. As an example, the 8 velocities can be calculated for the following 8 quadruplets of substrate concentrations: {a,b,p,q} = {100,100,0,0}, {1,100,0,0), {100,1,0,0}, {5,5,0,0}, {0,0,100,100}, {0,0,1,100}, {0,0,100,1}, and {0,0,5,5}. The first four quadruplets of velocities make the predominant contribution to vfexp, kIA, kB, and kIB. The second four quadruplets of velocities make the predominant contribution to vrexp, kIP, kIQ, and kP. The new parameter that is specfied arbitarily is kIB = 30.

To determine the values of 8 kinetic parameters, 8 velocities have to be measured.

```
vfrandABPQ2 /. vfexp → 1 /. vrexp → .5 /. kIA → 5 /. kIB → 30 /. kB → 20 /. kIP → 25 /. kIQ → 10 /.
    kP → 15 /. a → 100 /. b → 100 /. p → 0 /. q → 0 // N
```

```
0.80429
```

```
vfrandABPQ2 /. vfexp → 1 /. vrexp → .5 /. kIA → 5 /. kIB → 30 /. kB → 20 /. kIP → 25 /. kIQ → 10 /.
    kP → 15 /. a → 1 /. b → 100 /. p → 0 /. q → 0 // N
```

```
0.180723
```

```
vfrandABPQ2 /. vfexp → 1 /. vrexp → .5 /. kIA → 5 /. kIB → 30 /. kB → 20 /. kIP → 25 /. kIQ → 10 /.
    kP → 15 /. a → 100 /. b → 1 /. p → 0 /. q → 0 // N
```

```
0.0453858
```

```
vfrandABPQ2 /. vfexp → 1 /. vrexp → .5 /. kIA → 5 /. kIB → 30 /. kB → 20 /. kIP → 25 /. kIQ → 10 /.
    kP → 15 /. a → 5 /. b → 5 /. p → 0 /. q → 0 // N
```

```
0.103448
```

```
vfrandABPQ2 /. vfexp → 1 /. vrexp → .5 /. kIA → 5 /. kIB → 30 /. kB → 20 /. kIP → 25 /. kIQ → 10 /.
    kP → 15 /. a -> 0 /. b → 0 /. p → 100 /. q → 100 // N
```

-0.408163

```
vfrandABPQ2 /. vfexp → 1 /. vrexp → .5 /. kIA → 5 /. kIB → 30 /. kB → 20 /. kIP → 25 /. kIQ → 10 /.
    kP → 15 /. a → 0 /. b → 0 /. p → 1 /. q → 100 // N
```

-0.0284738

```
vfrandABPQ2 /. vfexp → 1 /. vrexp → .5 /. kIA → 5 /. kIB → 30 /. kB → 20 /. kIP → 25 /. kIQ → 10 /.
    kP → 15 /. a → 0 /. b → 0 /. p → 100 /. q → 1 // N
```

-0.0578035

```
vfrandABPQ2 /. vfexp → 1 /. vrexp → .5 /. kIA → 5 /. kIB → 30 /. kB → 20 /. kIP → 25 /. kIQ → 10 /.
    kP → 15 /. a → 0 /. b → 0 /. p → 5 /. q → 5 // N
```

-0.0446429

The following program uses Solve to calculate the 8 kinetic parameters from 8 measured velocities.

```
calckinparsrandABPQfr[v1_, a1_, b1_, p1_, q1_, v2_, a2_, b2_, p2_,
   q2_, v3_, a3_, b3_, p3_, q3_, v4_, a4_, b4_, p4_, q4_, v5_, a5_, b5_, p5_, q5_,
   v6_, a6_, b6_, p6_, q6_, v7_, a7_, b7_, p7_, q7_, v8_, a8_, b8_, p8_, q8_] :=
 Module[{}, (*This program calculates vfexp, kIA, kIB, kB, frexp, kIP, kIQ,
    and kP from 8 experimental quadruplets of velocities for random A + B =
    random P + Q products.  As an example,
    the following 8 quadruplets of velocities can be used:{100,100,0,0},{1,100,0,0},
    {100,1,0,0},{5,5,0,0},{0,0,100,100},{0,0,1,100},{0,0,100,1}, and {0,0,5,5}.*)
   Solve[{v1 == (a1 b1 kIB kIP kIQ kP vfexp - kB kIA kIB kIP p1 q1 vrexp) /
       (b1 kB kIA kIP kIQ kP + a1 b1 kIB kIP kIQ kP + a1 kB kIB kIP kIQ kP + kB kIA kIB kIP kIQ kP +
         kB kIA kIB kIQ kP p1 + kB kIA kIB kIP kP q1 + kB kIA kIB kIP p1 q1),
    v2 == (a2 b2 kIB kIP kIQ kP vfexp - kB kIA kIB kIP p2 q2 vrexp) /
       (b2 kB kIA kIP kIQ kP + a2 b2 kIB kIP kIQ kP + a2 kB kIB kIP kIQ kP + kB kIA kIB kIP kIQ kP +
         kB kIA kIB kIQ kP p2 + kB kIA kIB kIP kP q2 + kB kIA kIB kIP p2 q2),
    v3 == (a3 b3 kIB kIP kIQ kP vfexp - kB kIA kIB kIP p3 q3 vrexp) /
       (b3 kB kIA kIP kIQ kP + a3 b3 kIB kIP kIQ kP + a3 kB kIB kIP kIQ kP + kB kIA kIB kIP kIQ kP +
         kB kIA kIB kIQ kP p3 + kB kIA kIB kIP kP q3 + kB kIA kIB kIP p3 q3),
    v4 == (a4 b4 kIB kIP kIQ kP vfexp - kB kIA kIB kIP p4 q4 vrexp) /
       (b4 kB kIA kIP kIQ kP + a4 b4 kIB kIP kIQ kP + a4 kB kIB kIP kIQ kP + kB kIA kIB kIP kIQ kP +
         kB kIA kIB kIQ kP p4 + kB kIA kIB kIP kP q4 + kB kIA kIB kIP p4 q4),
    v5 == (a5 b5 kIB kIP kIQ kP vfexp - kB kIA kIB kIP p5 q5 vrexp) /
       (b5 kB kIA kIP kIQ kP + a5 b5 kIB kIP kIQ kP + a5 kB kIB kIP kIQ kP + kB kIA kIB kIP kIQ kP +
         kB kIA kIB kIQ kP p5 + kB kIA kIB kIP kP q5 + kB kIA kIB kIP p5 q5),
    v6 == (a6 b6 kIB kIP kIQ kP vfexp - kB kIA kIB kIP p6 q6 vrexp) /
       (b6 kB kIA kIP kIQ kP + a6 b6 kIB kIP kIQ kP + a6 kB kIB kIP kIQ kP + kB kIA kIB kIP kIQ kP +
         kB kIA kIB kIQ kP p6 + kB kIA kIB kIP kP q6 + kB kIA kIB kIP p6 q6),
    v7 == (a7 b7 kIB kIP kIQ kP vfexp - kB kIA kIB kIP p7 q7 vrexp) /
       (b7 kB kIA kIP kIQ kP + a7 b7 kIB kIP kIQ kP + a7 kB kIB kIP kIQ kP + kB kIA kIB kIP kIQ kP +
         kB kIA kIB kIQ kP p7 + kB kIA kIB kIP kP q7 + kB kIA kIB kIP p7 q7),
    v8 == (a8 b8 kIB kIP kIQ kP vfexp - kB kIA kIB kIP p8 q8 vrexp) /
       (b8 kB kIA kIP kIQ kP + a8 b8 kIB kIP kIQ kP + a8 kB kIB kIP kIQ kP + kB kIA kIB kIP kIQ kP +
         kB kIA kIB kIQ kP p8 + kB kIA kIB kIP kP q8 + kB kIA kIB kIP p8 q8)},
    {vfexp, vrexp, kIA, kIB, kB, kIP, kIQ, kP}]]

calckinparsrandABPQfr[.80429, 100, 100, 0, 0, .180723, 1, 100, 0,
  0, .0453858, 100, 1, 0, 0, .10345, 5, 5, 0, 0, -.408163, 0, 0, 100, 100,
  -.0284738, 0, 0, 1, 100, -.0578035, 0, 0, 100, 1, -.044643, 0, 0, 5, 5]
```

{{vfexp → 1., vrexp → 0.5, kIA → 4.99977,
  kIQ → 9.99993, kP → 15., kIP → 24.9998, kIB → 29.9983, kB → 20.0001}}

This yields the correct values for the kinetic parameters, as expected.

```
line1 = {vfexp, vrexp, kIA, kIQ, kP, kIP, kIB, kB} /. calckinparsrandABPQfr[.80429, 100, 100,
   0, 0, .180723, 1, 100, 0, 0, .0453858, 100, 1, 0, 0, .10345, 5, 5, 0, 0, -.408163, 0,
   0, 100, 100, -.0284738, 0, 0, 1, 100, -.0578035, 0, 0, 100, 1, -.044643, 0, 0, 5, 5]
```

```
{{1., 0.5, 4.99977, 9.99993, 15., 24.9998, 29.9983, 20.0001}}
```

Put in 5% errors in the measured velocities, one at a time.

```
calckinparsrandABPQfr[1.05 * .80429, 100, 100, 0, 0, .180723, 1,
   100, 0, 0, .0453858, 100, 1, 0, 0, .10345, 5, 5, 0, 0, -.408163, 0, 0, 100,
   100, -.0284738, 0, 0, 1, 100, -.0578035, 0, 0, 100, 1, -.044643, 0, 0, 5, 5]
```

```
{{vfexp → 1.06417, vrexp → 0.5, kIA → 5.03778,
   kIQ → 9.99993, kP → 15., kIP → 24.9998, kIB → 29.8563, kB → 21.3364}}
```

```
line2 = {vfexp, vrexp, kIA, kIQ, kP, kIP, kIB, kB} /.
   calckinparsrandABPQfr[1.05 * .80429, 100, 100, 0, 0, .180723, 1, 100, 0,
      0, .0453858, 100, 1, 0, 0, .10345, 5, 5, 0, 0, -.408163, 0, 0, 100, 100,
      -.0284738, 0, 0, 1, 100, -.0578035, 0, 0, 100, 1, -.044643, 0, 0, 5, 5]
```

```
{{1.06417, 0.5, 5.03778, 9.99993, 15., 24.9998, 29.8563, 21.3364}}
```

```
calckinparsrandABPQfr[.80429, 100, 100, 0, 0, 1.05 * .180723, 1,
   100, 0, 0, .0453858, 100, 1, 0, 0, .10345, 5, 5, 0, 0, -.408163, 0, 0, 100,
   100, -.0284738, 0, 0, 1, 100, -.0578035, 0, 0, 100, 1, -.044643, 0, 0, 5, 5]
```

```
{{vfexp → 0.997207, vrexp → 0.5, kIA → 5.07337,
   kIQ → 9.99993, kP → 15., kIP → 24.9998, kIB → 33.2097, kB → 19.9302}}
```

```
line3 = {vfexp, vrexp, kIA, kIQ, kP, kIP, kIB, kB} /.
   calckinparsrandABPQfr[.80429, 100, 100, 0, 0, 1.05 * .180723, 1, 100, 0,
      0, .0453858, 100, 1, 0, 0, .10345, 5, 5, 0, 0, -.408163, 0, 0, 100, 100,
      -.0284738, 0, 0, 1, 100, -.0578035, 0, 0, 100, 1, -.044643, 0, 0, 5, 5]
```

```
{{0.997207, 0.5, 5.07337, 9.99993, 15., 24.9998, 33.2097, 19.9302}}
```

```
calckinparsrandABPQfr[.80429, 100, 100, 0, 0, .180723, 1, 100, 0,
   0, 1.05 * .0453858, 100, 1, 0, 0, .10345, 5, 5, 0, 0, -.408163, 0, 0, 100,
   100, -.0284738, 0, 0, 1, 100, -.0578035, 0, 0, 100, 1, -.044643, 0, 0, 5, 5]
```

```
{{vfexp → 0.988968, vrexp → 0.5, kIA → 5.5905,
   kIQ → 9.99993, kP → 15., kIP → 24.9998, kIB → 32.2107, kB → 18.6761}}
```

```
line4 = {vfexp, vrexp, kIA, kIQ, kP, kIP, kIB, kB} /.
   calckinparsrandABPQfr[.80429, 100, 100, 0, 0, .180723, 1, 100, 0, 0,
      1.05 * .0453858, 100, 1, 0, 0, .10345, 5, 5, 0, 0, -.408163, 0, 0, 100, 100,
      -.0284738, 0, 0, 1, 100, -.0578035, 0, 0, 100, 1, -.044643, 0, 0, 5, 5]
```

```
{{0.988968, 0.5, 5.5905, 9.99993, 15., 24.9998, 32.2107, 18.6761}}
```

```
calckinparsrandABPQfr[.80429, 100, 100, 0, 0, .180723, 1, 100, 0,
   0, .0453858, 100, 1, 0, 0, 1.05 * .10345, 5, 5, 0, 0, -.408163, 0, 0, 100,
   100, -.0284738, 0, 0, 1, 100, -.0578035, 0, 0, 100, 1, -.044643, 0, 0, 5, 5]
```

```
{{vfexp → 1.00128, vrexp → 0.5, kIA → 4.33459,
   kIQ → 9.99993, kP → 15., kIP → 24.9998, kIB → 25.2088, kB → 20.1533}}
```

```
line5 = {vfexp, vrexp, kIA, kIQ, kP, kIP, kIB, kB} /.
   calckinparsrandABPQfr[.80429, 100, 100, 0, 0, .180723, 1, 100, 0, 0,
      .0453858, 100, 1, 0, 0, 1.05 * .10345, 5, 5, 0, 0, -.408163, 0, 0, 100, 100,
      -.0284738, 0, 0, 1, 100, -.0578035, 0, 0, 100, 1, -.044643, 0, 0, 5, 5]
```

```
{{1.00128, 0.5, 4.33459, 9.99993, 15., 24.9998, 25.2088, 20.1533}}
```

```
calckinparsrandABPQfr[.80429, 100, 100, 0, 0, .180723, 1, 100, 0,
  0, .0453858, 100, 1, 0, 0, .10345, 5, 5, 0, 0, 1.05 * - .408163, 0, 0, 100,
  100, - .0284738, 0, 0, 1, 100, - .0578035, 0, 0, 100, 1, - .044643, 0, 0, 5, 5]
```

```
{{vfexp → 1., vrexp → 0.531582, kIA → 4.99977,
  kIQ → 10.0335, kP → 15.9995, kIP → 24.962, kIB → 29.9983, kB → 20.0001}}
```

```
line6 = {vfexp, vrexp, kIA, kIQ, kP, kIP, kIB, kB} /.
  calckinparsrandABPQfr[.80429, 100, 100, 0, 0, .180723, 1, 100, 0, 0,
    .0453858, 100, 1, 0, 0, .10345, 5, 5, 0, 0, 1.05 * - .408163, 0, 0, 100, 100,
    - .0284738, 0, 0, 1, 100, - .0578035, 0, 0, 100, 1, - .044643, 0, 0, 5, 5]
```

```
{{1., 0.531582, 4.99977, 10.0335, 15.9995, 24.962, 29.9983, 20.0001}}
```

```
calckinparsrandABPQfr[.80429, 100, 100, 0, 0, .180723, 1, 100, 0,
  0, .0453858, 100, 1, 0, 0, .10345, 5, 5, 0, 0, - .408163, 0, 0, 100, 100,
  1.05 * - .0284738, 0, 0, 1, 100, - .0578035, 0, 0, 100, 1, - .044643, 0, 0, 5, 5]
```

```
{{vfexp → 1., vrexp → 0.495593, kIA → 4.99977,
  kIQ → 10.945, kP → 13.9866, kIP → 25.9329, kIB → 29.9983, kB → 20.0001}}
```

```
line7 = {vfexp, vrexp, kIA, kIQ, kP, kIP, kIB, kB} /. calckinparsrandABPQfr[.80429, 100, 100, 0,
    0, .180723, 1, 100, 0, 0, .0453858, 100, 1, 0, 0, .10345, 5, 5, 0, 0, - .408163, 0, 0,
    100, 100, 1.05 * - .0284738, 0, 0, 1, 100, - .0578035, 0, 0, 100, 1, - .044643, 0, 0, 5, 5]
```

```
{{1., 0.495593, 4.99977, 10.945, 13.9866, 25.9329, 29.9983, 20.0001}}
```

```
calckinparsrandABPQfr[.80429, 100, 100, 0, 0, .180723, 1, 100, 0,
  0, .0453858, 100, 1, 0, 0, .10345, 5, 5, 0, 0, - .408163, 0, 0, 100, 100,
  - .0284738, 0, 0, 1, 100, 1.05 * - .0578035, 0, 0, 100, 1, - .044643, 0, 0, 5, 5]
```

```
{{vfexp → 1., vrexp → 0.497819, kIA → 4.99977,
  kIQ → 10.1607, kP → 14.9128, kIP → 27.362, kIB → 29.9983, kB → 20.0001}}
```

```
line8 = {vfexp, vrexp, kIA, kIQ, kP, kIP, kIB, kB} /. calckinparsrandABPQfr[.80429, 100, 100, 0,
    0, .180723, 1, 100, 0, 0, .0453858, 100, 1, 0, 0, .10345, 5, 5, 0, 0, - .408163, 0, 0,
    100, 100, - .0284738, 0, 0, 1, 100, 1.05 * - .0578035, 0, 0, 100, 1, - .044643, 0, 0, 5, 5]
```

```
{{1., 0.497819, 4.99977, 10.1607, 14.9128, 27.362, 29.9983, 20.0001}}
```

```
calckinparsrandABPQfr[.80429, 100, 100, 0, 0, .180723, 1, 100, 0,
  0, .0453858, 100, 1, 0, 0, .10345, 5, 5, 0, 0, - .408163, 0, 0, 100, 100,
  - .0284738, 0, 0, 1, 100, - .0578035, 0, 0, 100, 1, 1.05 * - .044643, 0, 0, 5, 5]
```

```
{{vfexp → 1., vrexp → 0.500739, kIA → 4.99977,
  kIQ → 8.92709, kP → 15.1701, kIP → 21.9959, kIB → 29.9983, kB → 20.0001}}
```

```
line9 = {vfexp, vrexp, kIA, kIQ, kP, kIP, kIB, kB} /. calckinparsrandABPQfr[.80429, 100, 100, 0,
    0, .180723, 1, 100, 0, 0, .0453858, 100, 1, 0, 0, .10345, 5, 5, 0, 0, - .408163, 0, 0,
    100, 100, - .0284738, 0, 0, 1, 100, - .0578035, 0, 0, 100, 1, 1.05 * - .044643, 0, 0, 5, 5]
```

```
{{1., 0.500739, 4.99977, 8.92709, 15.1701, 21.9959, 29.9983, 20.0001}}
```

These estimated values for the kinetic parameters are summarized in Table 5.5 Notice that the order of the kinetic parameters is determined by *Mathematica*.

Table 5.5 Kinetic parameters for random A + B = random P + Q calculated from velocities at {100, 100, 0, 0}, {1, 100, 0, 0}, {100, 1, 0, 0}, {5, 5, 0, 0}, {0, 0, 100, 100}, {0, 0, 1, 100}, {0, 0,100, 1}, and {0, 0, 5, 5} when there are 5% errors in the velocities, one at a time.

```
TableForm[Round[{line1[[1]], line2[[1]], line3[[1]], line4[[1]], line5[[1]],
    line6[[1]], line7[[1]], line8[[1]], line9[[1]]}, 0.01], TableHeadings →
  {{"No errors", "1.05*v1", "1.05*v2", "1.05*v3", "1.05*v4", "1.05*v5", "1.05*v6",
    "1.05*v7", "1.05*v8"}, {"vfexp", "vrexp", "kIA", "kIQ", "kP", "kIP", "kIB", "kB"}}]
```

	vfexp	vrexp	kIA	kIQ	kP	kIP	kIB	kB
No errors	1.	0.5	5.	10.	15.	25.	30.	20.
1.05*v1	1.06	0.5	5.04	10.	15.	25.	29.86	21.34
1.05*v2	1.	0.5	5.07	10.	15.	25.	33.21	19.93
1.05*v3	0.99	0.5	5.59	10.	15.	25.	32.21	18.68
1.05*v4	1.	0.5	4.33	10.	15.	25.	25.21	20.15
1.05*v5	1.	0.53	5.	10.03	16.	24.96	30.	20.
1.05*v6	1.	0.5	5.	10.95	13.99	25.93	30.	20.
1.05*v7	1.	0.5	5.	10.16	14.91	27.36	30.	20.
1.05*v8	1.	0.5	5.	8.93	15.17	22.	30.	20.

Different effects of 5% errors in measured velocities would be obtained by using a different set of 8 quadruplets of substrate concentrations. A different set of 8 quadruplets will yield the correct values of kinetic parameters, but the effects of 5% errors in the measured velocities do change when the substrate concentrations are changed. The estimation of kinetic parameters is very dependent on the equipment for velocity neasurements because it is desirable to go to as low concentration of substrates as possible.

The random mechanism for the reverse reaction brings in 3 sides of the following thermodynamic cycle:

```
   kIQ = 10
E + Q  =  EQ
+         +
P         P
‖ kIP=25  ‖ kP = 15
EP + Q = EPQ
      kQ
```

Since kIQ kP = kIP kQ, it is possible to calculate kQ from the information in Table 5.5 when there are no experimental errors:

kQ = kIQ kP/kIP = 10*15/25 = 6

The effects of experimental errors on the values of kQ can be calculated by copying columns from Table 5.5 and repeating the calculations of kQ:

$$
\text{Round}\left[
\begin{matrix}
10.\char"0060 \\
10.\char"0060 \\
10.\char"0060 \\
10.\char"0060 \\
10.\char"0060 \\
10.03\char"0060 \\
10.950000000000001\char"0060 \\
10.16\char"0060 \\
8.93\char"0060
\end{matrix}
\; * \;
\begin{matrix}
15.\char"0060 \\
15.\char"0060 \\
15.\char"0060 \\
15.\char"0060 \\
15.\char"0060 \\
16.\char"0060 \\
13.99\char"0060 \\
14.91\char"0060 \\
15.17\char"0060
\end{matrix}
\; / \;
\begin{matrix}
25.\char"0060 \\
25.\char"0060 \\
25.\char"0060 \\
25.\char"0060 \\
25.\char"0060 \\
24.96\char"0060 \\
25.93\char"0060 \\
27.36\char"0060 \\
22.\char"0060
\end{matrix}
\; , 0.01
\right]
$$

{{6.}, {6.}, {6.}, {6.}, {6.}, {6.43}, {5.91}, {5.54}, {6.16}}

These values can be added to Table 5.5. They should be reported in an experimental article because they are equilibrium constants.

The random reaction in the forward direction brings in 3 sides of the following thermodynamic cycle:

```
       kIA = 5
E + A =  EA
+         +
B         B
|| kIB=30 || kB = 20
EB + A = EAB
      kA
```

Since kIA kB = kA kIB, it is possible to calculate kA from the information in Table 5.5 when there are no experimental errors:

kA = kIA kB/kIB = 5*20/30 = 3.33333

When there are errors, values of kA are given by

$$
\text{Round}\left[
\begin{array}{ccc}
5.\grave{} & 20.\grave{} & \overline{} \\
\overline{5.04\grave{}} & \overline{21.34\grave{}} & \overline{29.86\grave{}} \\
5.07\grave{} & 19.93\grave{} & 33.21\grave{} \\
5.59\grave{} & 18.68\grave{} & 32.21\grave{} \\
4.33\grave{} & * \; 20.150000000000002\grave{} & 25.21\grave{} \\
5.\grave{} & 20.\grave{} & 30.\grave{} \\
5.\grave{} & 20.\grave{} & 30.\grave{} \\
5.\grave{} & 20.\grave{} & 30.\grave{} \\
5.\grave{} & 20.\grave{} & 30.\grave{}
\end{array}
\middle/ \begin{array}{c} 30.\grave{} \end{array}, \; 0.01 \right]
$$

{{3.33}, {3.6}, {3.04}, {3.24}, {3.46}, {3.33}, {3.33}, {3.33}, {3.33}}

The effects of experimental errors on the values of kQ can be calculated by copying columns from Table 5.5 and repeating the calculations of kQ as described in connection with Table 5.4. These values for kA and kQ can be added to Table 5.5. Thus measurements of 8 velocities has yielded 10 parameters.

The values for the parameters in Table 5.5 can be used to calculate K' using the Haldane relation.

$$ K' = \frac{\text{vfexp kP kIQ}}{\text{vrexp kIA kB}} $$

Eight values of K' can be obtained by copying columns from Table 5.5.

$$
\left(
\begin{array}{ccc}
1.\grave{} & 15.\grave{} & 10.\grave{} \\
\overline{1.06\grave{}} & \overline{15.\grave{}} & \overline{10.\grave{}} \\
1.\grave{} & 15.\grave{} & 10.\grave{} \\
0.99\grave{} & 15.\grave{} & 10.\grave{} \\
1.\grave{} & * \; 15.\grave{} & * \; 10.\grave{} \\
1.\grave{} & 16.\grave{} & 10.03\grave{} \\
1.\grave{} & 13.99\grave{} & 10.950000000000001\grave{} \\
1.\grave{} & 14.91\grave{} & 10.16\grave{} \\
1.\grave{} & 15.17\grave{} & 8.93\grave{}
\end{array}
\right)
\middle/
\left(
\begin{array}{ccc}
0.5\grave{} & 5.\grave{} & 20.\grave{} \\
\overline{0.5\grave{}} & \overline{5.04\grave{}} & \overline{21.34\grave{}} \\
0.5\grave{} & 5.07\grave{} & 19.93\grave{} \\
0.5\grave{} & 5.59\grave{} & 18.68\grave{} \\
0.5\grave{} & * \; 4.33\grave{} & * \; 20.150000000000002\grave{} \\
0.53\grave{} & 5.\grave{} & 20.\grave{} \\
0.5\grave{} & 5.\grave{} & 20.\grave{} \\
0.5\grave{} & 5.\grave{} & 20.\grave{} \\
0.5\grave{} & 5.\grave{} & 20.\grave{}
\end{array}
\right)
$$

{{3.}, {2.95667}, {2.96897}, {2.84425}, {3.43842}, {3.02792}, {3.06381}, {3.02971}, {2.70936}}

These values could be added to Table 5.5 to show how 5% errors in the 8 velocities affect the apparent equilibrium constant calculated using the Haldane relation.

■ 5.3.3 Application of calckinparsrandABPQfr to data for ordered A + B = ordered P + Q

As recommended earlier [32,34,35], when the kinetics of A + B = P + Q is studied for the first time, there is an advantage in using a more general program first since the mechanism is unknown.

For ordered A + B = ordered P + Q, 6 velocities had to be measured (see Table 5.1), but to use calckinparsrandABPQfr two more velocities are needed. These velocities can be measured at {5,5,0,0} and {{0,0,5,5}. These velocities were calculated earlier in Section 5.1.1. When these velocities are put in calckinparsrandABPQfr, the following results are obtained:

```
calckinparsrandABPQfr[.82644, 100, 100, 0, 0, .454545, 1, 100,
 0, 0, .0454545, 100, 1, 0, 0, .11111, 5, 5, 0, 0, -.42919, 0, 0, 100,
 100, -.18868, 0, 0, 100, 1, -.028571, 0, 0, 1, 100, -.05000, 0, 0, 5, 5]
```

$\{\{vfexp \to 0.999991, vrexp \to 0.500009, kIA \to 5.00011, kIQ \to 9.99973,$
$kP \to 15.0006, kIP \to 9.93077 \times 10^6, kIB \to -3.4316 \times 10^6, kB \to 19.9998\}\}$

```
line1x = {vfexp, vrexp, kIA, kIQ, kP, kIP, kIB, kB} /. calckinparsrandABPQfr[.82644, 100,
    100, 0, 0, .454545, 1, 100, 0, 0, .0454545, 100, 1, 0, 0, .11111, 5, 5, 0, 0, -.42919,
    0, 0, 100, 100, -.18868, 0, 0, 100, 1, -.028571, 0, 0, 1, 100, -.05000, 0, 0, 5, 5]
```

$\{\{0.999991, 0.500009, 5.00011, 9.99973, 15.0006, 9.93077 \times 10^6, -3.4316 \times 10^6, 19.9998\}\}$

Note that the calculated parameters kIB and kIP are both unreasonable values for Michaelis constants. This shows that the mechanism is not random A + B = random P + Q because calckinparsrandABPQfr shows that the mechanism does not include EQ + P = EPQ and EB + A = EB. It is of interest to see the effects of 5% errors in measured velocities, one at a time.

```
calckinparsrandABPQfr[1.05 * .82644, 100, 100, 0, 0, .454545, 1,
 100, 0, 0, .0454545, 100, 1, 0, 0, .11111, 5, 5, 0, 0, -.42919, 0, 0, 100,
 100, -.18868, 0, 0, 100, 1, -.028571, 0, 0, 1, 100, -.05000, 0, 0, 5, 5]
```

$\{\{vfexp \to 1.06233, vrexp \to 0.500009, kIA \to 5.0371,$
$kIQ \to 9.99973, kP \to 15.0006, kIP \to 9.93077 \times 10^6, kIB \to 2089.17, kB \to 21.298\}\}$

Carrying more digits in the velocities shows that the values for kIP and kIB are actually infinite. This shows that the mechanism is ordered A + B = ordered P + Q.

```
line2x = {vfexp, vrexp, kIA, kIQ, kP, kIP, kIB, kB} /. calckinparsrandABPQfr[1.05 * .82644, 100,
    100, 0, 0, .454545, 1, 100, 0, 0, .0454545, 100, 1, 0, 0, .11111, 5, 5, 0, 0, -.42919,
    0, 0, 100, 100, -.18868, 0, 0, 100, 1, -.028571, 0, 0, 1, 100, -.05000, 0, 0, 5, 5]
```

$\{\{1.06233, 0.500009, 5.0371, 9.99973, 15.0006, 9.93077 \times 10^6, 2089.17, 21.298\}\}$

```
calckinparsrandABPQfr[.82644, 100, 100, 0, 0, 1.05 * .454545, 1,
 100, 0, 0, .0454545, 100, 1, 0, 0, .11111, 5, 5, 0, 0, -.42919, 0, 0, 100,
 100, -.18868, 0, 0, 100, 1, -.028571, 0, 0, 1, 100, -.05000, 0, 0, 5, 5]
```

$\{\{vfexp \to 0.998878, vrexp \to 0.500009, kIA \to 5.02936,$
$kIQ \to 9.99973, kP \to 15.0006, kIP \to 9.93077 \times 10^6, kIB \to -902.533, kB \to 19.972\}\}$

```
line3x = {vfexp, vrexp, kIA, kIQ, kP, kIP, kIB, kB} /. calckinparsrandABPQfr[.82644, 100, 100, 0,
    0, 1.05 * .454545, 1, 100, 0, 0, .0454545, 100, 1, 0, 0, .11111, 5, 5, 0, 0, -.42919,
    0, 0, 100, 100, -.18868, 0, 0, 100, 1, -.028571, 0, 0, 1, 100, -.05000, 0, 0, 5, 5]
```

$\{\{0.998878, 0.500009, 5.02936, 9.99973, 15.0006, 9.93077 \times 10^6, -902.533, 19.972\}\}$

```
calckinparsrandABPQfr[.82644, 100, 100, 0, 0, .454545, 1, 100, 0,
  0, 1.05 * .0454545, 100, 1, 0, 0, .11111, 5, 5, 0, 0, -.42919, 0, 0, 100,
  100, -.18868, 0, 0, 100, 1, -.028571, 0, 0, 1, 100, -.05000, 0, 0, 5, 5]
```

$\{\{\text{vfexp} \to 0.988975, \text{vrexp} \to 0.500009, \text{kIA} \to 5.58992,$
$\quad \text{kIQ} \to 9.99973, \text{kP} \to 15.0006, \text{kIP} \to 9.93077 \times 10^6, \text{kIB} \to -1894.55, \text{kB} \to 18.6779\}\}$

```
line4x = {vfexp, vrexp, kIA, kIQ, kP, kIP, kIB, kB} /. calckinparsrandABPQfr[.82644, 100, 100, 0,
  0, .454545, 1, 100, 0, 0, 1.05 * .0454545, 100, 1, 0, 0, .11111, 5, 5, 0, 0, -.42919,
  0, 0, 100, 100, -.18868, 0, 0, 100, 1, -.028571, 0, 0, 1, 100, -.05000, 0, 0, 5, 5]
```

$\{\{0.988975, 0.500009, 5.58992, 9.99973, 15.0006, 9.93077 \times 10^6, -1894.55, 18.6779\}\}$

```
calckinparsrandABPQfr[.82644, 100, 100, 0, 0, .454545, 1, 100, 0,
  0, .0454545, 100, 1, 0, 0, 1.05 * .11111, 5, 5, 0, 0, -.42919, 0, 0, 100,
  100, -.18868, 0, 0, 100, 1, -.028571, 0, 0, 1, 100, -.05000, 0, 0, 5, 5]
```

$\{\{\text{vfexp} \to 1.00118, \text{vrexp} \to 0.500009, \text{kIA} \to 4.38051,$
$\quad \text{kIQ} \to 9.99973, \text{kP} \to 15.0006, \text{kIP} \to 9.93077 \times 10^6, \text{kIB} \to 742.526, \text{kB} \to 20.1424\}\}$

```
line5x = {vfexp, vrexp, kIA, kIQ, kP, kIP, kIB, kB} /. calckinparsrandABPQfr[.82644, 100, 100, 0,
  0, .454545, 1, 100, 0, 0, .0454545, 100, 1, 0, 0, 1.05 * .11111, 5, 5, 0, 0, -.42919,
  0, 0, 100, 100, -.18868, 0, 0, 100, 1, -.028571, 0, 0, 1, 100, -.05000, 0, 0, 5, 5]
```

$\{\{1.00118, 0.500009, 4.38051, 9.99973, 15.0006, 9.93077 \times 10^6, 742.526, 20.1424\}\}$

```
calckinparsrandABPQfr[.82644, 100, 100, 0, 0, .454545, 1, 100, 0,
  0, .0454545, 100, 1, 0, 0, .11111, 5, 5, 0, 0, 1.05 * -.42919, 0, 0, 100,
  100, -.18868, 0, 0, 100, 1, -.028571, 0, 0, 1, 100, -.05000, 0, 0, 5, 5]
```

$\{\{\text{vfexp} \to 0.999991, \text{vrexp} \to 0.529953, \text{kIA} \to 5.00011,$
$\quad \text{kIQ} \to 10.0317, \text{kP} \to 15.9482, \text{kIP} \to 3240.41, \text{kIB} \to -3.4316 \times 10^6, \text{kB} \to 19.9998\}\}$

```
line6x = {vfexp, vrexp, kIA, kIQ, kP, kIP, kIB, kB} /. calckinparsrandABPQfr[.82644, 100, 100, 0,
  0, .454545, 1, 100, 0, 0, .0454545, 100, 1, 0, 0, .11111, 5, 5, 0, 0, 1.05 * -.42919,
  0, 0, 100, 100, -.18868, 0, 0, 100, 1, -.028571, 0, 0, 1, 100, -.05000, 0, 0, 5, 5]
```

$\{\{0.999991, 0.529953, 5.00011, 10.0317, 15.9482, 3240.41, -3.4316 \times 10^6, 19.9998\}\}$

```
calckinparsrandABPQfr[.82644, 100, 100, 0, 0, .454545, 1, 100,
  0, 0, .0454545, 100, 1, 0, 0, .11111, 5, 5, 0, 0, -.42919, 0, 0, 100, 100,
  1.05 * -.18868, 0, 0, 100, 1, -.028571, 0, 0, 1, 100, -.05000, 0, 0, 5, 5]
```

$\{\{\text{vfexp} \to 0.999991, \text{vrexp} \to 0.499339, \text{kIA} \to 5.00011,$
$\quad \text{kIQ} \to 10.049, \text{kP} \to 14.9738, \text{kIP} \to -1123.08, \text{kIB} \to -3.4316 \times 10^6, \text{kB} \to 19.9998\}\}$

```
line7x = {vfexp, vrexp, kIA, kIQ, kP, kIP, kIB, kB} /. calckinparsrandABPQfr[.82644, 100, 100,
  0, 0, .454545, 1, 100, 0, 0, .0454545, 100, 1, 0, 0, .11111, 5, 5, 0, 0, -.42919, 0, 0,
  100, 100, 1.05 * -.18868, 0, 0, 100, 1, -.028571, 0, 0, 1, 100, -.05000, 0, 0, 5, 5]
```

$\{\{0.999991, 0.499339, 5.00011, 10.049, 14.9738, -1123.08, -3.4316 \times 10^6, 19.9998\}\}$

```
calckinparsrandABPQfr[.82644, 100, 100, 0, 0, .454545, 1, 100,
  0, 0, .0454545, 100, 1, 0, 0, .11111, 5, 5, 0, 0, -.42919, 0, 0, 100, 100,
  -.18868, 0, 0, 100, 1, 1.05 * -.028571, 0, 0, 1, 100, -.05000, 0, 0, 5, 5]
```

$\{\{\text{vfexp} \to 0.999991, \text{vrexp} \to 0.495617, \text{kIA} \to 5.00011,$
$\quad \text{kIQ} \to 10.9414, \text{kP} \to 13.9905, \text{kIP} \to -3486.91, \text{kIB} \to -3.4316 \times 10^6, \text{kB} \to 19.9998\}\}$

```
line8x = {vfexp, vrexp, kIA, kIQ, kP, kIP, kIB, kB} /. calckinparsrandABPQfr[.82644, 100, 100,
    0, 0, .454545, 1, 100, 0, 0, .0454545, 100, 1, 0, 0, .11111, 5, 5, 0, 0, -.42919, 0, 0,
    100, 100, -.18868, 0, 0, 100, 1, 1.05 * -.028571, 0, 0, 1, 100, -.05000, 0, 0, 5, 5]
```

$$\{\{0.999991,\ 0.495617,\ 5.00011,\ 10.9414,\ 13.9905,\ -3486.91,\ -3.4316 \times 10^6,\ 19.9998\}\}$$

```
calckinparsrandABPQfr[.82644, 100, 100, 0, 0, .454545, 1, 100,
    0, 0, .0454545, 100, 1, 0, 0, .11111, 5, 5, 0, 0, -.42919, 0, 0, 100, 100,
    -.18868, 0, 0, 100, 1, -.028571, 0, 0, 1, 100, 1.05 * -.05000, 0, 0, 5, 5]
```

$$\{\{\text{vfexp} \to 0.999991,\ \text{vrexp} \to 0.500669,\ \text{kIA} \to 5.00011,$$
$$\text{kIQ} \to 9.04086,\ \text{kP} \to 15.1525,\ \text{kIP} \to 1037.02,\ \text{kIB} \to -3.4316 \times 10^6,\ \text{kB} \to 19.9998\}\}$$

```
line9x = {vfexp, vrexp, kIA, kIQ, kP, kIP, kIB, kB} /. calckinparsrandABPQfr[.82644, 100, 100,
    0, 0, .454545, 1, 100, 0, 0, .0454545, 100, 1, 0, 0, .11111, 5, 5, 0, 0, -.42919, 0, 0,
    100, 100, -.18868, 0, 0, 100, 1, -.028571, 0, 0, 1, 100, 1.05 * -.05000, 0, 0, 5, 5]
```

$$\{\{0.999991,\ 0.500669,\ 5.00011,\ 9.04086,\ 15.1525,\ 1037.02,\ -3.4316 \times 10^6,\ 19.9998\}\}$$

Table 5.6 Kinetic parameters for ordered A + B = ordered P + Q calculated by applying the more general program calckinparsrandABPQfr to velocities for the ordered mechanism

```
TableForm[Round[{line1x[[1]], line2x[[1]], line3x[[1]], line4x[[1]], line5x[[1]],
    line6x[[1]], line7x[[1]], line8x[[1]], line9x[[1]]}, 0.01], TableHeadings →
    {{"No errors", "1.05*v1", "1.05*v2", "1.05*v3", "1.05*v4", "1.05*v5", "1.05*v6",
    "1.05*v7", "1.05*v8"}, {"vfexp", "vrexp", "kIA", "kIQ", "kP", "kIP", "kIB", "kB"}}]
```

	vfexp	vrexp	kIA	kIQ	kP	kIP	kIB	kB
No errors	1.	0.5	5.	10.	15.	9.93077×10^6	-3.4316×10^6	20.
1.05*v1	1.06	0.5	5.04	10.	15.	9.93077×10^6	2089.17	21.3
1.05*v2	1.	0.5	5.03	10.	15.	9.93077×10^6	-902.53	19.97
1.05*v3	0.99	0.5	5.59	10.	15.	9.93077×10^6	-1894.55	18.68
1.05*v4	1.	0.5	4.38	10.	15.	9.93077×10^6	742.53	20.14
1.05*v5	1.	0.53	5.	10.03	15.95	3240.41	-3.4316×10^6	20.
1.05*v6	1.	0.5	5.	10.05	14.97	-1123.08	-3.4316×10^6	20.
1.05*v7	1.	0.5	5.	10.94	13.99	-3486.91	-3.4316×10^6	20.
1.05*v8	1.	0.5	5.	9.04	15.15	1037.02	-3.4316×10^6	20.

All of these kIP and kIB values are unreasonable, and so this shows that using the most general program can identify the mechanism and yield correct values for the kinetic parameters in the actual mechanism. Notice that the effects of 5% errors in measured velocities are highly variable; this is another sign that the data is not from a random A + B = random P + Q mechanism, but is from an ordered A + B = ordered P + Q mechanism.

■ 5.3.4 Application of calckinparsrandABPQfr to data for ordered A + B = random P + Q

Seven velocities were calculated in Section 5.2.1 using the rate equation for ordered A + B = random P + Q. In order to apply calckinparsrandABPQfr, it is necessary to have 8 velocities. As shown in Section 5.2.1, v = 0.1111 at {5,5,0,0}. Including this data makes it possible to calculate 8 kinetic parameters. The kinetic parameters are calculated as follows using 7 velocities.

```
calckinparsrandABPQfr[.82645, 100, 100, 0, 0, .454545, 1, 100,
    0, 0, .0454545, 100, 1, 0, 0, -.40816, 0, 0, 100, 100, -.028474, 0, 0,
    1, 100, -.057804, 0, 0, 100, 1, -.044643, 0, 0, 5, 5, .1111, 5, 5, 0, 0]
```

$$\{\{\text{vfexp} \to 1.,\ \text{vrexp} \to 0.499994,\ \text{kIA} \to 5.0013,$$
$$\text{kIQ} \to 10.0001,\ \text{kP} \to 14.9997,\ \text{kIP} \to 25.0004,\ \text{kIB} \to -414606.,\ \text{kB} \to 19.9998\}\}$$

Correct values are obtained for all of the kinetic parameters, except for kIB. The value of kIB is unreasonable; this is the program's way of saying there is no E + B = EB in the mechanism.

```
line1y = {vfexp, vrexp, kIA, kIQ, kP, kIP, kIB, kB} /. calckinparsrandABPQfr[.82645, 100,
    100, 0, 0, .454545, 1, 100, 0, 0, .0454545, 100, 1, 0, 0, -.40816, 0, 0, 100, 100,
    -.028474, 0, 0, 1, 100, -.057804, 0, 0, 100, 1, -.044643, 0, 0, 5, 5, .1111, 5, 5, 0, 0]
```

```
{{1., 0.499994, 5.0013, 10.0001, 14.9997, 25.0004, -414606., 19.9998}}
```

Calculate effects of 5% errors in velocities, one at a time.

```
calckinparsrandABPQfr[1.05 * .82645, 100, 100, 0, 0, .454545, 1,
    100, 0, 0, .0454545, 100, 1, 0, 0, -.40816, 0, 0, 100, 100, -.028474, 0,
    0, 1, 100, -.057804, 0, 0, 100, 1, -.044643, 0, 0, 5, 5, .1111, 5, 5, 0, 0]
```

```
{{vfexp → 1.06235, vrexp → 0.499994, kIA → 5.03829,
    kIQ → 10.0001, kP → 14.9997, kIP → 25.0004, kIB → 2098.87, kB → 21.2981}}
```

```
line2y = {vfexp, vrexp, kIA, kIQ, kP, kIP, kIB, kB} /. calckinparsrandABPQfr[1.05 * .82645,
    100, 100, 0, 0, .454545, 1, 100, 0, 0, .0454545, 100, 1, 0, 0, -.40816, 0, 0, 100, 100,
    -.028474, 0, 0, 1, 100, -.057804, 0, 0, 100, 1, -.044643, 0, 0, 5, 5, .1111, 5, 5, 0, 0]
```

```
{{1.06235, 0.499994, 5.03829, 10.0001, 14.9997, 25.0004, 2098.87, 21.2981}}
```

```
calckinparsrandABPQfr[.82645, 100, 100, 0, 0, 1.05 * .454545, 1,
    100, 0, 0, .0454545, 100, 1, 0, 0, -.40816, 0, 0, 100, 100, -.028474, 0,
    0, 1, 100, -.057804, 0, 0, 100, 1, -.044643, 0, 0, 5, 5, .1111, 5, 5, 0, 0]
```

```
{{vfexp → 0.998891, vrexp → 0.499994, kIA → 5.03055,
    kIQ → 10.0001, kP → 14.9997, kIP → 25.0004, kIB → -901.021, kB → 19.972}}
```

```
line3y = {vfexp, vrexp, kIA, kIQ, kP, kIP, kIB, kB} /. calckinparsrandABPQfr[.82645, 100, 100,
    0, 0, 1.05 * .454545, 1, 100, 0, 0, .0454545, 100, 1, 0, 0, -.40816, 0, 0, 100, 100,
    -.028474, 0, 0, 1, 100, -.057804, 0, 0, 100, 1, -.044643, 0, 0, 5, 5, .1111, 5, 5, 0, 0]
```

```
{{0.998891, 0.499994, 5.03055, 10.0001, 14.9997, 25.0004, -901.021, 19.972}}
```

```
calckinparsrandABPQfr[.82645, 100, 100, 0, 0, .454545, 1, 100,
    0, 0, 1.05 * .0454545, 100, 1, 0, 0, -.40816, 0, 0, 100, 100, -.028474, 0,
    0, 1, 100, -.057804, 0, 0, 100, 1, -.044643, 0, 0, 5, 5, .1111, 5, 5, 0, 0]
```

```
{{vfexp → 0.988987, vrexp → 0.499994, kIA → 5.59118,
    kIQ → 10.0001, kP → 14.9997, kIP → 25.0004, kIB → -1887.77, kB → 18.6779}}
```

```
line4y = {vfexp, vrexp, kIA, kIQ, kP, kIP, kIB, kB} /. calckinparsrandABPQfr[.82645, 100, 100,
    0, 0, .454545, 1, 100, 0, 0, 1.05 * .0454545, 100, 1, 0, 0, -.40816, 0, 0, 100, 100,
    -.028474, 0, 0, 1, 100, -.057804, 0, 0, 100, 1, -.044643, 0, 0, 5, 5, .1111, 5, 5, 0, 0]
```

```
{{0.988987, 0.499994, 5.59118, 10.0001, 14.9997, 25.0004, -1887.77, 18.6779}}
```

```
calckinparsrandABPQfr[.82645, 100, 100, 0, 0, .454545, 1, 100,
    0, 0, .0454545, 100, 1, 0, 0, 1.05 * -.40816, 0, 0, 100, 100, -.028474, 0,
    0, 1, 100, -.057804, 0, 0, 100, 1, -.044643, 0, 0, 5, 5, .1111, 5, 5, 0, 0]
```

```
{{vfexp → 1., vrexp → 0.531576, kIA → 5.0013,
    kIQ → 10.0337, kP → 15.9992, kIP → 24.9626, kIB → -414606., kB → 19.9998}}
```

```
line5y = {vfexp, vrexp, kIA, kIQ, kP, kIP, kIB, kB} /. calckinparsrandABPQfr[.82645, 100, 100,
    0, 0, .454545, 1, 100, 0, 0, .0454545, 100, 1, 0, 0, 1.05 * -.40816, 0, 0, 100, 100,
    -.028474, 0, 0, 1, 100, -.057804, 0, 0, 100, 1, -.044643, 0, 0, 5, 5, .1111, 5, 5, 0, 0]
```

```
{{1., 0.531576, 5.0013, 10.0337, 15.9992, 24.9626, -414606., 19.9998}}
```

```
calckinparsrandABPQfr[.82645, 100, 100, 0, 0, .454545, 1, 100,
    0, 0, .0454545, 100, 1, 0, 0, -.40816, 0, 0, 100, 100, 1.05 * -.028474, 0,
    0, 1, 100, -.057804, 0, 0, 100, 1, -.044643, 0, 0, 5, 5, .1111, 5, 5, 0, 0]
```

```
{{vfexp → 1., vrexp → 0.495588, kIA → 5.0013,
    kIQ → 10.9452, kP → 13.9863, kIP → 25.9334, kIB → -414606., kB → 19.9998}}
```

```
line6y = {vfexp, vrexp, kIA, kIQ, kP, kIP, kIB, kB} /.
   calckinparsrandABPQfr[.82645, 100, 100, 0, 0, .454545, 1, 100, 0, 0,
     .0454545, 100, 1, 0, 0, -.40816, 0, 0, 100, 100, 1.05 * -.028474, 0, 0, 1,
     100, -.057804, 0, 0, 100, 1, -.044643, 0, 0, 5, 5, .1111, 5, 5, 0, 0]
```

{{1., 0.495588, 5.0013, 10.9452, 13.9863, 25.9334, -414 606., 19.9998}}

```
calckinparsrandABPQfr[.82645, 100, 100, 0, 0, .454545, 1, 100,
   0, 0, .0454545, 100, 1, 0, 0, -.40816, 0, 0, 100, 100, -.028474, 0, 0, 1,
   100, 1.05 * -.057804, 0, 0, 100, 1, -.044643, 0, 0, 5, 5, .1111, 5, 5, 0, 0]
```

{{vfexp → 1., vrexp → 0.497814, kIA → 5.0013,
  kIQ → 10.1609, kP → 14.9125, kIP → 27.3626, kIB → -414 606., kB → 19.9998}}

```
line7y = {vfexp, vrexp, kIA, kIQ, kP, kIP, kIB, kB} /. calckinparsrandABPQfr[.82645, 100, 100, 0,
   0, .454545, 1, 100, 0, 0, .0454545, 100, 1, 0, 0, -.40816, 0, 0, 100, 100, -.028474,
   0, 0, 1, 100, 1.05 * -.057804, 0, 0, 100, 1, -.044643, 0, 0, 5, 5, .1111, 5, 5, 0, 0]
```

{{1., 0.497814, 5.0013, 10.1609, 14.9125, 27.3626, -414 606., 19.9998}}

```
calckinparsrandABPQfr[.82645, 100, 100, 0, 0, .454545, 1, 100,
   0, 0, .0454545, 100, 1, 0, 0, -.40816, 0, 0, 100, 100, -.028474, 0, 0, 1,
   100, -.057804, 0, 0, 100, 1, 1.05 * -.044643, 0, 0, 5, 5, .1111, 5, 5, 0, 0]
```

{{vfexp → 1., vrexp → 0.500734, kIA → 5.0013,
  kIQ → 8.92723, kP → 15.1698, kIP → 21.9964, kIB → -414 606., kB → 19.9998}}

```
line8y = {vfexp, vrexp, kIA, kIQ, kP, kIP, kIB, kB} /. calckinparsrandABPQfr[.82645, 100, 100, 0,
   0, .454545, 1, 100, 0, 0, .0454545, 100, 1, 0, 0, -.40816, 0, 0, 100, 100, -.028474,
   0, 0, 1, 100, -.057804, 0, 0, 100, 1, 1.05 * -.044643, 0, 0, 5, 5, .1111, 5, 5, 0, 0]
```

{{1., 0.500734, 5.0013, 8.92723, 15.1698, 21.9964, -414 606., 19.9998}}

```
calckinparsrandABPQfr[.82645, 100, 100, 0, 0, .454545, 1, 100,
   0, 0, .0454545, 100, 1, 0, 0, -.40816, 0, 0, 100, 100, -.028474, 0, 0, 1,
   100, -.057804, 0, 0, 100, 1, -.044643, 0, 0, 5, 5, 1.05 * .1111, 5, 5, 0, 0]
```

{{vfexp → 1.00119, vrexp → 0.499994, kIA → 4.38163,
  kIQ → 10.0001, kP → 14.9997, kIP → 25.0004, kIB → 743.971, kB → 20.1425}}

```
line9y = {vfexp, vrexp, kIA, kIQ, kP, kIP, kIB, kB} /. calckinparsrandABPQfr[.82645, 100, 100, 0,
   0, .454545, 1, 100, 0, 0, .0454545, 100, 1, 0, 0, -.40816, 0, 0, 100, 100, -.028474,
   0, 0, 1, 100, -.057804, 0, 0, 100, 1, -.044643, 0, 0, 5, 5, 1.05 * .1111, 5, 5, 0, 0]
```

{{1.00119, 0.499994, 4.38163, 10.0001, 14.9997, 25.0004, 743.971, 20.1425}}

Table 5.7 Kinetic parameters for ordered A + B = random P + Q calculated by applying the more general program calckinparsrandABPQfr to velocities for ordered A + B = random P + Q

```
TableForm[Round[{line1y[[1]], line2y[[1]], line3y[[1]], line4y[[1]], line5y[[1]],
    line6y[[1]], line7y[[1]], line8y[[1]], line9y[[1]]}, 0.01], TableHeadings →
  {{"No errors", "1.05*v1", "1.05*v2", "1.05*v3", "1.05*v4", "1.05*v5", "1.05*v6",
    "1.05*v7", "1.05*v8"}, {"vfexp", "vrexp", "kIA", "kIQ", "kP", "kIP", "kIB", "kB"}}]]
```

	vfexp	vrexp	kIA	kIQ	kP	kIP	kIB	kB
No errors	1.	0.5	5.	10.	15.	25.	-414606.	20.
1.05*v1	1.06	0.5	5.04	10.	15.	25.	2098.87	21.3
1.05*v2	1.	0.5	5.03	10.	15.	25.	-901.02	19.97
1.05*v3	0.99	0.5	5.59	10.	15.	25.	-1887.77	18.68
1.05*v4	1.	0.53	5.	10.03	16.	24.96	-414606.	20.
1.05*v5	1.	0.5	5.	10.95	13.99	25.93	-414606.	20.
1.05*v6	1.	0.5	5.	10.16	14.91	27.36	-414606.	20.
1.05*v7	1.	0.5	5.	8.93	15.17	22.	-414606.	20.
1.05*v8	1.	0.5	4.38	10.	15.	25.	743.97	20.14

The values for kIB are all unreasonable and are very sensitive to 5% errors, so E + B = EB is not in the mechanism.

■ 5.3.5 Table of velocities when calckinparsrandABPQfr is applied to velocity data for 3 mechanisms

It is of interest to make a table to show the 8 velocities for the 3 mechanisms.

velocity1 = {"(100,100,0,0)", 0.8264, 0.8265, 0.8043}

{(100,100,0,0), 0.8264, 0.8265, 0.8043}

velocity2 = {"(1,100,0,0)", 0.4545, 0.4545, 0.1807}

{(1,100,0,0), 0.4545, 0.4545, 0.1807}

velocity3 = {"(100,1,0,0)", 0.04545, 0.04545, 0.04539}

{(100,1,0,0), 0.04545, 0.04545, 0.04539}

velocity4 = {"(5,5,0,0)", 0.1111, 0.1111, 0.1035}

{(5,5,0,0), 0.1111, 0.1111, 0.1035}

velocity5 = {"(0,0,100,100)", -0.4292, -0.4082, -0.4082}

{(0,0,100,100), -0.4292, -0.4082, -0.4082}

velocity6 = {"(0,0,1,100)", -0.02857, -0.02847, -0.02847}

{(0,0,1,100), -0.02857, -0.02847, -0.02847}

velocity7 = {"(0,0,100,1)", -0.1887, -0.05780, -0.05780}

{(0,0,100,1), -0.1887, -0.0578, -0.0578}

velocity8 = {"(0,0,5,5)", -0.05000, -0.04464, -0.04464}

{(0,0,5,5), -0.05, -0.04464, -0.04464}

Table 5.8 Velocities used to calculate kinetic parameters for three mechanisms using the most general program calckinparsrand-ABPQfr

```
TableForm[
  {velocity1, velocity2, velocity3, velocity4, velocity5, velocity6, velocity7, velocity8},
  TableHeadings → {{""}, {"{a,b,p,q}", "ordAB=ordPQ", "ordAB=randPQ", "randAB=randPQ"}}]
```

{a,b,p,q}	ordAB=ordPQ	ordAB=randPQ	randAB=randPQ
(100,100,0,0)	0.8264	0.8265	0.8043
(1,100,0,0)	0.4545	0.4545	0.1807
(100,1,0,0)	0.04545	0.04545	0.04539
(5,5,0,0)	0.1111	0.1111	0.1035
(0,0,100,100)	− 0.4292	− 0.4082	− 0.4082
(0,0,1,100)	− 0.02857	− 0.02847	− 0.02847
(0,0,100,1)	− 0.1887	− 0.0578	− 0.0578
(0,0,5,5)	− 0.05	− 0.04464	− 0.04464

There are some small differences between the columns that are not really significant, but ordered A + B = ordered P + Q and ordered A + B = random P + Q differ mainly at {0,0,100,1}, where there is a 3-fold difference in velocities. The main difference between ordered A + B = random P + Q and random A + B = random P + Q is at {1,100,0,0}, where there is a 2.5-fold difference in velocities.

The effects of adding P and Q when the mechanism is ordered A + B = random P + Q can be discussed in the way they were in Sections 5.2.3 and 5.1.4. The estimates of pH-independent kinetic parameters can also be discussed (see preceeding chapter). Dead-end enzyme-substrate complexes like EAP and EBQ can also be added to the mechanism.

5.4 Appendix

▪ Alternate rate equation for ordered A + B = random P + Q

There is another way to write the mechanism that yields a different set of kinetic parameters and a different rapid-equilibrium rate equation. This alternate rate equation vfordABrandPQalt yields the same velocities as vfordABrandPQ, provided kQ is calculated using the thermodynamic cycle given after the complete mechanism. The equilibrium concentrations eab and epq are calculated using the following equilibria :

E + A = EA	kIA = e∗a/ea	
EA + B = EAB	kB = ea∗b/eab	
E + P = EP	kIP = e∗p/ep	
E + Q = EQ	kIQ = e∗q/eq	
EP + Q = EPQ	kQ = ep∗q/epq	This is the alternate reaction.

First, derive the expression for the equlibrium concentration eab.

```
Solve[{kIA == e * a / ea, kB == ea * b / eab, kIP == e * p / ep, kIQ == e * q / eq,
  kQ == ep * q / epq, et == e + ea + eab + ep + eq + epq}, {eab}, {e, ea, ep, eq, epq}]
```

{{eab → (a b et kIP kIQ kQ) / (a b kIP kIQ kQ + a kB kIP kIQ kQ +
    kB kIA kIP kIQ kQ + kB kIA kIQ kQ p + kB kIA kIP kQ q + kB kIA kIQ p q)}}

Second, derive the expression for epq.

```
Solve[{kIA == e * a / ea, kB == ea * b / eab, kIP == e * p / ep, kIQ == e * q / eq,
  kQ == ep * q / epq, et == e + ea + eab + ep + eq + epq}, {epq}, {e, ea, ep, eq, eab}]
```

{{epq → (et kB kIA kIQ p q) / (a b kIP kIQ kQ + a kB kIP kIQ kQ +
    kB kIA kIP kIQ kQ + kB kIA kIQ kQ p + kB kIA kIP kQ q + kB kIA kIQ p q)}}

The complete alternate mechanism is

E + A = EA $\quad$ kIA = e$*$a/ea = 5
EA + B = EAB $\quad$ kB = ea$*$b/eab = 20
EAB ⇌ EPQ $\quad$ v = kf$*$eab − kr$*$epq
E + P = EP $\quad$ kIP = e$*$p/ep = 25
E + Q = EQ $\quad$ kIQ = e$*$q/eq = 10
EP + Q = EPQ $\quad$ kQ = ep$*$q/epq = 6

The number of kinetic parameters is 7, and the number of denominator terms is 6. The number of reactants is 10. The number of components is 5. The number of equilibrium expressions is 5, and so $N' = C' + R'$ is 10 = 5 + 5. When the 5 reactions, excluding the rate-determining reaction are at equilibrium, EP + Q is also at equilibrium, as shown by the following cycle:

```
  kIQ = 10
E + Q =  EQ
+         +
P         P
|| kIP=25  || kP = 15
EP + Q = EPQ
      kQ
```

kQ = kIQkP/kIP = 10*15/25 = 6

Use v = kf*eab-kr*epq. To obtain the expression for the velocity, copy eab and subtract epq, and then change et in the expression for eab to vfexp and et in the expression for epq to vrexp.

```
(a b vfexp kIP kIQ kQ) / (a b kIP kIQ kQ + a kB kIP kIQ kQ + kB kIA kIP kIQ kQ +
    kB kIA kIQ kQ p + kB kIA kIP kQ q + kB kIA kIQ p q) - (vrexp kB kIA kIQ p q) /
    (a b kIP kIQ kQ + a kB kIP kIQ kQ + kB kIA kIP kIQ kQ + kB kIA kIQ kQ p + kB kIA kIP kQ q + kB kIA kIQ p q);
```

Since the denominators are the same, the velocity of ordered A + B = random P + Q for this mechanism can be written as

```
vfordABrandPQalt = (a b vfexp kIP kIQ kQ - vrexp kB kIA kIQ p q) /
    (a b kIP kIQ kQ + a kB kIP kIQ kQ + kB kIA kIP kIQ kQ + kB kIA kIQ kQ p + kB kIA kIP kQ q + kB kIA kIQ p q);
```

The rate equation is changed from vfordABrandPQ because there is no kP, but there is a kQ.

Use of the alternate rate equation to calculate velocities

vfordABrandPQalt is used to calculate the velocity for various sets of substrate concentrations. Since there are 7 kinetic parameters, 7 velocity measurements are required to determine the 7 kinetic parameters. This last quadruplet should be like {a,b,p,q} = {0,0,1,1}, but {0,0,5,5} is used to obtain a higher velocity: {100,100,0,0}, {1,100,0,0), {100,1,0,0}, {0,0,100,100}, {0,0,1,100}, {0,0,100,1}, and {0,0,5,5}

```
vfordABrandPQalt /. vfexp → 1 /. vrexp → .5 /. kIA → 5 /. kB → 20 /. kIP → 25 /. kIQ → 10 /. kQ → 6 /.
    a → 100 /. b → 100 /. p → 0 /. q → 0 // N

0.826446

vfordABrandPQalt /. vfexp → 1 /. vrexp → .5 /. kIA → 5 /. kB → 20 /. kIP → 25 /. kIQ → 10 /. kQ → 6 /.
    a → 1 /. b → 100 /. p → 0 /. q → 0 // N

0.454545

vfordABrandPQalt /. vfexp → 1 /. vrexp → .5 /. kIA → 5 /. kB → 20 /. kIP → 25 /. kIQ → 10 /. kQ → 6 /.
    a → 100 /. b → 1 /. p → 0 /. q → 0 // N

0.0454545
```

```
vfordABrandPQalt /. vfexp → 1 /. vrexp → .5 /. kIA → 5 /. kB → 20 /. kIP → 25 /. kIQ → 10 /. kQ → 6 /.
    a → 0 /. b → 0 /. p → 100 /. q → 100 // N
```

-0.408163

```
vfordABrandPQalt /. vfexp → 1 /. vrexp → .5 /. kIA → 5 /. kB → 20 /. kIP → 25 /. kIQ → 10 /. kQ → 6 /.
    a → 0 /. b → 0 /. p → 1 /. q → 100 // N
```

-0.0284738

```
vfordABrandPQalt /. vfexp → 1 /. vrexp → .5 /. kIA → 5 /. kB → 20 /. kIP → 25 /. kIQ → 10 /. kQ → 6 /.
    a → 0 /. b → 0 /. p → 100 /. q → 1 // N
```

-0.0578035

```
vfordABrandPQalt /. vfexp → 1 /. vrexp → .5 /. kIA → 5 /. kB → 20 /. kIP → 25 /. kIQ → 10 /. kQ → 6 /.
    a → 0 /. b → 0 /. p → 5 /. q → 5 // N
```

-0.0446429

As expected all these velocities are the same as in Section 5.2.2. The rate equations are different, but kQ has been calculated from the cycle after the alternate mechanism.

Calculation of kinetic parameters from measured velocities using the alternate rate equation

The 7 equations in the following program are given by vfordABrandPQalt with different subscripts.

```
calckinparsordABrandPQfralt[v1_, a1_, b1_, p1_, q1_, v2_,
    a2_, b2_, p2_, q2_, v3_, a3_, b3_, p3_, q3_, v4_, a4_, b4_, p4_, q4_, v5_,
    a5_, b5_, p5_, q5_, v6_, a6_, b6_, p6_, q6_, v7_, a7_, b7_, p7_, q7_] :=
  Module[{}, (*This program calculates vfexp, vrexp, kIA, kB,kIP,
    kIQ and kQ from 7 experimental velocities for ordered A + B = random P +
     Q. As an example, the following quadruplets of velocities can be used:{100,100,0,0},
  {0,0,100,100},{1,100,0,0},{100,1,0,0},{0,0,1,100},{0,0,100,1},{0,0,5,5}.*)
  Solve[{v1 == (a1 b1 vfexp kIP kIQ kQ - vrexp kB kIA kIQ p1 q1) / (a1 b1 kIP kIQ kQ +
        a1 kB kIP kIQ kQ + kB kIA kIP kIQ kQ + kB kIA kIQ kQ p1 + kB kIA kIP kQ q1 + kB kIA kIQ p1 q1),
    v2 == (a2 b2 vfexp kIP kIQ kQ - vrexp kB kIA kIQ p2 q2) / (a2 b2 kIP kIQ kQ + a2 kB kIP kIQ kQ +
        kB kIA kIP kIQ kQ + kB kIA kIQ kQ p2 + kB kIA kIP kQ q2 + kB kIA kIQ p2 q2),
    v3 == (a3 b3 vfexp kIP kIQ kQ - vrexp kB kIA kIQ p3 q3) / (a3 b3 kIP kIQ kQ + a3 kB kIP kIQ kQ +
        kB kIA kIP kIQ kQ + kB kIA kIQ kQ p3 + kB kIA kIP kQ q3 + kB kIA kIQ p3 q3),
    v4 == (a4 b4 vfexp kIP kIQ kQ - vrexp kB kIA kIQ p4 q4) / (a4 b4 kIP kIQ kQ + a4 kB kIP kIQ kQ +
        kB kIA kIP kIQ kQ + kB kIA kIQ kQ p4 + kB kIA kIP kQ q4 + kB kIA kIQ p4 q4),
    v5 == (a5 b5 vfexp kIP kIQ kQ - vrexp kB kIA kIQ p5 q5) / (a5 b5 kIP kIQ kQ + a5 kB kIP kIQ kQ +
        kB kIA kIP kIQ kQ + kB kIA kIQ kQ p5 + kB kIA kIP kQ q5 + kB kIA kIQ p5 q5),
    v6 == (a6 b6 vfexp kIP kIQ kQ - vrexp kB kIA kIQ p6 q6) / (a6 b6 kIP kIQ kQ + a6 kB kIP kIQ kQ +
        kB kIA kIP kIQ kQ + kB kIA kIQ kQ p6 + kB kIA kIP kQ q6 + kB kIA kIQ p6 q6),
    v7 == (a7 b7 vfexp kIP kIQ kQ - vrexp kB kIA kIQ p7 q7) / (a7 b7 kIP kIQ kQ + a7 kB kIP kIQ kQ +
        kB kIA kIP kIQ kQ + kB kIA kIQ kQ p7 + kB kIA kIP kQ q7 + kB kIA kIQ p7 q7)},
    {vfexp, vrexp, kIA, kB, kIP, kIQ, kQ}]]

calckinparsordABrandPQfralt[.82645, 100, 100, 0, 0, .454545, 1, 100, 0, 0, .0454545, 100, 1, 0,
    0, -.40816, 0, 0, 100, 100, -.028474, 0, 0, 1, 100, -.057804, 0, 0, 100, 1, -.044643, 0, 0, 5, 5]
```

{{vfexp → 1.00001, vrexp → 0.499994, kIA → 5.00003,
  kB → 20.0001, kIP → 25.0004, kQ → 5.99984, kIQ → 10.0001}}

```
line12alt = {vfexp, vrexp, kIA, kB, kIP, kQ, kIQ} /. calckinparsordABrandPQfralt[
    .82645, 100, 100, 0, 0, .454545, 1, 100, 0, 0, .0454545, 100, 1, 0, 0, -.40816, 0,
    0, 100, 100, -.028474, 0, 0, 1, 100, -.057804, 0, 0, 100, 1, -.044643, 0, 0, 5, 5]
```

{{1.00001, 0.499994, 5.00003, 20.0001, 25.0004, 5.99984, 10.0001}}

Notice the change in order of the rate parameters.

These are the expected values for the 7 kinetic parameters. Now put in 5% errors in the measured velocities, one at a time.

```
calckinparsordABrandPQfralt[1.05 * .82645, 100, 100, 0, 0,
 .454545, 1, 100, 0, 0, .0454545, 100, 1, 0, 0, -.40816, 0, 0, 100, 100,
 -.028474, 0, 0, 1, 100, -.057804, 0, 0, 100, 1, -.044643, 0, 0, 5, 5]
```

{{vfexp → 1.0618, vrexp → 0.499994, kIA → 5.29104,
  kB → 21.2361, kIP → 25.0004, kQ → 5.99984, kIQ → 10.0001}}

```
line22alt = {vfexp, vrexp, kIA, kB, kIP, kQ, kIQ} /. calckinparsordABrandPQfralt[
   1.05 * .82645, 100, 100, 0, 0, .454545, 1, 100, 0, 0, .0454545, 100, 1, 0, 0, -.40816,
   0, 0, 100, 100, -.028474, 0, 0, 1, 100, -.057804, 0, 0, 100, 1, -.044643, 0, 0, 5, 5]
```

{{1.0618, 0.499994, 5.29104, 21.2361, 25.0004, 5.99984, 10.0001}}

```
calckinparsordABrandPQfralt[.82645, 100, 100, 0, 0,
 1.05 * .454545, 1, 100, 0, 0, .0454545, 100, 1, 0, 0, -.40816, 0, 0, 100,
 100, -.028474, 0, 0, 1, 100, -.057804, 0, 0, 100, 1, -.044643, 0, 0, 5, 5]
```

{{vfexp → 1.00001, vrexp → 0.499994, kIA → 4.4474,
  kB → 20.106, kIP → 25.0004, kQ → 5.99984, kIQ → 10.0001}}

```
line32alt = {vfexp, vrexp, kIA, kB, kIP, kQ, kIQ} /. calckinparsordABrandPQfralt[.82645,
   100, 100, 0, 0, 1.05 * .454545, 1, 100, 0, 0, .0454545, 100, 1, 0, 0, -.40816, 0,
   0, 100, 100, -.028474, 0, 0, 1, 100, -.057804, 0, 0, 100, 1, -.044643, 0, 0, 5, 5]
```

{{1.00001, 0.499994, 4.4474, 20.106, 25.0004, 5.99984, 10.0001}}

```
calckinparsordABrandPQfralt[.82645, 100, 100, 0, 0, .454545,
 1, 100, 0, 0, 1.05 * .0454545, 100, 1, 0, 0, -.40816, 0, 0, 100, 100,
 -.028474, 0, 0, 1, 100, -.057804, 0, 0, 100, 1, -.044643, 0, 0, 5, 5]
```

{{vfexp → 0.989534, vrexp → 0.499994, kIA → 5.27937,
  kB → 18.7436, kIP → 25.0004, kQ → 5.99984, kIQ → 10.0001}}

```
line42alt = {vfexp, vrexp, kIA, kB, kIP, kQ, kIQ} /. calckinparsordABrandPQfralt[.82645,
   100, 100, 0, 0, .454545, 1, 100, 0, 0, 1.05 * .0454545, 100, 1, 0, 0, -.40816, 0,
   0, 100, 100, -.028474, 0, 0, 1, 100, -.057804, 0, 0, 100, 1, -.044643, 0, 0, 5, 5]
```

{{0.989534, 0.499994, 5.27937, 18.7436, 25.0004, 5.99984, 10.0001}}

```
calckinparsordABrandPQfralt[.82645, 100, 100, 0, 0, .454545,
 1, 100, 0, 0, .0454545, 100, 1, 0, 0, 1.05 * -.40816, 0, 0, 100, 100,
 -.028474, 0, 0, 1, 100, -.057804, 0, 0, 100, 1, -.044643, 0, 0, 5, 5]
```

{{vfexp → 1.00001, vrexp → 0.531576, kIA → 5.00003,
  kB → 20.0001, kIP → 24.9626, kQ → 6.43088, kIQ → 10.0337}}

```
line52alt = {vfexp, vrexp, kIA, kB, kIP, kQ, kIQ} /. calckinparsordABrandPQfralt[.82645,
   100, 100, 0, 0, .454545, 1, 100, 0, 0, .0454545, 100, 1, 0, 0, 1.05 * -.40816, 0,
   0, 100, 100, -.028474, 0, 0, 1, 100, -.057804, 0, 0, 100, 1, -.044643, 0, 0, 5, 5]
```

{{1.00001, 0.531576, 5.00003, 20.0001, 24.9626, 6.43088, 10.0337}}

```
calckinparsordABrandPQfralt[.82645, 100, 100, 0, 0,
 .454545, 1, 100, 0, 0, .0454545, 100, 1, 0, 0, -.40816, 0, 0, 100, 100,
 1.05 * -.028474, 0, 0, 1, 100, -.057804, 0, 0, 100, 1, -.044643, 0, 0, 5, 5]
```

{{vfexp → 1.00001, vrexp → 0.495588, kIA → 5.00003,
  kB → 20.0001, kIP → 25.9334, kQ → 5.9029, kIQ → 10.9452}}

```
line62alt = {vfexp, vrexp, kIA, kB, kIP, kQ, kIQ} /. calckinparsordABrandPQfralt[.82645,
   100, 100, 0, 0, .454545, 1, 100, 0, 0, .0454545, 100, 1, 0, 0, -.40816, 0, 0, 100,
   100, 1.05 * -.028474, 0, 0, 1, 100, -.057804, 0, 0, 100, 1, -.044643, 0, 0, 5, 5]
```

{{1.00001, 0.495588, 5.00003, 20.0001, 25.9334, 5.9029, 10.9452}}

```
calckinparsordABrandPQfralt[.82645, 100, 100, 0, 0, .454545,
 1, 100, 0, 0, .0454545, 100, 1, 0, 0, -.40816, 0, 0, 100, 100, -.028474,
 0, 0, 1, 100, 1.05 * -.057804, 0, 0, 100, 1, -.044643, 0, 0, 5, 5]
```

{{vfexp → 1.00001, vrexp → 0.497814, kIA → 5.00003,
  kB → 20.0001, kIP → 27.3626, kQ → 5.53763, kIQ → 10.1609}}

```
line72alt = {vfexp, vrexp, kIA, kB, kIP, kQ, kIQ} /. calckinparsordABrandPQfralt[.82645,
    100, 100, 0, 0, .454545, 1, 100, 0, 0, .0454545, 100, 1, 0, 0, -.40816, 0, 0, 100,
    100, -.028474, 0, 0, 1, 100, 1.05 * -.057804, 0, 0, 100, 1, -.044643, 0, 0, 5, 5]
```

{{1.00001, 0.497814, 5.00003, 20.0001, 27.3626, 5.53763, 10.1609}}

```
calckinparsordABrandPQfralt[.82645, 100, 100, 0, 0, .454545,
 1, 100, 0, 0, .0454545, 100, 1, 0, 0, -.40816, 0, 0, 100, 100, -.028474,
 0, 0, 1, 100, -.057804, 0, 0, 100, 1, 1.05 * -.044643, 0, 0, 5, 5]
```

{{vfexp → 1.00001, vrexp → 0.500734, kIA → 5.00003,
  kB → 20.0001, kIP → 21.9964, kQ → 6.15667, kIQ → 8.92723}}

```
line82alt = {vfexp, vrexp, kIA, kB, kIP, kQ, kIQ} /. calckinparsordABrandPQfralt[.82645,
    100, 100, 0, 0, .454545, 1, 100, 0, 0, .0454545, 100, 1, 0, 0, -.40816, 0, 0, 100,
    100, -.028474, 0, 0, 1, 100, -.057804, 0, 0, 100, 1, 1.05 * -.044643, 0, 0, 5, 5]
```

{{1.00001, 0.500734, 5.00003, 20.0001, 21.9964, 6.15667, 8.92723}}

The estimated values of kinetic parameters are summarized in the following table.

Table 5.9 Kinetic parameters for ordered A + B = random P + Q calculated using the alternate mechanism with velocities at $\{100, 100, 0, 0\}, \{1, 100, 0, 0), \{100, 1, 0, 0\}, \{0, 0, 100, 100\}, \{0, 0, 1, 100\}, \{0, 0, 100, 1\}$, and $\{0, 0, 5, 5\}$

```
TableForm[Round[{line12alt[[1]], line22alt[[1]], line32alt[[1]], line42alt[[1]],
    line52alt[[1]], line62alt[[1]], line72alt[[1]], line82alt[[1]]}, 0.01],
  TableHeadings → {{"No errors", "1.05*v1", "1.05*v2", "1.05*v3", "1.05*v4", "1.05*v5",
      "1.05*v6", "1.05*v7"}, {"vfexp", "vrexp", "kIA", "kB", "kIP", "kQ", "kIQ"}}]
```

	vfexp	vrexp	kIA	kB	kIP	kQ	kIQ
No errors	1.	0.5	5.	20.	25.	6.	10.
1.05*v1	1.06	0.5	5.29	21.24	25.	6.	10.
1.05*v2	1.	0.5	4.45	20.11	25.	6.	10.
1.05*v3	0.99	0.5	5.28	18.74	25.	6.	10.
1.05*v4	1.	0.53	5.	20.	24.96	6.43	10.03
1.05*v5	1.	0.5	5.	20.	25.93	5.9	10.95
1.05*v6	1.	0.5	5.	20.	27.36	5.54	10.16
1.05*v7	1.	0.5	5.	20.	22.	6.16	8.93

This table shows that the correct values of the kinetic parameters are obtained when there are no errors in the velocities. This table is different from Table 5.4 in that there is no column for kP and a new column for kQ. Changing the mechanism changes the rate equation, but when the correct dissociation constant is used for the new reaction in the mechanism, the velocities calculated at specific substrate concentrations are unchanged. Thus two investigators might derive rate equations for two different mechanisms and determine different values for the kinetic parameters, yet they would both be right. This supports the recommendation that when a mechanism provides 3 sides of a thermodynamic cycle, the additional equilibrium constant should be reported in publications.

5.5 Discussion

This chapter treats three general types of rapid-equilibrium rate equations for A + B = P + Q. It demonstrates the advantages in treating the forward and reverse reactions at the same time. The ordered A + B = random P + Q mechanism involves a thermodynamic cycle, and so an additional Michaelis constant can be estimated. The random A + B = random P + Q mechanism involves two thermodynamic cycles, and so two additional Michaelis constants can be estimated.

The rate equation for the random A + B = random P + Q reaction can be used to write a program to calculate kinetic parameters from eight velocity measurements when the kinetics are studied for the first time. This program distinguishes between the three types of mechanisms and give correct values of the kinetic parameter for the actual mechanism.

The three mechanisms have the same Haldane relation that yields the apparent equilibrium constant for the reaction that is catalyzed. When the apparent equilibrium constant is very large, the limiting velocity for the reverse reaction kr cannot be obtained from velocity measurements, but all six Michaelis constants can be determined by the study of product inhibition. When K' is known from thermodynamic measurements, the Haldane equation makes it possible to calculate kr. But it is not necessary to actually determnine K' from equilibrium measurements because standard transformed Gibbs energies of formation of reactants can be calculated from various enzyme-catalyzed reactions and can be used to calculate kr even for reactions with very large K'.

Chapter 6 A + B + C → Products

6.1 Introduction

6.2 Mechanism I

- 6.2.1 Derive the rate equation for mechanism I that involves 4 kinetic parameters

- 6.2.2 Estimate kinetic parameters from velocities

6.3 Mechanism II

- 6.3.1 Derive the rate equation for mechanism II that involves 5 kinetic parameters

- 6.3.2 Estimate kinetic parameters from velocities

6.4 Mechanism III

- 6.4.1 Derive the rate equation for mechanism III that involves 6 kinetic parameters

- 6.4.2 Estimate kinetic parameters from velocities

6.5. Mechanism IV

- 6.5.1 Derive the rate equation for mechanism IV that involves 7 kinetic parameters

- 6.5.2 Estimate kinetic parameters from velocities

6.6 Mechanism V

- 6.6.1 Derive the rate equation for mechanism V that involves 8 kinetic parameters

- 6.6.2 Estimate kinetic parameters from velocities

- 6.6.3 Use of calckinparsABC5 to determine the mechanism when A + B + C → products

6.7 Discussion

6.1 Introduction

There are 5 rapid-equilibrium rate equations for the enzyme-catalyzed reaction A + B + C → products, ranging from the completely-ordered mechanism to the completely-random mechanism [34]. The completely-ordered mechanism involves 4 kinetic parameters and the completely-random mechanism involves 8 kinetic parameters. Rapid-equilibrium rate equations are derived for each of these 5 mechanisms and are used to calculate velocities to test the use of the minimum number of velocities to estimate the kinetic parameters.

This chapter is about 5 different enzymes that catalyze the reaction A + B + C → products with different mechanisms. It is assumed that the only reactants involved are E, A, B, C, EA, EB, EC, EAB, EBC, EAC, and EABC. These 11 reactants are involved in 12 reactions.

E + A = EA	kA = e*a/ea = 5	Table 6.1	(6.1)
E + B = EB	kB = e*b/eb = 10	Table 6.2	(6.2)
E + C = EC	kC = e*c/ec = 15	Table 6.3	(6.3)
EA + B = EAB	kAB = ea*b/eab = 20	Table 6.1	(6.4)
EA + C = EAC	kAC = ea*c/eac = 25	Table 6.4	(6.5)
EB + A = EAB	kBA = eb*a/eab	Table 6.2	(6.6)
EB + C = EBC	kBC = ea*c/eac = 30	Table 6.5	(6.7)
EC + A = EAC	kCA = ec*a/eac	Table 6.4	(6.8)
EC + B = EBC	kCB = ec*b/ebc	Table 6.5	(6.9)
EAB + C = EABC	kABC = eab*c/eabc = 35	Table 6.1	(6.10)
EAC + B = EABC	kACB = eac*b/eabc	Table 6.4	(6.11)
EBC + A = EABC	kBCA = ebc*a/eabc	Table 6.5	(6.12)

The symbols for reactants have A, B, and C in alphabetical order, but the symbols for equilibrium constants indicate the order in which the substrates are bound by the enzyme. These 12 reactions are not independent. Actually, only 7 reactions are independent. Thus when arbitrary values are assigned to the equilibrium constants of 7 of these reactions, the equilibrium constants for the other 5 reactions can be calculated. The objective of this chapter is to show that these 5 dependent equilibrium constants can be obtained by making the minimum number of velocity measurements for the 5 mechanisms. For these calculations, the values of equilibrium constants shown for seven of these reactions can be used to calculate the values for the remaining five reactions. The column of table numbers indicate where each equilibrium constant appears first.

Linear algebra can be used to clarify that 7 independent equilibrium constants can be assigned arbitrary values. Linear algebra also clarifies the significance of 4 components. For more information on linear algebra see reference [23].

First, construct the stoichiometric number matrix for the whole system of 12 reactions. This matrix contains the stoichiometric numbers of 11 reactants in 12 biochemical reactions. There is a row in the matrix for each reactant, and a column for each enzyme-catalyzed reaction.

```
line1 = {-1, -1, -1, 0, 0, 0, 0, 0, 0, 0, 0, 0}

{-1, -1, -1, 0, 0, 0, 0, 0, 0, 0, 0, 0}

line2 = {-1, 0, 0, 0, 0, -1, 0, -1, 0, 0, 0, -1}

{-1, 0, 0, 0, 0, -1, 0, -1, 0, 0, 0, -1}

line3 = {0, -1, 0, -1, 0, 0, 0, 0, -1, 0, -1, 0}

{0, -1, 0, -1, 0, 0, 0, 0, -1, 0, -1, 0}

line4 = {0, 0, -1, 0, -1, 0, -1, 0, 0, -1, 0, 0}

{0, 0, -1, 0, -1, 0, -1, 0, 0, -1, 0, 0}
```

```
line5 = {1, 0, 0, -1, -1, 0, 0, 0, 0, 0, 0, 0}

{1, 0, 0, -1, -1, 0, 0, 0, 0, 0, 0, 0}

line6 = {0, 1, 0, 0, 0, -1, -1, 0, 0, 0, 0, 0}

{0, 1, 0, 0, 0, -1, -1, 0, 0, 0, 0, 0}

line7 = {0, 0, 1, 0, 0, 0, 0, -1, -1, 0, 0, 0}

{0, 0, 1, 0, 0, 0, 0, -1, -1, 0, 0, 0}

line8 = {0, 0, 0, 1, 0, 1, 0, 0, 0, -1, 0, 0}

{0, 0, 0, 1, 0, 1, 0, 0, 0, -1, 0, 0}

line9 = {0, 0, 0, 0, 0, 0, 1, 0, 1, 0, 0, -1}

{0, 0, 0, 0, 0, 0, 1, 0, 1, 0, 0, -1}

line10 = {0, 0, 0, 0, 1, 0, 0, 1, 0, 0, -1, 0}

{0, 0, 0, 0, 1, 0, 0, 1, 0, 0, -1, 0}

line11 = {0, 0, 0, 0, 0, 0, 0, 0, 0, 1, 1, 1}

{0, 0, 0, 0, 0, 0, 0, 0, 0, 1, 1, 1}

snMat = {line1, line2, line3, line4, line5, line6, line7, line8, line9, line10, line11}

{{-1, -1, -1, 0, 0, 0, 0, 0, 0, 0, 0, 0},
 {-1, 0, 0, 0, 0, -1, 0, -1, 0, 0, 0, -1}, {0, -1, 0, -1, 0, 0, 0, 0, -1, 0, -1, 0},
 {0, 0, -1, 0, -1, 0, -1, 0, 0, -1, 0, 0}, {1, 0, 0, -1, -1, 0, 0, 0, 0, 0, 0, 0},
 {0, 1, 0, 0, 0, -1, -1, 0, 0, 0, 0, 0}, {0, 0, 1, 0, 0, 0, 0, -1, -1, 0, 0, 0},
 {0, 0, 0, 1, 0, 1, 0, 0, 0, -1, 0, 0}, {0, 0, 0, 0, 0, 0, 1, 0, 1, 0, 0, -1},
 {0, 0, 0, 0, 1, 0, 0, 1, 0, 0, -1, 0}, {0, 0, 0, 0, 0, 0, 0, 0, 0, 1, 1, 1}}
```

The stoichiometric number matrix ν, which is 11×12, is given by

```
TableForm[snMat,
  TableHeadings → {{"E", "A", "B", "C", "EA", "EB", "EC", "EAB", "EBC", "EAC", "EABC"},
    {1, 2, 3, 4, 5, 6, 7, 8, 9, 10, 11, 12}}]
```

	1	2	3	4	5	6	7	8	9	10	11	12
E	-1	-1	-1	0	0	0	0	0	0	0	0	0
A	-1	0	0	0	0	-1	0	-1	0	0	0	-1
B	0	-1	0	-1	0	0	0	0	-1	0	-1	0
C	0	0	-1	0	-1	0	-1	0	0	-1	0	0
EA	1	0	0	-1	-1	0	0	0	0	0	0	0
EB	0	1	0	0	0	-1	-1	0	0	0	0	0
EC	0	0	1	0	0	0	0	-1	-1	0	0	0
EAB	0	0	0	1	0	1	0	0	0	-1	0	0
EBC	0	0	0	0	0	0	1	0	1	0	0	-1
EAC	0	0	0	0	1	0	0	1	0	0	-1	0
EABC	0	0	0	0	0	0	0	0	0	1	1	1

The number of reactions that are independent can be obtained by row-reducing the stoichiometric number matrix.

TableForm[RowReduce[snMat]]

1	0	0	0	0	1	0	1	0	0	0	1
0	1	0	0	0	-1	0	0	1	0	0	-1
0	0	1	0	0	0	0	-1	-1	0	0	0
0	0	0	1	0	1	0	0	0	0	1	1
0	0	0	0	1	0	0	1	0	0	-1	0
0	0	0	0	0	0	1	0	1	0	0	-1
0	0	0	0	0	0	0	0	0	1	1	1
0	0	0	0	0	0	0	0	0	0	0	0
0	0	0	0	0	0	0	0	0	0	0	0
0	0	0	0	0	0	0	0	0	0	0	0
0	0	0	0	0	0	0	0	0	0	0	0

This shows there are 7 independent reactions that can be assigned arbitrary equilibrium constants. The other 5 reactions are dependent, but their equilibrium constants can be calculated from the values of equilibrium constants shown in equations 6.1-6.12. The fact that there are 7 independent reactions is shown more clearly when the stoichiometric matrix is transposed first.

TableForm[RowReduce[Transpose[snMat]]]

1	0	0	0	0	0	0	-1	-1	-1	2
0	1	0	0	0	0	0	0	1	0	-1
0	0	1	0	0	0	0	0	0	1	-1
0	0	0	1	0	0	0	1	0	0	-1
0	0	0	0	1	0	0	-1	0	-1	1
0	0	0	0	0	1	0	-1	-1	0	1
0	0	0	0	0	0	1	0	-1	-1	1
0	0	0	0	0	0	0	0	0	0	0
0	0	0	0	0	0	0	0	0	0	0
0	0	0	0	0	0	0	0	0	0	0
0	0	0	0	0	0	0	0	0	0	0
0	0	0	0	0	0	0	0	0	0	0

The stoichiometric number matrix v' is related to the conservation matrix $\mathbf{A}'$ by

$$\mathbf{A}'v' = \mathbf{0}$$

The $\mathbf{0}$ is the corresponding zero matrix. The conservation matrix has the dimensions $C'{\times}N'$, where C' is the number of components and N' is the number of different reactants. The stoichiometric number matrix, v' has the dimensions $N'{\times}R'$. Here R' is the number of reactions, but after this section, R' will be used for the number of independent reactions. The order of multiplication can be changed by using the transposes A'^{T} and v'^{T}.

$$v'^{\mathrm{T}}A'^{\mathrm{T}} = \mathbf{0}$$

This equation makes it possible to calculate a basis for the conservation matrix from the transpose of the stoichiometric number matrix by using the operation NullSpace in *Mathematica*.

conservMat = NullSpace[Transpose[snMat]]

{{-2, 1, 1, 1, -1, -1, -1, 0, 0, 0, 1}, {1, 0, -1, 0, 1, 0, 1, 0, 0, 1, 0},
  {1, -1, 0, 0, 0, 1, 1, 0, 1, 0, 0}, {1, 0, 0, -1, 1, 1, 0, 1, 0, 0, 0}}

```
TableForm[conservMat]
```

```
-2   1    1    1    -1  -1  -1   0  0  0  1
 1   0   -1    0     1   0   1   0  0  1  0
 1  -1    0    0     0   1   1   0  1  0  0
 1   0    0   -1     1   1   0   1  0  0  0
```

```
TableForm[conservMat, TableHeadings →
   {{1, 2, 3, 4}, {"E", "A", "B", "C", "EA", "EB", "EC", "EAB", "EBC", "EAC", "EABC"}}]
```

	E	A	B	C	EA	EB	EC	EAB	EBC	EAC	EABC
1	-2	1	1	1	-1	-1	-1	0	0	0	1
2	1	0	-1	0	1	0	1	0	0	1	0
3	1	-1	0	0	0	1	1	0	1	0	0
4	1	0	0	-1	1	1	0	1	0	0	0

This shows that there are 4 components. The row-reduced form of this matrix shows how the other reactants are made up of E, A, B, and C.

```
TableForm[RowReduce[NullSpace[Transpose[snMat]]], TableHeadings →
   {{"E", "A", "B", "C"}, {"E", "A", "B", "C", "EA", "EB", "EC", "EAB", "EBC", "EAC", "EABC"}}]
```

	E	A	B	C	EA	EB	EC	EAB	EBC	EAC	EABC
E	1	0	0	0	1	1	1	1	1	1	1
A	0	1	0	0	1	0	0	1	0	1	1
B	0	0	1	0	0	1	0	1	1	0	1
C	0	0	0	1	0	0	1	0	1	1	1

There are 4 components and 11 reactants.

The conservation matrix can be used to find the stoichiometric number matrix by use of NullSpace.

```
NullSpace[conservMat]
```

```
{{-1, -1, -1, -1, 0, 0, 0, 0, 0, 0, 1},
 {-1, -1, 0, -1, 0, 0, 0, 0, 0, 1, 0}, {-1, 0, -1, -1, 0, 0, 0, 0, 1, 0, 0},
 {-1, -1, -1, 0, 0, 0, 0, 1, 0, 0, 0}, {-1, 0, 0, -1, 0, 0, 1, 0, 0, 0, 0},
 {-1, 0, -1, 0, 0, 1, 0, 0, 0, 0, 0}, {-1, -1, 0, 0, 1, 0, 0, 0, 0, 0, 0}}
```

```
TableForm[RowReduce[NullSpace[conservMat]], TableHeadings →
   {{1, 2, 3, 4, 5, 6, 7}, {"E", "A", "B", "C", "EA", "EB", "EC", "EAB", "EBC", "EAC", "EABC"}}]
```

	E	A	B	C	EA	EB	EC	EAB	EBC	EAC	EABC
1	1	0	0	0	0	0	0	-1	-1	-1	2
2	0	1	0	0	0	0	0	0	1	0	-1
3	0	0	1	0	0	0	0	0	0	1	-1
4	0	0	0	1	0	0	0	1	0	0	-1
5	0	0	0	0	1	0	0	-1	0	-1	1
6	0	0	0	0	0	1	0	-1	-1	0	1
7	0	0	0	0	0	0	1	0	-1	-1	1

This shows that there are 7 independent reactions for which equilibrium constants can be arbitrarily assigned. This stoichiometric number matrix contains only the independent reactions. This matrix is $R' \times N'$; here R' is used for the number of independent reactions.

6.2 Mechanism I

■ 6.2.1 Derive the rate equation for mechanism I that involves 4 kinetic parameters

The completely-ordered mechanism is

E + A = EA	kA = e*a/ea
EA + B = EAB	kAB = ea*b/eab
EAB + C = EABC	kABC = eab*c/eabc
EABC → products	v = kf*eab = kf*et*(eab/et) = kfet*(eab/et)
kf	

The number N' of reactants is 7. The number of components C' is 4 (E, A, B, and C). The number of independent equilibrium equations R' is 3, and so $N' = C' + R'$ is $7 = 4 + 3$. To calculate the equilibrium concentration eabc using Solve in *Mathematica*, three equilibrium equations and one conservation equation are used.

```
Solve[{kA == e * a / ea, kAB == ea * b / eab, kABC == eab * c / eabc, et == e + ea + eab + eabc},
  {eabc}, {e, ea, eab}]
```

$$\left\{\left\{eabc \rightarrow \frac{a\,b\,c\,et}{a\,b\,c + a\,b\,kABC + a\,kAB\,kABC + kA\,kAB\,kABC}\right\}\right\}$$

In order to obtain the expression for the velocity v1 for the completely-ordered mechanism, it is necessary to replace et with vfexp, which is equal to kf*et. The rapid-equilibrium rate equation for mechanism I is

$$v1 = \frac{a\,b\,c\,vfexp}{a\,b\,c + a\,b\,kABC + a\,kAB\,kABC + kA\,kAB\,kABC};$$

This function for v1 makes it possible to calculate the velocity for any set of 4 kinetic parameters and 3 reactant concentrations.

EABC is formed in the same reaction in all five mechanisms. This is arbitrary because all the reactions up to EABC → products are at equilibrium. The velocity of A + B + C → products with the arbitrary kinetic parameters (see equations 6.1-6.12) is given by

```
v1funconc = v1 /. vfexp → 1 /. kA → 5 /. kAB → 20 /. kABC → 35
```

$$\frac{a\,b\,c}{3500 + 700\,a + 35\,a\,b + a\,b\,c}$$

In order to calculate the 4 kinetic parameters from the minimum number of velocity measurements, the velocities are calculated using the following 4 triplets of substrate concentrations {a,b,c}: {200,200,200}, {0.5,200,200}, {200,0.5,200}, and {200,200,0.5}. There are reasons for using these particular triplets of concentrations. The first triplet is at the highest practical concentrations a, b, and c because this velocity measurement will make the predominant contribution to vfexp. The second triplet has the lowest practical concentration a because this velocity measurement makes the predominant contribution to kA. The third triplet has the lowest practical concentration b because this velocity measurement makes the predominant contribution to kAB. The fourth triplet has the lowest practical concentration c because this velocity measurement makes the predominant contribution to kABC.

The velocities for the 4 triplets of substrate concentrations are as follows:

```
v1funconc /. a → 200 /. b → 200 /. c → 200 // N
```

```
0.838267
```

```
v1funconc /. a → .5 /. b → 200 /. c → 200 // N

0.731261

v1funconc /. a → 200 /. b → .5 /. c → 200 // N

0.11976

v1funconc /. a → 200 /. b → 200 /. c → .5 // N

0.0127918
```

These velocities are now considered to be measured velocities, and they are used in the program in the next section. The values of the 4 kinetic parameters are obtained by solving the following 4 simultaneous equations.

$$(*0.8383 == \frac{8000000\ \text{vfexp}}{8000000 + 40000\ \text{kABC} + 200\ \text{kAB kABC} + \text{kA kAB kABC}},$$

$$0.7313 == \frac{20000.`\ \text{vfexp}}{20000.` + 100.`\ \text{kABC} + 0.5`\ \text{kAB kABC} + \text{kA kAB kABC}},$$

$$0.1198 == \frac{20000.`\ \text{vfexp}}{20000.` + 100.`\ \text{kABC} + 200\ \text{kAB kABC} + \text{kA kAB kABC}}, \quad 0.01279 == \frac{20000.`\ \text{vfexp}}{20000.` + 40000\ \text{kABC} + 200\ \text{kAB kABC} + \text{kA kAB kABC}} *)$$

The (*...*) is used to show *Mathematica* that it should not do anything with these equations now.

■ 6.2.2 Estimate kinetic parameters from velocities

In 1979, Duggleby [9] pointed out that the values of kinetic parameters for an enzyme-catalyzed reaction can be calculated from as many velocity measurements as there are kinetic parameters. This concept to has been applied to ordered A + B → products, random A + B → products, and to A + B = P + Q (see Chapters 4 and 5).

The following program uses Solve to estimate the kinetic parameters from four velocity measurements for the completely ordered mechanism.

```
calckinparsABC1[v1_, a1_, b1_, c1_, v2_, a2_, b2_, c2_, v3_, a3_, b3_, c3_, v4_, a4_, b4_, c4_] :=
 Module[{}, (*This program calculates vfexp, kA,
   kAB and kABC from four experimental velocities for the completely-
     ordered mechanism of A + B + C → products at four triplets of substrate
     concentrations on the assumption that the mechanism is ordered, with A bound first,
   B bound second, and C bound last.  The first velocity is at high [A],
   high [B], and high [C], the second velocity is at low [A],
   high [B] and high [C], the third velocity is at high [A], high [B] and low [C],
   and the fourth velocity is at high [A], high [B], and low[C].*)
  Solve[{v1 == (a1 b1 c1 vfexp) / (a1 b1 c1 + a1 b1 kABC + a1 kAB kABC + kA kAB kABC),
    v2 == (a2 b2 c2 vfexp) / (a2 b2 c2 + a2 b2 kABC + a2 kAB kABC + kA kAB kABC),
    v3 == (a3 b3 c3 vfexp) / (a3 b3 c3 + a3 b3 kABC + a3 kAB kABC + kA kAB kABC), v4 ==
     (a4 b4 c4 vfexp) / (a4 b4 c4 + a4 b4 kABC + a4 kAB kABC + kA kAB kABC)}, {vfexp, kA, kAB, kABC}]]
```

The 4 kinetic parameters are calculated from 4 measured velocities as follows:

```
calckinparsABC1[.8383, 200, 200, 200, .7313,
 .5, 200, 200, .1198, 200, .5, 200, .01279, 200, 200, .5]

{{vfexp → 1.00008, kA → 5.00121, kAB → 19.9884, kABC → 35.0096}}
```

These values are correct. The following line puts these values in a suitable form for making a table.

```
line11 = {vfexp, kA, kAB, kABC} /. calckinparsABC1[.8383, 200,
   200, 200, .7313, .5, 200, 200, .1198, 200, .5, 200, .01279, 200, 200, .5]

{{1.00008, 5.00121, 19.9884, 35.0096}}
```

The effects of 5% errors in velocities, one at a time, can be calculated.

```
calckinparsABC1[1.05 * .8383, 200, 200, 200,
  .7313, .5, 200, 200, .1198, 200, .5, 200, .01279, 200, 200, .5]
```

{{vfexp → 1.06047, kA → 6.62888, kAB → 19.9884, kABC → 37.1238}}

```
line12 = {vfexp, kA, kAB, kABC} /. calckinparsABC1[1.05 * .8383,
    200, 200, 200, .7313, .5, 200, 200, .1198, 200, .5, 200, .01279, 200, 200, .5]
```

{{1.06047, 6.62888, 19.9884, 37.1238}}

```
calckinparsABC1[.8383, 200, 200, 200, 1.05 * .7313,
  .5, 200, 200, .1198, 200, .5, 200, .01279, 200, 200, .5]
```

{{vfexp → 1.00008, kA → 3.1064, kAB → 20.1748, kABC → 35.0096}}

```
line13 = {vfexp, kA, kAB, kABC} /. calckinparsABC1[.8383, 200, 200,
    200, 1.05 * .7313, .5, 200, 200, .1198, 200, .5, 200, .01279, 200, 200, .5]
```

{{1.00008, 3.1064, 20.1748, 35.0096}}

```
calckinparsABC1[.8383, 200, 200, 200, .7313, .5,
  200, 200, 1.05 * .1198, 200, .5, 200, .01279, 200, 200, .5]
```

{{vfexp → 1.00008, kA → 5.30321, kAB → 18.7434, kABC → 35.2088}}

```
line14 = {vfexp, kA, kAB, kABC} /. calckinparsABC1[.8383, 200, 200,
    200, .7313, .5, 200, 200, 1.05 * .1198, 200, .5, 200, .01279, 200, 200, .5]
```

{{1.00008, 5.30321, 18.7434, 35.2088}}

```
calckinparsABC1[.8383, 200, 200, 200, .7313, .5,
  200, 200, .1198, 200, .5, 200, 1.05 * .01279, 200, 200, .5]
```

{{vfexp → 0.990829, kA → 5.00121, kAB → 21.114, kABC → 32.8368}}

```
line15 = {vfexp, kA, kAB, kABC} /. calckinparsABC1[.8383, 200, 200,
    200, .7313, .5, 200, 200, .1198, 200, .5, 200, 1.05 * .01279, 200, 200, .5]
```

{{0.990829, 5.00121, 21.114, 32.8368}}

Table 6.1 Effects of 5% errors in 4 measured velocities on the 4 kinetic parameters for the completely-ordered mechanism I.

```
TableForm[Round[{line11[[1]], line12[[1]], line13[[1]], line14[[1]], line15[[1]]}, .01],
  TableHeadings →
   {{"No errors", "1.05*v1", "1.05*v2", "1.05*v3", "1.05*v4"}, {"vfexp", "kA", "kAB", "kABC"}}]
```

	vfexp	kA	kAB	kABC
No errors	1.	5.	19.99	35.01
1.05*v1	1.06	6.63	19.99	37.12
1.05*v2	1.	3.11	20.17	35.01
1.05*v3	1.	5.3	18.74	35.21
1.05*v4	0.99	5.	21.11	32.84

This mechanism yields 3 equilibrium constants.

6.3 Mechanism II

■ 6.3.1 Derive the rate equation for mechanism II that involves 5 kinetic parameters

Mechanism II is

E + A = EA   kA = e*a/ea
E + B = EB   kB = e*b/eb   New reaction
EA + B = EAB   kAB = ea*b/eab
EAB + C = EABC   kABC = eab*c/eabc
EABC → products

The number R' of reactants is 8. The number of components C' is 4 (E, A, B, and C). The number of independent equilibrium expressions R' is 4, and so $N' = C' + R'$ is 8 = 4 + 4. To calculate the equilibrium concentration eabc, there are 4 equilibrium equations and one conservation equation.

```
Solve[{kA == e * a / ea, kB == e * b / eb, kAB == ea * b / eab,
   kABC == eab * c / eabc, et == e + ea + eb + eab + eabc}, {eabc}, {e, ea, eb, eab}]
```

$$\left\{\left\{eabc \to \frac{a\,b\,c\,et\,kB}{b\,kA\,kAB\,kABC + a\,b\,c\,kB + a\,b\,kABC\,kB + a\,kAB\,kABC\,kB + kA\,kAB\,kABC\,kB}\right\}\right\}$$

In order to obtain the expression for the velocity v2, it is necessary to replace et with vfexp.

$$v2 = \frac{a\,b\,c\,vfexp\,kB}{b\,kA\,kAB\,kABC + a\,b\,c\,kB + a\,b\,kABC\,kB + a\,kAB\,kABC\,kB + kA\,kAB\,kABC\,kB};$$

This function for the velocity makes it possible to calculate the velocities for specified values of the kinetic parameters and substrate concentrations and to write a program to estimate the kinetic parameters from the minimum number of velocity measurements. The rate equation with the kinetic parameters specified in Section 6.1 is given by

```
v2funconc = v2 /. vfexp → 1 /. kA → 5 /. kB → 10 /. kAB -> 20 /. kABC → 35
```

$$\frac{10\,a\,b\,c}{35\,000 + 7000\,a + 3500\,b + 350\,a\,b + 10\,a\,b\,c}$$

In order to determine the 5 kinetic parameters with the minimum number of velocities, the velocities are calculated using the following triplets of substrate concentrations {a,b,c}: {200,200,200}, {0.5,200,200}, {200,0.5,200}, {200,200,0.5}, and {0.5,0.5.200}, {5,5,50}.

```
v2funconc /. a → 200 /. b → 200 /. c → 200 // N
```
0.832163

```
v2funconc /. a → .5 /. b → 200 /. c → 200 // N
```
0.205444

```
v2funconc /. a → 200 /. b → .5 /. c → 200 // N
```
0.119635

```
v2funconc /. a → 200 /. b → 200 /. c → .5 // N
```
0.0122436

```
v2funconc /. a → .5 /. b → .5 /. c → 200 // N

0.0122436
```

▪ 6.3.2 Estimate kinetic parameters from velocities

The following program calculates the 5 kinetic parameters in mechanism II from 5 measured velocities.

```
calckinparsABC2[v1_, a1_, b1_, c1_, v2_, a2_, b2_, c2_, v3_, a3_, b3_, c3_, v4_, a4_,
  b4_, c4_, v5_, a5_, b5_, c5_] := Module[{}, (*This program calculates vfexp,
  kA, kB, kAB,and kABC from 5 experimental velocities for A + B + C →
   products at 5 triplets of substrate concentrations for mechanism 2.*)
  Solve[{v1 == a1 b1 c1 vfexp kB
                 / (b1 kA kAB kABC + a1 b1 c1 kB + a1 b1 kABC kB + a1 kAB kABC kB + kA kAB kABC kB),
   v2 == a2 b2 c2 vfexp kB
                 / (b2 kA kAB kABC + a2 b2 c2 kB + a2 b2 kABC kB + a2 kAB kABC kB + kA kAB kABC kB),
   v3 == a3 b3 c3 vfexp kB
                 / (b3 kA kAB kABC + a3 b3 c3 kB + a3 b3 kABC kB + a3 kAB kABC kB + kA kAB kABC kB),
   v4 == a4 b4 c4 vfexp kB
                 / (b4 kA kAB kABC + a4 b4 c4 kB + a4 b4 kABC kB + a4 kAB kABC kB + kA kAB kABC kB),
   v5 == a5 b5 c5 vfexp kB
                 / (b5 kA kAB kABC + a5 b5 c5 kB + a5 b5 kABC kB + a5 kAB kABC kB + kA kAB kABC kB)},
  {vfexp, kA, kAB, kABC, kB}]]

calckinparsABC2[.8322, 200, 200, 200, .2054, .5, 200,
 200, .1196, 200, .5, 200, .01224, 200, 200, .5, .01224, .5, .5, 200]

{{vfexp → 1.00011, kA → 4.99973, kAB → 20.0011, kB → 9.99998, kABC → 35.0145}}

line21 = {vfexp, kA, kAB, kB, kABC} /. calckinparsABC2[.8322, 200, 200, 200, .2054,
    .5, 200, 200, .1196, 200, .5, 200, .01224, 200, 200, .5, .01224, .5, .5, 200]

{{1.00011, 4.99973, 20.0011, 9.99998, 35.0145}}
```

Calculate the effects of 5 % errors in velocities, one by one.

```
calckinparsABC2[1.05 * .8322, 200, 200, 200, .2054, .5, 200,
 200, .1196, 200, .5, 200, .01224, 200, 200, .5, .01224, .5, .5, 200]

{{vfexp → 1.06098, kA → 4.95494, kAB → 20.1819, kB → 9.83029, kABC → 37.115}}

line22 = {vfexp, kA, kAB, kB, kABC} /. calckinparsABC2[1.05 * .8322, 200, 200, 200,
    .2054, .5, 200, 200, .1196, 200, .5, 200, .01224, 200, 200, .5, .01224, .5, .5, 200]

{{1.06098, 4.95494, 20.1819, 9.83029, 37.115}}

calckinparsABC2[.8322, 200, 200, 200, 1.05 * .2054, .5, 200,
 200, .1196, 200, .5, 200, .01224, 200, 200, .5, .01224, .5, .5, 200]

{{vfexp → 1.00011, kA → 5.01679, kAB → 19.9331, kB → 10.7486, kABC → 35.131}}

line23 = {vfexp, kA, kAB, kB, kABC} /. calckinparsABC2[.8322, 200, 200, 200, 1.05 * .2054,
    .5, 200, 200, .1196, 200, .5, 200, .01224, 200, 200, .5, .01224, .5, .5, 200]

{{1.00011, 5.01679, 19.9331, 10.7486, 35.131}}
```

```
calckinparsABC2[.8322, 200, 200, 200, .2054, .5, 200, 200,
   1.05 * .1196, 200, .5, 200, .01224, 200, 200, .5, .01224, .5, .5, 200]
```

{{vfexp → 1.00011, kA → 5.33306, kAB → 18.751, kB → 10.06, kABC → 35.2146}}

```
line24 = {vfexp, kA, kAB, kB, kABC} /. calckinparsABC2[.8322, 200, 200, 200, .2054,
   .5, 200, 200, 1.05 * .1196, 200, .5, 200, .01224, 200, 200, .5, .01224, .5, .5, 200]
```

{{1.00011, 5.33306, 18.751, 10.06, 35.2146}}

```
calckinparsABC2[.8322, 200, 200, 200, .2054, .5, 200, 200,
   .1196, 200, .5, 200, 1.05 * .01224, 200, 200, .5, .01224, .5, .5, 200]
```

{{vfexp → 0.990456, kA → 4.99973, kAB → 21.1808, kB → 9.99998, kABC → 32.7448}}

```
line25 = {vfexp, kA, kAB, kB, kABC} /. calckinparsABC2[.8322, 200, 200, 200, .2054,
   .5, 200, 200, .1196, 200, .5, 200, 1.05 * .01224, 200, 200, .5, .01224, .5, .5, 200]
```

{{0.990456, 4.99973, 21.1808, 9.99998, 32.7448}}

```
calckinparsABC2[.8322, 200, 200, 200, .2054, .5, 200, 200,
   .1196, 200, .5, 200, .01224, 200, 200, .5, 1.05 * .01224, .5, .5, 200]
```

{{vfexp → 1.00011, kA → 4.71397, kAB → 20.0318, kB → 9.41529, kABC → 35.0096}}

```
line26 = {vfexp, kA, kAB, kB, kABC} /. calckinparsABC2[.8322, 200, 200, 200, .2054,
   .5, 200, 200, .1196, 200, .5, 200, .01224, 200, 200, .5, 1.05 * .01224, .5, .5, 200]
```

{{1.00011, 4.71397, 20.0318, 9.41529, 35.0096}}

Table 6.2 Effects of 5% errors in velocities on the kinetic parameters for mechanism II

```
TableForm[
  Round[{line21[[1]], line22[[1]], line23[[1]], line24[[1]], line25[[1]], line26[[1]]}, 0.01],
  TableHeadings → {{"No errors", "1.05*v1", "1.05*v2", "1.05*v3", "1.05*v4", "1.05*v5"},
    {"vfexp", "kA", "kAB", "kB", "kABC"}}]
```

	vfexp	kA	kAB	kB	kABC
No errors	1.	5.	20.	10.	35.01
1.05*v1	1.06	4.95	20.18	9.83	37.12
1.05*v2	1.	5.02	19.93	10.75	35.13
1.05*v3	1.	5.33	18.75	10.06	35.21
1.05*v4	0.99	5.	21.18	10.	32.74
1.05*v5	1.	4.71	20.03	9.42	35.01

Mechanism II yields the equilibrium constants for three sides of the following cycle:

```
      kA
E + A =  EA
+         +
B         B
|| kB     || kAB
EB + A = EAB
     kBA
```

This cycle indicates that kA kAB = kB kBA, and so kBA = kA kAB/kB

```
5 * 20 / 10
```

10

when there are no errors. When there are errors, kBA is given by

```
Round[{4.955 * 20.18 / 9.830, 5.017 * 19.93 / 9.83,
   5.33 * 18.75 / 10.06, 5.00 * 21.18 / 10, 4.714 * 20.03 / 9.415}, 0.01]
```

{10.17, 10.17, 9.93, 10.59, 10.03}

Thus the measurement of 5 velocities actually yields 4 equilibrium constants directly and one equilibrium constant indirectly. kBA is not in the rate equation, but its value is determined by the fact that when the 4 reactions in mechanism II are at equilibrium, the reaction EB + A = EAB is necessarily at equilibrium.

6.4 Mechanism III

■ 6.4.1 Derive the rate equation for mechanism III that involves 6 kinetic parameters

Mechanism III is

E + A = EA	kA = e*a/ea
E + B = EB	kB = e*b/eb
E + C = EC	kC = e*c/ec New reaction
EA + B = EAB	kAB = ea*b/eab
EAB + C = EABC	kABC = eab*c/eabc
EABC → products	

The number R' of reactants is 9. The number of components C' is 4 (E, A, B, and C). The number of independent equilibrium relations R' is 5, and so $N' = C' + R'$ is 9 = 4 + 5. To calculate the equilibrium concentration eabc, there are 5 equilibrium equations and one conservation equation.

```
Solve[{kA == e * a / ea, kB == e * b / eb, kC == e * c / ec, kAB == ea * b / eab,
   kABC == eab * c / eabc, et == e + ea + eb + ec + eab + eabc}, {eabc}, {e, ea, eb, ec, eab}]
```

{{eabc → (a b c et kB kC) / (c kA kAB kABC kB + b kA kAB kABC kC +
   a b c kB kC + a b kABC kB kC + a kAB kABC kB kC + kA kAB kABC kB kC)}}

In order to obtain the expression for the velocity v3, it is necessary to replace et with vfexp.

```
v3 = (a b c vfexp kB kC) /
   (c kA kAB kABC kB + b kA kAB kABC kC + a b c kB kC + a b kABC kB kC + a kAB kABC kB kC + kA kAB kABC kB kC);
```

This function for v3 makes it possible to calculate velocities at triplets of substrate concentrations and arbitrary values of kinetic parameters. This function is used to write a program to calculate the 6 kinetic parameters from 6 velocity measurements

The velocity with the specific kinetic parameters is given by

```
v3funconc = v3 /. vfexp → 1 /. kA → 5 /. kB → 10 /. kC → 15 /. kAB -> 20 /. kABC → 35
```

$$\frac{150 \, a \, b \, c}{525\,000 + 105\,000 \, a + 52\,500 \, b + 5250 \, a \, b + 35\,000 \, c + 150 \, a \, b \, c}$$

In order to calculate the 6 kinetic parameters from the minimum number of measured velocities, the velocities are calculated using the following triplets of substrate concentrations {a,b,c}: {200,200,200}, {0.5,200,200},{200,0.5,200},{200,200,0.5}, {5,5,50}, and {5,5,5}. The concentrations of A and B in the sixth triplet are chosen to be low so that unoccupied enzymatic sites are available to C.

```
v3funconc /. a → 200 /. b → 200 /. c → 200 // N
```

0.828143

```
v3funconc /. a → .5 /. b → 200 /. c → 200 // N
```

0.138873

```
v3funconc /. a → 200 /. b → .5 /. c → 200 // N
```

0.0935271

```
v3funconc /. a → 200 /. b → 200 /. c → .5 // N
```

0.0122428

```
v3funconc /. a → 5 /. b → 5 /. c → 50 // N
```

0.0554529

```
v3funconc /. a → 5 /. b → 5 /. c → 5 // N
```

0.0114504

■ 6.4.2 Estimate 6 kinetic parameters from 6 velocities

The following program calculates the kinetic parameters in mechanism III from 6 measured velocities.

```
calckinparsABC3[v1_, a1_, b1_, c1_, v2_, a2_, b2_, c2_, v3_, a3_,
  b3_, c3_, v4_, a4_, b4_, c4_, v5_, a5_, b5_, c5_, v6_, a6_, b6_, c6_] :=
Module[{}, (*This program calculates vfexp, kA, kAB, kABC, kB,
  and kC from 6 experimental velocities for A + B + C →
  products at 6 triplets of substrate concentrations for mechanism III.*)
 Solve[{v1 == (a1 b1 c1 vfexp kB kC) / (c1 kA kAB kABC kB + b1 kA kAB kABC kC +
        a1 b1 c1 kB kC + a1 b1 kABC kB kC + a1 kAB kABC kB kC + kA kAB kABC kB kC),
    v2 == (a2 b2 c2 vfexp kB kC) / (c2 kA kAB kABC kB + b2 kA kAB kABC kC + a2 b2 c2 kB kC +
        a2 b2 kABC kB kC + a2 kAB kABC kB kC + kA kAB kABC kB kC),
    v3 == (a3 b3 c3 vfexp kB kC) / (c3 kA kAB kABC kB + b3 kA kAB kABC kC + a3 b3 c3 kB kC +
        a3 b3 kABC kB kC + a3 kAB kABC kB kC + kA kAB kABC kB kC),
    v4 == (a4 b4 c4 vfexp kB kC) / (c4 kA kAB kABC kB + b4 kA kAB kABC kC + a4 b4 c4 kB kC +
        a4 b4 kABC kB kC + a4 kAB kABC kB kC + kA kAB kABC kB kC),
    v5 == (a5 b5 c5 vfexp kB kC) / (c5 kA kAB kABC kB + b5 kA kAB kABC kC + a5 b5 c5 kB kC +
        a5 b5 kABC kB kC + a5 kAB kABC kB kC + kA kAB kABC kB kC),
    v6 == (a6 b6 c6 vfexp kB kC) / (c6 kA kAB kABC kB + b6 kA kAB kABC kC + a6 b6 c6 kB kC +
        a6 b6 kABC kB kC + a6 kAB kABC kB kC + kA kAB kABC kB kC)}, {vfexp, kA, kAB, kABC, kB, kC}]]

calckinparsABC3[.82814, 200, 200, 200, .13887, .5, 200, 200, .093527,
  200, .5, 200, .012243, 200, 200, .5, .055453, 5, 5, 50, .01145, 5, 5, 5]
```

{{vfexp → 0.999992, kA → 5.00033, kAB → 20.0008, kB → 10.0001, kC → 15.0018, kABC → 34.9988}}

```
line31 = {vfexp, kA, kAB, kB, kC, kABC} /. calckinparsABC3[.82814, 200, 200, 200, .13887, .5, 200,
  200, .093527, 200, .5, 200, .012243, 200, 200, .5, .055453, 5, 5, 50, .01145, 5, 5, 5]
```

{{0.999992, 5.00033, 20.0008, 10.0001, 15.0018, 34.9988}}

Calculate the effects of 5% errors in the measured velocities.

```
calckinparsABC3[1.05 * .82814, 200, 200, 200, .13887, .5, 200, 200,
  .093527, 200, .5, 200, .012243, 200, 200, .5, .055453, 5, 5, 50, .01145, 5, 5, 5]
```

{{vfexp → 1.0612, kA → 4.90405, kAB → 20.1397, kB → 9.74471, kC → 14.7176, kABC → 37.1246}}

```
line32 = {vfexp, kA, kAB, kB, kC, kABC} /. calckinparsABC3[1.05 * .82814, 200, 200, 200, .13887,
  .5, 200, 200, .093527, 200, .5, 200, .012243, 200, 200, .5, .055453, 5, 5, 50, .01145, 5, 5, 5]
```

{{1.0612, 4.90405, 20.1397, 9.74471, 14.7176, 37.1246}}

```
calckinparsABC3[.82814, 200, 200, 200, 1.05 * .13887, .5, 200, 200,
   .093527, 200, .5, 200, .012243, 200, 200, .5, .055453, 5, 5, 50, .01145, 5, 5, 5]
```

{{vfexp → 0.999992, kA → 5.25879, kAB → 19.8755, kB → 11.6804, kC → 15.7573, kABC → 35.1751}}

```
line33 = {vfexp, kA, kAB, kB, kC, kABC} /. calckinparsABC3[.82814, 200, 200, 200, 1.05 * .13887,
   .5, 200, 200, .093527, 200, .5, 200, .012243, 200, 200, .5, .055453, 5, 5, 50, .01145, 5, 5, 5]
```

{{0.999992, 5.25879, 19.8755, 11.6804, 15.7573, 35.1751}}

```
calckinparsABC3[.82814, 200, 200, 200, .13887, .5, 200, 200, 1.05 * .093527,
   200, .5, 200, .012243, 200, 200, .5, .055453, 5, 5, 50, .01145, 5, 5, 5]
```

{{vfexp → 0.999992, kA → 5.80867, kAB → 18.3676, kB → 10.7883, kC → 16.1236, kABC → 35.2605}}

```
line34 = {vfexp, kA, kAB, kB, kC, kABC} /. calckinparsABC3[.82814, 200, 200, 200, .13887, .5, 200,
   200, 1.05 * .093527, 200, .5, 200, .012243, 200, 200, .5, .055453, 5, 5, 50, .01145, 5, 5, 5]
```

{{0.999992, 5.80867, 18.3676, 10.7883, 16.1236, 35.2605}}

```
calckinparsABC3[.82814, 200, 200, 200, .13887, .5, 200, 200, .093527,
   200, .5, 200, 1.05 * .012243, 200, 200, .5, .055453, 5, 5, 50, .01145, 5, 5, 5]
```

{{vfexp → 0.990332, kA → 5.07, kAB → 21.1814, kB → 10.1395, kC → 15.2267, kABC → 32.7287}}

```
line35 = {vfexp, kA, kAB, kB, kC, kABC} /. calckinparsABC3[.82814, 200, 200, 200, .13887, .5, 200,
   200, .093527, 200, .5, 200, 1.05 * .012243, 200, 200, .5, .055453, 5, 5, 50, .01145, 5, 5, 5]
```

{{0.990332, 5.07, 21.1814, 10.1395, 15.2267, 32.7287}}

```
calckinparsABC3[.82814, 200, 200, 200, .13887, .5, 200, 200, .093527,
   200, .5, 200, .012243, 200, 200, .5, 1.05 * .055453, 5, 5, 50, .01145, 5, 5, 5]
```

{{vfexp → 0.999395, kA → 4.66674, kAB → 20.8437, kB → 9.02816, kC → 16.1415, kABC → 34.7364}}

```
line36 = {vfexp, kA, kAB, kB, kC, kABC} /. calckinparsABC3[.82814, 200, 200, 200, .13887, .5, 200,
   200, .093527, 200, .5, 200, .012243, 200, 200, .5, 1.05 * .055453, 5, 5, 50, .01145, 5, 5, 5]
```

{{0.999395, 4.66674, 20.8437, 9.02816, 16.1415, 34.7364}}

```
calckinparsABC3[.82814, 200, 200, 200, .13887, .5, 200, 200, .093527,
   200, .5, 200, .012243, 200, 200, .5, .055453, 5, 5, 50, 1.05 * .01145, 5, 5, 5]
```

{{vfexp → 1.00028, kA → 4.35842, kAB → 19.6822, kB → 8.83292, kC → 12.2965, kABC → 35.1123}}

```
line37 = {vfexp, kA, kAB, kB, kC, kABC} /. calckinparsABC3[.82814, 200, 200, 200, .13887, .5, 200,
   200, .093527, 200, .5, 200, .012243, 200, 200, .5, .055453, 5, 5, 50, 1.05 * .01145, 5, 5, 5]
```

{{1.00028, 4.35842, 19.6822, 8.83292, 12.2965, 35.1123}}

Table 6.3 Effects of 5% errors in 6 velocities for mechanism III

```
TableForm[Round[{line31[[1]], line32[[1]],
   line33[[1]], line34[[1]], line35[[1]], line36[[1]], line37[[1]]}, 0.01],
  TableHeadings → {{"No errors", "1.05*v1", "1.05*v2", "1.05*v3", "1.05*v4",
    "1.05*v5", "1.05*v6"}, {"vfexp", "kA", "kAB", "kB", "kC", "kABC"}}]
```

	vfexp	kA	kAB	kB	kC	kABC
No errors	1.	5.	20.	10.	15.	35.
1.05*v1	1.06	4.9	20.14	9.74	14.72	37.12
1.05*v2	1.	5.26	19.88	11.68	15.76	35.18
1.05*v3	1.	5.81	18.37	10.79	16.12	35.26
1.05*v4	0.99	5.07	21.18	10.14	15.23	32.73
1.05*v5	1.	4.67	20.84	9.03	16.14	34.74
1.05*v6	1.	4.36	19.68	8.83	12.3	35.11

Thus mechanism III yields 5 equilibrium constants directly. Two more eqilibrum constants can be obtained from thermodynamic cycles. They are kCA and kACB.

6.5 Mechanism IV

■ 6.5.1 Derive the rate equation for mechanism IV that involves 7 kinetic parameters

Mechanism IV is

E + A = EA	kA = e*a/ea	
E + B = EB	kB = e*b/eb	
E + C = EC	kC = e*c/ec	
EA + B = EAB	kAB = ea*b/eab	
EA + C = EAC	kAC = ea*c/eac	New reaction
EAB + C = EABC	kABC = eab*c/eabc	
EABC → products		

The number N' of reactants is 10. The number of components C' is 4 (E, A, B, and C). The number of independent reactions R' is 6, and so $N' = C' + R'$ is 10 = 4 + 6. To calculate the equilibrium concentration eabc, there are 6 equilibrium equations and one conservation equation.

```
Solve[{kA ⩵ e * a / ea, kB ⩵ e * b / eb, kC ⩵ e * c / ec, kAB ⩵ ea * b / eab, kAC ⩵ ea * c / eac,
  kABC ⩵ eab * c / eabc, et ⩵ e + ea + eb + ec + eab + eac + eabc}, {eabc}, {e, ea, eb, ec, eab, eac}]

{{eabc → (a b c et kAC kB kC) / (c kA kAB kABC kAC kB + b kA kAB kABC kAC kC + a c kAB kABC kB kC +
    a b c kAC kB kC + a b kABC kAC kB kC + a kAB kABC kAC kB kC + kA kAB kABC kAC kB kC)}}

v4 = (a b c vfexp kAC kB kC) / (c kA kAB kABC kAC kB + b kA kAB kABC kAC kC + a c kAB kABC kB kC +
   a b c kAC kB kC + a b kABC kAC kB kC + a kAB kABC kAC kB kC + kA kAB kABC kAC kB kC) ;

v4funconc = v4 /. vfexp → 1 /. kA → 5 /. kB → 10 /. kC → 15 /. kAB -> 20 /. kAC → 25 /. kABC → 35
```

$$\frac{3750\ a\ b\ c}{13\,125\,000 + 2\,625\,000\ a + 1\,312\,500\ b + 131\,250\ a\ b + 875\,000\ c + 105\,000\ a\ c + 3750\ a\ b\ c}$$

In order to calculate the 7 kinetic parameters with the minimum number of velocities, the velocities are calculated using the following pairs of substrate concentrations {a,b,c}: {200,200,200}, {0.5,200,200},{200,0.5,200},{200,200,0.5}, {5,5,50}, {5,5,5}, and {200,5,5}.

```
v4funconc /. a → 200 /. b → 200 /. c → 200 // N
```

0.742104

```
v4funconc /. a → .5 /. b → 200 /. c → 200 // N
```

0.136224

```
v4funconc /. a → 200 /. b → .5 /. c → 200 // N
```

0.0149943

```
v4funconc /. a → 200 /. b → 200 /. c → .5 // N
```

0.0122218

```
v4funconc /. a → 5 /. b → 5 /. c → 50 // N
```

0.0423131

```
v4funconc /. a → 5 /. b → 5 /. c → 5 // N
```

0.0107604

```
v4funconc /. a → 200 /. b → 5 /. c → 5 // N
```

0.0233191

▪ 6.5.2 Estimate kinetic parameters from velocities

The following program calculates 7 kinetic paarameters from 7 measured velocities.

```
calckinparsABC4[v1_, a1_, b1_, c1_, v2_, a2_, b2_, c2_, v3_, a3_, b3_, c3_,
  v4_, a4_, b4_, c4_, v5_, a5_, b5_, c5_, v6_, a6_, b6_, c6_, v7_, a7_, b7_, c7_] :=
 Module[{}, (*This program calculates vfexp, kA, kAB, kABC, kB,
  kC, and kAC from 7 experimental velocities for A + B + C →
   products at 7 triplets of substrate concentrations for mechanism IV.*)
  Solve[{v1 == (a1 b1 c1 vfexp kAC kB kC) /
     (c1 kA kAB kABC kAC kB + b1 kA kAB kABC kAC kC + a1 c1 kAB kABC kB kC + a1 b1 c1 kAC kB kC +
       a1 b1 kABC kAC kB kC + a1 kAB kABC kAC kB kC + kA kAB kABC kAC kB kC), v2 ==
    (a2 b2 c2 vfexp kAC kB kC) / (c2 kA kAB kABC kAC kB + b2 kA kAB kABC kAC kC + a2 c2 kAB kABC kB kC +
       a2 b2 c2 kAC kB kC + a2 b2 kABC kAC kB kC + a2 kAB kABC kAC kB kC + kA kAB kABC kAC kB kC), v3 ==
    (a3 b3 c3 vfexp kAC kB kC) / (c3 kA kAB kABC kAC kB + b3 kA kAB kABC kAC kC + a3 c3 kAB kABC kB kC +
       a3 b3 c3 kAC kB kC + a3 b3 kABC kAC kB kC + a3 kAB kABC kAC kB kC + kA kAB kABC kAC kB kC), v4 ==
    (a4 b4 c4 vfexp kAC kB kC) / (c4 kA kAB kABC kAC kB + b4 kA kAB kABC kAC kC + a4 c4 kAB kABC kB kC +
       a4 b4 c4 kAC kB kC + a4 b4 kABC kAC kB kC + a4 kAB kABC kAC kB kC + kA kAB kABC kAC kB kC), v5 ==
    (a5 b5 c5 vfexp kAC kB kC) / (c5 kA kAB kABC kAC kB + b5 kA kAB kABC kAC kC + a5 c5 kAB kABC kB kC +
       a5 b5 c5 kAC kB kC + a5 b5 kABC kAC kB kC + a5 kAB kABC kAC kB kC + kA kAB kABC kAC kB kC), v6 ==
    (a6 b6 c6 vfexp kAC kB kC) / (c6 kA kAB kABC kAC kB + b6 kA kAB kABC kAC kC + a6 c6 kAB kABC kB kC +
       a6 b6 c6 kAC kB kC + a6 b6 kABC kAC kB kC + a6 kAB kABC kAC kB kC + kA kAB kABC kAC kB kC),
   v7 == (a7 b7 c7 vfexp kAC kB kC) / (c7 kA kAB kABC kAC kB + b7 kA kAB kABC kAC kC +
       a7 c7 kAB kABC kB kC + a7 b7 c7 kAC kB kC + a7 b7 kABC kAC kB kC + a7 kAB kABC kAC kB kC +
       kA kAB kABC kAC kB kC)}, {vfexp, kA, kAB, kABC, kB, kC, kAC}]]
```

```
calckinparsABC4[.7421, 200, 200, 200, .1362, .5, 200, 200, .01500, 200, .5,
  200, .01222, 200, 200, .5, .04231, 5, 5, 50, .01076, 5, 5, 5, .02332, 200, 5, 5]
```

```
{{vfexp → 0.999961, kA → 5.00009, kAB → 19.9962,
  kB → 9.99925, kC → 14.9934, kABC → 35.0047, kAC → 25.0114}}
```

```
line41 = {vfexp, kA, kAB, kB, kC, kABC, kAC} /.
   calckinparsABC4[.7421, 200, 200, 200, .1362, .5, 200, 200, .01500, 200, .5,
     200, .01222, 200, 200, .5, .04231, 5, 5, 50, .01076, 5, 5, 5, .02332, 200, 5, 5]
```

{{0.999961, 5.00009, 19.9962, 9.99925, 14.9934, 35.0047, 25.0114}}

Calculate the effects of 5% errors in velocities.

```
calckinparsABC4[1.05 * .7421, 200, 200, 200, .1362, .5, 200, 200, .01500, 200,
   .5, 200, .01222, 200, 200, .5, .04231, 5, 5, 50, .01076, 5, 5, 5, .02332, 200, 5, 5]
```

{{vfexp → 1.06885, kA → 4.94221, kAB → 20.0397,
   kB → 9.76344, kC → 14.7558, kABC → 37.4163, kAC → 25.0499}}

```
line42 = {vfexp, kA, kAB, kB, kC, kABC, kAC} /.
   calckinparsABC4[1.05 * .7421, 200, 200, 200, .1362, .5, 200, 200, .01500, 200, .5,
     200, .01222, 200, 200, .5, .04231, 5, 5, 50, .01076, 5, 5, 5, .02332, 200, 5, 5]
```

{{1.06885, 4.94221, 20.0397, 9.76344, 14.7558, 37.4163, 25.0499}}

```
calckinparsABC4[.7421, 200, 200, 200, 1.05 * .1362, .5, 200, 200, .01500, 200,
   .5, 200, .01222, 200, 200, .5, .04231, 5, 5, 50, .01076, 5, 5, 5, .02332, 200, 5, 5]
```

{{vfexp → 0.999961, kA → 5.26363, kAB → 19.8686,
   kB → 11.716, kC → 15.7634, kABC → 35.1844, kAC → 24.9793}}

```
line43 = {vfexp, kA, kAB, kB, kC, kABC, kAC} /.
   calckinparsABC4[.7421, 200, 200, 200, 1.05 * .1362, .5, 200, 200, .01500, 200, .5,
     200, .01222, 200, 200, .5, .04231, 5, 5, 50, .01076, 5, 5, 5, .02332, 200, 5, 5]
```

{{0.999961, 5.26363, 19.8686, 11.716, 15.7634, 35.1844, 24.9793}}

```
calckinparsABC4[.7421, 200, 200, 200, .1362, .5, 200, 200, 1.05 * .01500, 200,
   .5, 200, .01222, 200, 200, .5, .04231, 5, 5, 50, .01076, 5, 5, 5, .02332, 200, 5, 5]
```

{{vfexp → 0.991867, kA → 4.94247, kAB → 20.2294,
   kB → 10.2379, kC → 14.4869, kABC → 34.7213, kAC → 26.9116}}

```
line44 = {vfexp, kA, kAB, kB, kC, kABC, kAC} /.
   calckinparsABC4[.7421, 200, 200, 200, .1362, .5, 200, 200, 1.05 * .01500, 200, .5,
     200, .01222, 200, 200, .5, .04231, 5, 5, 50, .01076, 5, 5, 5, .02332, 200, 5, 5]
```

{{0.991867, 4.94247, 20.2294, 10.2379, 14.4869, 34.7213, 26.9116}}

```
calckinparsABC4[.7421, 200, 200, 200, .1362, .5, 200, 200, .01500, 200, .5, 200,
   1.05 * .01222, 200, 200, .5, .04231, 5, 5, 50, .01076, 5, 5, 5, .02332, 200, 5, 5]
```

{{vfexp → 0.990044, kA → 4.92954, kAB → 21.5136,
   kB → 9.99925, kC → 14.9934, kABC → 32.6741, kAC → 25.4148}}

```
line45 = {vfexp, kA, kAB, kB, kC, kABC, kAC} /.
   calckinparsABC4[.7421, 200, 200, 200, .1362, .5, 200, 200, .01500, 200, .5, 200,
     1.05 * .01222, 200, 200, .5, .04231, 5, 5, 50, .01076, 5, 5, 5, .02332, 200, 5, 5]
```

{{0.990044, 4.92954, 21.5136, 9.99925, 14.9934, 32.6741, 25.4148}}

```
calckinparsABC4[.7421, 200, 200, 200, .1362, .5, 200, 200, .01500, 200, .5, 200,
   .01222, 200, 200, .5, 1.05 * .04231, 5, 5, 50, .01076, 5, 5, 5, .02332, 200, 5, 5]
```

{{vfexp → 0.999961, kA → 5.00009, kAB → 20.0882,
   kB → 9.16012, kC → 17.3808, kABC → 34.8444, kAC → 24.869}}

```
line46 = {vfexp, kA, kAB, kB, kC, kABC, kAC} /.
   calckinparsABC4[.7421, 200, 200, 200, .1362, .5, 200, 200, .01500, 200, .5, 200,
     .01222, 200, 200, .5, 1.05 * .04231, 5, 5, 50, .01076, 5, 5, 5, .02332, 200, 5, 5]
```

{{0.999961, 5.00009, 20.0882, 9.16012, 17.3808, 34.8444, 24.869}}

```
calckinparsABC4[.7421, 200, 200, 200, .1362, .5, 200, 200, .01500, 200, .5, 200,
  .01222, 200, 200, .5, .04231, 5, 5, 50, 1.05 * .01076, 5, 5, 5, .02332, 200, 5, 5]
```

{{vfexp → 0.999961, kA → 4.15149, kAB → 20.0516,
  kB → 8.57431, kC → 11.8599, kABC → 35.0532, kAC → 25.1721}}

```
line47 = {vfexp, kA, kAB, kB, kC, kABC, kAC} /.
  calckinparsABC4[.7421, 200, 200, 200, .1362, .5, 200, 200, .01500, 200, .5, 200,
    .01222, 200, 200, .5, .04231, 5, 5, 50, 1.05 * .01076, 5, 5, 5, .02332, 200, 5, 5]
```

{{0.999961, 4.15149, 20.0516, 8.57431, 11.8599, 35.0532, 25.1721}}

```
calckinparsABC4[.7421, 200, 200, 200, .1362, .5, 200, 200, .01500, 200, .5, 200,
  .01222, 200, 200, .5, .04231, 5, 5, 50, .01076, 5, 5, 5, 1.05 * .02332, 200, 5, 5]
```

{{vfexp → 1.0013, kA → 5.84315, kAB → 18.2836,
  kB → 10.7664, kC → 16.237, kABC → 35.3206, kAC → 22.8202}}

```
line48 = {vfexp, kA, kAB, kB, kC, kABC, kAC} /.
  calckinparsABC4[.7421, 200, 200, 200, .1362, .5, 200, 200, .01500, 200, .5, 200,
    .01222, 200, 200, .5, .04231, 5, 5, 50, .01076, 5, 5, 5, 1.05 * .02332, 200, 5, 5]
```

{{1.0013, 5.84315, 18.2836, 10.7664, 16.237, 35.3206, 22.8202}}

Table 6.4 Effects of 5% errors in 7 velocities for mechanism IV

```
TableForm[Round[{line41[[1]], line42[[1]], line43[[1]],
  line44[[1]], line45[[1]], line46[[1]], line47[[1]], line48[[1]]}, 0.01],
  TableHeadings → {{"No errors", "1.05*v1", "1.05*v2", "1.05*v3", "1.05*v4", "1.05*v5",
    "1.05*v6", "1.05*v7"}, {"vfexp", "kA", "kAB", "kB", "kC", "kABC", "kAC"}}]
```

	vfexp	kA	kAB	kB	kC	kABC	kAC
No errors	1.	5.	20.	10.	14.99	35.	25.01
1.05*v1	1.07	4.94	20.04	9.76	14.76	37.42	25.05
1.05*v2	1.	5.26	19.87	11.72	15.76	35.18	24.98
1.05*v3	0.99	4.94	20.23	10.24	14.49	34.72	26.91
1.05*v4	0.99	4.93	21.51	10.	14.99	32.67	25.41
1.05*v5	1.	5.	20.09	9.16	17.38	34.84	24.87
1.05*v6	1.	4.15	20.05	8.57	11.86	35.05	25.17
1.05*v7	1.	5.84	18.28	10.77	16.24	35.32	22.82

Mechanism IV yields the equilibrium constants for three sides of the following cycle:

```
        kA
E + A  =  EA
+          +
C          C
‖ kC       ‖ kAC
EC + A = EAC
       kCA
```

This cycle indicates that kA kAC = kC kCA, and so kCA = kA kAC/kC =

```
5 * 25 / 15 // N
```

8.33333

when there are no errors. When there are 5% errors, kCA is given by

$$\text{Round}\left[\begin{array}{ccc} 5.\text{`} & 25.01\text{`} & 14.99\text{`} \\ \hline 4.94\text{`} & 25.05\text{`} & 14.76\text{`} \\ 5.26\text{`} & 24.98\text{`} & 15.76\text{`} \\ 4.94\text{`} & 26.91\text{`} & 14.49\text{`} \\ 4.93\text{`} & 25.41\text{`} & 14.99\text{`} \\ 5.\text{`} & 24.87\text{`} & 17.38\text{`} \\ 4.15\text{`} & 25.17\text{`} & 11.86\text{`} \\ 5.84\text{`} & 22.82\text{`} & 16.240000000000002\text{`} \end{array}\right., 0.01\right]$$

{{8.34}, {8.38}, {8.34}, {9.17}, {8.36}, {7.15}, {8.81}, {8.21}}

Mechanism IV also yields the equilibrium constants for three sides of the following cycle:

```
          kAB
EA + B  =   EAB
+           +
C           C
|| kAC      || kABC
EAC + B = EABC
        kACB
```

This cycle indicates that kAB kABC = kAC kACB, and so kACB = kAB kABC/kAC =

20 * 35 / 25 // N

28.

when there are no errors. When there are 5% errors, kACB is given by

$$\text{Round}\left[\begin{array}{cc} 20.\text{`} & 35.\text{`} \\ \hline 20.04\text{`} & 37.42\text{`} \\ 19.87\text{`} & 35.18\text{`} \\ 20.23\text{`} & 34.72\text{`} \\ 21.51\text{`} & 32.67\text{`} \\ 20.09\text{`} & 34.84\text{`} \\ 20.05\text{`} & 35.050000000000004\text{`} \\ 18.28\text{`} & 35.32\text{`} \end{array}\right. * \left/\begin{array}{c} 25.01\text{`} \\ \hline 25.05\text{`} \\ 24.98\text{`} \\ 26.91\text{`} \\ 25.41\text{`} \\ 24.87\text{`} \\ 25.17\text{`} \\ 22.82\text{`} \end{array}\right., 0.01\right]$$

{{27.99}, {29.94}, {27.98}, {26.1}, {27.66}, {28.14}, {27.92}, {28.29}}

Thus mechanism IV yields 6 equilibrium constants directly, but kCA, and kACB can be obtained indirectly, so that 8 equilibrium constants are obtained.

6.6 Mechanism V

■ 6.6.1 Derive the rate equation for mechanism V that involves 8 kinetic parameters

The mechanism is

E + A = EA	kA = e*a/ea	
E + B = EB	kB = e*b/eb	
E + C = EC	kC = e*c/ec	
EA + B = EAB	kAB = ea*b/eab	
EA + C = EAC	kAC = ea*c/eac	
EB + C = EBC	kBC = ea*c/eac	New reaction
EAB + C = EABC	kABC = eab*c/eabc	
EABC → products	v = kf*eabc	

The number N' of reactants is 11. The number of components C' is 4 (E, A, B, and C). The number of independent equilibrium expressions R' is 7, and so $N' = C' + R'$ is 11 = 4 + 7.

This mechanism leads to the most general rapid-equilibrium rate equation for A + B + C → products, but this rate equation does not involve all of the equilibrium constants in equations 1-12. However, this rate equation makes it possible to obtain 7 equilibrium constants from 8 velocity measurements. These 7 equilibrium constants make it possible to obtain the remaining 5 equilibrium constants.

To calculate the equilibrium concentration eabc, there are 6 equilibrium equations and one conservation equation.

```
Solve[{kA == e * a / ea, kB == e * b / eb, kC == e * c / ec,
  kAB == ea * b / eab, kAC == ea * c / eac, kBC == eb * c / ebc, kABC == eab * c / eabc,
  et == e + ea + eb + ec + eab + eac + ebc + eabc}, {eabc}, {e, ea, eb, ec, eab, eac, ebc}]
```

```
{{eabc → (a b c et kAC kB kBC kC) /
    (c kA kAB kABC kAC kB kBC + b c kA kAB kABC kAC kC + b kA kAB kABC kAC kBC kC + a c kAB kABC kB kBC kC +
      a b c kAC kB kBC kC + a b kABC kAC kB kBC kC + a kAB kABC kAC kB kBC kC + kA kAB kABC kAC kB kBC kC)}}
```

```
v5 = (a b c vfexp kAC kB kBC kC) /
    (c kA kAB kABC kAC kB kBC + b c kA kAB kABC kAC kC + b kA kAB kABC kAC kBC kC + a c kAB kABC kB kBC kC +
      a b c kAC kB kBC kC + a b kABC kAC kB kBC kC + a kAB kABC kAC kB kBC kC + kA kAB kABC kAC kB kBC kC);
```

This velocity function is used to write a program to estimate 8 kinetic parameters from 8 velocities. The 8 arbitrarily chosen kinetic parameters can be inserted into the rate equation.

```
v5funconc =
  v5 /. vfexp → 1 /. kA → 5 /. kB → 10 /. kC → 15 /. kAB -> 20 /. kAC → 25 /. kBC → 30 /. kABC → 35
```

$$(112\,500\ a\,b\,c) / (393\,750\,000 + 78\,750\,000\ a + 39\,375\,000\ b +$$
$$3\,937\,500\ a\,b + 26\,250\,000\ c + 3\,150\,000\ a\,c + 1\,312\,500\ b\,c + 112\,500\ a\,b\,c)$$

v5 can be put in a more familiar form by dividing the numerator and denominator by a b c kAC kB kBC kC. The numerator yields vfexp, and the denominator yields

```
(c kA kAB kABC kAC kB kBC / (a b c kAC kB kBC kC) + b c kA kAB kABC kAC kC / (a b c kAC kB kBC kC) +
  b kA kAB kABC kAC kBC kC / (a b c kAC kB kBC kC) + a c kAB kABC kB kBC kC / (a b c kAC kB kBC kC) +
  a b c kAC kB kBC kC / (a b c kAC kB kBC kC) + a b kABC kAC kB kBC kC / (a b c kAC kB kBC kC) +
  a kAB kABC kAC kB kBC kC / (a b c kAC kB kBC kC) + kA kAB kABC kAC kB kBC kC / (a b c kAC kB kBC kC))
```

$$1 + \frac{kABC}{c} + \frac{kAB\ kABC}{b\ c} + \frac{kA\ kAB\ kABC}{a\ b\ c} + \frac{kAB\ kABC}{b\ kAC} + \frac{kA\ kAB\ kABC}{a\ c\ kB} + \frac{kA\ kAB\ kABC}{a\ kB\ kBC} + \frac{kA\ kAB\ kABC}{a\ b\ kC}$$

The velocity for mechanism V is given by

$$\mathtt{vfexp} \Big/ \left(1 + \frac{\mathtt{kABC}}{\mathtt{c}} + \frac{\mathtt{kAB\ kABC}}{\mathtt{b\ c}} + \frac{\mathtt{kA\ kAB\ kABC}}{\mathtt{a\ b\ c}} + \frac{\mathtt{kAB\ kABC}}{\mathtt{b\ kAC}} + \frac{\mathtt{kA\ kAB\ kABC}}{\mathtt{a\ c\ kB}} + \frac{\mathtt{kA\ kAB\ kABC}}{\mathtt{a\ kB\ kBC}} + \frac{\mathtt{kA\ kAB\ kABC}}{\mathtt{a\ b\ kC}}\right)$$

$$\frac{\mathtt{vfexp}}{1 + \frac{\mathtt{kABC}}{\mathtt{c}} + \frac{\mathtt{kAB\ kABC}}{\mathtt{b\ c}} + \frac{\mathtt{kA\ kAB\ kABC}}{\mathtt{a\ b\ c}} + \frac{\mathtt{kAB\ kABC}}{\mathtt{b\ kAC}} + \frac{\mathtt{kA\ kAB\ kABC}}{\mathtt{a\ c\ kB}} + \frac{\mathtt{kA\ kAB\ kABC}}{\mathtt{a\ kB\ kBC}} + \frac{\mathtt{kA\ kAB\ kABC}}{\mathtt{a\ b\ kC}}}$$

In order to calculate the 8 kinetic parameters with the minimum number of velocities, the velocities are calculated using the following triplets of substrate concentrations {a,b,c}: {200,200,200}, {0.5,200,200},{200,0.5,200},{200,200,0.5}, {5,5,50}, {5,5,5}, {200,5,5}, and {5,200,5}.

```
v5funconc /. a → 200 /. b → 200 /. c → 200 // N
```

0.711311

```
v5funconc /. a → .5 /. b → 200 /. c → 200 // N
```

0.0326007

```
v5funconc /. a → 200 /. b → .5 /. c → 200 // N
```

0.0149812

```
v5funconc /. a → 200 /. b → 200 /. c → .5 // N
```

0.0122131

```
v5funconc /. a → 5 /. b → 5 /. c → 50 // N
```

0.0385109

```
v5funconc /. a → 5 /. b → 5 /. c → 5 // N
```

0.0104969

```
v5funconc /. a → 200 /. b → 5 /. c → 5 // N
```

0.0232874

```
v5funconc /. a → 5 /. b → 200 /. c → 5 // N
```

0.0383044

These 8 velocity measurements make it possible to calculate 13 kinetic parameters.

■ 6.6.2 Estimate kinetic parameters from velocities

The following program calculates 8 kinetic parameters from 8 measured velocities.

```
calckinparsABC5[v1_, a1_, b1_, c1_, v2_, a2_, b2_, c2_, v3_, a3_, b3_, c3_, v4_, a4_, b4_,
   c4_, v5_, a5_, b5_, c5_, v6_, a6_, b6_, c6_, v7_, a7_, b7_, c7_, v8_, a8_, b8_, c8_] :=
  Module[{}, (*This program calculates vfexp, kA, kAB, kABC, kB, kC,
   kAC, and kBC from 8 experimental velocities for A + B + C →
    products at 8 triplets of substrate concentrations for mechanism V.*)
   Solve[{v1 == (a1 b1 c1 vfexp kAC kB kBC kC) / (c1 kA kAB kABC kAC kB kBC + b1 c1 kA kAB kABC kAC kC +
        b1 kA kAB kABC kAC kBC kC + a1 c1 kAB kABC kB kBC kC + a1 b1 c1 kAC kB kBC kC +
        a1 b1 kABC kAC kB kBC kC + a1 kAB kABC kAC kB kBC kC + kA kAB kABC kAC kB kBC kC),
     v2 == (a2 b2 c2 vfexp kAC kB kBC kC) / (c2 kA kAB kABC kAC kB kBC + b2 c2 kA kAB kABC kAC kC +
        b2 kA kAB kABC kAC kBC kC + a2 c2 kAB kABC kB kBC kC + a2 b2 c2 kAC kB kBC kC +
        a2 b2 kABC kAC kB kBC kC + a2 kAB kABC kAC kB kBC kC + kA kAB kABC kAC kB kBC kC),
     v3 == (a3 b3 c3 vfexp kAC kB kBC kC) / (c3 kA kAB kABC kAC kB kBC + b3 c3 kA kAB kABC kAC kC +
        b3 kA kAB kABC kAC kBC kC + a3 c3 kAB kABC kB kBC kC + a3 b3 c3 kAC kB kBC kC +
        a3 b3 kABC kAC kB kBC kC + a3 kAB kABC kAC kB kBC kC + kA kAB kABC kAC kB kBC kC),
     v4 == (a4 b4 c4 vfexp kAC kB kBC kC) / (c4 kA kAB kABC kAC kB kBC + b4 c4 kA kAB kABC kAC kC +
        b4 kA kAB kABC kAC kBC kC + a4 c4 kAB kABC kB kBC kC + a4 b4 c4 kAC kB kBC kC +
        a4 b4 kABC kAC kB kBC kC + a4 kAB kABC kAC kB kBC kC + kA kAB kABC kAC kB kBC kC),
     v5 == (a5 b5 c5 vfexp kAC kB kBC kC) / (c5 kA kAB kABC kAC kB kBC + b5 c5 kA kAB kABC kAC kC +
        b5 kA kAB kABC kAC kBC kC + a5 c5 kAB kABC kB kBC kC + a5 b5 c5 kAC kB kBC kC +
        a5 b5 kABC kAC kB kBC kC + a5 kAB kABC kAC kB kBC kC + kA kAB kABC kAC kB kBC kC),
     v6 == (a6 b6 c6 vfexp kAC kB kBC kC) / (c6 kA kAB kABC kAC kB kBC + b6 c6 kA kAB kABC kAC kC +
        b6 kA kAB kABC kAC kBC kC + a6 c6 kAB kABC kB kBC kC + a6 b6 c6 kAC kB kBC kC +
        a6 b6 kABC kAC kB kBC kC + a6 kAB kABC kAC kB kBC kC + kA kAB kABC kAC kB kBC kC),
     v7 == (a7 b7 c7 vfexp kAC kB kBC kC) / (c7 kA kAB kABC kAC kB kBC + b7 c7 kA kAB kABC kAC kC +
        b7 kA kAB kABC kAC kBC kC + a7 c7 kAB kABC kB kBC kC + a7 b7 c7 kAC kB kBC kC +
        a7 b7 kABC kAC kB kBC kC + a7 kAB kABC kAC kB kBC kC + kA kAB kABC kAC kB kBC kC),
     v8 == (a8 b8 c8 vfexp kAC kB kBC kC) / (c8 kA kAB kABC kAC kB kBC + b8 c8 kA kAB kABC kAC kC +
        b8 kA kAB kABC kAC kBC kC + a8 c8 kAB kABC kB kBC kC + a8 b8 c8 kAC kB kBC kC +
        a8 b8 kABC kAC kB kBC kC + a8 kAB kABC kAC kB kBC kC + kA kAB kABC kAC kB kBC kC)},
    {vfexp, kA, kAB, kABC, kB, kC, kAC, kBC}]]
```

```
calckinparsABC5[.7113, 200, 200, 200, .03260, .5, 200, 200, .01498, 200, .5, 200, .01221,
  200, 200, .5, .03851, 5, 5, 50, .01050, 5, 5, 5, .02329, 200, 5, 5, .03830, 5, 200, 5]
```

```
{{vfexp → 1.00005, kA → 4.99602, kAB → 19.9886,
  kB → 9.98836, kC → 14.979, kABC → 35.0129, kBC → 30.004, kAC → 24.9909}}
```

These values are correct.

```
line51 = {vfexp, kA, kAB, kB, kC, kABC, kBC, kAC} /.
  calckinparsABC5[.7113, 200, 200, 200, .03260, .5, 200, 200, .01498, 200, .5, 200, .01221,
    200, 200, .5, .03851, 5, 5, 50, .01050, 5, 5, 5, .02329, 200, 5, 5, .03830, 5, 200, 5]
```

```
{{1.00005, 4.99602, 19.9886, 9.98836, 14.979, 35.0129, 30.004, 24.9909}}
```

Calculate effects of 5% errors in measured velocities on the calculated kinetic parameters.

```
calckinparsABC5[1.05 * .7113, 200, 200, 200, .03260, .5, 200, 200, .01498, 200, .5, 200, .01221,
  200, 200, .5, .03851, 5, 5, 50, .01050, 5, 5, 5, .02329, 200, 5, 5, .03830, 5, 200, 5]
```

```
{{vfexp → 1.07226, kA → 4.96341, kAB → 20.0204,
  kB → 9.90209, kC → 14.8182, kABC → 37.5615, kBC → 30.083, kAC → 25.0272}}
```

```
line52 = {vfexp, kA, kAB, kB, kC, kABC, kBC, kAC} /.
  calckinparsABC5[1.05 * .7113, 200, 200, 200, .03260, .5, 200, 200, .01498, 200, .5, 200,
    .01221, 200, 200, .5, .03851, 5, 5, 50, .01050, 5, 5, 5, .02329, 200, 5, 5, .03830, 5, 200, 5]
```

```
{{1.07226, 4.96341, 20.0204, 9.90209, 14.8182, 37.5615, 30.083, 25.0272}}
```

```
calckinparsABC5[.7113, 200, 200, 200, 1.05 * .03260, .5, 200, 200, .01498, 200, .5, 200, .01221,
  200, 200, .5, .03851, 5, 5, 50, .01050, 5, 5, 5, .02329, 200, 5, 5, .03830, 5, 200, 5]
```

```
{{vfexp → 0.99621, kA → 4.96782, kAB → 20.0023,
  kB → 9.82526, kC → 14.6547, kABC → 34.8594, kBC → 32.4783, kAC → 25.0115}}
```

```
line53 = {vfexp, kA, kAB, kB, kC, kABC, kBC, kAC} /.
    calckinparsABC5[.7113, 200, 200, 200, 1.05 * .03260, .5, 200, 200, .01498, 200, .5, 200,
    .01221, 200, 200, .5, .03851, 5, 5, 50, .01050, 5, 5, 5, .02329, 200, 5, 5, .03830, 5, 200, 5]
```

{{0.99621, 4.96782, 20.0023, 9.82526, 14.6547, 34.8594, 32.4783, 25.0115}}

```
calckinparsABC5[.7113, 200, 200, 200, .03260, .5, 200, 200, 1.05 * .01498, 200, .5, 200, .01221,
    200, 200, .5, .03851, 5, 5, 50, .01050, 5, 5, 5, .02329, 200, 5, 5, .03830, 5, 200, 5]
```

{{vfexp → 0.991735, kA → 4.87773, kAB → 20.2522,
  kB → 9.86275, kC → 14.2865, kABC → 34.6803, kBC → 30.1302, kAC → 26.9003}}

```
line54 = {vfexp, kA, kAB, kB, kC, kABC, kBC, kAC} /.
    calckinparsABC5[.7113, 200, 200, 200, .03260, .5, 200, 200, 1.05 * .01498, 200, .5, 200,
    .01221, 200, 200, .5, .03851, 5, 5, 50, .01050, 5, 5, 5, .02329, 200, 5, 5, .03830, 5, 200, 5]
```

{{0.991735, 4.87773, 20.2522, 9.86275, 14.2865, 34.6803, 30.1302, 26.9003}}

```
calckinparsABC5[.7113, 200, 200, 200, .03260, .5, 200, 200, .01498, 200, .5, 200, 1.05 * .01221,
    200, 200, .5, .03851, 5, 5, 50, .01050, 5, 5, 5, .02329, 200, 5, 5, .03830, 5, 200, 5]
```

{{vfexp → 0.989869, kA → 4.85128, kAB → 21.5479,
  kB → 9.56066, kC → 14.7425, kABC → 32.6209, kBC → 31.0221, kAC → 25.4047}}

```
line55 = {vfexp, kA, kAB, kB, kC, kABC, kBC, kAC} /.
    calckinparsABC5[.7113, 200, 200, 200, .03260, .5, 200, 200, .01498, 200, .5, 200, 1.05 * .01221,
    200, 200, .5, .03851, 5, 5, 50, .01050, 5, 5, 5, .02329, 200, 5, 5, .03830, 5, 200, 5]
```

{{0.989869, 4.85128, 21.5479, 9.56066, 14.7425, 32.6209, 31.0221, 25.4047}}

```
calckinparsABC5[.7113, 200, 200, 200, .03260, .5, 200, 200, .01498, 200, .5, 200, .01221, 200,
    200, .5, 1.05 * .03851, 5, 5, 50, .01050, 5, 5, 5, .02329, 200, 5, 5, .03830, 5, 200, 5]
```

{{vfexp → 1.00095, kA → 5.26096, kAB → 19.9602,
  kB → 10.5317, kC → 18.6381, kABC → 35.0491, kBC → 29.4702, kAC → 24.7987}}

```
line56 = {vfexp, kA, kAB, kB, kC, kABC, kBC, kAC} /.
    calckinparsABC5[.7113, 200, 200, 200, .03260, .5, 200, 200, .01498, 200, .5, 200, .01221,
    200, 200, .5, 1.05 * .03851, 5, 5, 50, .01050, 5, 5, 5, .02329, 200, 5, 5, .03830, 5, 200, 5]
```

{{1.00095, 5.26096, 19.9602, 10.5317, 18.6381, 35.0491, 29.4702, 24.7987}}

```
calckinparsABC5[.7113, 200, 200, 200, .03260, .5, 200, 200, .01498, 200, .5, 200, .01221, 200,
    200, .5, .03851, 5, 5, 50, 1.05 * .01050, 5, 5, 5, .02329, 200, 5, 5, .03830, 5, 200, 5]
```

{{vfexp → 0.999716, kA → 4.03013, kAB → 20.0927,
  kB → 8.01955, kC → 11.4881, kABC → 34.9847, kBC → 30.461, kAC → 25.1688}}

```
line57 = {vfexp, kA, kAB, kB, kC, kABC, kBC, kAC} /.
    calckinparsABC5[.7113, 200, 200, 200, .03260, .5, 200, 200, .01498, 200, .5, 200, .01221,
    200, 200, .5, .03851, 5, 5, 50, 1.05 * .01050, 5, 5, 5, .02329, 200, 5, 5, .03830, 5, 200, 5]
```

{{0.999716, 4.03013, 20.0927, 8.01955, 11.4881, 34.9847, 30.461, 25.1688}}

```
calckinparsABC5[.7113, 200, 200, 200, .03260, .5, 200, 200, .01498, 200, .5, 200, .01221, 200,
    200, .5, .03851, 5, 5, 50, .01050, 5, 5, 5, 1.05 * .02329, 200, 5, 5, .03830, 5, 200, 5]
```

{{vfexp → 1.00143, kA → 5.85113, kAB → 18.2695,
  kB → 10.8187, kC → 16.2564, kABC → 35.3374, kBC → 29.8682, kAC → 22.797}}

```
line58 = {vfexp, kA, kAB, kB, kC, kABC, kBC, kAC} /.
    calckinparsABC5[.7113, 200, 200, 200, .03260, .5, 200, 200, .01498, 200, .5, 200, .01221,
    200, 200, .5, .03851, 5, 5, 50, .01050, 5, 5, 5, 1.05 * .02329, 200, 5, 5, .03830, 5, 200, 5]
```

{{1.00143, 5.85113, 18.2695, 10.8187, 16.2564, 35.3374, 29.8682, 22.797}}

```
calckinparsABC5[.7113, 200, 200, 200, .03260, .5, 200, 200, .01498, 200, .5, 200, .01221, 200,
   200, .5, .03851, 5, 5, 50, .01050, 5, 5, 5, .02329, 200, 5, 5, 1.05 * .03830, 5, 200, 5]
```

{{vfexp → 1.00089, kA → 5.2419, kAB → 19.8695,
  kB → 11.5765, kC → 15.7539, kABC → 35.2101, kBC → 26.7451, kAC → 24.9572}}

```
line59 = {vfexp, kA, kAB, kB, kC, kABC, kBC, kAC} /.
   calckinparsABC5[.7113, 200, 200, 200, .03260, .5, 200, 200, .01498, 200, .5, 200, .01221,
      200, 200, .5, .03851, 5, 5, 50, .01050, 5, 5, 5, .02329, 200, 5, 5, 1.05 * .03830, 5, 200, 5]
```

{{1.00089, 5.2419, 19.8695, 11.5765, 15.7539, 35.2101, 26.7451, 24.9572}}

Table 6.5 Effects of 5 % errors in 8 velocities for mechanism V

```
TableForm[Round[{line51[[1]], line52[[1]], line53[[1]], line54[[1]], line55[[1]],
   line56[[1]], line57[[1]], line58[[1]], line59[[1]]}, 0.01], TableHeadings →
   {{"No errors", "1.05*v1", "1.05*v2", "1.05*v3", "1.05*v4", "1.05*v5", "1.05*v6",
      "1.05*v7", "1.05*v8"}, {"vfexp", "kA", "kAB", "kB", "kC", "kABC", "kBC", "kAC"}}]
```

	vfexp	kA	kAB	kB	kC	kABC	kBC	kAC
No errors	1.	5.	19.99	9.99	14.98	35.01	30.	24.99
1.05*v1	1.07	4.96	20.02	9.9	14.82	37.56	30.08	25.03
1.05*v2	1.	4.97	20.	9.83	14.65	34.86	32.48	25.01
1.05*v3	0.99	4.88	20.25	9.86	14.29	34.68	30.13	26.9
1.05*v4	0.99	4.85	21.55	9.56	14.74	32.62	31.02	25.4
1.05*v5	1.	5.26	19.96	10.53	18.64	35.05	29.47	24.8
1.05*v6	1.	4.03	20.09	8.02	11.49	34.98	30.46	25.17
1.05*v7	1.	5.85	18.27	10.82	16.26	35.34	29.87	22.8
1.05*v8	1.	5.24	19.87	11.58	15.75	35.21	26.75	24.96

As mentioned after mechanism V, these 7 equilibrium constants make it possible to calculate the other 5 equilibrium constants in mechanism V.

The rate equation for mechanism V yields the equilibrium constants for three sides of five thermodynamic cycles. The first one of the five is

$$
\begin{array}{ccc}
& kB & \\
E + B & = & EB \\
+ & & + \\
C & & C \\
\| kC & & \| kBC \\
EC + B & = & EBC \\
& kCB &
\end{array}
$$

When there are no errors, this cycle indicates that kB kBC = kC kCB, and so kCB = kB kBC/kC

```
10 * 30 / 15 // N
```

20.

When there are errors, kCB = kB*kBC/kC can be calculated by copying columns from Table 6.5 and multiplying and dividing them.

$$\text{rowkCB} = \text{Round}\left[\begin{array}{c} 9.99` \\ \hline 9.9` \\ 9.83` \\ 9.86` \\ 9.56` \\ 10.53` \\ 8.02` \\ 10.82` \\ 11.58` \end{array} * \begin{array}{c} 30.` \\ \hline 30.080000000000002` \\ 32.480000000000004` \\ 30.13` \\ 31.02` \\ 29.47` \\ 30.46` \\ 29.87` \\ 26.75` \end{array} \middle/ \begin{array}{c} 14.98` \\ \hline 14.82` \\ 14.65` \\ 14.290000000000001` \\ 14.74` \\ 18.64` \\ 11.49` \\ 16.26` \\ 15.75` \end{array} , 0.01\right]$$

{{20.01}, {20.09}, {21.79}, {20.79}, {20.12}, {16.65}, {21.26}, {19.88}, {19.67}}

Mechanism V yields the equilibrium constants for three sides of the following cycle 2:

```
         kA
E + A  =   EA
+          +
B          B
‖ kB      ‖ kAB
EB + A = EAB
      kBA
```

When there are no errors, this cycle indicates that kA kAB = kB kBA, and so kBA = kA kAB/kB

5 * 20 / 10 // N

10.

When there are errors, kBA = kA*kAB/kB is given by

$$\text{rowkBA} = \text{Round}\left[\begin{array}{c} 5.` \\ \hline 4.96` \\ 4.97` \\ 4.88` \\ 4.850000000000005` \\ 5.26` \\ 4.03` \\ 5.850000000000005` \\ 5.24` \end{array} * \begin{array}{c} 19.990000000000002` \\ \hline 20.02` \\ 20.` \\ 20.25` \\ 21.55` \\ 19.96` \\ 20.09` \\ 18.27` \\ 19.87` \end{array} \middle/ \begin{array}{c} 9.99` \\ \hline 9.9` \\ 9.83` \\ 9.86` \\ 9.56` \\ 10.53` \\ 8.02` \\ 10.82` \\ 11.58` \end{array} , 0.01\right]$$

{{10.01}, {10.03}, {10.11}, {10.02}, {10.93}, {9.97}, {10.1}, {9.88}, {8.99}}

These are the kBA values calculated from Table 6.5.

The rate equation for mechanism V yields the equilibrium constants for three sides of cycle 3.

```
         kA
E + A  =   EA
+          +
C          C
‖ kC      ‖ kAC
EC + A = EAC
      kCA
```

When there are no errors, this cycle indicates that kA kAC = kC kCA, and so kCA = kA kAC/kC

```
5 * 25 / 15 // N
```

```
8.33333
```

When there are errors, kCA = kA*kAC/kC is given by

$$
\text{rowkCA} = \text{Round}\left[
\begin{array}{ccc}
5.\grave{} & 24.990000000000002\grave{} & 14.98\grave{} \\
\hline
4.96\grave{} & 25.03\grave{} & 14.82\grave{} \\
4.97\grave{} & 25.01\grave{} & 14.65\grave{} \\
4.88\grave{} & 26.900000000000002\grave{} & 14.290000000000001\grave{} \\
4.8500000000000005\grave{} & 25.400000000000002\grave{} & 14.74\grave{} \\
5.26\grave{} & 24.8\grave{} & 18.64\grave{} \\
4.03\grave{} & 25.17\grave{} & 11.49\grave{} \\
5.8500000000000005\grave{} & 22.8\grave{} & 16.26\grave{} \\
5.24\grave{} & 24.96\grave{} & 15.75\grave{}
\end{array}
\ast \Big/ \quad , 0.01\right]
$$

{{8.34}, {8.38}, {8.48}, {9.19}, {8.36}, {7.}, {8.83}, {8.2}, {8.3}}

These are the kCA values calculated from Table 6.5. The rate equation for mechanism V also yields the equilibrium constants for three sides of cycle 4.

```
       kBA
EB + A  =   EAB
+           +
C           C
|| kBC      || kABC
EBC + A = EABC
       kBCA
```

This cycle indicates that kBA kABC = kB CkBCA, and so kBCA = kB AkABC/kBC

```
10 * 35 / 30 // N
```

```
11.6667
```

kBA is given by cycle 4:

{{10.01}, {10.03}, {10.11}, {10.02}, {10.93}, {9.97}, {10.1}, {9.88}, {8.99}}

kBCA is given by

```
rowkBCA =
```

$$
\text{Round}\Big[\{\{10.01\grave{}\}, \{10.03\grave{}\}, \{10.11\grave{}\}, \{10.02\grave{}\}, \{10.93\grave{}\}, \{9.97\grave{}\}, \{10.1\grave{}\}, \{9.88\grave{}\}, \{8.99\grave{}\}\} \ast
$$

$$
\begin{array}{cc}
35.01\grave{} & 30.\grave{} \\
\hline
37.56\grave{} & 30.080000000000002\grave{} \\
34.86\grave{} & 32.480000000000004\grave{} \\
34.68\grave{} & 30.13\grave{} \\
32.62\grave{} & 31.02\grave{} \\
35.050000000000004\grave{} & 29.47\grave{} \\
34.980000000000004\grave{} & 30.46\grave{} \\
35.34\grave{} & 29.87\grave{} \\
35.21\grave{} & 26.75\grave{}
\end{array}
\Big/ \quad , 0.01\Big]
$$

{{11.68}, {12.52}, {10.85}, {11.53}, {11.49}, {11.86}, {11.6}, {11.69}, {11.83}}

The rate equation for mechanism V yields the equilibrium constants for three sides of cycle 5:

```
         kAB
EA + B  =  EAB
+          +
C          C
|| kAC     || kABC
EAC + B = EACB
      kACB
```

When there are no errors, this cycle indicates that kAB kABC = kAC kACB, and so kACB = kAB kABC/kAC.

When there are no errors

20 * 35 / 25 // N

28.

$$
\text{rowkACB} = \text{Round}\left[
\begin{array}{ccc}
\underline{19.990000000000002`} & 35.01` & \underline{24.990000000000002`} \\
20.02` & 37.56` & 25.03` \\
20.` & 34.86` & 25.01` \\
20.25` & 34.68` & 26.900000000000002` \\
21.55` & *\ 32.62` & 25.400000000000002` \\
19.96` & 35.050000000000004` & 24.8` \\
20.09` & 34.980000000000004` & 25.17` \\
18.27` & 35.34` & 22.8` \\
19.87` & 35.21` & 24.96`
\end{array}
,\ 0.01\right]
$$

{{28.01}, {30.04}, {27.88}, {26.11}, {27.68}, {28.21}, {27.92}, {28.32}, {28.03}}

Table 6.6 summarizes the 5 additional kinetic parameters calculated from velocities of mechanism V.

Table 6.6 Effects of 5% errors in velocities on 5 dependent equilibrium constants

```
TableForm[{rowkCB, rowkBA, rowkCA, rowkBCA, rowkACB},
  TableHeadings → {{"kCB", "kBA", "kCA", "kBCA", "kACB"}, {"No errors", "1.05*v1",
   "1.05*v2", "1.05*v3", "1.05*v4", "1.05*v5", "1.05*v6", "1.05*v7", "1.05*v8"}}]
```

	No errors	1.05*v1	1.05*v2	1.05*v3	1.05*v4	1.05*v5	1.05*v6	1.05*v7	1.05*v8
kCB	20.01	20.09	21.79	20.79	20.12	16.65	21.26	19.88	19.67
kBA	10.01	10.03	10.11	10.02	10.93	9.97	10.1	9.88	8.99
kCA	8.34	8.38	8.48	9.19	8.36	7.	8.83	8.2	8.3
kBCA	11.68	12.52	10.85	11.53	11.49	11.86	11.6	11.69	11.83
kACB	28.01	30.04	27.88	26.11	27.68	28.21	27.92	28.32	28.03

Thus mechanism V yields 7 equilibrium constants directly from the measurement of 8 velocities and 5 more equilibrium constants indirectly. Thus the equilibrium constants for the 12 reactions in Section 6.1 can all be determined from velocity measurements when the A + B + C → products reaction is completely random. This table has been rotated, partly as a convenience and partly to emphasize that these kinetic parameters have been calculated from the values of kinetic parameters in the rapid-equilibrium rate equation for mechanism V.

■ 6.6.3 Use of calckinparsABC5 to determine the mechanism when A + B + C → products is studied for the first time

Earlier [32] it was shown that more general programs can be used to determine the mechanism for simpler mechanisms and to evaluate the kinetic parameters. This is demonstrated here by applying calckinparsABC5 to velocities calculated using mechanisms I, II, III, and IV. To make these tests, velocities are calculated at the triplets of substrate concentrations that were used to test calckinparsABC5 in the preceding section.

```
v1funconc /. a → 200 /. b → 200 /. c → 200 // N
```

0.838267

```
v1funconc /. a → .5 /. b → 200 /. c → 200 // N
```

0.731261

```
v1funconc /. a → 200 /. b → .5 /. c → 200 // N
```

0.11976

```
v1funconc /. a → 200 /. b → 200 /. c → .5 // N
```

0.0127918

```
v1funconc /. a → 5 /. b → 5 /. c → 50 // N
```

0.136986

```
v1funconc /. a → 5 /. b → 5 /. c → 5 // N
```

0.015625

```
v1funconc /. a → 200 /. b → 5 /. c → 5 // N
```

0.027248

```
v1funconc /. a → 5 /. b → 200 /. c → 5 // N
```

0.106383

```
calckinparsABC5[.8383, 200, 200, 200, .7313, .5, 200, 200, .1198, 200, .5, 200, .01279,
  200, 200, .5, .1370, 5, 5, 50, .01563, 5, 5, 5, .02725, 200, 5, 5, .1064, 5, 200, 5]
```

{{vfexp → 1.00007, kA → 4.99742, kAB → 19.9941,
  kB → -62 695.7, kC → 80 596., kABC → 35.0088, kBC → -532.674, kAC → -630 467.}}

```
line61 = {vfexp, kA, kAB, kB, kC, kABC, kBC, kAC} /.
  calckinparsABC5[.8383, 200, 200, 200, .7313, .5, 200, 200, .1198, 200, .5, 200, .01279,
    200, 200, .5, .1370, 5, 5, 50, .01563, 5, 5, 5, .02725, 200, 5, 5, .1064, 5, 200, 5]
```

{{1.00007, 4.99742, 19.9941, -62 695.7, 80 596., 35.0088, -532.674, -630 467.}}

Calculate effects of 5 % errors in velocities

```
calckinparsABC5[1.05 * .8383, 200, 200, 200, .7313, .5, 200, 200, .1198, 200, .5, 200, .01279,
  200, 200, .5, .1370, 5, 5, 50, .01563, 5, 5, 5, .02725, 200, 5, 5, .1064, 5, 200, 5]
```

{{vfexp → 1.06068, kA → 4.96974, kAB → 20.021,
  kB → 2876.55, kC → 2658.85, kABC → 37.1477, kBC → 74.6276, kAC → 46 689.8}}

```
line62 = {vfexp, kA, kAB, kB, kC, kABC, kBC, kAC} /.
   calckinparsABC5[1.05 * .8383, 200, 200, 200, .7313, .5, 200, 200, .1198, 200, .5, 200,
    .01279, 200, 200, .5, .1370, 5, 5, 50, .01563, 5, 5, 5, .02725, 200, 5, 5, .1064, 5, 200, 5]
```

{{1.06068, 4.96974, 20.021, 2876.55, 2658.85, 37.1477, 74.6276, 46 689.8}}

```
calckinparsABC5[.8383, 200, 200, 200, 1.05 * .7313, .5, 200, 200, .1198, 200, .5, 200, .01279,
  200, 200, .5, .1370, 5, 5, 50, .01563, 5, 5, 5, .02725, 200, 5, 5, .1064, 5, 200, 5]
```

{{vfexp → 0.999899, kA → 4.99617, kAB → 19.9947, kB → 30 176.3,
  kC → 16 259.3, kABC → 35.0019, kBC → -3.38544, kAC → -355 557.}}

```
line63 = {vfexp, kA, kAB, kB, kC, kABC, kBC, kAC} /.
   calckinparsABC5[.8383, 200, 200, 200, 1.05 * .7313, .5, 200, 200, .1198, 200, .5, 200,
    .01279, 200, 200, .5, .1370, 5, 5, 50, .01563, 5, 5, 5, .02725, 200, 5, 5, .1064, 5, 200, 5]
```

{{0.999899, 4.99617, 19.9947, 30 176.3, 16 259.3, 35.0019, -3.38544, -355 557.}}

```
calckinparsABC5[.8383, 200, 200, 200, .7313, .5, 200, 200, 1.05 * .1198, 200, .5, 200, .01279,
  200, 200, .5, .1370, 5, 5, 50, .01563, 5, 5, 5, .02725, 200, 5, 5, .1064, 5, 200, 5]
```

{{vfexp → 0.999024, kA → 4.98248, kAB → 20.027,
  kB → -118 027., kC → 3200.2, kABC → 34.9669, kBC → 5.7626, kAC → -3326.83}}

```
line64 = {vfexp, kA, kAB, kB, kC, kABC, kBC, kAC} /.
   calckinparsABC5[.8383, 200, 200, 200, .7313, .5, 200, 200, 1.05 * .1198, 200, .5, 200,
    .01279, 200, 200, .5, .1370, 5, 5, 50, .01563, 5, 5, 5, .02725, 200, 5, 5, .1064, 5, 200, 5]
```

{{0.999024, 4.98248, 20.027, -118 027., 3200.2, 34.9669, 5.7626, -3326.83}}

```
calckinparsABC5[.8383, 200, 200, 200, .7313, .5, 200, 200, .1198, 200, .5, 200, 1.05 * .01279,
  200, 200, .5, .1370, 5, 5, 50, .01563, 5, 5, 5, .02725, 200, 5, 5, .1064, 5, 200, 5]
```

{{vfexp → 0.990349, kA → 4.85916, kAB → 21.479, kB → 353.349,
  kC → 11 941.7, kABC → 32.7243, kBC → -199.289, kAC → -14 141.1}}

```
line65 = {vfexp, kA, kAB, kB, kC, kABC, kBC, kAC} /.
   calckinparsABC5[.8383, 200, 200, 200, .7313, .5, 200, 200, .1198, 200, .5, 200, 1.05 * .01279,
    200, 200, .5, .1370, 5, 5, 50, .01563, 5, 5, 5, .02725, 200, 5, 5, .1064, 5, 200, 5]
```

{{0.990349, 4.85916, 21.479, 353.349, 11 941.7, 32.7243, -199.289, -14 141.1}}

```
calckinparsABC5[.8383, 200, 200, 200, .7313, .5, 200, 200, .1198, 200, .5, 200, .01279,
  200, 200, .5, 1.05 * .1370, 5, 5, 50, .01563, 5, 5, 5, .02725, 200, 5, 5, .1064, 5, 200, 5]
```

{{vfexp → 1.00033, kA → 5.07182, kAB → 19.9861,
  kB → -11 457.9, kC → -350.885, kABC → 35.019, kBC → -6.08604, kAC → 14 085.4}}

```
line66 = {vfexp, kA, kAB, kB, kC, kABC, kBC, kAC} /.
   calckinparsABC5[.8383, 200, 200, 200, .7313, .5, 200, 200, .1198, 200, .5, 200, .01279,
    200, 200, .5, 1.05 * .1370, 5, 5, 50, .01563, 5, 5, 5, .02725, 200, 5, 5, .1064, 5, 200, 5]
```

{{1.00033, 5.07182, 19.9861, -11 457.9, -350.885, 35.019, -6.08604, 14 085.4}}

```
calckinparsABC5[.8383, 200, 200, 200, .7313, .5, 200, 200, .1198, 200, .5, 200, .01279,
  200, 200, .5, .1370, 5, 5, 50, 1.05 * .01563, 5, 5, 5, .02725, 200, 5, 5, .1064, 5, 200, 5]
```

{{vfexp → 0.999849, kA → 4.34763, kAB → 20.064, kB → 1406.87,
  kC → 341.238, kABC → 34.9899, kBC → -48.8616, kAC → -15 390.3}}

```
line67 = {vfexp, kA, kAB, kB, kC, kABC, kBC, kAC} /.
   calckinparsABC5[.8383, 200, 200, 200, .7313, .5, 200, 200, .1198, 200, .5, 200, .01279,
    200, 200, .5, .1370, 5, 5, 50, 1.05 * .01563, 5, 5, 5, .02725, 200, 5, 5, .1064, 5, 200, 5]
```

{{0.999849, 4.34763, 20.064, 1406.87, 341.238, 34.9899, -48.8616, -15 390.3}}

```
calckinparsABC5[.8383, 200, 200, 200, .7313, .5, 200, 200, .1198, 200, .5, 200, .01279,
   200, 200, .5, .1370, 5, 5, 50, .01563, 5, 5, 5, 1.05 * .02725, 200, 5, 5, .1064, 5, 200, 5]
```

$\{\{\text{vfexp} \to 1.00125, \text{kA} \to 5.71929, \text{kAB} \to 18.5229,$
$\text{kB} \to -3025.11, \text{kC} \to -3289.49, \text{kABC} \to 35.2861, \text{kBC} \to -205.81, \text{kAC} \to 2782.98\}\}$

```
line68 = {vfexp, kA, kAB, kB, kC, kABC, kBC, kAC} /.
   calckinparsABC5[.8383, 200, 200, 200, .7313, .5, 200, 200, .1198, 200, .5, 200, .01279,
      200, 200, .5, .1370, 5, 5, 50, .01563, 5, 5, 5, 1.05 * .02725, 200, 5, 5, .1064, 5, 200, 5]
```

$\{\{1.00125, 5.71929, 18.5229, -3025.11, -3289.49, 35.2861, -205.81, 2782.98\}\}$

```
calckinparsABC5[.8383, 200, 200, 200, .7313, .5, 200, 200, .1198, 200, .5, 200, .01279,
   200, 200, .5, .1370, 5, 5, 50, .01563, 5, 5, 5, .02725, 200, 5, 5, 1.05 * .1064, 5, 200, 5]
```

$\{\{\text{vfexp} \to 1.00037, \text{kA} \to 5.08585, \text{kAB} \to 19.951, \text{kB} \to -293.397,$
$\text{kC} \to -13\,770.4, \text{kABC} \to 35.0798, \text{kBC} \to -200.576, \text{kAC} \to 1.75434 \times 10^6\}\}$

```
line69 = {vfexp, kA, kAB, kB, kC, kABC, kBC, kAC} /.
   calckinparsABC5[.8383, 200, 200, 200, .7313, .5, 200, 200, .1198, 200, .5, 200, .01279,
      200, 200, .5, .1370, 5, 5, 50, .01563, 5, 5, 5, .02725, 200, 5, 5, 1.05 * .1064, 5, 200, 5]
```

$\{\{1.00037, 5.08585, 19.951, -293.397, -13\,770.4, 35.0798, -200.576, 1.75434 \times 10^6\}\}$

Table 6.7 Effects of 5% errors in 8 velocities when calckinparsABC5 is applied to velocities for the completely-ordered mechanism I

```
TableForm[Round[{line61[[1]], line62[[1]], line63[[1]], line64[[1]], line65[[1]],
   line66[[1]], line67[[1]], line68[[1]], line69[[1]]}, 0.01], TableHeadings ->
   {{"No errors", "1.05*v1", "1.05*v2", "1.05*v3", "1.05*v4", "1.05*v5", "1.05*v6",
      "1.05*v7", "1.05*v8"}, {"vfexp", "kA", "kAB", "kB", "kC", "kABC", "kBC", "kAC"}}]
```

	vfexp	kA	kAB	kB	kC	kABC	kBC	kAC
No errors	1.	5.	19.99	-62 695.7	80 596.	35.01	-532.67	-630 467.
1.05*v1	1.06	4.97	20.02	2876.55	2658.85	37.15	74.63	46 689.8
1.05*v2	1.	5.	19.99	30 176.3	16 259.3	35.	-3.39	-355 557.
1.05*v3	1.	4.98	20.03	-118 027.	3200.2	34.97	5.76	-3326.83
1.05*v4	0.99	4.86	21.48	353.35	11 941.7	32.72	-199.29	-14 141.1
1.05*v5	1.	5.07	19.99	-11 457.9	-350.88	35.02	-6.09	14 085.5
1.05*v6	1.	4.35	20.06	1406.87	341.24	34.99	-48.86	-15 390.3
1.05*v7	1.	5.72	18.52	-3025.11	-3289.49	35.29	-205.81	2782.98
1.05*v8	1.	5.09	19.95	-293.4	-13 770.4	35.08	-200.58	1.75434×10^6

This table shows that unreasonable values are obtained for kB, kC, kBC, and kAC. These parameters are not only unreasonable, but they are also extremely dependent on experimental errors in velocities. Thus these calculations show that the mechanism is the completely-ordered mechanism I. This calculation gives values for vfexp, kA, kAB, and kABC with the usual effects for 5% errors.

Similar calculations will now be made velocities calculated from mechanisms II, III, and IV. Five velocities are calculated for mechanism II.

```
v2funconc /. a -> 200 /. b -> 200 /. c -> 200 // N
```

0.832163

```
v2funconc /. a -> .5 /. b -> 200 /. c -> 200 // N
```

0.205444

```
v2funconc /. a -> 200 /. b -> .5 /. c -> 200 // N
```

0.119635

```
v2funconc /. a → 200 /. b → 200 /. c → .5 // N
```

0.0122436

```
v2funconc /. a → 5 /. b → 5 /. c → 50 // N
```

0.114943

```
v2funconc /. a → 5 /. b → 5 /. c → 5 // N
```

0.0128205

```
v2funconc /. a → 200 /. b → 5 /. c → 5 // N
```

0.0269906

```
v2funconc /. a → 5 /. b → 200 /. c → 5 // N
```

0.042735

```
calckinparsABC5[.8322, 200, 200, 200, .2054, .5, 200, 200, .1196, 200, .5, 200, .01224,
  200, 200, .5, .1149, 5, 5, 50, .01282, 5, 5, 5, .02699, 200, 5, 5, .04274, 5, 200, 5]
```

{{vfexp → 1.00013, kA → 5.00123, kAB → 19.991,
  kB → 10.0057, kC → 47 300., kABC → 35.0176, kBC → 406 716., kAC → 621 795.}}

```
line71 = {vfexp, kA, kAB, kB, kC, kABC, kBC, kAC} /.
  calckinparsABC5[.8322, 200, 200, 200, .2054, .5, 200, 200, .1196, 200, .5, 200, .01224,
    200, 200, .5, .1149, 5, 5, 50, .01282, 5, 5, 5, .02699, 200, 5, 5, .04274, 5, 200, 5]
```

{{1.00013, 5.00123, 19.991, 10.0057, 47 300., 35.0176, 406 716., 621 795.}}

Calculate effects of 5% errors in velocities.

```
calckinparsABC5[1.05 * .8322, 200, 200, 200, .2054, .5, 200, 200, .1196, 200, .5, 200, .01224,
  200, 200, .5, .1149, 5, 5, 50, .01282, 5, 5, 5, .02699, 200, 5, 5, .04274, 5, 200, 5]
```

{{vfexp → 1.06122, kA → 4.97334, kAB → 20.0182,
  kB → 9.9318, kC → 2582.31, kABC → 37.1738, kBC → 20 526.5, kAC → 40 365.2}}

```
line72 = {vfexp, kA, kAB, kB, kC, kABC, kBC, kAC} /.
  calckinparsABC5[1.05 * .8322, 200, 200, 200, .2054, .5, 200, 200, .1196, 200, .5, 200, .01224,
    200, 200, .5, .1149, 5, 5, 50, .01282, 5, 5, 5, .02699, 200, 5, 5, .04274, 5, 200, 5]
```

{{1.06122, 4.97334, 20.0182, 9.9318, 2582.31, 37.1738, 20 526.5, 40 365.2}}

```
calckinparsABC5[.8322, 200, 200, 200, 1.05 * .2054, .5, 200, 200, .1196, 200, .5, 200, .01224,
  200, 200, .5, .1149, 5, 5, 50, .01282, 5, 5, 5, .02699, 200, 5, 5, .04274, 5, 200, 5]
```

{{vfexp → 0.999515, kA → 4.99675, kAB → 19.9932,
  kB → 9.97952, kC → 5104.2, kABC → 34.9931, kBC → - 2887.31, kAC → - 362 598.}}

```
line73 = {vfexp, kA, kAB, kB, kC, kABC, kBC, kAC} /.
  calckinparsABC5[.8322, 200, 200, 200, 1.05 * .2054, .5, 200, 200, .1196, 200, .5, 200, .01224,
    200, 200, .5, .1149, 5, 5, 50, .01282, 5, 5, 5, .02699, 200, 5, 5, .04274, 5, 200, 5]
```

{{0.999515, 4.99675, 19.9932, 9.97952, 5104.2, 34.9931, - 2887.31, - 362 598.}}

```
calckinparsABC5[.8322, 200, 200, 200, .2054, .5, 200, 200, 1.05 * .1196, 200, .5, 200, .01224,
  200, 200, .5, .1149, 5, 5, 50, .01282, 5, 5, 5, .02699, 200, 5, 5, .04274, 5, 200, 5]
```

{{vfexp → 0.999077, kA → 4.98625, kAB → 20.024, kB → 9.98993,
  kC → 3110.54, kABC → 34.9756, kBC → - 79 731.6, kAC → - 3357.03}}

```
line74 = {vfexp, kA, kAB, kB, kC, kABC, kBC, kAC} /.
    calckinparsABC5[.8322, 200, 200, 200, .2054, .5, 200, 200, 1.05 * .1196, 200, .5, 200, .01224,
       200, 200, .5, .1149, 5, 5, 50, .01282, 5, 5, 5, .02699, 200, 5, 5, .04274, 5, 200, 5]
```

{{0.999077, 4.98625, 20.024, 9.98993, 3110.54, 34.9756, -79 731.6, -3357.03}}

```
calckinparsABC5[.8322, 200, 200, 200, .2054, .5, 200, 200, .1196, 200, .5, 200, 1.05 * .01224,
   200, 200, .5, .1149, 5, 5, 50, .01282, 5, 5, 5, .02699, 200, 5, 5, .04274, 5, 200, 5]
```

{{vfexp → 0.989971, kA → 4.85679, kAB → 21.5464, kB → 9.57829,
  kC → 10 438.8, kABC → 32.6312, kBC → -7141.75, kAC → -14 159.1}}

```
line75 = {vfexp, kA, kAB, kB, kC, kABC, kBC, kAC} /.
    calckinparsABC5[.8322, 200, 200, 200, .2054, .5, 200, 200, .1196, 200, .5, 200, 1.05 * .01224,
       200, 200, .5, .1149, 5, 5, 50, .01282, 5, 5, 5, .02699, 200, 5, 5, .04274, 5, 200, 5]
```

{{0.989971, 4.85679, 21.5464, 9.57829, 10 438.8, 32.6312, -7141.75, -14 159.1}}

```
calckinparsABC5[.8322, 200, 200, 200, .2054, .5, 200, 200, .1196, 200, .5, 200, .01224, 200,
   200, .5, 1.05 * .1149, 5, 5, 50, .01282, 5, 5, 5, .02699, 200, 5, 5, .04274, 5, 200, 5]
```

{{vfexp → 1.00043, kA → 5.08993, kAB → 19.9815,
  kB → 10.1876, kC → -295.872, kABC → 35.0297, kBC → 5692.13, kAC → 11 343.8}}

```
line76 = {vfexp, kA, kAB, kB, kC, kABC, kBC, kAC} /.
    calckinparsABC5[.8322, 200, 200, 200, .2054, .5, 200, 200, .1196, 200, .5, 200, .01224,
       200, 200, .5, 1.05 * .1149, 5, 5, 50, .01282, 5, 5, 5, .02699, 200, 5, 5, .04274, 5, 200, 5]
```

{{1.00043, 5.08993, 19.9815, 10.1876, -295.872, 35.0297, 5692.13, 11 343.8}}

```
calckinparsABC5[.8322, 200, 200, 200, .2054, .5, 200, 200, .1196, 200, .5, 200, .01224, 200,
   200, .5, .1149, 5, 5, 50, 1.05 * .01282, 5, 5, 5, .02699, 200, 5, 5, .04274, 5, 200, 5]
```

{{vfexp → 0.999855, kA → 4.20958, kAB → 20.0763,
  kB → 8.38947, kC → 270.678, kABC → 34.9945, kBC → -6601.45, kAC → -13 222.1}}

```
line77 = {vfexp, kA, kAB, kB, kC, kABC, kBC, kAC} /.
    calckinparsABC5[.8322, 200, 200, 200, .2054, .5, 200, 200, .1196, 200, .5, 200, .01224,
       200, 200, .5, .1149, 5, 5, 50, 1.05 * .01282, 5, 5, 5, .02699, 200, 5, 5, .04274, 5, 200, 5]
```

{{0.999855, 4.20958, 20.0763, 8.38947, 270.678, 34.9945, -6601.45, -13 222.1}}

```
calckinparsABC5[.8322, 200, 200, 200, .2054, .5, 200, 200, .1196, 200, .5, 200, .01224, 200,
   200, .5, .1149, 5, 5, 50, .01282, 5, 5, 5, 1.05 * .02699, 200, 5, 5, .04274, 5, 200, 5]
```

{{vfexp → 1.00132, kA → 5.7308, kAB → 18.5061,
  kB → 10.7223, kC → -3350.83, kABC → 35.2976, kBC → 51 209., kAC → 2728.75}}

```
line78 = {vfexp, kA, kAB, kB, kC, kABC, kBC, kAC} /.
    calckinparsABC5[.8322, 200, 200, 200, .2054, .5, 200, 200, .1196, 200, .5, 200, .01224,
       200, 200, .5, .1149, 5, 5, 50, .01282, 5, 5, 5, 1.05 * .02699, 200, 5, 5, .04274, 5, 200, 5]
```

{{1.00132, 5.7308, 18.5061, 10.7223, -3350.83, 35.2976, 51 209., 2728.75}}

```
calckinparsABC5[.8322, 200, 200, 200, .2054, .5, 200, 200, .1196, 200, .5, 200, .01224, 200,
   200, .5, .1149, 5, 5, 50, .01282, 5, 5, 5, .02699, 200, 5, 5, 1.05 * .04274, 5, 200, 5]
```

{{vfexp → 1.00088, kA → 5.2215, kAB → 19.8842,
  kB → 11.4157, kC → -5390.4, kABC → 35.1943, kBC → 2116.26, kAC → 143 214.}}

```
line79 = {vfexp, kA, kAB, kB, kC, kABC, kBC, kAC} /.
    calckinparsABC5[.8322, 200, 200, 200, .2054, .5, 200, 200, .1196, 200, .5, 200, .01224,
       200, 200, .5, .1149, 5, 5, 50, .01282, 5, 5, 5, .02699, 200, 5, 5, 1.05 * .04274, 5, 200, 5]
```

{{1.00088, 5.2215, 19.8842, 11.4157, -5390.4, 35.1943, 2116.26, 143 214.}}

Table 6.8 Effects of 5% errors in 8 velocities when calckinparsABC5 is applied to velocities for mechanism II

```
TableForm[Round[{line71[[1]], line72[[1]], line73[[1]], line74[[1]], line75[[1]],
    line76[[1]], line77[[1]], line78[[1]], line79[[1]]}, 0.01], TableHeadings →
  {{"No errors", "1.05*v1", "1.05*v2", "1.05*v3", "1.05*v4", "1.05*v5", "1.05*v6",
    "1.05*v7", "1.05*v8"}, {"vfexp", "kA", "kAB", "kB", "kC", "kABC", "kBC", "kAC"}}]
```

	vfexp	kA	kAB	kB	kC	kABC	kBC	kAC
No errors	1.	5.	19.99	10.01	47 300.	35.02	406 716.	621 795.
1.05*v1	1.06	4.97	20.02	9.93	2582.31	37.17	20 526.5	40 365.2
1.05*v2	1.	5.	19.99	9.98	5104.2	34.99	−2887.31	−362 598.
1.05*v3	1.	4.99	20.02	9.99	3110.54	34.98	−79 731.6	−3357.03
1.05*v4	0.99	4.86	21.55	9.58	10 438.8	32.63	−7141.75	−14 159.1
1.05*v5	1.	5.09	19.98	10.19	−295.87	35.03	5692.13	11 343.8
1.05*v6	1.	4.21	20.08	8.39	270.68	34.99	−6601.45	−13 222.1
1.05*v7	1.	5.73	18.51	10.72	−3350.83	35.3	51 209.	2728.75
1.05*v8	1.	5.22	19.88	11.42	−5390.4	35.19	2116.26	143 214.

Unreasonable values of kC, kBC, and kAC are obtained. In contrast with the preceding table, the expected values for kB are obtained.

Similar calculations are now made with velocities from mechanisms III.

v3funconc /. a → 200 /. b → 200 /. c → 200 // N

0.828143

v3funconc /. a → .5 /. b → 200 /. c → 200 // N

0.138873

v3funconc /. a → 200 /. b → .5 /. c → 200 // N

0.0935271

v3funconc /. a → 200 /. b → 200 /. c → .5 // N

0.0122428

v3funconc /. a → 5 /. b → 5 /. c → 50 // N

0.0554529

v3funconc /. a → 5 /. b → 5 /. c → 5 // N

0.0114504

v3funconc /. a → 200 /. b → 5 /. c → 5 // N

0.0268216

v3funconc /. a → 5 /. b → 200 /. c → 5 // N

0.0423131

**calckinparsABC5[.8281, 200, 200, 200, .1389, .5, 200, 200, .09353, 200, .5, 200, .01224,
  200, 200, .5, .05545, 5, 5, 50, .01145, 5, 5, 5, .02682, 200, 5, 5, .04231, 5, 200, 5]**

$\{\{$vfexp → 0.999978, kA → 4.99989, kAB → 19.9951, kB → 9.99994,
    kC → 14.9984, kABC → 35.0084, kBC → −408 271., kAC → −1.88772 × $10^6\}\}$

```
line81 = {vfexp, kA, kAB, kB, kC, kABC, kBC, kAC} /.
    calckinparsABC5[.8281, 200, 200, 200, .1389, .5, 200, 200, .09353, 200, .5, 200, .01224,
      200, 200, .5, .05545, 5, 5, 50, .01145, 5, 5, 5, .02682, 200, 5, 5, .04231, 5, 200, 5]
```

$$\{\{0.999978, 4.99989, 19.9951, 9.99994, 14.9984, 35.0084, -408271., -1.88772 \times 10^6\}\}$$

Calculate effects of 5% errors in velocities.

```
calckinparsABC5[1.05 * .8281, 200, 200, 200, .1389, .5, 200, 200, .09353, 200, .5, 200, .01224,
  200, 200, .5, .05545, 5, 5, 50, .01145, 5, 5, 5, .02682, 200, 5, 5, .04231, 5, 200, 5]
```

$\{\{vfexp \rightarrow 1.06137, kA \rightarrow 4.97187, kAB \rightarrow 20.0224,$
$kB \rightarrow 9.92574, kC \rightarrow 14.86, kABC \rightarrow 37.175, kBC \rightarrow 22708.1, kAC \rightarrow 43951.9\}\}$

```
line82 = {vfexp, kA, kAB, kB, kC, kABC, kBC, kAC} /.
    calckinparsABC5[1.05 * .8281, 200, 200, 200, .1389, .5, 200, 200, .09353, 200, .5, 200, .01224,
      200, 200, .5, .05545, 5, 5, 50, .01145, 5, 5, 5, .02682, 200, 5, 5, .04231, 5, 200, 5]
```

$\{\{1.06137, 4.97187, 20.0224, 9.92574, 14.86, 37.175, 22708.1, 43951.9\}\}$

```
calckinparsABC5[.8281, 200, 200, 200, 1.05 * .1389, .5, 200, 200, .09353, 200, .5, 200, .01224,
  200, 200, .5, .05545, 5, 5, 50, .01145, 5, 5, 5, .02682, 200, 5, 5, .04231, 5, 200, 5]
```

$\{\{vfexp \rightarrow 0.999075, kA \rightarrow 4.99327, kAB \rightarrow 19.9983,$
$kB \rightarrow 9.9613, kC \rightarrow 14.9213, kABC \rightarrow 34.9723, kBC \rightarrow -1932., kAC \rightarrow -143154.\}\}$

```
line83 = {vfexp, kA, kAB, kB, kC, kABC, kBC, kAC} /.
    calckinparsABC5[.8281, 200, 200, 200, 1.05 * .1389, .5, 200, 200, .09353, 200, .5, 200, .01224,
      200, 200, .5, .05545, 5, 5, 50, .01145, 5, 5, 5, .02682, 200, 5, 5, .04231, 5, 200, 5]
```

$\{\{0.999075, 4.99327, 19.9983, 9.9613, 14.9213, 34.9723, -1932., -143154.\}\}$

```
calckinparsABC5[.8281, 200, 200, 200, .1389, .5, 200, 200, 1.05 * .09353, 200, .5, 200, .01224,
  200, 200, .5, .05545, 5, 5, 50, .01145, 5, 5, 5, .02682, 200, 5, 5, .04231, 5, 200, 5]
```

$\{\{vfexp \rightarrow 0.998637, kA \rightarrow 4.98075, kAB \rightarrow 20.0373,$
$kB \rightarrow 9.9798, kC \rightarrow 14.884, kABC \rightarrow 34.9548, kBC \rightarrow -46247.3, kAC \rightarrow -2608.92\}\}$

```
line84 = {vfexp, kA, kAB, kB, kC, kABC, kBC, kAC} /.
    calckinparsABC5[.8281, 200, 200, 200, .1389, .5, 200, 200, 1.05 * .09353, 200, .5, 200, .01224,
      200, 200, .5, .05545, 5, 5, 50, .01145, 5, 5, 5, .02682, 200, 5, 5, .04231, 5, 200, 5]
```

$\{\{0.998637, 4.98075, 20.0373, 9.9798, 14.884, 34.9548, -46247.3, -2608.92\}\}$

```
calckinparsABC5[.8281, 200, 200, 200, .1389, .5, 200, 200, .09353, 200, .5, 200, 1.05 * .01224,
  200, 200, .5, .05545, 5, 5, 50, .01145, 5, 5, 5, .02682, 200, 5, 5, .04231, 5, 200, 5]
```

$\{\{vfexp \rightarrow 0.989826, kA \rightarrow 4.85548, kAB \rightarrow 21.5509,$
$kB \rightarrow 9.57287, kC \rightarrow 14.7624, kABC \rightarrow 32.6225, kBC \rightarrow -6909.18, kAC \rightarrow -13750.1\}\}$

```
line85 = {vfexp, kA, kAB, kB, kC, kABC, kBC, kAC} /.
    calckinparsABC5[.8281, 200, 200, 200, .1389, .5, 200, 200, .09353, 200, .5, 200, 1.05 * .01224,
      200, 200, .5, .05545, 5, 5, 50, .01145, 5, 5, 5, .02682, 200, 5, 5, .04231, 5, 200, 5]
```

$\{\{0.989826, 4.85548, 21.5509, 9.57287, 14.7624, 32.6225, -6909.18, -13750.1\}\}$

```
calckinparsABC5[.8281, 200, 200, 200, .1389, .5, 200, 200, .09353, 200, .5, 200, .01224, 200,
  200, .5, 1.05 * .05545, 5, 5, 50, .01145, 5, 5, 5, .02682, 200, 5, 5, .04231, 5, 200, 5]
```

$\{\{vfexp \rightarrow 1.00061, kA \rightarrow 5.18378, kAB \rightarrow 19.9754,$
$kB \rightarrow 10.377, kC \rightarrow 17.4081, kABC \rightarrow 35.0335, kBC \rightarrow 2803.69, kAC \rightarrow 5590.67\}\}$

```
line86 = {vfexp, kA, kAB, kB, kC, kABC, kBC, kAC} /.
    calckinparsABC5[.8281, 200, 200, 200, .1389, .5, 200, 200, .09353, 200, .5, 200, .01224,
      200, 200, .5, 1.05 * .05545, 5, 5, 50, .01145, 5, 5, 5, .02682, 200, 5, 5, .04231, 5, 200, 5]
```

$\{\{1.00061, 5.18378, 19.9754, 10.377, 17.4081, 35.0335, 2803.69, 5590.67\}\}$

```
calckinparsABC5[.8281, 200, 200, 200, .1389, .5, 200, 200, .09353, 200, .5, 200, .01224, 200,
    200, .5, .05545, 5, 5, 50, 1.05 * .01145, 5, 5, 5, .02682, 200, 5, 5, .04231, 5, 200, 5]
```

$\{\{vfexp \to 0.999674, kA \to 4.11402, kAB \to 20.0906,$
$\quad kB \to 8.19276, kC \to 11.7811, kABC \to 34.9826, kBC \to -5729.34, kAC \to -11500.3\}\}$

```
line87 = {vfexp, kA, kAB, kB, kC, kABC, kBC, kAC} /.
    calckinparsABC5[.8281, 200, 200, 200, .1389, .5, 200, 200, .09353, 200, .5, 200, .01224,
        200, 200, .5, .05545, 5, 5, 50, 1.05 * .01145, 5, 5, 5, .02682, 200, 5, 5, .04231, 5, 200, 5]
```

$\{\{0.999674, 4.11402, 20.0906, 8.19276, 11.7811, 34.9826, -5729.34, -11500.3\}\}$

```
calckinparsABC5[.8281, 200, 200, 200, .1389, .5, 200, 200, .09353, 200, .5, 200, .01224, 200,
    200, .5, .05545, 5, 5, 50, .01145, 5, 5, 5, 1.05 * .02682, 200, 5, 5, .04231, 5, 200, 5]
```

$\{\{vfexp \to 1.00118, kA \to 5.73426, kAB \to 18.5006,$
$\quad kB \to 10.7208, kC \to 16.1072, kABC \to 35.2901, kBC \to 68005.8, kAC \to 2727.62\}\}$

```
line88 = {vfexp, kA, kAB, kB, kC, kABC, kBC, kAC} /.
    calckinparsABC5[.8281, 200, 200, 200, .1389, .5, 200, 200, .09353, 200, .5, 200, .01224,
        200, 200, .5, .05545, 5, 5, 50, .01145, 5, 5, 5, 1.05 * .02682, 200, 5, 5, .04231, 5, 200, 5]
```

$\{\{1.00118, 5.73426, 18.5006, 10.7208, 16.1072, 35.2901, 68005.8, 2727.62\}\}$

```
calckinparsABC5[.8281, 200, 200, 200, .1389, .5, 200, 200, .09353, 200, .5, 200, .01224, 200,
    200, .5, .05545, 5, 5, 50, .01145, 5, 5, 5, .02682, 200, 5, 5, 1.05 * .04231, 5, 200, 5]
```

$\{\{vfexp \to 1.00074, kA \to 5.22239, kAB \to 19.8872,$
$\quad kB \to 11.4245, kC \to 15.6999, kABC \to 35.1869, kBC \to 2117.83, kAC \to 204230.\}\}$

```
line89 = {vfexp, kA, kAB, kB, kC, kABC, kBC, kAC} /.
    calckinparsABC5[.8281, 200, 200, 200, .1389, .5, 200, 200, .09353, 200, .5, 200, .01224,
        200, 200, .5, .05545, 5, 5, 50, .01145, 5, 5, 5, .02682, 200, 5, 5, 1.05 * .04231, 5, 200, 5]
```

$\{\{1.00074, 5.22239, 19.8872, 11.4245, 15.6999, 35.1869, 2117.83, 204230.\}\}$

Table 6.9 Effects of 5% errors in 8 velocities when calckinparsABC5 is applied to velocities for mechanism III

```
TableForm[Round[{line81[[1]], line82[[1]], line83[[1]], line84[[1]], line85[[1]],
    line86[[1]], line87[[1]], line88[[1]], line89[[1]]}, 0.01], TableHeadings →
    {{"No errors", "1.05*v1", "1.05*v2", "1.05*v3", "1.05*v4", "1.05*v5", "1.05*v6",
        "1.05*v7", "1.05*v8"}, {"vfexp", "kA", "kAB", "kB", "kC", "kABC", "kBC", "kAC"}}]
```

	vfexp	kA	kAB	kB	kC	kABC	kBC	kAC
No errors	1.	5.	20.	10.	15.	35.01	-408271.	-1.88772×10^6
1.05*v1	1.06	4.97	20.02	9.93	14.86	37.18	22708.1	43951.9
1.05*v2	1.	4.99	20.	9.96	14.92	34.97	-1932.	-143154.
1.05*v3	1.	4.98	20.04	9.98	14.88	34.95	-46247.3	-2608.92
1.05*v4	0.99	4.86	21.55	9.57	14.76	32.62	-6909.18	-13750.1
1.05*v5	1.	5.18	19.98	10.38	17.41	35.03	2803.69	5590.67
1.05*v6	1.	4.11	20.09	8.19	11.78	34.98	-5729.34	-11500.3
1.05*v7	1.	5.73	18.5	10.72	16.11	35.29	68005.8	2727.62
1.05*v8	1.	5.22	19.89	11.42	15.7	35.19	2117.83	204230.

Unreasonable values of kBC and kAC are obtained. In contrast with the preceding table, the expected values for kB and kC are obtained.

Similar calculations are now made with velocities from mechanisms IV.

```
v4funconc /. a → 200 /. b → 200 /. c → 200 // N
```

0.742104

```
v4funconc /. a → .5 /. b → 200 /. c → 200 // N
```

0.136224

```
v4funconc /. a → 200 /. b → .5 /. c → 200 // N
```

0.0149943

```
v4funconc /. a → 200 /. b → 200 /. c → .5 // N
```

0.0122218

```
v4funconc /. a → 5 /. b → 5 /. c → 50 // N
```

0.0423131

```
v4funconc /. a → 5 /. b → 5 /. c → 5 // N
```

0.0107604

```
v4funconc /. a → 200 /. b → 5 /. c → 5 // N
```

0.0233191

```
v4funconc /. a → 5 /. b → 200 /. c → 5 // N
```

0.0420639

```
calckinparsABC5[.7421, 200, 200, 200, .1362, .5, 200, 200, .01499, 200, .5, 200, .01222,
  200, 200, .5, .04231, 5, 5, 50, .01076, 5, 5, 5, .02332, 200, 5, 5, .04206, 5, 200, 5]
```

{{vfexp → 1.00008, kA → 5.00179, kAB → 19.9925,
  kB → 10.0012, kC → 15.0035, kABC → 35.0094, kBC → 579995., kAC → 24.9863}}

```
line91 = {vfexp, kA, kAB, kB, kC, kABC, kBC, kAC} /.
    calckinparsABC5[.7421, 200, 200, 200, .1362, .5, 200, 200, .01499, 200, .5, 200, .01222,
      200, 200, .5, .04231, 5, 5, 50, .01076, 5, 5, 5, .02332, 200, 5, 5, .04206, 5, 200, 5]
```

{{1.00008, 5.00179, 19.9925, 10.0012, 15.0035, 35.0094, 579995., 24.9863}}

Calculate effects of 5% errors in velocities.

```
calckinparsABC5[1.05 * .7421, 200, 200, 200, .1362, .5, 200, 200, .01499, 200, .5, 200, .01222,
  200, 200, .5, .04231, 5, 5, 50, .01076, 5, 5, 5, .02332, 200, 5, 5, .04206, 5, 200, 5]
```

{{vfexp → 1.06909, kA → 4.97052, kAB → 20.023,
  kB → 9.9184, kC → 14.8491, kABC → 37.4447, kBC → 18672.1, kAC → 25.0211}}

```
line92 = {vfexp, kA, kAB, kB, kC, kABC, kBC, kAC} /.
    calckinparsABC5[1.05 * .7421, 200, 200, 200, .1362, .5, 200, 200, .01499, 200, .5, 200, .01222,
      200, 200, .5, .04231, 5, 5, 50, .01076, 5, 5, 5, .02332, 200, 5, 5, .04206, 5, 200, 5]
```

{{1.06909, 4.97052, 20.023, 9.9184, 14.8491, 37.4447, 18672.1, 25.0211}}

```
calckinparsABC5[.7421, 200, 200, 200, 1.05 * .1362, .5, 200, 200, .01499, 200, .5, 200, .01222,
  200, 200, .5, .04231, 5, 5, 50, .01076, 5, 5, 5, .02332, 200, 5, 5, .04206, 5, 200, 5]
```

{{vfexp → 0.999157, kA → 4.99504, kAB → 19.9958,
  kB → 9.96176, kC → 14.9248, kABC → 34.9725, kBC → -1909.89, kAC → 24.9912}}

```
line93 = {vfexp, kA, kAB, kB, kC, kABC, kBC, kAC} /.
    calckinparsABC5[.7421, 200, 200, 200, 1.05 * .1362, .5, 200, 200, .01499, 200, .5, 200, .01222,
      200, 200, .5, .04231, 5, 5, 50, .01076, 5, 5, 5, .02332, 200, 5, 5, .04206, 5, 200, 5]
```

{{0.999157, 4.99504, 19.9958, 9.96176, 14.9248, 34.9725, -1909.89, 24.9912}}

```
calckinparsABC5[.7421, 200, 200, 200, .1362, .5, 200, 200, 1.05 * .01499, 200, .5, 200, .01222,
  200, 200, .5, .04231, 5, 5, 50, .01076, 5, 5, 5, .02332, 200, 5, 5, .04206, 5, 200, 5]
```

{{vfexp → 0.991771, kA → 4.88351, kAB → 20.256,
  kB → 9.87561, kC → 14.3103, kABC → 34.677, kBC → -8485.76, kAC → 26.8936}}

```
line94 = {vfexp, kA, kAB, kB, kC, kABC, kBC, kAC} /.
  calckinparsABC5[.7421, 200, 200, 200, .1362, .5, 200, 200, 1.05 * .01499, 200, .5, 200, .01222,
    200, 200, .5, .04231, 5, 5, 50, .01076, 5, 5, 5, .02332, 200, 5, 5, .04206, 5, 200, 5]
```

{{0.991771, 4.88351, 20.256, 9.87561, 14.3103, 34.677, -8485.76, 26.8936}}

```
calckinparsABC5[.7421, 200, 200, 200, .1362, .5, 200, 200, .01499, 200, .5, 200, 1.05 * .01222,
  200, 200, .5, .04231, 5, 5, 50, .01076, 5, 5, 5, .02332, 200, 5, 5, .04206, 5, 200, 5]
```

{{vfexp → 0.989907, kA → 4.85709, kAB → 21.551,
  kB → 9.5734, kC → 14.7671, kABC → 32.6193, kBC → -7097.29, kAC → 25.3997}}

```
line95 = {vfexp, kA, kAB, kB, kC, kABC, kBC, kAC} /.
  calckinparsABC5[.7421, 200, 200, 200, .1362, .5, 200, 200, .01499, 200, .5, 200, 1.05 * .01222,
    200, 200, .5, .04231, 5, 5, 50, .01076, 5, 5, 5, .02332, 200, 5, 5, .04206, 5, 200, 5]
```

{{0.989907, 4.85709, 21.551, 9.5734, 14.7671, 32.6193, -7097.29, 25.3997}}

```
calckinparsABC5[.7421, 200, 200, 200, .1362, .5, 200, 200, .01499, 200, .5, 200, .01222, 200,
  200, .5, 1.05 * .04231, 5, 5, 50, .01076, 5, 5, 5, .02332, 200, 5, 5, .04206, 5, 200, 5]
```

{{vfexp → 1.0009, kA → 5.2429, kAB → 19.9667,
  kB → 10.4956, kC → 18.2857, kABC → 35.0423, kBC → 2115.85, kAC → 24.8113}}

```
line96 = {vfexp, kA, kAB, kB, kC, kABC, kBC, kAC} /.
  calckinparsABC5[.7421, 200, 200, 200, .1362, .5, 200, 200, .01499, 200, .5, 200, .01222, 200,
    200, .5, 1.05 * .04231, 5, 5, 50, .01076, 5, 5, 5, .02332, 200, 5, 5, .04206, 5, 200, 5]
```

{{1.0009, 5.2429, 19.9667, 10.4956, 18.2857, 35.0423, 2115.85, 24.8113}}

```
calckinparsABC5[.7421, 200, 200, 200, .1362, .5, 200, 200, .01499, 200, .5, 200, .01222, 200,
  200, .5, .04231, 5, 5, 50, 1.05 * .01076, 5, 5, 5, .02332, 200, 5, 5, .04206, 5, 200, 5]
```

{{vfexp → 0.999755, kA → 4.05917, kAB → 20.0942,
  kB → 8.07925, kC → 11.5899, kABC → 34.9819, kBC → -5514.83, kAC → 25.1598}}

```
line97 = {vfexp, kA, kAB, kB, kC, kABC, kBC, kAC} /.
  calckinparsABC5[.7421, 200, 200, 200, .1362, .5, 200, 200, .01499, 200, .5, 200, .01222,
    200, 200, .5, .04231, 5, 5, 50, 1.05 * .01076, 5, 5, 5, .02332, 200, 5, 5, .04206, 5, 200, 5]
```

{{0.999755, 4.05917, 20.0942, 8.07925, 11.5899, 34.9819, -5514.83, 25.1598}}

```
calckinparsABC5[.7421, 200, 200, 200, .1362, .5, 200, 200, .01499, 200, .5, 200, .01222, 200,
  200, .5, .04231, 5, 5, 50, .01076, 5, 5, 5, 1.05 * .02332, 200, 5, 5, .04206, 5, 200, 5]
```

{{vfexp → 1.00146, kA → 5.85614, kAB → 18.2754,
  kB → 10.8305, kC → 16.2799, kABC → 35.3334, kBC → 46558.8, kAC → 22.7958}}

```
line98 = {vfexp, kA, kAB, kB, kC, kABC, kBC, kAC} /.
  calckinparsABC5[.7421, 200, 200, 200, .1362, .5, 200, 200, .01499, 200, .5, 200, .01222,
    200, 200, .5, .04231, 5, 5, 50, .01076, 5, 5, 5, 1.05 * .02332, 200, 5, 5, .04206, 5, 200, 5]
```

{{1.00146, 5.85614, 18.2754, 10.8305, 16.2799, 35.3334, 46558.8, 22.7958}}

```
calckinparsABC5[.7421, 200, 200, 200, .1362, .5, 200, 200, .01499, 200, .5, 200, .01222, 200,
  200, .5, .04231, 5, 5, 50, .01076, 5, 5, 5, .02332, 200, 5, 5, 1.05 * .04206, 5, 200, 5]
```

{{vfexp → 1.00084, kA → 5.22565, kAB → 19.884,
  kB → 11.435, kC → 15.7093, kABC → 35.1889, kBC → 2084.14, kAC → 24.9556}}

```
line99 = {vfexp, kA, kAB, kB, kC, kABC, kBC, kAC} /.
   calckinparsABC5[.7421, 200, 200, 200, .1362, .5, 200, 200, .01499, 200, .5, 200, .01222,
      200, 200, .5, .04231, 5, 5, 50, .01076, 5, 5, 5, .02332, 200, 5, 5, 1.05 * .04206, 5, 200, 5]
```

{{1.00084, 5.22565, 19.884, 11.435, 15.7093, 35.1889, 2084.14, 24.9556}}

Table 6.10 Effects of 5% errors in 8 velocities when calckinparsABC5 is applied to velocities for mechanism IV

```
TableForm[Round[{line91[[1]], line92[[1]], line93[[1]], line94[[1]], line95[[1]],
   line96[[1]], line97[[1]], line98[[1]], line99[[1]]}, 0.01], TableHeadings →
   {{"No errors", "1.05*v1", "1.05*v2", "1.05*v3", "1.05*v4", "1.05*v5", "1.05*v6",
      "1.05*v7", "1.05*v8"}, {"vfexp", "kA", "kAB", "kB", "kC", "kABC", "kBC", "kAC"}}]
```

	vfexp	kA	kAB	kB	kC	kABC	kBC	kAC
No errors	1.	5.	19.99	10.	15.	35.01	579 995.	24.99
1.05*v1	1.07	4.97	20.02	9.92	14.85	37.44	18 672.1	25.02
1.05*v2	1.	5.	20.	9.96	14.92	34.97	−1909.89	24.99
1.05*v3	0.99	4.88	20.26	9.88	14.31	34.68	−8485.76	26.89
1.05*v4	0.99	4.86	21.55	9.57	14.77	32.62	−7097.29	25.4
1.05*v5	1.	5.24	19.97	10.5	18.29	35.04	2115.85	24.81
1.05*v6	1.	4.06	20.09	8.08	11.59	34.98	−5514.83	25.16
1.05*v7	1.	5.86	18.28	10.83	16.28	35.33	46 558.8	22.8
1.05*v8	1.	5.23	19.88	11.44	15.71	35.19	2084.14	24.96

Unreasonable values of kBC are obtained. These calculations support the generalization that programs for more general mechanisms can be applied to velocities for simpler mechanisms to identify the simpler mechanism and obtain correct values for kinetic parameters in the simpler mechanism. Application of less general programs to data for more complicated mechanisms involves dropping some measured velocities and does not provide interesting information.

6.7 Discussion

The calculations in this chapter support the following conclusions:

1. Computer programs using the minimum number of velocities can be used to estimate kinetic parameters for 5 mechanisms of the reaction A + B + C → products.

2. Equilibrium constants can be calculated for reactions not included in the mechanism used to derive the rate equation.

3. A more general computer program can be used to identify the mechanism when it is applied to velocities for a simpler mechanism, as pointed out earlier [32].

The estimations of equilibrium constants for the 5 mechanisms can be summarized as follows:

I   kA, kAB, kABC

II   kA, kAB, kABC, kB, (kBA)

III   kA, kAB, kABC, kB, kC, (kCA, kACB)

IV   kA, kAB, kABC, kB, kC, (kCA, kACB)

V   kA, kAB, kABC, kB, kC, kBC, kAC, (kCB, kBA, kCA, kBCA, kACB)

The equilibrium constants in parentheses are calculated from thermodynamic cycles. Thus mechanism I yields 3 equilibrium constants, mechanism II yields 5 equilibrium constants, mechanism III yields 7 equilibrium constants, mechanism IV yields 8 equilibrium constants, and mechanism V yields 12 equilibrium constants. Mechanism V involves 7 independent biochemical reactions.

When A, B, C, E, and enzyme-substrate complexes have pKs, the rapid-equilibrium rate equations become much more complicated. When the kinetic parameters depend on the pH, it is probably better to use the programs discussed here to determine the pH-dependent kinetic parameters at a number of pHs. Plotting these parameters and certain ratios versus pH makes it possible to determine the pH-independent kinetic parameters [26].

Chapter 7 Ordered O + mR → Products

7.1 Introduction

7.2 Mechanism I for ordered O + mR → products in which mR is bound first

- 7.2.1 Derivation of the rate equation when mR is bound first

- 7.2.2 Use of Solve to estimate the kinetic parameters with the minimum number of velocity measurements for mechanism I for ordered O + mR → products

---

7.3 Effects of pH on mechanism I

- 7.3.1 Mechanism I with pKs and the consumption of hydrogen ions

- 7.3.2 Plots of kinetic parameters for a nitrate reductase reaction with mechanism I

- 7.3.3 3D plots of rates for the nitrate reductase type of reaction with mechanism I

7.4 Mechanism II for ordered O + 2R → products in which O is bound first

- 7.4.1 Derivation of the rate equation when O is bound first

- 7.4.2 Use of Solve to estimate the kinetic parameters with the minimum number of velocity measurements for mechanism II for ordered O + 2R → products

---

7.5 Effects of pH on mechanism II

- 7.5.1 Mechanism II with pKs and the consumption of hydrogen ions

- 7.5.2 Plots of kinetic parameters for a nitrate reductase reaction with mechanism II

- 7.5.3 3D plots of rates for the nitrate reductase type of reaction with mechanism II

7.6 Discussion

7.1 Introduction

The reaction is represented as Oxidant + mReductant → products because this type of reaction is frequently encountered with oxidoreductase reactions. The stoichiometic number m can be in the range 1-6. If $m = 1$, the rapid-equilibrium kinetics are represented by ordered A + B → products (see Chapter 3). Here are two examples of the types of reactions that can be represented by O + mR → products [Web2]:

EC 1.9.6.1 nitrate reductase (cytochrome).

$$\text{nitrate} + 2\text{cytochromec}_\text{red} = \text{nitrite} + 2\text{cytochromec}_\text{ox} + H_2O \tag{1}$$

A chemical reference reaction is

$$NO_3^- + 2Fe^{2+} + 2H^+ = NO_2^- + 2Fe^{3+} + H_2O \tag{2}$$

where cytochrome c_red is represented by Fe^{2+} and cytochrome c_ox is represented by Fe^{3+}. For this reaction, $m = 2$ and the $n = 2$.

The number of reductant molecules involved can be as large as 6, as shown by ferridoxin-nitrite reductase.

EC 1.7.7.1 ferredoxin-nitrite reductase

$$\text{nitrite} + 6\,\text{ferredoxin}_\text{red} = \text{ammonia} + 6\,\text{ferredoxin}_\text{ox} + 2H_2O \tag{3}$$

A chemical reference reaction is

$$NO_2^- + 6\,Fe^{2+} + 8H^+ = NH_4^+ + 6\,Fe^{3+} + 2H_2O \tag{4}$$

For this reaction, $m = 6$ and $n = 8$. It is possible that $n = 7$ or 9 because n can only be determined from kinetic measurements.

Section 1.7 discusses the change in binding of hydrogen ions $\Delta_r N_H$ in a biochemical reaction. This is a thermodynamic property of the reaction that is catalyzed. $\Delta_r N_H$ can be calculated in two ways. When the apparent equilibrium constant K' of an enzyme-catalyzed reaction is known as a function of pH, the change in binding of hydrogen ions in an enzyme-catalyzed reaction can be calculated using

$$\Delta_r N_H = -\text{dlog}K'/\text{dpH} \tag{5}$$

The values of $\Delta_r N_H$ at pHs 5, 6, 7, 8, and 9 have been calculated [23] for about 200 enzyme-catalyzed reactions.

The second way to calculate $\Delta_r N_H$ only requires knowing the pKs of the reactants in the pH range of interest and their molecular formulas. The average number of hydrogen atoms in a biochemical reactant is given by

$$\overline{N}_H(i) = \sum r_j N_H(j) \tag{6}$$

where r_j is the equilibrium mole fraction of species j in reactant i and $N_H(j)$ is the number of hydrogen atoms in species j. The change in binding of hydrogen ions is given by

$$\Delta_r N_H = \sum v_i ' \overline{N}_H(i) \tag{7}$$

where v_i' is the stoichiometric number for reactant i in the catalyzed reaction. The values of $\Delta_r N_H$ at pHs 5, 6, 7, 8, and 9 have been calculated using equation 7 for about 200 enzyme-catalyzed reactions, and they are in agreement with the values calculated using equation 5. Values of $\Delta_r N_H$ at a desired pH can be calculated for most of the reactions in the EC list [Web2] because at least approximate pKs are known for most substrates.

7.2 Mechanism I for ordered O + mR → products in which mR is bound first

▪ 7.2.1 Derivation of the rate equation when mR is bound first

There are two possible ordered mechanisms for O + mR → products with different orders of binding of 2R and O [29]. In mechanism I, mR are bound first:

E + mR = ER$_m$	$K_{IRm}{}^m = [E][R]^m / [ER_m]$	kIRm^m = e*r^m/erm kIRm = 20	(8)
ER$_m$ + O = ER$_m$O	$K_O = [ER_m][O] / [ER_m O]$	kO = erm*o/ermo = 5	(9)
ER$_m$O → products	v = kf*ermo = kf*et*(ermo/et) = kfet*(ermo/et)	kfet = 1	(10)

This mechanism applies at a specified pH. The total concentration of enzymatic sites is given by

$$et == e + erm + ermo \tag{11}$$

The values of the kinetic parameters shown here have been arbitrarily chosen for test calculations. The rapid-equilibrium rate equation for mechanism I can be derived using Solve.

```
Solve[{kIRm^m == e * r^m / erm, kO == erm * o / ermo, et == e + erm + ermo}, {ermo}, {e, erm}]
```

$$\left\{\left\{ermo \to \frac{et\, o\, r^m}{kIRm^m\, kO + kO\, r^m + o\, r^m}\right\}\right\}$$

The expression for the velocity is obtained by multiplying ermo by kf.

This rate equation is named vI.

$$vI = \frac{kfet\, o\, r^m}{kIRm^m\, kO + kO\, r^m + o\, r^m};$$

The symbol vI indicates that mR is bound first.

This rate equation can also be written in text.

$$v = \frac{kfet}{1 + \frac{K_O}{[O]}\left(1 + \frac{K_{IRm}^m}{[R]^m}\right)} \tag{12}$$

Calculation of velocities when m = 2 and the following values are chosen for the kinetic parameters: kIRm = 20, kO = 5, and kfet = 1

```
vI /. kIRm → 20 /. kO → 5 /. kfet → 1 /. m → 2
```

$$\frac{o\,r^2}{2000 + 5\,r^2 + o\,r^2}$$

This velocity can be shown as a function of [O] and [R] by making a 3D plot with *m* = 2.

```
plot1 = Plot3D[Evaluate[vI /. kfet -> 1 /. kO → 5 /. kIRm → 20 /. m → 2],
    {o, 0.00001, 20}, {r, 0.00001, 50}, ViewPoint → {-2, -2, 1}, PlotRange -> {0, 1},
    Lighting → {{"Ambient", GrayLevel[1]}}, AxesLabel → {"[O]", "[R]"}, PlotLabel → "vIm2"];

plot2 = Plot3D[Evaluate[D[vI /. kfet -> 1 /. kO → 5 /. kIRm → 20 /. m → 2, o]],
    {o, 0.00001, 20}, {r, 0.00001, 50}, ViewPoint → {-2, -2, 1},
    PlotRange -> {0, .2}, Lighting → {{"Ambient", GrayLevel[1]}},
    AxesLabel → {"[O]", "[R]"}, PlotLabel → "dvIm2/d[O]"];

plot3 = Plot3D[Evaluate[D[vI /. kfet -> 1 /. kO → 5 /. kIRm → 20 /. m → 2, r]],
    {o, 0.00001, 20}, {r, 0.00001, 50}, ViewPoint → {-2, -2, 1},
    PlotRange -> {0, .06}, Lighting → {{"Ambient", GrayLevel[1]}},
    AxesLabel → {"[O]", "[R]"}, PlotLabel → "dvIm2/d[R]"];

figlord = GraphicsArray[{{plot1, plot2, plot3}}]
```

Fig. 7.1 (a) Velocity for mechanism I with *m* = 2, kfet = 1, kIR2 = 20, and kO = 5. Notice that the velocity versus [R] at constant [P] are sigmoid. (b) dvIm2/d[O]. This plot shows that the velocity is most sensitive to kO at high [R] and low [O]. (c) dvOm2/d[R]. This plot shows that the velocity is most sensitive to kIRm at high [O] and low [R], but not too low.

Calculation of velocities when m = 3 and the following values are chosen for the kinetic parameters: kIRm = 20, kO = 5, and kfet = 1

The velocity when *m* = 3 can be shown as a function of [O] and [R] by making a 3D plot .

```
vI /. kIRm → 20 /. kO → 5 /. kfet → 1 /. m → 3
```

$$\frac{o\,r^3}{40\,000 + 5\,r^3 + o\,r^3}$$

This velocity can be shown as a function of [O] and [R] by making a 3D plot.

```
plot4 = Plot3D[Evaluate[vI /. kfet -> 1 /. kO → 5 /. kIRm → 20 /. m → 3],
    {o, 0.00001, 20}, {r, 0.00001, 50}, ViewPoint → {-2, -2, 1}, PlotRange -> {0, 1},
    Lighting → {{"Ambient", GrayLevel[1]}}, AxesLabel → {"[O]", "[R]"}, PlotLabel → "vIm3"];

plot5 = Plot3D[Evaluate[D[vI /. kfet -> 1 /. kO → 5 /. kIRm → 20 /. m → 3, o]],
    {o, 0.00001, 20}, {r, 0.00001, 50}, ViewPoint → {-2, -2, 1},
    PlotRange -> {0, .2}, Lighting → {{"Ambient", GrayLevel[1]}},
    AxesLabel → {"[O]", "[R]"}, PlotLabel → "dvIm3/d[O]"];

plot6 = Plot3D[Evaluate[D[vI /. kfet -> 1 /. kO → 5 /. kIRm → 20 /. m → 3, r]],
    {o, 0.00001, 20}, {r, 0.00001, 50}, ViewPoint → {-2, -2, 1},
    PlotRange -> {0, .06}, Lighting → {{"Ambient", GrayLevel[1]}},
    AxesLabel → {"[O]", "[R]"}, PlotLabel → "dvIm/d[R]"];
```

```
fig2ord = GraphicsArray[{{plot4, plot5, plot6}}]
```

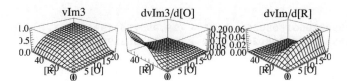

Fig. 7.2 (a) Velocity for mechanism I with *m* = 3, kfet = 1, kIR2 = 20, and kO = 5. The sigmoid character of this plot is increased. (b) dvIm3/d[O]. This plot shows that the velocity is most sensitive to kO at high [R] and low [O]. (c) dvOm3/d[R]. This plot shows that the velocity is most sensitive to kIRm at high [O] and low [R], but not too low.

■ **7.2.2 Use of Solve to estimate the kinetic parameters with the minimum number of velocity measurements for mechanism I for ordered O + mR → products**

Since there are 3 kinetic parameters, 3 velocities must be measured [30,32]. In the first test, velocities are calculated at the following pairs of substrate concentrations: {o,r} at {100,100}, {1,100}, {100,1} with *m* = 2.

```
vI /. kIRm → 20 /. kO → 5 /. kfet → 1 /. o → 100 /. r → 100 /. m → 2 // N
```

```
0.95057
```

```
vI /. kIRm → 20 /. kO → 5 /. kfet → 1 /. o → 1 /. r → 100 /. m → 2 // N
```

```
0.16129
```

```
vI /. kIRm → 20 /. kO → 5 /. kfet → 1 /. o → 100 /. r → 1 /. m → 2 // N
```

```
0.0475059
```

The velocity at {5,5} is used in the next chapter (Section 8.1.3).

```
vI /. kIRm → 20 /. kO → 5 /. kfet → 1 /. o → 5 /. r → 5 /. m → 2 // N
```

```
0.0555556
```

The following program was written in *Mathematica* to calculate the values of kfet, kO, and kIR2 from three measured velocities at three pairs of substrate concentrations on the assumption that the mechanism is ordered, with 2R being bound first.

```
calckinparsordORI[v1_, o1_, r1_, v2_, o2_, r2_, v3_, o3_, r3_] :=
 Module[{}, (*This program calculates kfet, kO,
   and kIR2 from three experimental velocities for O + 2R → products at three pairs
     of substrate concentrations on the assumption that the mechanism is ordered,
   with 2R bound first.  The first velocity is at high [O] and high [R],
   the second velocity is at low [O] and high [R],
   and the third velocity is at high [O] and low [R].*)
```

$$
\text{Solve}\left[\left\{v1 == \frac{\text{kfet o1 r1}^2}{\text{kIR2}^2 \text{ kO} + \text{kO r1}^2 + \text{o1 r1}^2}, \; v2 == \frac{\text{kfet o2 r2}^2}{\text{kIR2}^2 \text{ kO} + \text{kO r2}^2 + \text{o2 r2}^2},\right.\right.
$$

$$
\left.\left. v3 == \frac{\text{kfet o3 r3}^2}{\text{kIR2}^2 \text{ kO} + \text{kO r3}^2 + \text{o3 r3}^2}\right\}, \{\text{kfet, kIR2, kO}\}\right]\right]
$$

```
calckinparsordORI[.9506, 100, 100, .1613, 1, 100, .04751, 100, 1]
```

```
{{kfet → 1.00003, kO → 4.99982, kIR2 → -19.9998}, {kfet → 1.00003, kO → 4.99982, kTR2 → 19.9998}}
```

There are two solutions. The second is correct. The -20 is mathematically correct because the rate constant involved is $kIR2^2$. The dimensions of the output are calculated.

```
Dimensions[
 {{kfet → 1.0000293964188307`, kO → 4.9998220601808026`, kIR2 → -19.99976631994697`},
  {kfet → 1.0000293964188307`, kO → 4.9998220601808026`, kIR2 → 19.99976631994697`}}]
```

{2, 3}

The positive value of kIR2 is obtained as follows:

```
Take[calckinparsordORI[.9506, 100, 100, .1613, 1, 100, .04751, 100, 1], {2}]
```

{{kfet → 1.00003, kO → 4.99982, kIR2 → 19.9998}}

```
line1 =
 {kfet, kO, kIR2} /. Take[calckinparsordORI[.9506, 100, 100, .1613, 1, 100, .04751, 100, 1], {2}]
```

{{1.00003, 4.99982, 19.9998}}

Put in 5% errors in velocities, one at a time.

```
calckinparsordORI[1.05 * .9506, 100, 100, .1613, 1, 100, .04751, 100, 1]
```

{{kfet → 1.05333, kO → 5.31907, kIR2 → -19.9252}, {kfet → 1.05333, kO → 5.31907, kIR2 → 19.9252}}

```
line2 = {kfet, kO, kIR2} /.
 Take[calckinparsordORI[1.05 * .9506, 100, 100, .1613, 1, 100, .04751, 100, 1], {2}]
```

{{1.05333, 5.31907, 19.9252}}

```
calckinparsordORI[.9506, 100, 100, 1.05 * .1613, 1, 100, .04751, 100, 1]
```

{{kfet → 0.997056, kO → 4.68763, kIR2 → -20.6243},
 {kfet → 0.997056, kO → 4.68763, kIR2 → 20.6243}}

```
line3 = {kfet, kO, kIR2} /.
 Take[calckinparsordORI[.9506, 100, 100, 1.05 * .1613, 1, 100, .04751, 100, 1], {2}]
```

{{0.997056, 4.68763, 20.6243}}

```
calckinparsordORI[.9506, 100, 100, .1613, 1, 100, 1.05 * .04751, 100, 1]
```

{{kfet → 1.00003, kO → 5.00985, kIR2 → -19.4726}, {kfet → 1.00003, kO → 5.00985, kIR2 → 19.4726}}

```
line4 = {kfet, kO, kIR2} /.
 Take[calckinparsordORI[.9506, 100, 100, .1613, 1, 100, 1.05 * .04751, 100, 1], {2}]
```

{{1.00003, 5.00985, 19.4726}}

Table 7.1 Values of kinetic constants for mechanism I calculated using velocities at {100,100}, {1,100}, {100,1}

```
TableForm[Round[{line1[[1]], line2[[1]], line3[[1]], line4[[1]]}, 0.01],
 TableHeadings -> {{"No errors", "1.05*v1", "1.05*v2", "1.05*v3"}, {"kfet", "kO", "kIR2"}}]
```

	kfet	kO	kIR2
No errors	1.	5.	20.
1.05*v1	1.05	5.32	19.93
1.05*v2	1.	4.69	20.62
1.05*v3	1.	5.01	19.47

The accuracy of the estimates of kinetic parameters can be improved by using a wider range of substrate concentrations.

calckinparsordORI was written for $m = 2$, but it is possible to write a similar program in which m can be specified. The property m is not a kinetic parameter, but rather a property of the reaction being catalyzed.

The following program was written in *Mathematica* to calculate the values of kfet, kO, and kIRm from three measured velocities at three pairs of substrate concentrations on the assumption that the mechanism is ordered, with R being bound first and m being specified.

```
calckinparsordORIm2[m_, v1_, o1_, r1_, v2_, o2_, r2_, v3_, o3_, r3_] :=
 Module[{}, (*This program calculates kfet, kO,
   and kIRm from three experimental velocities for O + mR →
    products at three pairs of substrate concentrations on the assumption that
    the mechanism is ordered.  The first velocity is at high [O] and high [R],
    the second velocity is at low [O] and high [R],
    and the third velocity is at high [O] and low [R](but not too low).*)
```

$$\text{Solve}\left[\left\{v1 == \frac{\text{kfet } o1 \; r1^m}{\text{kIRm}^m \; kO + kO \; r1^m + o1 \; r1^m}, \; v2 == \frac{\text{kfet } o2 \; r2^m}{\text{kIRm}^m \; kO + kO \; r2^m + o2 \; r2^m},\right.\right.$$

$$\left.\left. v3 == \frac{\text{kfet } o3 \; r3^m}{\text{kIRm}^m \; kO + kO \; r3^m + o3 \; r3^m}\right\}, \{\text{kfet}, \text{kIRm}, kO\}\right]\right]$$

This program is applied to the velocities calculated above.

```
calckinparsordORIm2[2, .9506, 100, 100, .1613, 1, 100, .04751, 100, 1]
```

{{kfet → 1.00003, kO → 4.99982, kIRm → -19.9998}, {kfet → 1.00003, kO → 4.99982, kIRm → 19.9998}}

There are two solutions. The second is correct. The -20 is correct mathematically because the rate constant involved is $kIR2^2$.

7.3 Effects of pH on mechanism I

■ **7.3.1 Mechanism I with p*K*s and the consumption of hydrogen ions**

The effects of pH on the velocity of the forward reaction can be discussed by including two p*K*s for the enzymatic site, each enzyme-substrate complex, but no p*K*s for R and O. This mechanism also includes the consumption of hydrogen ions in the rate-determining reaction This mechanism is presumably adequate to discuss reactions like nitrite reductase (EC 1.9.6.1) and ferredodin-nitrate reductase (EC 1.7.7.1). However, the consumption of hydrogen ions in the rate-determining reaction is only one possibility. The other possibilities are that a hydrogen ion can be consumed in the reaction prior to the rate-determining reaction of in the first reaction in the mechanism [37].

$$
\begin{array}{ll}
E^- & ER_m{}^- \\
\| & \| \\
HE + mR \;=\; HER_m \qquad & K_{HERm}^m = \dfrac{[HE]\,[R]^m}{[HER_m]} \\
\| & \| \\
H_2\,E^+ & H_2\,ER_m{}^+
\end{array}
\tag{13}
$$

$$
\begin{array}{ll}
ER_m{}^- & ER_m\,O^- \\
\| & \| \\
HER_m + O \;=\; HER_m\,O \qquad & K_{HERmO} = \dfrac{[HER_m]\,[O]}{[HER_m\,O]} \\
\| & \| \\
H_2\,ER_m{}^+ & H_2\,ER_m\,O^+
\end{array}
\tag{14}
$$

$$
\begin{array}{l}
ER_m\,O^- \\
\| \qquad\qquad k_f \\
|\,n\,|\;H^+ + HER_m\,O \;\rightarrow \\
\| \\
H_2\,ER_m\,O^+
\end{array}
\tag{15}
$$

k_f is the rate constant for the rate-determining reaction. The absolute value signs have to be put on n in reaction 15 because n is the number of hydrogen ions produced; in this case, n is negative. The chemical equilibrium constant expressions in equations 13 and 14 are always obeyed. More information on the use of mechanism like this is given in reference [33]

When $n = 0$, the limiting velocity of the forward reaction is given by

$$
V_f \;=\; \frac{k_f\,[E]_t}{1 + 10^{pH - pK1ERmO} + 10^{-pH + K2ERmO}}
\tag{16}
$$

where $[E]_t$ is the concentration of enzymatic sites, bound and unbound. Note $pK_1 > pK_2$. When n is not equal to zero, k_f is replaced $k_f [H^+]^{-n} = k_f 10^{npH}$ since $[H^+] = 10^{-pH}$. Thus V_{fexp} is given by

$$V_{fexp} = \frac{k_f 10^{npH} [E]_t}{1 + 10^{pH-pK1ERmO} + 10^{-pH+K2ERmO}} = 10^{npH} V_f \tag{17}$$

Since n is negative for the reactions considered here, V_{fexp} increases rapidly when the pH is reduced. The effect of npH is very large because n is in the exponent. The pH dependencies of the two Michaelis constants are given by

$$K_{IRm}^m = \frac{K_{HERm}^m \left(1 + 10^{pH-pK1E} + 10^{-pH+K2E}\right)}{1 + 10^{pH-pK1ERm} + 10^{-pH+K2ERm}} \tag{18}$$

$$K_O = \frac{K_{HERmO} \left(1 + 10^{pH-pK1ERm} + 10^{-pH+K2ERm}\right)}{1 + 10^{pH-pK1ERmO} + 10^{-pH+K2ERmO}} \tag{19}$$

These expressions for the 3 kinetic parameters can be substituted in the rate equation in Section 7.2.1.

The rate equation can also be written as

$$v = \frac{V_{fexp}}{1 + \frac{K_O}{[O]} \left(1 + \frac{K_{IRm}^m}{[R]^m}\right)} = \frac{10^{npH} V_f}{1 + \frac{K_O}{[O]} \left(1 + \frac{K_{IRm}^m}{[R]^m}\right)} \tag{20}$$

There has to be a 10^{npH} factor here because the apparent equilibrium constant K' has to change very rapidly with pH over the whole range of pH, and this cannot be accomplished with pKs. The values of V_{fexp}, K_O, and K_{Rm}^m can be determined at several pH values. Since the pH dependencies of K_O and K_{IRm}^m each involve five unknowns, it is convenient to determine the kinetic parameters in equations 18 and 19 by plotting V_f/K_O and $V_f/K_O K_{IRm}^m$ versus pH as shown by the following two equations:

$$V_f/K_O = \frac{k_f [E]_t}{K_{HERmO} \left(1 + 10^{pH-pK1ERm} + 10^{-pH+pK2ERm}\right)} \tag{21}$$

$$V_f/K_O K_{IRm}^m = \frac{k_f [E]_t}{K_{HERmO} K_{HERm}^m \left(1 + 10^{pH-pK1E} + 10^{-pH+pK2E}\right)} \tag{22}$$

The following computer program **calcpropsRmO** derives the expressions for the pH dependencies of the kinetic parameters for Mechanism I and the rapid-equilibrium velocity when m and n are known.

```
calcpropsRmO[pK1e_, pK2e_, pK1erm_,
   pK2erm_, pK1ermo_, pK2ermo_, kfEt_, kerm_, kermo_, m_, n_] :=
Module[{efactor, ermfactor, ermofactor,
    vf, krm, krmo, vfexp, v},
  (*This program derives kinetic parameters for the initial velocity of the forward
      reaction mR+O→products.  The reductant R binds first.  The oxidant O binds
      second.  This program also derives the rate equation for the forward reaction.  o
      and r are the concentrations of the oxidant and the reductant.  m is the
      stoichiometric number of R molecules needed to reduce the oxidant in the
      biochemical reaction.  n is the number of hydrogen ions produced in the rate-
     determining step.  The output is a list of seven functions.*)
  efactor = 1 + 10^pK2e-pH + 10^pH-pK1e;
  ermfactor = 1 + 10^pK2erm-pH + 10^pH-pK1erm;
  ermofactor = 1 + 10^pK2ermo-pH + 10^pH-pK1ermo;
  vf = kfEt / ermofactor;
  vfexp = (10 ^ (n * pH)) * vf;
  krm = kerm * (efactor / ermfactor) ^ (1 / m);
  krmo = kermo * ermfactor / ermofactor;
  v = vfexp / (1 + (krmo / o) * (1 + (krm / r) ^m));
  {vfexp, vf, krm, krmo, vf / krmo, vf / (krmo * krm^m), v}]
```

■ 7.3.2 Plots of kinetic parameters for a nitrate reductase reaction with mechanism I

To test the equations that are derived by using calcpropsRmO, the following values are assumed for the kinetic parameters:

pK1e	7.5	pK2e	6.5
pK1erm	7.0	pK2erm	6.0
pK1ermo	8.0	pK2ermo	7.0
kfEt	10^(2*7.5)=10^15		
kerm	1	kermo	2
m	2	n	-2

The pKs, chemical equilibrium constants (kerm and kermo), and rate constant (kfEt) used here are arbitrary. These calculations can be used for any enzyme-catalyzed reaction mR + O -> products with Mechanism I. An example of a reaction that might follow Mechanism I is nitrate reductase (EC 1.9.6.1). Note kfEt is taken to be 10^(2*7.5) so that V_{fexp} will be of the order of 1 at pH 7.5.

The seven mathematical functions for properties of the ordered mechanism when mR is bound first are derived by putting the input of properties in calcpropsRmO[pK1e_, pK2e_, pK1erm_, pK2erm_, pK1ermo_, pK2ermo_, kfEt_, kerm_, kermo_, m_, n_]

rxIpH = calcpropsRmO[7.5, 6.5, 7.0, 6.0, 8.0, 7.0, 10 ^ (2 * 7.5), 1, 2, 2, -2]

$$\left\{ \frac{1.\times 10^{15}\ 10^{-2\,\text{pH}}}{1+10^{7.-\text{pH}}+10^{-8.+\text{pH}}},\ \frac{1.\times 10^{15}}{1+10^{7.-\text{pH}}+10^{-8.+\text{pH}}},\ \sqrt{\frac{1+10^{6.5-\text{pH}}+10^{-7.5+\text{pH}}}{1+10^{6.-\text{pH}}+10^{-7.+\text{pH}}}}, \right.$$

$$\frac{2\left(1+10^{6.-\text{pH}}+10^{-7.+\text{pH}}\right)}{1+10^{7.-\text{pH}}+10^{-8.+\text{pH}}},\ \frac{5.\times 10^{14}}{1+10^{6.-\text{pH}}+10^{-7.+\text{pH}}},\ \frac{5.\times 10^{14}}{1+10^{6.5-\text{pH}}+10^{-7.5+\text{pH}}},$$

$$\left. \frac{1.\times 10^{15}\ 10^{-2\,\text{pH}}}{\left(1+10^{7.-\text{pH}}+10^{-8.+\text{pH}}\right)\left(1+\frac{2\left(1+10^{6.-\text{pH}}+10^{-7.+\text{pH}}\right)\left(1+\frac{1+10^{6.5-\text{pH}}+10^{-7.5+\text{pH}}}{\left(1+10^{6.-\text{pH}}+10^{-7.+\text{pH}}\right)r^2}\right)}{\left(1+10^{7.-\text{pH}}+10^{-8.+\text{pH}}\right)\text{o}}\right)} \right\}$$

Note that $V_f = 10^{2 \, pH} \, V_{fexp}$. Note the square root in the third function (K_{Rm}).

   Plots of the functions of pH for the two limiting velocities, two Michaelis constants, and two ratios that are useful for calculating pKs and chemical equilibrium constants from bell-shaped plots are constructed as follows:

```
plot1a = Plot[Evaluate[Log[10, rxIpH[[1]]]],
    {pH, 5, 9}, AxesLabel → {"pH", "logV_fexp"}, DisplayFunction → Identity];

plot2a = Plot[rxIpH[[2]] * 10^-14, {pH, 5, 9},
    AxesLabel → {"pH", "V_f×10^-14"}, DisplayFunction → Identity];

plot3a = Plot[rxIpH[[3]], {pH, 5, 9}, AxesLabel → {"pH", "K_IR2"}, DisplayFunction → Identity];

plot4a = Plot[rxIpH[[4]], {pH, 5, 9}, AxesLabel → {"pH", "K_O"}, DisplayFunction → Identity];

plot5a = Plot[rxIpH[[5]] * 10^-14, {pH, 5, 9},
    AxesLabel → {"pH", " (V_f × 10^-14) / K_O "}, DisplayFunction → Identity];

plot6a = Plot[rxIpH[[6]] * 10^-14, {pH, 5, 9},
    AxesLabel → {"pH", " (V_f × 10^-14) / (K_O K_IR2^2) "}, DisplayFunction → Identity];
```

figlMN = Show[GraphicsGrid[{{plot1a, plot2a}, {plot3a, plot4a}, {plot5a, plot6a}}]]

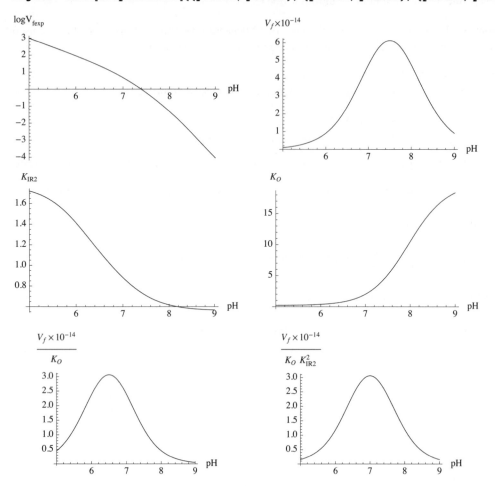

Fig. 7.3 Plots of pH dependencies of kinetic parameters when $m = 2$ and $n = -2$, as in the nitrate reductase reaction (EC 1.9.6.1). The second plot yields $pK_{1\ ER2O}$, $pK_{2\ ER2O}$, and kfEt. The fifth plot yields $pK_{1\ ER2}$, $pK_{2\ ER2}$, and K_{ER2O}. The sixth plot yields $pK_{1\ E}$, $pK_{2\ E}$, and K_{ER2}.

The reaction velocity v is given by

rxIpH[[7]]

$$
\frac{1. \times 10^{15}\ 10^{-2\ pH}}{\left(1 + 10^{7.-pH} + 10^{-8.+pH}\right)\left(1 + \dfrac{2\left(1+10^{6.-pH}+10^{-7.+pH}\right)\left(1+\dfrac{1+10^{6.5-pH}+10^{-7.5+pH}}{\left(1+10^{6.-pH}+10^{-7.+pH}\right) r^2}\right)}{\left(1+10^{7.-pH}+10^{-8.+pH}\right) o}\right)}
$$

This function can be treated like an experimental system in the sense that v can be calculated at any [R], [O], and pH. The rate equations at pHs 5, 6, 7, 8, and 9 are given by

rxIpH[[7]] /. pH → {5.0, 6.0, 7.0, 8.0, 9.0}

$$
\left\{\frac{990.089}{1+\dfrac{0.218018\left(1+\frac{2.9633}{r^2}\right)}{o}},\ \frac{90.8265}{1+\dfrac{0.381471\left(1+\frac{1.9971}{r^2}\right)}{o}},\ \frac{4.7619}{1+\dfrac{2.\left(1+\frac{0.77736}{r^2}\right)}{o}},\ \frac{0.047619}{1+\dfrac{10.4857\left(1+\frac{0.380917}{r^2}\right)}{o}},\ \frac{0.0000908265}{1+\dfrac{18.3471\left(1+\frac{0.323026}{r^2}\right)}{o}}\right\}
$$

Note that the Michaelis constant K_O is equal to $0.21, 0.38, 2.0, 10.5$, and 18.4 at these pHs. The Michaelis constant K_{IRm} at these pHs is given by

```
{2.96, 2.00, .78, .38, .32}^.5

{1.72047, 1.41421, 0.883176, 0.616441, 0.565685}
```

The easiest property to calculate from the rate equation is V_{fexp} because it can be obtained by raising [R] and [O] to infinity.

```
rxIpH[[7]] /. r → ∞ /. o → ∞
```

$$\frac{1. \times 10^{15}\ 10^{-2\ pH}}{1 + 10^{7. - pH} + 10^{-8. + pH}}$$

This agrees with the expression in rx1961[[1]].

V_{fexp} is very sensitive to a change of even $\Delta pH = 0.1$.

Calculation of % errors in V_{fexp} introduced by $\Delta pH = 0.1$ at pHs $5, 6, 7, 8$, and 9

```
rxIpH[[7]] /. r → ∞ /. o → ∞ /. pH → {5.0, 6.0, 7.0, 8.0, 9.0}

{990.089, 90.8265, 4.7619, 0.047619, 0.0000908265}

rxIpH[[7]] /. r → ∞ /. o → ∞ /. pH → {5.1, 6.1, 7.1, 8.1, 9.1}

{784.44, 70.4518, 3.28586, 0.0269829, 0.0000464035}

((rxIpH[[7]] /. r → ∞ /. o → ∞ /. pH → {5.0, 6.0, 7.0, 8.0, 9.0}) /
    (rxIpH[[7]] /. r → ∞ /. o → ∞ /. pH → {5.1, 6.1, 7.1, 8.1, 9.1}) - 1) * 100

{26.216, 28.9201, 44.9212, 76.4785, 95.7321}
```

The percentage error in V_{fexp} caused by an error of 0.1 in the pH for $O + 2R \rightarrow$ is 26% at pH 5, 29% at pH 6, 45% at pH 7, 76% at pH 8, 96% at pH 9.

■ 7.3.3 3D plots of rates for the nitrate reductase type of reaction with mechanism I

An overview of the rate equation can be obtained by plotting v at a specified pH versus [O] and [R] in three-dimensions.

```
plot1bnew5 = Plot3D[Evaluate[rxIpH[[7]]/100 /. pH → 5.`], {r, 0.0001`, 5}, {o, 0.0001`, 5},
    ViewPoint → {-2, -2, 1}, AxesLabel → {"    [R]", "[O]    ", "  v/100"}, PlotRange → {0, 12},
    Lighting → {{"Ambient", GrayLevel[1]}}, PlotLabel → "pH 5.0", DisplayFunction → Identity];

plot2bnew6 = Plot3D[Evaluate[rxIpH[[7]]/10 /. pH → 6.`], {r, 0.0001`, 5}, {o, 0.0001`, 10},
    ViewPoint → {-2, -2, 1}, AxesLabel → {"    [R]", "[O]    ", "v/10"}, PlotRange → {0, 10},
    Lighting → {{"Ambient", GrayLevel[1]}}, PlotLabel → "pH 6.0", DisplayFunction → Identity];

plot3bnew7 = Plot3D[Evaluate[rxIpH[[7]] /. pH → 7.`], {r, 0.0001`, 5}, {o, 0.0001`, 10},
    ViewPoint → {-2, -2, 1}, AxesLabel → {"    [R]", "[O]    ", "v"}, PlotRange → {0, 5.5`},
    Lighting → {{"Ambient", GrayLevel[1]}}, PlotLabel → "pH 7.0", DisplayFunction → Identity];

plot4bnew8 = Plot3D[Evaluate[rxIpH[[7]] 100 /. pH → 8.`], {r, 0.0001`, 5}, {o, 0.0001`, 50},
    ViewPoint → {-2, -2, 1}, AxesLabel → {"    [R]", "[O]    ", "v×100"}, PlotRange → {0, 5},
    Lighting → {{"Ambient", GrayLevel[1]}}, PlotLabel → "pH 8.0", DisplayFunction → Identity];
```

```
fig2MNnew = GraphicsGrid[{{plot1bnew5, plot2bnew6}, {plot3bnew7, plot4bnew8}}]
```

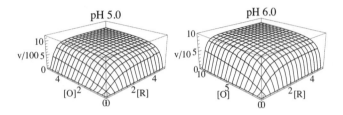

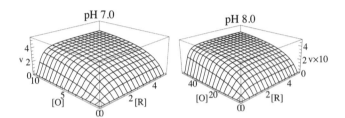

Fig. 7.4 Plots of velocities v of the catalyzed reaction 2R + O → products as functions of [O] and [R] at four pHs for mechanism I. The number n of hydrogen ions produced in the rate-determining reaction for this ordered mechanism is -2 and $m = 2$. Note that plots of v versus [R] at constant [O] are sigmoid. Note that the scale of the ordinate changes by a factor of 10^4 between pH 5 and pH 8.

These plots show the large effect of the pH on v for given [O] and [R]. This is due to the $10^{-2 \, pH}$ factor in the rate equation for the forward reaction. The plots show the sigmoid shape of the plots of v versus [R] at constant [O]. A sigmoid plot of v versus the concentration is usually taken to be an indication of allosterism. This can arise when there is positive cooperativity between active sites of a polymeric enzyme. But a sigmoid plot can have other origins. In this case, it is caused by the stoichiometric number of R in the biochemical equation for the forward reaction: O + 2R → products. The effects of changing the pH by 0.1 are very large over the whole range of pH and substrate concentrations.

7.4 Mechanism II for ordered O + 2R → products in which O is bound first

▪ 7.4.1 Derivation of the rate equation when O is bound first

In mechanism II, O is bound first [29].

E + O = EO	$K_{IO} = [E][O] / [EO]$	kIO = e*o/eo = 20		(23)
EO + 2R = ER$_2$O	$K_{R2}^2 = [EO][R]^2 / [ER_2 O]$	kR2^2 = eo*r^2/er2o	kR2 = 10	(24)
ER$_2$O → products	v = kf*er2o = kf*et*(er2o/et) = kfet*(er2o/et)	kfet = 1		(25)

The values of the 3 kinetic parameters have been chosen arbitrarily for test calculations. The total concentration of enzymatic sites is given by

et = e + eo + er2o   (26)

The rapid-equilibrium rate equation for mechanism II can be derived using Solve.

```
Solve[{kIO == e * o / eo, kR2^2 == eo * r^2 / er2o, et == e + eo + er2o}, {er2o}, {e, eo}]
```

$$\left\{\left\{er2o \to \frac{et\ o\ r^2}{kIO\ kR2^2 + kR2^2\ o + o\ r^2}\right\}\right\}$$

The rapid-equilibrium rate equation for mechanism II is given by

$$vII = \frac{kfet\ o\ r^2}{kIO\ kR2^2 + kR2^2\ o + o\ r^2}$$

$$\frac{kfet\ o\ r^2}{kIO\ kR2^2 + kR2^2\ o + o\ r^2}$$

This rate equation can also be written in a more familiar way in text.

$$vII = \frac{V_{fexp}}{1 + \frac{K_{Rm}^m}{[R]^m} + \frac{K_{IO}\ K_{Rm}^m}{[O]\ [R]^m}} = \frac{V_{fexp}}{1 + \frac{K_{Rm}^m}{[R]^m}\left(1 + \frac{K_{IO}}{[O]}\right)} \tag{27}$$

Ths rate equation can be compared with equation 12. These two rate equations are definitely different, and so the order of binding is significant. This is different from ordered A + B → products, where A and B play the same roles.

Substitute the kinetic parameters for mechanism II:

```
vII /. kfet -> 1 /. kIO → 20 /. kR2 → 10
```

$$\frac{o\ r^2}{2000 + 100\ o + o\ r^2}$$

The velocity for mechanism II is shown in a 3D plot.

```
plotII1 = Plot3D[Evaluate[vII /. kfet -> 1 /. kIO → 20 /. kR2 → 10],
    {o, 0.00001, 1}, {r, 0.00001, 200}, ViewPoint → {-2, -2, 1}, PlotRange -> {0, 1},
    Lighting → {{"Ambient", GrayLevel[1]}}, AxesLabel → {"[O]", "[R]"}, PlotLabel -> "vII"];

plotII2 = Plot3D[Evaluate[D[vII /. kfet -> 1 /. kIO → 20 /. kR2 → 10, o]],
    {o, 0.00001, 1}, {r, 0.00001, 200}, ViewPoint → {-2, -2, 1},
    PlotRange -> {0, 5}, Lighting → {{"Ambient", GrayLevel[1]}},
    AxesLabel → {"[O]", "[R]"}, PlotLabel -> "dvII/d[O]"];

plotII3 = Plot3D[Evaluate[D[vII /. kfet -> 1 /. kIO → 20 /. kR2 → 10, r]],
    {o, 0.00001, 1}, {r, 0.00001, 200}, ViewPoint → {-2, -2, 1},
    PlotRange -> {0, .02}, Lighting → {{"Ambient", GrayLevel[1]}},
    AxesLabel → {"[O]", "[R]"}, PlotLabel -> "dvII/d[R]"];

fig3ord = GraphicsArray[{{plotII1, plotII2, plotII3}}]
```

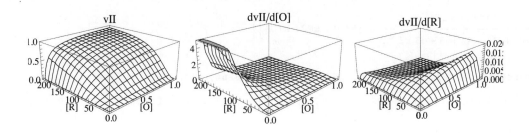

Fig. 7.5 (a) Velocity for mechanism II with kfet = 1, kR2 = 10, and kIO = 20. (b) dvII/d[O]. (c) dvII/d[R].

Since there are 3 kinetic parameters, 3 velocities must be measured. In the test of the estimation of kinetic parameters, the following pairs of velocities are measured: {o,r} at {100,100}, {1,100}, {100,1}.

```
vII /. kfet -> 1 /. kIO → 20 /. kR2 → 10
```

$$\frac{o\, r^2}{2000 + 100\, o + o\, r^2}$$

```
vII /. kfet -> 1 /. kIO → 20 /. kR2 → 10 /. o → 100 /. r → 100 // N
```

```
0.988142
```

```
vII /. kfet -> 1 /. kIO → 20 /. kR2 → 10 /. o → 1 /. r → 100 // N
```

```
0.826446
```

```
vII /. kfet -> 1 /. kIO → 20 /. kR2 → 10 /. o → 100 /. r → 1 // N
```

```
0.00826446
```

The last velocity is quite low, but the enzyme concentration can be increased, or a higher [R] can be used.

The following velocity is needed in the next chapter.

```
vII /. kfet -> 1 /. kIO → 20 /. kR2 → 10 /. o → 5 /. r → 5 // N
```

```
0.047619
```

■ **7.4.2 Use of Solve to estimate the kinetic parameters with the minimum number of velocity measurements for mechanism II for ordered O + 2R → products**

The following program was written in *Mathematica* to calculate the values of kfet, kIO, and kR2 from three measured velocities at three pairs of substrate concentrations on the assumption that the mechanism is ordered (that is reactions 23-25), with O being bound first.

```
calckinparsordORII[v1_, o1_, r1_, v2_, o2_, r2_, v3_, o3_, r3_] :=
Module[{}, (*This program calculates kfet, kIO,
    and kR2 from three experimental velocities for O + 2R → products at three pairs
      of substrate concentrations on the assumption that the mechanism is ordered,
    with O bound first.  The first velocity is at high [O] and high [R],
    the second velocity is at low [O] and high [R],
    and the third velocity is at high [O] and low [R].*)
```

$$\text{Solve}\Big[\Big\{v1 == \frac{\text{kfet}\, o1\, r1^2}{\text{kIO}\, \text{kR2}^2 + \text{kR2}^2\, o1 + o1\, r1^2},\ v2 == \frac{\text{kfet}\, o2\, r2^2}{\text{kIO}\, \text{kR2}^2 + \text{kR2}^2\, o2 + o2\, r2^2},$$

$$v3 == \frac{\text{kfet}\, o3\, r3^2}{\text{kIO}\, \text{kR2}^2 + \text{kR2}^2\, o3 + o3\, r3\,\hat{}\,2}\Big\},\ \{\text{kfet, kR2, kIO}\}\Big]\Big]$$

```
calckinparsordORII[0.9881, 100, 100, 0.8264, 1, 100, 0.008264, 100, 1]
```

```
{{kfet → 0.999957, kIO → 20.0016, kR2 → -10.}, {kfet → 0.999957, kIO → 20.0016, kR2 → 10.}}
```

There are two solutions, and the second is reasonable.

```
Take[calckinparsordORII[0.9881, 100, 100, 0.8264, 1, 100, 0.008264, 100, 1], {2}]
```

$\{\{kfet \to 0.999957, kIO \to 20.0016, kR2 \to 10.\}\}$

Put in 5% errors in velocities, one at a time.

```
line5 = {kfet, kIO, kR2} /.
  Take[calckinparsordORII[0.9881, 100, 100, 0.8264, 1, 100, 0.008264, 100, 1], {2}]
{{kfet → 0.9999573597381077`, kIO → 20.00161661884169`, kR2 → 10.`}}
```

$\{\{0.999957, 20.0016, 10.\}\}$

$\{\{kfet \to 0.999957, kIO \to 20.0016, kR2 \to 10.\}\}$

```
calckinparsordORII[1.05 * 0.9881, 100, 100, 0.8264, 1, 100, 0.008264, 100, 1]
```

$\{\{kfet \to 1.05059, kIO \to 26.1286, kR2 \to -10.\}, \{kfet \to 1.05059, kIO \to 26.1286, kR2 \to 10.\}\}$

```
Take[calckinparsordORII[1.05 * 0.9881, 100, 100, 0.8264, 1, 100, 0.008264, 100, 1], {2}]
```

$\{\{kfet \to 1.05059, kIO \to 26.1286, kR2 \to 10.\}\}$

```
line6 = {kfet, kIO, kR2} /.
  Take[calckinparsordORII[1.05 * 0.9881, 100, 100, 0.8264, 1, 100, 0.008264, 100, 1], {2}]
```

$\{\{1.05059, 26.1286, 10.\}\}$

```
calckinparsordORII[0.9881, 100, 100, 1.05 * 0.8264, 1, 100, 0.008264, 100, 1]
```

$\{\{kfet \to 0.999957, kIO \to 13.4014, kR2 \to -10.2869\},$
$\{kfet \to 0.999957, kIO \to 13.4014, kR2 \to 10.2869\}\}$

```
line7 = {kfet, kIO, kR2} /.
  Take[calckinparsordORII[0.9881, 100, 100, 1.05 * 0.8264, 1, 100, 0.008264, 100, 1], {2}]
{{kfet → 0.9999573597381077`, kIO → 13.401444374391012`, kR2 → -10.286893777013052`},
 {kfet → 0.9999573597381077`, kIO → 13.401444374391012`, kR2 → 10.286893777013052`}}
```

$\{\{0.999957, 13.4014, 10.2869\}\}$

$\{\{kfet \to 0.999957, kIO \to 13.4014, kR2 \to -10.2869\},$
$\{kfet \to 0.999957, kIO \to 13.4014, kR2 \to 10.2869\}\}$

```
calckinparsordORII[0.9881, 100, 100, 0.8264, 1, 100, 1.05 * 0.008264, 100, 1]
```

$\{\{kfet \to 0.999381, kIO \to 21.2247, kR2 \to -9.7048\}, \{kfet \to 0.999381, kIO \to 21.2247, kR2 \to 9.7048\}\}$

```
line8 = {kfet, kIO, kR2} /.
  Take[calckinparsordORII[0.9881, 100, 100, 0.8264, 1, 100, 1.05 * 0.008264, 100, 1], {2}]
```

$\{\{0.999381, 21.2247, 9.7048\}\}$

Table 7.2 Values of kinetic constants for mechanism II using velocities at $\{100,100\}, \{1,100\}, \{100,1\}$

```
TableForm[Round[{line5[[1]], line6[[1]], line7[[1]], line8[[1]]}, 0.01], TableHeadings ->
  {{"No errors", "1.05*v1", "1.05*v2", "1.05*v3", "1.05*v4"}, {"kfet", "kIO", "kR2"}}]
```

	kfet	kIO	kR2
No errors	1.	20.	10.
1.05*v1	1.05	26.13	10.
1.05*v2	1.	13.4	10.29
1.05*v3	1.	21.22	9.7

Velocities at $\{200,100\}, \{0.1,200\}, \{200,0.1\}$ could be used to improve the accuracy of the estimations.

Velocities calculated using mechanism I are put into calckinparsordORII

> `calckinparsordORII[.9506, 100, 100, .1613, 1, 100, .04751, 100, 1]`

> {{kfet → 0.952411, kIO → -104., kR2 → 0. - 21.8214 i},
> {kfet → 0.952411, kIO → -104., kR2 → 0. + 21.8214 i}}

This is the program's way to say that the data were not determined using vI; in other words, mechanism II does not apply.

Velocities calculated using mechanism II are put into calckinparsordORI

> `calckinparsordORI[0.9881, 100, 100, 0.8264, 1, 100, 0.008264, 100, 1]`

> {{kfet → 0.990057, kO → -0.990099, kIR2 → 0. - 109.545 i},
> {kfet → 0.990057, kO → -0.990099, kIR2 → 0. + 109.545 i}}

This shows mechanisms I and II are clearly distinguishable by using 3 velocity measurements.

7.5 Effects of pH on mechanism II

▪ 7.5.1 Mechanism II with p*K*s and the consumption of hydrogen ions

The following extension of mechanism II introduces the p*K*s and chemical equilibrium constants:

$$
\begin{array}{cc}
E^- & EO^- \\
\| & \| \\
HE + O = & HEO \qquad\qquad K_{HEO} = \dfrac{[HE][O]}{[HEO]} \qquad\qquad (28) \\
\| & \| \\
H_2E^+ & H_2EO^+
\end{array}
$$

$$
\begin{array}{cc}
EO^- & ER_mO^- \\
\| & \| \\
HEO + mR = & HER_mO \qquad\qquad K^m_{HERmO} = \dfrac{[HEO][R]^m}{[HER_mO]} \qquad\qquad (29) \\
\| & \| \\
H_2EO^+ & H_2ER_mO^+
\end{array}
$$

$$
\begin{array}{c}
ER_mO^- \\
\| \qquad k_f \\
|\,n\,|\,H^+ + HER_mO \rightarrow \qquad\qquad\qquad (30) \\
\| \\
H_2ER_mO^+
\end{array}
$$

The following program has been written to estimate the kinetic parameters when *m* and *n* are known.

```
calcpropsORm[pK1e_, pK2e_, pK1eo_, pK2eo_, pK1ermo_, pK2ermo_, kfEt_, keo_, kermo_, m_, n_] :=
Module[{efactor, eofactor, ermofactor,
    vf, ko, krmo, vfexp, v},
   (*This program estimates kinetic parameters for the initial velocity of
      the forward reaction mR+O→.  The oxidant O binds first.  The reductant
      O binds second. This program also derives the rate equation for the
      forward reaction.  o and r are the concentrations of the oxidant and the
      reductant.  m is the number of R molecules needed to reduce the oxidant in the
      biochemical reaction.  n is the number of hydrogen ions produced in the rate-
      determining step.  The output is a list of seven functions.*)
   efactor = 1 + 10^(pK2e-pH) + 10^(pH-pK1e);
   eofactor = 1 + 10^(pK2eo-pH) + 10^(pH-pK1eo);
   ermofactor = 1 + 10^(pK2ermo-pH) + 10^(pH-pK1ermo);
   vf = kfEt / ermofactor;
   vfexp = (10^(n * pH)) * vf;
   ko = keo * (efactor / eofactor);
   krmo = kermo * (eofactor / ermofactor) ^ (1 / m);
   v = vfexp / (1 + ((krmo / r) ^m) * (1 + (ko / o)));
   {vfexp, vf, ko, krmo, vf / (krmo^m), vf / ((krmo^m) * ko), v}]
```

■ 7.5.2 Plots of kinetic parameters for the nitrate reductase type of reaction with mechanism II

To test the equations that are derived by this program, the following values are assumed for the kinetic parameters:

pK1e	7.5	pK2e	6.5	
pK1eo	7.0	pK2eo	6.0	(These values were previously assigned to pK1erm and pK2erm.)
pK1ermo	8.0	pK2ermo	7.0	
kfEt	10^(2*7.5)=10^15			
keo	1	kermo	2	
m	2	n	-2	

The pKs, chemical equilibrium rate constants, and rate constant used in these calculations are arbitrary.
Note kfEt for EC 1.9.61 is taken to be $10^{(2*7.5)}$ for nitrate reductase so that V_{fexp} will be of the order of 1 at pH 7.5.

The seven mathematical functions for nitrate reductase are derived as follows:

rxIOpH = calcpropsORm[7.5, 6.5, 7.0, 6.0, 8.0, 7.0, 10^(2 * 7.5), 1, 2, 2, -2]

$$\left\{ \frac{1. \times 10^{15} \, 10^{-2\,\text{pH}}}{1 + 10^{7.-\text{pH}} + 10^{-8.+\text{pH}}}, \; \frac{1. \times 10^{15}}{1 + 10^{7.-\text{pH}} + 10^{-8.+\text{pH}}}, \; \frac{1 + 10^{6.5-\text{pH}} + 10^{-7.5+\text{pH}}}{1 + 10^{6.-\text{pH}} + 10^{-7.+\text{pH}}}, \right.$$

$$2\sqrt{\frac{1 + 10^{6.-\text{pH}} + 10^{-7.+\text{pH}}}{1 + 10^{7.-\text{pH}} + 10^{-8.+\text{pH}}}}, \; \frac{2.5 \times 10^{14}}{1 + 10^{6.-\text{pH}} + 10^{-7.+\text{pH}}}, \; \frac{2.5 \times 10^{14}}{1 + 10^{6.5-\text{pH}} + 10^{-7.5+\text{pH}}},$$

$$\left. \frac{1. \times 10^{15} \, 10^{-2\,\text{pH}}}{\left(1 + 10^{7.-\text{pH}} + 10^{-8.+\text{pH}}\right) \left(1 + \frac{4 \left(1+10^{6.-\text{pH}}+10^{-7.+\text{pH}}\right)\left(1 + \frac{1+10^{6.5-\text{pH}}+10^{-7.5+\text{pH}}}{\left(1+10^{6.-\text{pH}}+10^{-7.+\text{pH}}\right) o}\right)}{\left(1+10^{7.-\text{pH}}+10^{-8.+\text{pH}}\right) r^2} \right)} \right\}$$

The form of the rate equation shows that O is bound first. This sequence in binding brings in new properties, for example pKs of EO, and new chemical equilibrium constants like K_{HEO} and K_{HERmO}. The rate equations at specified pHs are readily calculated.

```
rxIOpH[[7]] /. pH → {5, 6, 7, 8, 9}
```

$$\left\{ \frac{990.089}{1 + \frac{0.436035\left(1 + \frac{2.9633}{o}\right)}{r^2}}, \frac{90.8265}{1 + \frac{0.762943\left(1 + \frac{1.9971}{o}\right)}{r^2}}, \frac{4.7619}{1 + \frac{4.\left(1 + \frac{0.77736}{o}\right)}{r^2}}, \frac{0.047619}{1 + \frac{20.9714\left(1 + \frac{0.380917}{o}\right)}{r^2}}, \frac{0.0000908265}{1 + \frac{36.6943\left(1 + \frac{0.323026}{o}\right)}{r^2}} \right\}$$

This shows that the Michaelis constants K_{Rm} are given by

```
{.43, .76, 4.0, 20.97, 36.7}^.5
```

```
{0.655744, 0.87178, 2., 4.5793, 6.05805}
```

V_{fexp} is readily calculated as follows :

```
rxIOpH[[7]] /. o → ∞ /. r → ∞
```

$$\frac{1. \times 10^{15}\, 10^{-2\, pH}}{1 + 10^{7.-pH} + 10^{-8.+pH}}$$

Plot the functions of pH for the first six functions.

```
plot1a = Plot[Evaluate[Log[10, rxIOpH[[1]]]],
    {pH, 5, 9}, AxesLabel → {"pH", "logV_fexp"}, DisplayFunction → Identity];

plot2a = Plot[rxIOpH[[2]] * 10^-14, {pH, 5, 9},
    AxesLabel → {"pH", "V_f×10^-14"}, DisplayFunction → Identity];

plot3a = Plot[rxIOpH[[3]], {pH, 5, 9}, AxesLabel → {"pH", "K_IO"}, DisplayFunction → Identity];

plot4a = Plot[rxIOpH[[4]], {pH, 5, 9}, AxesLabel → {"pH", "K_R2"}, DisplayFunction → Identity];

plot5a = Plot[rxIOpH[[5]] * 10^-14, {pH, 5, 9},
    AxesLabel → {"pH", "V_f×10^-14/K_R2"}, DisplayFunction → Identity];

plot6a = Plot[rxIOpH[[6]] * 10^-14, {pH, 5, 9},
    AxesLabel → {"pH", "V_f×10^-14/K_IO K_R2^2"}, DisplayFunction → Identity];
```

fig1MOfirst = GraphicsGrid[{{plot1a, plot2a}, {plot3a, plot4a}, {plot5a, plot6a}}]

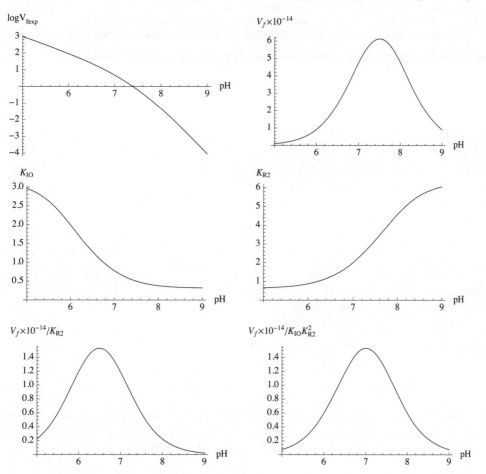

Fig. 7.6 Plots of pH dependencies of kinetic parameters when $m = 2$ and $n = -2$ in Mechanism II, as for example EC 1.9.6.1.

▪ 7.5.3 3D plots of rates for the nitrate reductase type of reaction with mechanism II

An overview of the rate equation can be obtained by plotting v at a specified pH versus [O] and [R] in three-dimensions.

```
plot1bnew5 = Plot3D[Evaluate[rxIOpH[[7]]/100 /. pH → 5.`], {r, 0.0001`, 5}, {o, 0.0001`, 2},
    ViewPoint → {-2, -2, 1}, AxesLabel → {"    [R]", "[O]    ", "  v/100  "}, PlotRange → {0, 12},
    Lighting → {{"Ambient", GrayLevel[1]}}, PlotLabel → "pH 5.0", DisplayFunction → Identity];

plot2bnew6 = Plot3D[Evaluate[rxIOpH[[7]]/10 /. pH → 6.`], {r, 0.0001`, 5}, {o, 0.0001`, 2},
    ViewPoint → {-2, -2, 1}, AxesLabel → {"    [R]", "[O]    ", "  v/10"}, PlotRange → {0, 10},
    Lighting → {{"Ambient", GrayLevel[1]}}, PlotLabel → "pH 6.0", DisplayFunction → Identity];

plot3bnew7 = Plot3D[Evaluate[rxIOpH[[7]] /. pH → 7.`], {r, 0.0001`, 20}, {o, 0.0001`, 0.5`},
    ViewPoint → {-2, -2, 1}, AxesLabel → {"    [R]", "[O]    ", "v"}, PlotRange → {0, 5.5`},
    Lighting → {{"Ambient", GrayLevel[1]}}, PlotLabel → "pH 7.0", DisplayFunction → Identity];
```

```
plot4bnew8 = Plot3D[Evaluate[rxIOpH[[7]] 10 /. pH → 8.`],
    {r, 0.0001`, 50}, {o, 0.0001`, 0.5`}, ViewPoint → {-2, -2, 1},
    AxesLabel → {"    [R]", "[O]      ", "  v×10    "}, PlotRange → {0, 1},
    Lighting → {{"Ambient", GrayLevel[1]}}, PlotLabel → "pH 8.0", DisplayFunction → Identity];

fig2MNnew = GraphicsGrid[{{plot1bnew5, plot2bnew6}, {plot3bnew7, plot4bnew8}}]
```

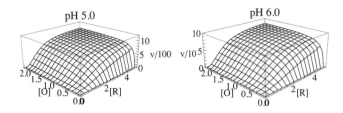

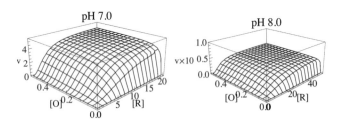

Fig. 7.7 Plots of initial velocities v of the catalyzed reaction 2R + O → as functions of [O] and [R] at four pHs for Mechanism II. The number *n* of hydrogen ions consumed in the rate-determining reaction for this ordered mechanism is -2 and *m* = 2. Note that plots of v versus [R] at constant [O] are sigmoid. Note the change in ordinate by a factor of 1000.

7.6 Discussion

This chapter discusses two ordered mechanisms for reductase reactions of the type O + *m*R. In the first mechanism *m*R is bound first and in the second O is bound first. The effects of pH when the enzymatic site and enzyme-substrate complexes each have two p*K*s and hydrogen ions are consumed in the rate-determining are discussed.

In general, rapid - equilibrium rate equations have fewer terms in the denominator than steady - state rate equations, but often rapid - equilibrium rate equations do represent the experimental data. When they do, they are to be preferred because the Michaelis constants are then equilibrium constants and can be interpreted using thermodynamics. More reactions can be included in rapid - equilibrium rate equations than are discussed here. An example is provided by dead - end complexes. But any additional reactions in rapid - equilibrium mechanisms must be mathematically independent.

There are two different sources of pH effects in rapid-equilibrium enzyme kinetics. Acid dissociations of substrates, the enzymatic site, and enzyme-substrate complexes introduce pH effects, but these effects extend over a range of about 2 units of pH. The other source of pH effects in kinetics is consumption of hydrogen ions. This chapter has discussed the consumption of hydogen ions in the rate-determining reaction, but there are other possibilities.

The rate equation vI for O + mR → products at a specified pH when mR is bound first is given by vI. When mR is bound first, the output from the program rxIpH is a list of 7 functions, and the rate equation is rxIpH[[7]]. When O is bound first, the rate equation at a specified pH is named vII. When O is bound first and the enzymatic site and enzyme-substrate complexes have two pKs, the rate equation is rxIOpH[[7]].

Chapter 8 Random O + *m*R → Products

8.1 Derivation of the rapid-equilibrium rate equation for random O + *m*R → products

- 8.1.1 Mechanism and rate equation for random O + *m*R → products

- 8.1.2 Estimation of kinetic parameters when $m = 2$

- 8.1.3 Application of the program for random O + 2R → products to velocities for mechanisms I and II

8.2 Effects of pH for random O + *m*R → products

- 8.2.1 Mechanism and rate equation for random O + *m*R → products when the enzymatic site and enzyme-substrate complexes each have two pKs using the factor method

- 8.2.2 Plots of pH dependencies of kinetic parameters for the random mechanism when $m = 2$ and $n = -2$ and the enzyme-substrate complexes each have two pKs

- 8.2.3 3D plots for the nitrate reductase type of reaction (O + 2R → products)

8.3 Estimation of kinetic parameters for random O + 3R → products

---

8.4 Discussion

8.1 Derivation of the rapid-equilibrium rate equation for random O + mR → products

This chapter is very much like the preceding chapter, but there is an additional kinetic parameter in the rate equation at a specified pH; that is, there are four kinetic parameters at a specified pH. Estimating these four parameters makes it possible to calculate a fifth kinetic parameter by using a thermodynamic cycle. This fifth kinetic parameter should be reported because it is an equlibrium constant and another investigator may report its value because of the use of a rate equation involving this parameter.

- ### 8.1.1 Mechanism and rate equation for random O + mR → products

The random mechanism can be written as

$E + mR = ER_m$	$K_{IRm}^m = [E] [R]^m / [ER_m]$	(1)
$E + O = EO$	$K_{IO} = [E][O]/[EO]$	(2)
$ER_m + O = ER_mO$	$K_O = [ER_m] [O] / [ER_m O]$	(3)
$ER_mO →$ products	$v = kf×[ER_mO] = kf*et*([ER_mO]/et) = kfet*([ER_mO]/et)$	(4)

But it can also be written as

$E + mR = ER_m$	$K_{IRm}^m = [E] [R]^m / [ER_m]$	(5)
$E + O = EO$	$K_{IO} = [E][O]/[EO]$	(6)
$EO + mR = ER_mO$	$K_{Rm}^m = [EO] [R]^m / [ER_m O]$	(7)
$ER_mO →$ products	$v = kf×[ER_mO] = kfet*([ER_mO]/et)$	(8)

Rapid-equilibrium rate equations can be derived from both mechanisms. The first mechanism yields K_{IRm}^m, K_{IO}, K_O, and kfet directly, and K_{Rm}^m indirectly. The second mechanism yields K_{IRm}^m, K_{IO}, K_{Rm}^m, and kfet directly and K_O indirectly.

Four equilibrium constants are not independent because they are related by the following thermodynamic cycle:

$$
\begin{array}{ccc}
& K_{IRm}^m & \\
E + mR & = & ER_m \\
+ & & + \\
O & & O \\
\| K_{IO} & & \| K_O \\
EO + mR & = & ER_mO \\
& K_{Rm}^m &
\end{array}
\tag{9}
$$

The four equilibrium constants in this thermodynamic cycle are related by

$$K_{IRm}^m K_O = K_{Rm}^m K_{IO} \tag{10}$$

When three of the reactions in this cycle are at equilibrium, the fourth also has to be at equilibrium. Since these four equilibrium constants are not independent, it is necessary to omit one of these reactions from the random mechanism. The reaction EO + mR = ER_mO is arbitrarily omitted in the first calculations done here. O and mR play similar roles in the random mechanism, and so it is not possible to tell which binds first. The conservation of enzymatic sites is represented by

$$et == e + erm + eo + ermo \tag{11}$$

The rapid-equilibrium rate equation can be derived by using Solve.

```
Solve[{kIRm^m == e * r^m / erm, kO == erm * o / ermo, kIO == e * o / eo, et == e + erm + eo + ermo},
  {ermo}, {e, eo, erm}]
```

$$\left\{\left\{ermo \rightarrow \frac{et\ kIO\ o\ r^m}{kIO\ kIRm^m\ kO + kIRm^m\ kO\ o + kIO\ kO\ r^m + kIO\ o\ r^m}\right\}\right\}$$

The rate equation for mechanism 1 - 4 is given by

$$vORrand = \frac{kfet\ kIO\ o\ r^m}{kIO\ kIRm^m\ kO + kIRm^m\ kO\ o + kIO\ kO\ r^m + kIO\ o\ r^m}$$

$$\frac{kfet\ kIO\ o\ r^m}{kIO\ kIRm^m\ kO + kIRm^m\ kO\ o + kIO\ kO\ r^m + kIO\ o\ r^m}$$

This rate equation is important because it can be used to calculate rate equations for any value of *m*. The number m is not a kinetic parameter because it is a property of the reaction catalyzed. When *m* is specified, it is necessary to change the symbols kIRm and kRm. When *m* = 1, kIRm has to be changed to kIR and kRm has to be changed to kR. The following four rate equations are used in this chapter.

```
vORrand1 = vORrand /. m → 1 /. kIRm → kIR
```

$$\frac{kfet\ kIO\ o\ r}{kIO\ kIR\ kO + kIR\ kO\ o + kIO\ kO\ r + kIO\ o\ r}$$

```
vORrand2 = vORrand /. m → 2 /. kIRm → kIR2
```

$$\frac{kfet\ kIO\ o\ r^2}{kIO\ kIR2^2\ kO + kIR2^2\ kO\ o + kIO\ kO\ r^2 + kIO\ o\ r^2}$$

```
vORrand3 = vORrand /. m → 3 /. kIRm → kIR3
```

$$\frac{kfet\ kIO\ o\ r^3}{kIO\ kIR3^3\ kO + kIR3^3\ kO\ o + kIO\ kO\ r^3 + kIO\ o\ r^3}$$

```
vORrand6 = vORrand /. m → 6 /. kIRm → kIR6
```

$$\frac{kfet\ kIO\ o\ r^6}{kIO\ kIR6^6\ kO + kIR6^6\ kO\ o + kIO\ kO\ r^6 + kIO\ o\ r^6}$$

vORrand1 can be rearranged into the following more familiar form:

$$v = \frac{V_{fexp}}{1 + \dfrac{K_O}{[O]} + \dfrac{K_O\, K_{IRm}^m}{K_{IO}\,[R]^m} + \dfrac{K_{IRm}^m\, K_O}{[R]^m\,[O]}} \tag{12}$$

Rate equation 12 can also be written in the form

$$v = \frac{V_{fexp}}{1 + \dfrac{K_O}{[O]} + \dfrac{K_O\, K_{IRm}^m}{K_{IO}\,[R]^m}\left(1 + \dfrac{K_{IO}}{[O]}\right)} \tag{13}$$

This rate equation makes it possible to estimate V_{fexp}, K_O, K_{IO}, and K_{IRm} by measuring four velocities. Equation 10 shows that this equation can also be written as

$$v = \frac{V_{fexp}}{1 + \dfrac{K_O}{[O]} + \dfrac{K_{Rm}^m}{[R]^m}\left(1 + \dfrac{K_{IO}}{[O]}\right)} \tag{14}$$

Rate equation vORrand can be used to calculate the velocity of O + mR → products for any value of m and any pair of substrate concentrations {o,r}. For the first calculation when $m = 1$, the following kinetic parameters are chosen arbtrarily: kO = 5, kIO = 15, kIR = 20, and kfet = 1.

```
vORrand1 /. kO → 5 /. kIO → 15 /. kIR → 20 /. kfet → 1
```

$$\frac{15\,o\,r}{1500 + 100\,o + 75\,r + 15\,o\,r}$$

This rate equation will not be used in this chapter because it is the same as the rate equation for random A + B → products.

Calculation of velocities when *m* = 2

```
vORrand2
```

$$\frac{kfet\ kIO\ o\ r^2}{kIO\ kIR2^2\ kO + kIR2^2\ kO\ o + kIO\ kO\ r^2 + kIO\ o\ r^2}$$

```
vel2 = vORrand2 /. kO → 5 /. kIO → 15 /. kIR2 → 20 /. kfet → 1
```

$$\frac{15\,o\,r^2}{30\,000 + 2000\,o + 75\,r^2 + 15\,o\,r^2}$$

It is of interest to plot this velocity as a function of [O] and [R], and also plot the partial derivatives of the velocity with respect to the substrate concentrations.

```
plota2 =
  Plot3D[vel2, {o, 0.00001, 20}, {r, 0.00001, 50}, ViewPoint → {-2, -2, 1}, PlotRange -> {0, 1},
    Lighting → {{"Ambient", GrayLevel[1]}}, AxesLabel → {"[O]", "[R]"}, PlotLabel → "v"];

plotb2 = Plot3D[Evaluate[D[vel2, o]], {o, 0.00001, 20},
    {r, 0.00001, 50}, ViewPoint → {-2, -2, 1}, PlotRange -> {0, 0.2},
    Lighting → {{"Ambient", GrayLevel[1]}}, AxesLabel → {"[O]", "[R]"}, PlotLabel → "dv/d[O]"];

plotc2 = Plot3D[Evaluate[D[vel2, r]], {o, 0.00001, 20},
    {r, 0.00001, 50}, ViewPoint → {-2, -2, 1}, PlotRange -> {0, 0.05},
    Lighting → {{"Ambient", GrayLevel[1]}}, AxesLabel → {"[O]", "[R]"}, PlotLabel → "dv/d[R]"];
```

```
plotd2 = Plot3D[Evaluate[D[vel2, o, r]], {o, 0.00001, 20}, {r, 0.00001, 50},
    ViewPoint → {-2, -2, 1}, PlotRange -> {0, 0.007}, Lighting → {{"Ambient", GrayLevel[1]}},
    AxesLabel → {"[O]", "[R]"}, PlotLabel → "d(dv/d[O])/d[R]"];
```

```
fig81 = GraphicsArray[{{plota2, plotb2}, {plotc2, plotd2}}]
```

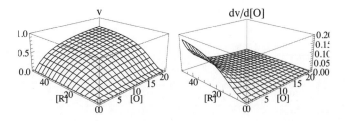

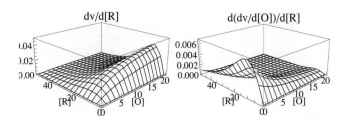

Fig. 8.1 (a) v versus [R] and [O] for the random mechanism for O + 2R → products. (b) dv/d[O] versus [R] and [O] for the random mechanism. (c) dv/d[R] versus [R] and [O] for the random mechanism. (d) d (dv/d[O])/d[R] versus [R] and [O] for the random mechanism.

Figure 8.1 (b) shows that the velocity is most sensitive to [O] at low [O] and high [R]. Figure 8.1 (c) shows that the velocity is most sensitive to [R] at high [O] and [R] of the order of K_{IR}. At lower values of [R] the sensitivity drops off quickly Figure 8.1 (d) shows that the velocity is most sensitive to [O] and [R] together at low [O] and [R] of the order of K_{IR2}. At low [R], very low concentrations of R are to be avoided in estimating kinetic parameters.

Calculation of velocities when $m = 3$

```
vel3 = vORrand3 /. kO → 5 /. kIO → 15 /. kIR3 → 20 /. kfet → 1
```

$$\frac{15\, o\, r^3}{600\,000 + 40\,000\, o + 75\, r^3 + 15\, o\, r^3}$$

It is of interest to plot this velocity as a function of [O] and [R], and to also plot the derivatives of this velocity with respect to the substrate concentrations.

```
plota3 = Plot3D[vel3, {o, 0.00001, 20}, {r, 0.00001, 50},
    ViewPoint → {-2, -2, 1}, PlotRange -> {0, 1}, AxesLabel → {"[O]", "[R]"},
    Lighting → {{"Ambient", GrayLevel[1]}}, PlotLabel → "v"];
```

```
plotb3 = Plot3D[Evaluate[D[vel3, o]], {o, 0.00001, 20},
    {r, 0.00001, 50}, ViewPoint → {-2, -2, 1}, PlotRange -> {0, 0.2},
    Lighting → {{"Ambient", GrayLevel[1]}}, AxesLabel → {"[O]", "[R]"}, PlotLabel → "dv/d[O]"];
```

```
plotc3 = Plot3D[Evaluate[D[vel3, r]], {o, 0.00001, 20},
    {r, 0.00001, 50}, ViewPoint → {-2, -2, 1}, PlotRange -> {0, 0.05},
    Lighting → {{"Ambient", GrayLevel[1]}}, AxesLabel → {"[O]", "[R]"}, PlotLabel → "dv/d[R]"];

plotd3 = Plot3D[Evaluate[D[vel3, o, r]], {o, 0.00001, 20}, {r, 0.00001, 50},
    ViewPoint → {-2, -2, 1}, PlotRange -> {0, 0.01}, Lighting → {{"Ambient", GrayLevel[1]}},
    AxesLabel → {"[O]", "[R]"}, PlotLabel → "d(dv/d[O])/d[R]"];

fig82 = GraphicsArray[{{plota3, plotb3}, {plotc3, plotd3}}]
```

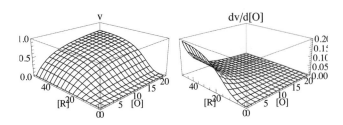

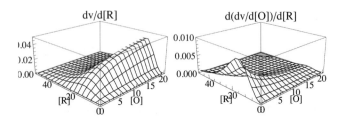

Fig. 8.2 (a) v versus [R] and [O] for the random mechanism for O + 3R → products. (b) dv/d[O] versus [R] and [O] for the random mechanism. (c) dv/d[R] versus [R] and [O] for the random mechanism. (d) d (dv/d[O])/d[R] versus [R] and [O] for the random mechanism.

Calculation of velocities for *m* = 6

As mentioned in Section 7.1, ferredoxin-nitrite reductase is an example of $m = 6$:

nitrite + 6ferredoxin$_{red}$ = ammonia + 6ferredoxin$_{ox}$ + 2 H$_2$ O.

```
vel6 = vORrand6 /. kO → 5 /. kIO → 15 /. kIR6 → 20 /. kfet → 1
```

$$\frac{15\ o\ r^6}{4\,800\,000\,000 + 320\,000\,000\ o + 75\ r^6 + 15\ o\ r^6}$$

It is of interest to plot this velocity as a function of [O] and [R], and to also plot the derivatives of the velocity with respect to the substrate concentrations.

```
plota6 =
    Plot3D[vel6, {o, 0.00001, 20}, {r, 0.00001, 50}, ViewPoint → {-2, -2, 1}, PlotRange -> {0, 1},
      AxesLabel → {"[O]", "[R]"}, Lighting → {{"Ambient", GrayLevel[1]}}, PlotLabel → v];
```

```
plotb6 = Plot3D[Evaluate[D[vel6, o]], {o, 0.00001, 20},
   {r, 0.00001, 50}, ViewPoint → {-2, -2, 1}, PlotRange -> {0, 0.2},
   Lighting → {{"Ambient", GrayLevel[1]}}, AxesLabel → {"[O]", "[R]"}, PlotLabel → "dv/d[O]"];

plotc6 = Plot3D[Evaluate[D[vel6, r]], {o, 0.00001, 20}, {r, 0.00001, 50},
   ViewPoint → {-2, -2, 1}, PlotRange -> {0, 0.07}, AxesLabel → {"[O]", "[R]"},
   Lighting → {{"Ambient", GrayLevel[1]}}, PlotLabel → "dv/d[R]"];

plotd6 = Plot3D[Evaluate[D[vel6, o, r]], {o, 0.00001, 20}, {r, 0.00001, 50},
   ViewPoint → {-2, -2, 1}, PlotRange -> {0, 0.015}, AxesLabel → {"[O]", "[R]"},
   Lighting → {{"Ambient", GrayLevel[1]}}, PlotLabel → "d(dv/d[O])/d[R]"];

fig86 = GraphicsArray[{{plota6, plotb6}, {plotc6, plotd6}}]
```

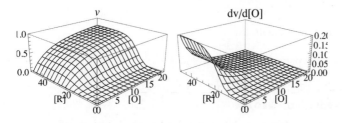

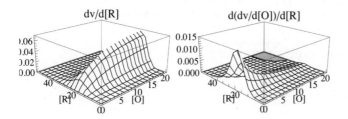

Fig. 8.3 (a) v versus [R] and [O] for the random mechanism for O + 6R → products. (b) dv/d[O] versus [R] and [O] for the random mechanism. (c) dv/d[R] versus [R] and [O] for the random mechanism. (d) d (dv/d[O])/d[R] versus [R] and [O] for the random mechanism.

■ 8.1.2 Estimation of kinetic parameters when *m* = 2

The program in this section was written to estimate the values of kfet, kIR2, kO, and kIO from four measured velocities at four pairs of substrate concentrations on the assumption that the mechanism is random.

When *m* = 2, the rate equation that is used to construct the program to estimate the kinetic parameters is

vORrand2

$$\frac{kfet \ kIO \ o \ r^2}{kIO \ kIR2^2 \ kO + kIR2^2 \ kO \ o + kIO \ kO \ r^2 + kIO \ o \ r^2}$$

This expression for the velocity is used to write a computer program to estimate the kinetic parameters using the minimum number of velocity measurements. When the chosen parameters are inserted, the velocity is given by

vel2

$$\frac{15 \ o \ r^2}{30\,000 + 2000 \ o + 75 \ r^2 + 15 \ o \ r^2}$$

```
vel2 /. o → 100 /. r → 100 // N
```

```
0.938673
```

```
vel2 /. o → 1 /. r → 100 // N
```

```
0.160944
```

```
vel2 /. o → 100 /. r → 5 // N
```

```
0.139211
```

```
vel2 /. o → 5 /. r → 5 // N
```

```
0.0428571
```

The four velocities are considered to be experimental data for test calculations.

The following program is written to estimate four kinetic parameters from 4 velocities.

```
calckinparsrandOR2[v1_, o1_, r1_, v2_, o2_, r2_, v3_, o3_, r3_, v4_, o4_, r4_] :=
 Module[{}, (*This program calculates kfet, kO, kIO,
   and kIR2 from four experimental velocities for O + 2R → products at 4 pairs of
      substrate concentrations on the assumption that the mechanism is random and 2
      molecules of R are involved.  The first velocity is at high [O] and high [R],
   the second velocity is at low [O] and high [R],  the third velocity is at
   high [O] and low [R], and the 4th velocity is at medium [O] and medium [R].*)
```

$$\text{Solve}\Big[\Big\{v1 == \frac{\text{kfet kIO o1 r1}^2}{\text{kIO kIR2}^2 \text{kO} + \text{kIR2}^2 \text{kO o1} + \text{kIO kO r1}^2 + \text{kIO o1 r1}^2},$$

$$v2 == \frac{\text{kfet kIO o2 r2}^2}{\text{kIO kIR2}^2 \text{kO} + \text{kIR2}^2 \text{kO o2} + \text{kIO kO r2}^2 + \text{kIO o2 r2}^2},$$

$$v3 == \frac{\text{kfet kIO o3 r3}^2}{\text{kIO kIR2}^2 \text{kO} + \text{kIR2}^2 \text{kO o3} + \text{kIO kO r3}^2 + \text{kIO o3 r3}^2},$$

$$v4 == \frac{\text{kfet kIO o4 r4}^2}{\text{kIO kIR2}^2 \text{kO} + \text{kIR2}^2 \text{kO o4} + \text{kIO kO r4}^2 + \text{kIO o4 r4}^2}\Big\}, \{\text{kfet, kIR2, kO, kIO}\}\Big]\Big]$$

```
calckinparsrandOR2[.9387, 100, 100, .1609, 1, 100, .1392, 100, 5, 0.04286, 5, 5][[2]]
```

```
{kfet → 1.00005, kIO → 14.9955, kO → 5.00204, kIR2 → 19.9948}
```

```
row1 = {kfet, kIO, kO, kIR2} /.
   calckinparsrandOR2[.9387, 100, 100, .1609, 1, 100, .1392, 100, 5, 0.04286, 5, 5][[2]]
```

```
{1.00005, 14.9955, 5.00204, 19.9948}
```

These values are correct, but it is necessary to take experimental errors into account.

```
calckinparsrandOR2[1.05 * .9387, 100, 100, .1609, 1, 100, .1392, 100, 5, 0.04286, 5, 5][[2]]
```

```
{kfet → 1.05421, kIO → 14.8429, kO → 5.32707, kIR2 → 19.8865}
```

```
row2 = {kfet, kIO, kO, kIR2} /.
   calckinparsrandOR2[1.05 * .9387, 100, 100, .1609, 1, 100, .1392, 100, 5, 0.04286, 5, 5][[2]]
```

```
{1.05421, 14.8429, 5.32707, 19.8865}
```

```
calckinparsrandOR2[.9387, 100, 100, 1.05 * .1609, 1, 100, .1392, 100, 5, 0.04286, 5, 5][[2]]

{kfet → 0.997061, kIO → 15.0601, kO → 4.68828, kIR2 → 20.6608}

row3 = {kfet, kIO, kO, kIR2} /.
    calckinparsrandOR2[.9387, 100, 100, 1.05 * .1609, 1, 100, .1392, 100, 5, 0.04286, 5, 5][[2]]

{0.997061, 15.0601, 4.68828, 20.6608}

calckinparsrandOR2[.9387, 100, 100, .1609, 1, 100, 1.05 * .1392, 100, 5, 0.04286, 5, 5][[2]]

{kfet → 0.999148, kIO → 16.447, kO → 4.99302, kIR2 → 20.2283}

row4 = {kfet, kIO, kO, kIR2} /.
    calckinparsrandOR2[.9387, 100, 100, .1609, 1, 100, 1.05 * .1392, 100, 5, 0.04286, 5, 5][[2]]

{0.999148, 16.447, 4.99302, 20.2283}

calckinparsrandOR2[.9387, 100, 100, .1609, 1, 100, .1392, 100, 5, 1.05 * 0.04286, 5, 5][[2]]

{kfet → 1.0002, kIO → 13.7454, kO → 5.01743, kIR2 → 19.22}

row5 = {kfet, kIO, kO, kIR2} /.
    calckinparsrandOR2[.9387, 100, 100, .1609, 1, 100, .1392, 100, 5, 1.05 * 0.04286, 5, 5][[2]]

{1.0002, 13.7454, 5.01743, 19.22}
```

Table 8.1 Values of kinetic constants for mechanism 1-4 using velocities at {100,100}, {1,100}, {100,5},{5,5}}

```
TableForm[Round[{row1, row2, row3, row4, row5}, 0.01], TableHeadings ->
    {{"No errors", "1.05*v1", "1.05*v2", "1.05*v3", "1.05*v4"}, {"kfet", "kIO", "kO", "kIR2"}}]
```

	kfet	kIO	kO	kIR2
No errors	1.	15.	5.	19.99
1.05*v1	1.05	14.84	5.33	19.89
1.05*v2	1.	15.06	4.69	20.66
1.05*v3	1.	16.45	4.99	20.23
1.05*v4	1.	13.75	5.02	19.22

The fifth kinetic parameter, kR2, can be calculated by using equation 10, which is $K_{IR2}^2 K_O = K_{R2}^2 K_{IO}$.

$$K_{R2} = \left(K_{IR2}^2 K_O / K_{IO} \right)^{.5}$$

$$\left(\left(\begin{array}{c} 19.990000000000002` \\ 19.89` \\ 20.66` \\ 20.23` \\ 19.22` \end{array} \right)^{\wedge}2 * \left( \begin{array}{c} 5.` \\ 5.33` \\ 4.69` \\ 4.99` \\ 5.020000000000005` \end{array} \middle/ \begin{array}{c} 15.` \\ 14.84` \\ 15.06` \\ 16.45` \\ 13.75` \end{array} \right) \right)^{\wedge}.5$$

{{11.5412}, {11.9201}, {11.5293}, {11.142}, {11.6133}}

This list can be added to Table 8.1 as a column. Each line in the table is augmented with kRm.

```
lin1a =
{1.0000496469268365`, 14.99549203084873`, 5.002035687976978`, 19.99480160865908`, 11.54};

lin2a =
 {1.0542064498320818`, 14.842917426539504`, 5.327070614775765`, 19.88652120368042`, 11.92}

{1.05421, 14.8429, 5.32707, 19.8865, 11.92}
```

```
lin3a = {0.9970613756812245`, 15.060138194712566`,
    4.688276701169326`, 20.660751825254437`, 11.53}

{0.997061, 15.0601, 4.68828, 20.6608, 11.53}

lin4a = {0.9991478758551077`, 16.446976314380162`,
    4.993016589305692`, 20.228271639706215`, 11.14}

{0.999148, 16.447, 4.99302, 20.2283, 11.14}

lin5a =
    {1.0001962384372636`, 13.74541368296546`, 5.01742733083862`, 19.22001767415077`, 11.61}

{1.0002, 13.7454, 5.01743, 19.22, 11.61}
```

Table 8.2 Values of five kinetic constants for mechanism 1-4 using velocities at {100,100}, {1,100}, {100,5}, {5,5} and the thermodynamic cycle

```
TableForm[Round[{lin1a, lin2a, lin3a, lin4a, lin5a}, 0.01],
   TableHeadings -> {{"No errors", "1.05*v1", "1.05*v2", "1.05*v3", "1.05*v4"},
      {"kfet", "kIO", "kO", "kIR2", "kR2"}}]
```

	kfet	kIO	kO	kIR2	kR2
No errors	1.	15.	5.	19.99	11.54
1.05*v1	1.05	14.84	5.33	19.89	11.92
1.05*v2	1.	15.06	4.69	20.66	11.53
1.05*v3	1.	16.45	4.99	20.23	11.14
1.05*v4	1.	13.75	5.02	19.22	11.61

■ 8.1.3 Application of the program for random O + 2R → products to velocities for ordered mechanisms I and II

One more velocity has to be measured using vI, and {5,5} is chosen for this fourth velocity. Section 7.2.1 shows that this velocity is 0.05556. vI velocities are for $m = 2$.

```
Take[calckinparsrandOR2[0.9506, 100, 100, 0.1613, 1, 100, 0.04751, 100, 1, 0.05556, 5, 5], {2}]
```

$$\{\{kfet \to 1.00003, kIO \to -2.22708 \times 10^7, kO \to 4.99982, kIR2 \to 19.9998\}\}$$

The values for kO and kIR2 are correct, but the value of kIO is unreasonable. This is the program's way of showing that this data does not include the reaction E + O = EO.

One more velocity has to be measured using vII, and {5,5} is chosen for this fourth velocity. See Section 7.2.2.

```
Take[calckinparsrandOR2[0.9506, 100, 100, 0.1613, 1, 100, 0.04751, 100, 1, 0.05556, 5, 5], {2}]
```

$$\{\{kfet \to 1.00003, kIO \to -2.22708 \times 10^7, kO \to 4.99982, kIR2 \to 19.9998\}\}$$

This is the program's way of showing that this data is not for the random mechanism. vI and v2 are equal because the thermodynamic cycle was used to calculate the new kinetic parameter in the second mechanism.

8.2 Effects of pH for random O + mR → products

- ### 8.2.1 Mechanism and rate equation for random O + mR → products when the enzymatic site and enzyme-substrate complexes each have two pKs using the factor method

The effects of pH on the velocity of the forward reaction can be discussed by including two pKs for the enzymatic site, each enzyme-substrate complex, and the consumption of hydrogen ions in the rate-determining reaction:

$$
\begin{array}{ll}
E^- & ER_m^- \\
\| & \| \\
HE + mR = HER_m & \quad K_{HERm}^m = \dfrac{[HE][R]^m}{[HER_m]} \\
\| & \| \\
H_2 E^+ & H_2 ER_m^+
\end{array}
\tag{15}
$$

$$
\begin{array}{ll}
E^- & EO^- \\
\| & \| \\
HE + O = HEO & \quad K_{HEO} = \dfrac{[HE][O]}{[HEO]} \\
\| & \| \\
H_2 E^+ & H_2 EO^+
\end{array}
\tag{16}
$$

$$
\begin{array}{ll}
ER_m^- & ER_m O^- \\
\| & \| \\
HER_m + O = HER_m O & \quad K_{HERmO} = \dfrac{[HER_m][O]}{[HER_m O]} \\
\| & \| \\
H_2 ER_m^+ & H_2 ER_m O^+
\end{array}
\tag{17}
$$

$$
\begin{array}{l}
\quad\quad ER_m O^- \\
\quad\quad \| \quad\quad k_f \\
|n| H^+ + HER_m O \ \text{-> products} \\
\quad\quad \| \\
\quad\quad H_2 ER_m O^+
\end{array}
\tag{18}
$$

The pH dependencies of the three Michaelis constants are given by

$$
K_{IRm}^m = \frac{K_{HERm}^m \left(1 + 10^{pH-pK1E} + 10^{-pH+K2E}\right)}{1 + 10^{pH-pK1ERm} + 10^{-pH+K2ERm}}
\tag{19}
$$

$$K_{IO} = \frac{K_{HEO}\left(1 + 10^{pH-pK1E} + 10^{-pH+K2E}\right)}{1 + 10^{pH-pK1EO} + 10^{-pH+K2EO}} \tag{20}$$

$$K_O = \frac{K_{HERmO}\left(1 + 10^{pH-pK1ERm} + 10^{-pH+K2ERm}\right)}{1 + 10^{pH-pK1ERmO} + 10^{-pH+K2ERmO}} \tag{21}$$

Since equations 19 to 21 each involve five unknowns, it is convenient to determine the kinetic parameters in them by plotting V_f / K_O, $V_f / K_O\ K_{IRm}^m$, and $V_f\ K_{IO} / K_O\ K_{IRm}^m$ versus pH [33].

$$V_f/K_O = \frac{k_f\,[E]_t}{K_{HERmO}\left(1 + 10^{pH-pK1ERm} + 10^{-pH+pK2ERm}\right)} \tag{22}$$

$$V_f/K_O\ K_{IRm}^m = \frac{k_f\,[E]_t}{K_{HERmO}\ K_{HERm}^m\left(1 + 10^{pH-pK1E} + 10^{-pH+pK2E}\right)} \tag{23}$$

$$V_f\ K_{IO}/K_O\ K_{IRm}^m = \frac{k_f\,[E]_t\ K_{HEO}}{K_{HERmO}\ K_{HERm}^m\left(1 + 10^{pH-pK1EO} + 10^{-pH+pK2EO}\right)} \tag{24}$$

The kinetic parameters in these three equations can all be estimated from bell-shaped plots

- ### 8.2.2 Plots of pH dependencies of kinetic parameters for the random mechanism when *m* = 2 and *n* = -2 and enzyme-substrate complexes have each have two pKs

```
calcpropsrandom[pK1e_, pK2e_, pK1erm_, pK2erm_, pK1ermo_,
  pK2ermo_, kfEt_, kerm_, kermo_, m_, n_, pK1eo_, pK2eo_, keo_] :=
Module[{efactor, ermfactor, ermofactor, eofactor,
   vf, krm, krmo, ko, vfexp, v},
  (*This program derives kinetic parameters for the initial velocity of the forward
    reaction mR+O→ and several useful ratios.  The binding is random.  This
    program also derives the rate equation for the forward reaction.  o and
    r are the concentrations of the oxidant and the reductant.  m is the
    stoichiometric number of R molecules needed to reduce the oxidant in the
    biochemical reaction.  n is the number of hydrogen ions produced in the rate-
    determining step.  The output is a list of nine functions.*)
  efactor = 1 + 10^pK2e-pH + 10^pH-pK1e;
  ermfactor = 1 + 10^pK2erm-pH + 10^pH-pK1erm;
  ermofactor = 1 + 10^pK2ermo-pH + 10^pH-pK1ermo;
  eofactor = 1 + 10^pK2eo-pH + 10^pH-pK1eo;
  vf = kfEt / ermofactor;
  vfexp = (10^(n*pH)) * vf;
  krm = kerm * (efactor / ermfactor) ^ (1 / m);
  ko = keo * (efactor / eofactor);
  krmo = kermo * ermfactor / ermofactor;
  v = vfexp / (1 + krmo / o + krmo * (krm^m) / (ko * r^m) + (krmo * (krm / r) ^m) / o);
  {vfexp, vf, krm, ko, krmo, vf / ko, vf / (ko * krm^m), (vf * ko) / (krmo * (krm^m)), v}]
```

This mechanism involves one more Michaelis constant, 2 more p*K*s, and one more chemical equilibrium constant than Mechanisms I and II for ordered O + *m*R → products (Sections 7.3 and 7.4). The rate equation involves one more term in the denominator.

To test the equations that are derived by this program, the following values are chosen for the pH-independent kinetic parameters:

pK1e	7.5	pK2e	6.5
pK1erm	7.0	pK2erm	6.0
pK1ermo	8.0	pK2ermo	7.0
kfEt	10^(2*7.5)=10^15		
kerm	1	kermo	2
m	2	n	-2
pK1eo	7.0	pK2eo	6.0
keo	1		

The nine mathematical functions for the random mechanism are derived as follows: The 1961 is the EC number for nitrate reductase.

```
rx1961random =
  calcpropsrandom[7.5, 6.5, 7.0, 6.0, 8.0, 7.0, 10^(2 * 7.5), 1, 2, 2, -2, 7.0, 6.0, 1]
```

$$
\left\{
\frac{1.\times10^{15}\,10^{-2\,\mathrm{pH}}}{1+10^{7.-\mathrm{pH}}+10^{-8.+\mathrm{pH}}},\;
\frac{1.\times10^{15}}{1+10^{7.-\mathrm{pH}}+10^{-8.+\mathrm{pH}}},\;
\sqrt{\frac{1+10^{6.5-\mathrm{pH}}+10^{-7.5+\mathrm{pH}}}{1+10^{6.-\mathrm{pH}}+10^{-7.+\mathrm{pH}}}},\right.
$$

$$
\frac{1+10^{6.5-\mathrm{pH}}+10^{-7.5+\mathrm{pH}}}{1+10^{6.-\mathrm{pH}}+10^{-7.+\mathrm{pH}}},\;
\frac{2\left(1+10^{6.-\mathrm{pH}}+10^{-7.+\mathrm{pH}}\right)}{1+10^{7.-\mathrm{pH}}+10^{-8.+\mathrm{pH}}},\;
\frac{1.\times10^{15}\left(1+10^{6.-\mathrm{pH}}+10^{-7.+\mathrm{pH}}\right)}{\left(1+10^{7.-\mathrm{pH}}+10^{-8.+\mathrm{pH}}\right)\left(1+10^{6.5-\mathrm{pH}}+10^{-7.5+\mathrm{pH}}\right)},
$$

$$
\frac{1.\times10^{15}\left(1+10^{6.-\mathrm{pH}}+10^{-7.+\mathrm{pH}}\right)^{2}}{\left(1+10^{7.-\mathrm{pH}}+10^{-8.+\mathrm{pH}}\right)\left(1+10^{6.5-\mathrm{pH}}+10^{-7.5+\mathrm{pH}}\right)^{2}},\;
\frac{5.\times10^{14}}{1+10^{6.-\mathrm{pH}}+10^{-7.+\mathrm{pH}}},
$$

$$
\left.\frac{1.\times10^{15}\,10^{-2\,\mathrm{pH}}}{\left(1+10^{7.-\mathrm{pH}}+10^{-8.+\mathrm{pH}}\right)\left(1+\frac{2\left(1+10^{6.-\mathrm{pH}}+10^{-7.+\mathrm{pH}}\right)}{\left(1+10^{7.-\mathrm{pH}}+10^{-8.+\mathrm{pH}}\right)\mathrm{o}}+\frac{2\left(1+10^{6.-\mathrm{pH}}+10^{-7.+\mathrm{pH}}\right)}{\left(1+10^{7.-\mathrm{pH}}+10^{-8.+\mathrm{pH}}\right)\mathrm{r}^{2}}+\frac{2\left(1+10^{6.5-\mathrm{pH}}+10^{-7.5+\mathrm{pH}}\right)}{\left(1+10^{7.-\mathrm{pH}}+10^{-8.+\mathrm{pH}}\right)\mathrm{o}\,\mathrm{r}^{2}}\right)}\right\}
$$

Plots of the functions of pH for the two limiting velocities, three Michaelis constants, and three ratios that are useful for calculating pKs and chemical equilibrium constants from bell-shaped plots.

```
plot1b = Plot[Evaluate[Log[10, rx1961random[[1]]]],
    {pH, 5, 9}, AxesLabel → {"pH", "logV_fexp"}, DisplayFunction → Identity];

plot2b = Plot[rx1961random[[2]] * 10^-14, {pH, 5, 9},
    AxesLabel → {"pH", "V_f×10^-14"}, DisplayFunction → Identity];

plot3b =
  Plot[rx1961random[[3]], {pH, 5, 9}, AxesLabel → {"pH", "K_IR2"}, DisplayFunction → Identity];

plot4b =
  Plot[rx1961random[[4]], {pH, 5, 9}, AxesLabel → {"pH", "K_IO"}, DisplayFunction → Identity];

plot5b =
  Plot[rx1961random[[5]], {pH, 5, 9}, AxesLabel → {"pH", "K_O"}, DisplayFunction → Identity];

plot6b = Plot[rx1961random[[6]] * 10^-14, {pH, 5, 9},
    AxesLabel → {"pH", "V_f×10^-14/K_O"}, DisplayFunction → Identity];

plot7b = Plot[rx1961random[[7]] * 10^-14, {pH, 5, 9},
    AxesLabel → {"pH", "V_f/K_OK_IR2^2"}, DisplayFunction → Identity];
```

```
plot8b = Plot[rx1961random[[8]] * 10^-14, {pH, 5, 9},
    AxesLabel → {"pH", "V_f K_IO/K_O K²_IR2"}, DisplayFunction → Identity];

fig1MN = GraphicsGrid[{{plot1b, plot2b}, {plot3b, plot4b}, {plot5b, plot6b}, {plot7b, plot8b}}]
```

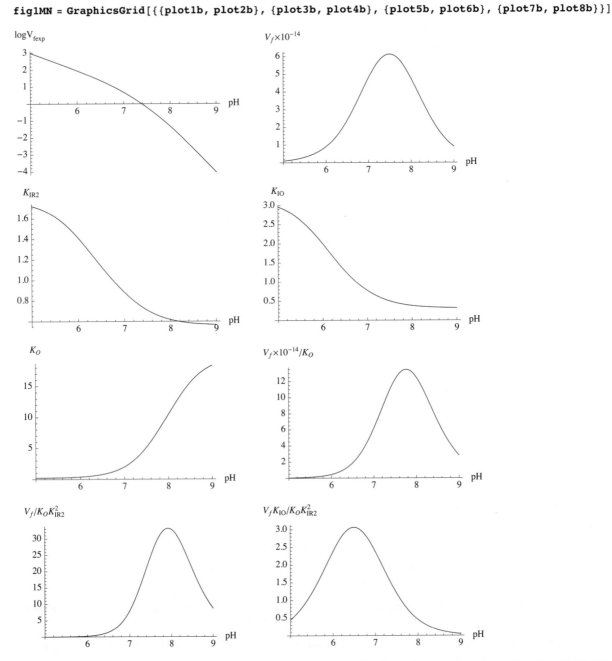

Fig. 8.4 Plots of pH dependencies of kinetic parameters for the random mechanism when *m* = 2 and *n* = -2, as in EC 1.9.6.1. The second plot yields $pK_{1\,ER2O}$, $pK2_{ER2O}$, and kfEt. The fifth plot yields $pK_{1\,ER2}$, $pK_{2\,ER2}$, and K_{ER2O}. The sixth plot yields $pK_{1\,E}$, $pK_{2\,E}$, and K_{ER2}.

The reaction velocity v is given by

rx1961random[[9]]

$$\frac{1. \times 10^{15}\ 10^{-2\,pH}}{\left(1 + 10^{7.-pH} + 10^{-8.+pH}\right)\left(1 + \frac{2\,\left(1+10^{6.-pH}+10^{-7.+pH}\right)}{\left(1+10^{7.-pH}+10^{-8.+pH}\right)o} + \frac{2\,\left(1+10^{6.-pH}+10^{-7.+pH}\right)}{\left(1+10^{7.-pH}+10^{-8.+pH}\right)r^2} + \frac{2\,\left(1+10^{6.5-pH}+10^{-7.5+pH}\right)}{\left(1+10^{7.-pH}+10^{-8.+pH}\right)o\,r^2}\right)}$$

This function can be treated like an experimental system in the sense that v can be calculated at any [R], [O], and pH for a specified set of pH-independent kinetic parameters. The rate equations at pHs 5, 6, 7, 8, and 9 are

rx1961random[[9]] /. pH → {5, 6, 7, 8, 9}

$$\left\{\frac{990.089}{1 + \frac{0.218018}{o} + \frac{0.218018}{r^2} + \frac{0.646052}{o\,r^2}}, \frac{90.8265}{1 + \frac{0.381471}{o} + \frac{0.381471}{r^2} + \frac{0.761835}{o\,r^2}},\right.$$

$$\frac{4.7619}{1 + \frac{2.}{o} + \frac{2.}{r^2} + \frac{1.55472}{o\,r^2}}, \frac{0.047619}{1 + \frac{10.4857}{o} + \frac{10.4857}{r^2} + \frac{3.99419}{o\,r^2}}, \left.\frac{0.0000908265}{1 + \frac{18.3471}{o} + \frac{18.3471}{r^2} + \frac{5.9266}{o\,r^2}}\right\}$$

The easiest property to calculate is V_{fexp} because it can be obtained by raising [R] and [O] to infinity.

rx1961random[[9]] /. r → ∞ /. o → ∞

$$\frac{1. \times 10^{15}\ 10^{-2\,pH}}{1 + 10^{7.-pH} + 10^{-8.+pH}}$$

rx1961random[[9]] /. r → ∞ /. o → ∞ /. pH → {5, 6, 7, 8, 9}

{990.089, 90.8265, 4.7619, 0.047619, 0.0000908265}

K_{IR2} at pH 7 is given by

rx1961random[[3]] /. pH → 7.

0.88168

K_{IR2}^2 at pH 7 is given by

(rx1961random[[3]] /. pH → 7.)^2

0.77736

K_{IO} at pH 7 is given by

rx1961random[[4]] /. pH → 7.

0.77736

K_O at pH 7 is given by

rx1961random[[5]] /. pH → 7.

2.

rx1961random[[9]] /. pH → 7.

$$\frac{4.7619}{1 + \frac{2.}{o} + \frac{2.}{r^2} + \frac{1.55472}{o\,r^2}}$$

8.2.3 3D plots for the nitrate reductase type of reaction (O + 2R → products)

An overview of rate equation rx1961random[[9]] can be obtained by plotting rx1961random[[9]] at 4 different pHs versus [O] and [R] in three-dimensions.

```
plot1c = Plot3D[Evaluate[ rx1961random[[9]] /. pH → 5.`],
                          ─────────────────
                                100
    {r, 0.0001`, 20}, {o, 0.0001`, 1}, ViewPoint → {-2, -2, 1},
    AxesLabel → {"    [R]", "[O]      ", "v/100   "}, PlotRange → {0, 12},
    Lighting → {{"Ambient", GrayLevel[1]}}, PlotLabel → "pH 5.0", DisplayFunction → Identity];

plot2c = Plot3D[Evaluate[ rx1961random[[9]] /. pH → 6.`], {r, 0.0001`, 10}, {o, 0.0001`, 3},
                          ─────────────────
                                10
    ViewPoint → {-2, -2, 1}, AxesLabel → {"    [R]", "[O]     ", "  v/10"}, PlotRange → {0, 10},
    Lighting → {{"Ambient", GrayLevel[1]}}, PlotLabel → "pH 6.0", DisplayFunction → Identity];

plot3c = Plot3D[Evaluate[rx1961random[[9]] /. pH → 7.`], {r, 0.0001`, 5}, {o, 0.0001`, 10},
    ViewPoint → {-2, -2, 1}, AxesLabel → {"    [R]", "[O]      ", "v"}, PlotRange → {0, 5.5`},
    Lighting → {{"Ambient", GrayLevel[1]}}, PlotLabel → "pH 7.0", DisplayFunction → Identity];

plot4c = Plot3D[Evaluate[rx1961random[[9]] 100 /. pH → 8.`], {r, 0.0001`, 10}, {o, 0.0001`, 50},
    ViewPoint → {-2, -2, 1}, AxesLabel → {"    [R]", "[O]      ", "vx100"}, PlotRange → {0, 5},
    Lighting → {{"Ambient", GrayLevel[1]}}, PlotLabel → "pH 8.0", DisplayFunction → Identity];

fig2new = Show[GraphicsGrid[{{plot1c, plot2c}, {plot3c, plot4c}}]]
```

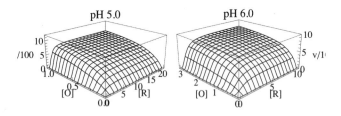

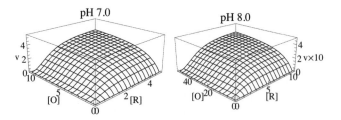

Fig. 8.5 Plots of initial velocities of the catalyzed reaction 2R + O → products as functions of [O] and [R] at four pHs for Mechanism III. The number n of hydrogen ions produced in the rate-determining reaction for the ordered mechanism is -2. Note the change in scale of the ordinate is 10^4 over the range pH 5 to pH 8.

This mechanism will not be applied to the ferredoxin-nitrite reductase reaction (EC 1.7.7.1) because $n = 8$ and the sensitivity to pH is so great.

8.3 Estimation of kinetic parameters for random O + 3R → products

This section is based on vORrand3 (see Section 8.1.1).

vORrand3

$$\frac{\text{kfet kIO o } r^3}{\text{kIO kIR3}^3 \text{ kO + kIR3}^3 \text{ kO o + kIO kO } r^3 + \text{kIO o } r^3}$$

Since there are 4 kinetic parameters, 4 velocities must be measured. In the first test, the following pairs of velocities are calculated: {o,r} at {100,100}, {1,100}, {100,5}, {5,5}.

vel3 = vORrand3 /. kfet → 1 /. kIR3 → 20 /. kIO → 15 /. kO → 5

$$\frac{15 \text{ o } r^3}{600\,000 + 40\,000 \text{ o} + 75 \, r^3 + 15 \text{ o } r^3}$$

vel3 /. o → 100 /. r → 100 // N

0.949607

vel3 /. o → 1 /. r → 100 // N

0.16549

vel3 /. o → 100 /. r → 5 // N

0.0390879

vel3 /. o → 5 /. r → 5 // N

0.0114504

These four velocities can be used to estimate the four kinetic parameters by use of the following program that is based on vORrand3.

```
calckinparsrandOR3[v1_, o1_, r1_, v2_, o2_, r2_, v3_, o3_, r3_, v4_, o4_, r4_] :=
Module[{}, (*This program calculates kfet, kO,kIO,
  and kIR3 from four experimental velocities for O + 3R →
   products at 4 pairs of substrate concentrations on the assumption that
    the mechanism is random. The first velocity is at high [O] and high [R],
   the second velocity is at low [O] and high [R],  the third velocity is at
   high [O] and low [R], and the 4th velocity is at medium [O] and medium [R].*)
```

$$\text{Solve}\Big[\Big\{v1 == \frac{\text{kfet kIO o1 r1}^3}{\text{kIO kIR3}^3 \text{ kO + kIR3}^3 \text{ kO o1 + kIO kO r1}^3 + \text{kIO o1 r1}^3},$$

$$v2 == \frac{\text{kfet kIO o2 r2}^3}{\text{kIO kIR3}^3 \text{ kO + kIR3}^3 \text{ kO o2 + kIO kO r2}^3 + \text{kIO o2 r2}^3},$$

$$v3 == \frac{\text{kfet kIO o3 r3}^3}{\text{kIO kIR3}^3 \text{ kO + kIR3}^3 \text{ kO o3 + kIO kO r3}^3 + \text{kIO o3 r3}^3},$$

$$v4 == \frac{\text{kfet kIO o4 r4}^3}{\text{kIO kIR3}^3 \text{ kO + kIR3}^3 \text{ kO o4 + kIO kO r4}^3 + \text{kIO o4 r4}^3}\Big\}, \{\text{kfet, kIR3, kO, kIO}\}\Big]\Big]$$

```
calckinparsrandOR3[.9496, 100, 100, .1655, 1, 100, .03909, 100, 5, 0.01145, 5, 5]
```

{{kfet → 0.999988, kIO → 15.0022, kO → 4.99955, kIR3 → -10.0005 - 17.3214 i},
 {kfet → 0.999988, kIO → 15.0022, kO → 4.99955, kIR3 → -10.0005 + 17.3214 i},
 {kfet → 0.999988, kIO → 15.0022, kO → 4.99955, kIR3 → 20.001}}

```
Dimensions[calckinparsrandOR3[.9496, 100, 100, .1655, 1, 100, .03909, 100, 5, 0.01145, 5, 5]]
```

{3, 4}

Since only the third output is physically realistic, it can be obtained as follows:

```
Take[calckinparsrandOR3[.9496, 100, 100, .1655, 1, 100, .03909, 100, 5, 0.01145, 5, 5], {3}]
```

{{kfet → 0.999988, kIO → 15.0022, kO → 4.99955, kIR3 → 20.001}}

```
ln1 = {kfet, kIR3, kO, kIO} /.
    Take[calckinparsrandOR3[.9496, 100, 100, .1655, 1, 100, .03909, 100, 5, 0.01145, 5, 5], {3}]
```

{{0.999988, 20.001, 4.99955, 15.0022}}

Put in 5% errors, one at a time.

```
Take[calckinparsrandOR3[1.05 * .9496,
    100, 100, .1655, 1, 100, .03909, 100, 5, 0.01145, 5, 5], {3}]
```

{{kfet → 1.05335, kIO → 14.9643, kO → 5.3197, kIR3 → 19.9328}}

```
ln2 = {kfet, kIR3, kO, kIO} /. Take[
    calckinparsrandOR3[1.05 * .9496, 100, 100, .1655, 1, 100, .03909, 100, 5, 0.01145, 5, 5], {3}]
```

{{1.05335, 19.9328, 5.3197, 14.9643}}

```
Take[calckinparsrandOR3[.9496, 100, 100,
    1.05 * .1655, 1, 100, .03909, 100, 5, 0.01145, 5, 5], {3}]
```

{{kfet → 0.997089, kIO → 15.0179, kO → 4.69524, kIR3 → 20.4105}}

```
ln3 = {kfet, kIR3, kO, kIO} /. Take[
    calckinparsrandOR3[.9496, 100, 100, 1.05 * .1655, 1, 100, .03909, 100, 5, 0.01145, 5, 5], {3}]
```

{{0.997089, 20.4105, 4.69524, 15.0179}}

```
Take[calckinparsrandOR3[.9496, 100, 100,
    .1655, 1, 100, 1.05 * .03909, 100, 5, 0.01145, 5, 5], {3}]
```

{{kfet → 0.999827, kIO → 16.2816, kO → 4.99795, kIR3 → 20.1348}}

```
ln4 = {kfet, kIR3, kO, kIO} /. Take[
    calckinparsrandOR3[.9496, 100, 100, .1655, 1, 100, 1.05 * .03909, 100, 5, 0.01145, 5, 5], {3}]
```

{{0.999827, 20.1348, 4.99795, 16.2816}}

```
Take[calckinparsrandOR3[.9496, 100, 100,
    .1655, 1, 100, .03909, 100, 5, 1.05 * 0.01145, 5, 5], {3}]
```

{{kfet → 1.00002, kIO → 13.834, kO → 5.00243, kIR3 → 19.5305}}

```
ln5 = {kfet, kIR3, kO, kIO} /. Take[
    calckinparsrandOR3[.9496, 100, 100, .1655, 1, 100, .03909, 100, 5, 1.05 * 0.01145, 5, 5], {3}]
```

{{1.00002, 19.5305, 5.00243, 13.834}}

Table 8.3 Values of kinetic parameters for random O + 3R → products obtained using {100,100}, {1,100}, {100,1}, and {5,5} and testing the effects of 5% errors in the measured velocities

```
TableForm[Round[{ln1[[1]], ln2[[1]], ln3[[1]], ln4[[1]], ln5[[1]]}, 0.01], TableHeadings →
   {{"No errors", "1.05*v1", "1.05*v2", "1.05*v3", "1.05*v4"}, {"kfet", "kIR3", "kO", "kIO"}}]
```

	kfet	kIR3	kO	kIO
No errors	1.	20.	5.	15.
1.05*v1	1.05	19.93	5.32	14.96
1.05*v2	1.	20.41	4.7	15.02
1.05*v3	1.	20.13	5.	16.28
1.05*v4	1.	19.53	5.	13.83

The fifth kinetic parameter, kRm, can be calculated by using equation 10 for any possible mechanism for random O + mR → products, which is $K_{IR3}^3 K_O = K_{R3}^3 K_{IO}$.

8.4 Discussion

The general rate equation vORrand for random O + mR → products is derived using Solve, and it is used to derive more specialized rate equations vORrand1, vORrand2, vORrand3, and vORrand6 for m = 1, 2, 3, and 6. The number of reductant molecules m is not a kinetic parameter because it is a property of the reaction being catalyzed. 3D plots are presented for m - 2, 3, and 6. The estimation of kinetic parameters using the minimum number of velocity measurements is demonstrated for m = 2, and a fifth paremeter is calculated using a thermodynamic cycle. When the program calckinparsrandOR2 is applied to velocity data for the two ordered mechanisms (in the preceeding chapter), unreasonable values are obtained for kIO, but correct values are obtained for the other parameters. This illustrates again the use of a more general program to identify the type of mechanism.

The effects of pH when the enzymatic site and the enzyme-substrate complexes each have two pKs and hydrogen ions are consumed in the rate-determining reaction are discussed. Plots of the pH dependencies of the kinetc parameters are presented, and it is shown that all the pH-independent parameters can be calculated from bell-shaped plots.

Thermodynamics is very different from kinetics because when the equilibrium composition of a system is calculated, this composition will satisify ALL possible equilibrium constant expressions that can be obtained by adding and subtracting the reactions in the independent set, not simply the set of equilibrium constant expressions used to calculate the equilibrium composition. This is very different from the situation in steady-state enzyme kinetics where every possible reaction in the mechanism must be included.

Chapter 9 Inhibition and activation of A → products

9.1 Introduction

9.2 Competitive inhibition

9.3 Uncompetitive inhibition

9.4 Mixed inhibition

9.5 Essential activation

9.6 Essential activation when two, three, and four molecules of activator are bound in a single reaction

- 9.6.1 Two molecules of activator X are bound

- 9.6.2 Three molecules of activator X are bound

- 9.6.3 Four molecules of activator X are bound

9.7 Mixed activation when one X is bound

9.8 Mixed activation when two X are bound in a single reaction

9.9 Derivation of the simplest rate equation that is sigmoid in both A and X

9.10 Derivation of the rate equation when 2A is bound in two reactions and 2X is bound in one reaction

- 9.10.1 Positive cooperativity with respect to A

- 9.10.2 Negative cooperativity with respect to A

9.11 Discussion

9.1 Introduction

Inhibition and activation lead to a large number of rapid-equilibrium rate equations for A → products. Inhibition can be competitive, uncompetitive, or mixed. Activation can be essential or mixed. This chapter uses the terminology of Cornish-Bowden [21].

Product inhibition has been described earlier (Section 2.3.2), where it was pointed out that the extent of inhibition is determined by the Michaelis constant of the product in the rate equation that includes the reverse reaction.

Inhibition and activation by hydrogen ions have already been described in previous chapters.

9.2 Competitive inhibition

The mechanism for competitive inhibition of A → products is

E + A = EA   kA = e*a/ea
E + I = EI   kI = e*i/ei
EA → products   v = kf*ea = kf*et*(ea/et) = kfet*(ea/et)

The rapid-equilibrium rate equation can be derived by use of Solve.

    Solve[{kA == e * a / ea, kI == e * i / ei, et == e + ea + ei}, {ea}, {e, ei}]

$$\left\{\left\{ea \to \frac{a\,et\,kI}{i\,kA + a\,kI + kA\,kI}\right\}\right\}$$

When a competitive inhibitor is present, the velocity is named vcomp.

    vcomp = $\dfrac{a\,kfet\,kI}{i\,kA + a\,kI + kA\,kI}$

$$\frac{a\,kfet\,kI}{i\,kA + a\,kI + kA\,kI}$$

For test calculations, the following kinetic parameters are chosen:

    vcomp /. kfet → 1 /. kA → 10 /. kI → 20

$$\frac{20\,a}{200 + 20\,a + 10\,i}$$

The characteristics of vcomp are shown by a 3D plot.

    plot1a = Plot3D[Evaluate[vcomp /. kfet → 1 /. kA → 10 /. kI → 20],
       {a, 0.00001, 50}, {i, 0.00001, 100}, ViewPoint → {-2, -2, 1}, PlotRange -> {0, 1},
       AxesLabel → {"[A]", "[I]"}, Lighting -> {{"Ambient", GrayLevel[1]}}, PlotLabel → "vcomp"];

    plot1b = Plot3D[Evaluate[D[vcomp /. kfet → 1 /. kA → 10 /. kI → 20, a]], {a, 0.00001, 50},
       {i, 0.00001, 100}, ViewPoint → {-2, -2, 1}, PlotRange -> {0, .1}, AxesLabel → {"[A]", "[I]"},
       Lighting -> {{"Ambient", GrayLevel[1]}}, PlotLabel → "dvcomp/d[A]"];

    plot1c = Plot3D[Evaluate[D[vcomp /. kfet → 1 /. kA → 10 /. kI → 20, i]],
       {a, 0.00001, 50}, {i, 0.00001, 100}, ViewPoint → {-2, -2, 1},
       PlotRange -> {-.02, .001}, Lighting -> {{"Ambient", GrayLevel[1]}},
       AxesLabel → {"[A]", "[I]"}, PlotLabel → "dvcomp/d[I]"];

```
plot1d = Plot3D[Evaluate[D[vcomp /. kfet → 1 /. kA → 10 /. kI → 20, a, i]],
    {a, 0.00001, 50}, {i, 0.00001, 100}, ViewPoint → {-2, -2, 1},
    PlotRange -> {-.005, .005}, AxesLabel → {"[A]", "[I]"},
    Lighting -> {{"Ambient", GrayLevel[1]}}, PlotLabel → "d(dvcomp/d[A])/d[I]"];

fig1vcomp = GraphicsArray[{{plot1a, plot1b}, {plot1c, plot1d}}]
```

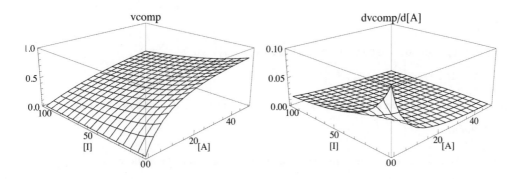

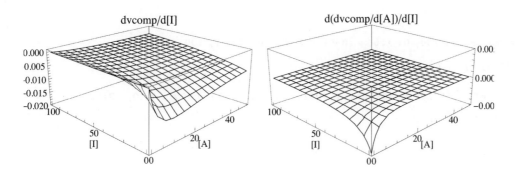

Fig. 9.1 Plots of vcomp, ∂vcomp$/\partial$[A], ∂vcomp$/\partial$[I], and $\partial(\partial$vcomp$/\partial$[A]$)/\partial$[I] versus [A] and [I] for competitive inhibition of A $\rightarrow$ products when kfet = 1, kA = 10, and kI = 20.

Since there are three kinetic parameters, three velocities have to be measured.

vcomp /. kfet → 1 /. kA → 10 /. kI → 20 /. a → 100 /. i → 0 // N

0.909091

vcomp /. kfet → 1 /. kA → 10 /. kI → 20 /. a → 1 /. i → 0 // N

0.0909091

vcomp /. kfet → 1 /. kA → 10 /. kI → 20 /. a → 1 /. i → 100 // N

0.0163934

The rate equation vcomp can be used to write a program to estimate the 3 kinetic parameters from 3 velocity measurements.

```
calckinparsAcomp[v1_, a1_, i1_, v2_, a2_, i2_, v3_, a3_, i3_] :=
 Module[{}, (*This program calculates kfet, kA,
   and kI from 3 experimental velocities for A → products with a
     competitive inhibitor I.  The first velocity is at high [A] and zero [I],
     the second velocity is at low [A] and zero [I], and the third velocity is at low [A]
     and high [I]. High and low are with respect to the two dissociation constants.*)
```

$$\text{Solve}\left[\left\{v1 == \frac{a1\,kfet\,kI}{i1\,kA + a1\,kI + kA\,kI},\ v2 == \frac{a2\,kfet\,kI}{i2\,kA + a2\,kI + kA\,kI},\ v3 == \frac{a3\,kfet\,kI}{i3\,kA + a3\,kI + kA\,kI}\right\},\right.$$

```
  {kfet, kA, kI}]]
```

```
calckinparsAcomp[.9091, 100, 0, .09091, 1, 0, .01639, 1, 100]
```

```
{{kfet → 1.00001, kI → 19.9946, kA → 10.}}
```

```
row1 = {kfet, kA, kI} /. calckinparsAcomp[.9091, 100, 0, .09091, 1, 0, .01639, 1, 100]
```

```
{{1.00001, 10., 19.9946}}
```

These values are correct. The effects of 5% errors in the velocities, one by one, are calculated as follows:

```
calckinparsAcomp[1.05 * .9091, 100, 0, .09091, 1, 0, .01639, 1, 100]
```

```
{{kfet → 1.05588, kI → 20.1004, kA → 10.6145}}
```

```
row2 = {kfet, kA, kI} /. calckinparsAcomp[1.05 * .9091, 100, 0, .09091, 1, 0, .01639, 1, 100]
```

```
{{1.05588, 10.6145, 20.1004}}
```

```
calckinparsAcomp[.9091, 100, 0, 1.05 * .09091, 1, 0, .01639, 1, 100]
```

```
{{kfet → 0.994747, kI → 18.7404, kA → 9.42105}}
```

```
row3 = {kfet, kA, kI} /. calckinparsAcomp[.9091, 100, 0, 1.05 * .09091, 1, 0, .01639, 1, 100]
```

```
{{0.994747, 9.42105, 18.7404}}
```

```
calckinparsAcomp[.9091, 100, 0, .09091, 1, 0, 1.05 * .01639, 1, 100]
```

```
{{kfet → 1.00001, kI → 21.2278, kA → 10.}}
```

```
row4 = {kfet, kA, kI} /. calckinparsAcomp[.9091, 100, 0, .09091, 1, 0, 1.05 * .01639, 1, 100]
```

```
{{1.00001, 10., 21.2278}}
```

Table 9.1 Estimation of kinetic parameters for competitive inhibition of A → products using velocities at {a,i} = {100,0}, {1,0}, {1,100}

```
TableForm[Round[{row1[[1]], row2[[1]], row3[[1]], row4[[1]]}, 0.01],
 TableHeadings -> {{"No errors", "1.05*v1", "1.05*v2", "1.05*v3"}, {"kfet", "kA", "kI"}}]
```

	kfet	kA	kI
No errors	1.	10.	19.99
1.05*v1	1.06	10.61	20.1
1.05*v2	0.99	9.42	18.74
1.05*v3	1.	10.	21.23

9.3 Uncompetitive inhibition

An uncompetitive inhibitor is bound by an enzyme-substrate complex. The mechanism for uncompetitive inhibition of A → products is

E + A = EA   kA = e*a/ea
EA + I = EAI   kuI = ea*i/eai
EA → products   v = kf*ea

The rapid-equilibrium rate equation can be derived by use of Solve.

Solve[{kA == e * a / ea, kuI == ea * i / eai, et == e + ea + eai}, {ea}, {e, eai}]

$$\left\{\left\{ea \to \frac{a\ et\ kuI}{a\ i + a\ kuI + kA\ kuI}\right\}\right\}$$

When an uncompetitive inhibitor is present, the velocity is given by vuncomp.

$$vuncomp = \frac{a\ kfet\ kuI}{a\ i + a\ kuI + kA\ kuI}$$

$$\frac{a\ kfet\ kuI}{a\ i + a\ kuI + kA\ kuI}$$

For test calculations, kfet = 1, kA = 10, and kuI = 30 are chosen.

vuncomp /. kfet → 1 /. kA → 10 /. kuI → 30

$$\frac{30\ a}{300 + 30\ a + a\ i}$$

The characteristics of vuncomp are shown by a 3D plot.

```
plot2a = Plot3D[Evaluate[vuncomp /. kfet → 1 /. kA → 10 /. kuI → 30],
    {a, 0.00001, 50}, {i, 0.00001, 100}, ViewPoint → {-2, -2, 1}, PlotRange -> {0, 1},
    AxesLabel → {"[A]", "[I]"}, Lighting -> {{"Ambient", GrayLevel[1]}}, PlotLabel → "vuncomp"];

plot2b = Plot3D[Evaluate[D[vuncomp /. kfet → 1 /. kA → 10 /. kuI → 30, a]], {a, 0.00001, 50},
    {i, 0.00001, 100}, ViewPoint → {-2, -2, 1}, PlotRange -> {0, .15}, AxesLabel → {"[A]", "[I]"},
    Lighting -> {{"Ambient", GrayLevel[1]}}, PlotLabel → "dvuncomp/d[A]"];

plot2c = Plot3D[Evaluate[D[vuncomp /. kfet → 1 /. kA → 10 /. kuI → 30, i]],
    {a, 0.00001, 50}, {i, 0.00001, 100}, ViewPoint → {-2, -2, 1},
    PlotRange -> {0, -.03}, AxesLabel → {"[A]", "[I]"},
    Lighting -> {{"Ambient", GrayLevel[1]}}, PlotLabel → "dvuncomp/d[I]"];

plot2d = Plot3D[Evaluate[D[vuncomp /. kfet → 1 /. kA → 10 /. kuI → 30, a, i]],
    {a, 0.00001, 50}, {i, 0.00001, 100}, ViewPoint → {-2, -2, 1},
    PlotRange -> {-.001, .001}, AxesLabel → {"[A]", "[I]"},
    Lighting -> {{"Ambient", GrayLevel[1]}}, PlotLabel → "d(dvuncomp/d[A])/d[I]"];
```

```
fig2vuncomp = GraphicsArray[{{plot2a, plot2b}, {plot2c, plot2d}}]
```

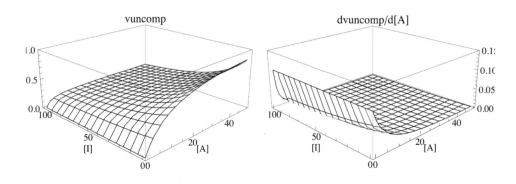

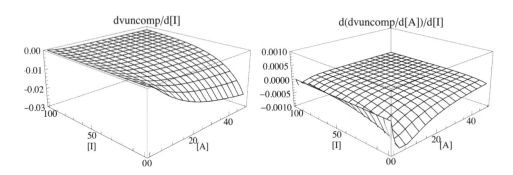

Fig. 9.2 Plots of vuncomp, ∂vuncomp/∂[A], ∂vuncomp/∂[I], and $\partial(\partial$vuncomp/∂[A])/∂[I] versus [A] and [I] for uncompetitive inhibition of A → products when kfet = 1, kA = 10, and kuI = 30.

Adding an uncompetitive inhibitor always inhibits, but Michaelis-Menten plots are obtained at each inhibitor concentration.

Rate equation vuncomp can be used to write a program to estimate the 3 kinetic parameters from 3 velocity measurements.

```
vuncomp /. kfet → 1 /. kA → 10 /. kuI → 30 /. a → 100 /. i → 0 // N

0.909091

vuncomp /. kfet → 1 /. kA → 10 /. kuI → 30 /. a → 1 /. i → 0 // N

0.0909091

vuncomp /. kfet → 1 /. kA → 10 /. kuI → 30 /. a → 1 /. i → 100 // N

0.0697674
```

```
calckinparsAucomp[v1_, a1_, i1_, v2_, a2_, i2_, v3_, a3_, i3_] :=
  Module[{}, (*This program calculates kfet, kA,
    and kuI from 3 experimental velocities for A → products with an
      uncompetitive inhibitor I.  The first velocity is at high [A] and zero [I],
      the second velocity is at low [A] and zero [I], and the third is at low [A] and
      high [I]. High and low are with respect to the two dissociation constants.*)
```

$$\text{Solve}\left[\left\{v1 == \frac{a1\ kfet\ kuI}{a1\ i1 + a1\ kuI + kA\ kuI},\ v2 == \frac{a2\ kfet\ kuI}{a2\ i2 + a2\ kuI + kA\ kuI},\right.\right.$$
$$\left.\left. v3 == \frac{a3\ kfet\ kuI}{a3\ i3 + a3\ kuI + kA\ kuI}\right\},\ \{kfet,\ kA,\ kuI\}\right]\right]$$

```
calckinparsAucomp[.9091, 100, 0, .09091, 1, 0, .06977, 1, 100]
```

{{kfet → 1.00001, kuI → 30.0034, kA → 10.}}

```
ln1 = {kfet, kA, kuI} /. calckinparsAucomp[.9091, 100, 0, .09091, 1, 0, .06977, 1, 100]
```

{{1.00001, 10., 30.0034}}

These values are correct, but the effects of errors in measurements of velocities have to be considered.

```
calckinparsAucomp[1.05 * .9091, 100, 0, .09091, 1, 0, .06977, 1, 100]
```

{{kfet → 1.05588, kuI → 28.416, kA → 10.6145}}

```
ln2 = {kfet, kA, kuI} /. calckinparsAucomp[1.05 * .9091, 100, 0, .09091, 1, 0, .06977, 1, 100]
```

{{1.05588, 10.6145, 28.416}}

```
calckinparsAucomp[.9091, 100, 0, 1.05 * .09091, 1, 0, .06977, 1, 100]
```

{{kfet → 0.994747, kuI → 26.0657, kA → 9.42105}}

```
ln3 = {kfet, kA, kuI} /. calckinparsAucomp[.9091, 100, 0, 1.05 * .09091, 1, 0, .06977, 1, 100]
```

{{0.994747, 9.42105, 26.0657}}

```
calckinparsAucomp[.9091, 100, 0, .09091, 1, 0, 1.05 * .06977, 1, 100]
```

{{kfet → 1.00001, kuI → 37.7297, kA → 10.}}

```
ln4 = {kfet, kA, kuI} /. calckinparsAucomp[.9091, 100, 0, .09091, 1, 0, 1.05 * .06977, 1, 100]
```

{{1.00001, 10., 37.7297}}

Table 9.2 Estimation of kinetic parameters for uncompetitive inhibition of A → products using velocities at {a,i} = {100,0}, {1,0}, {1,100}

```
TableForm[Round[{ln1[[1]], ln2[[1]], ln3[[1]], ln4[[1]]}, 0.01],
  TableHeadings -> {{"No errors", "1.05*v1", "1.05*v2", "1.05*v3"}, {"kfet", "kA", "kuI"}}]
```

	kfet	kA	kuI
No errors	1.	10.	30.
1.05*v1	1.06	10.61	28.42
1.05*v2	0.99	9.42	26.07
1.05*v3	1.	10.	37.73

Other pairs of concentrations can be used to improve the accuracy of the determination of the kinetic parameters.

9.4 Mixed inhibition

The mechanism for mixed inhibition is

E + A = EA	kA = e*a/ea
E + I = EI	kI = e*i/ei
EA + I = EAI	kuI = ea*i/eai
EA → products	v = kf*ea

The rapid-equilibrium rate equation can be derived by use of Solve.

```
Solve[{kA == e * a / ea, kI == e * i / ei, kuI == ea * i / eai, et == e + ea + ei + eai}, {ea}, {e, ei, eai}]
```

$$\left\{\left\{ea \to \frac{a\ et\ kI\ kuI}{a\ i\ kI + i\ kA\ kuI + a\ kI\ kuI + kA\ kI\ kuI}\right\}\right\}$$

$$vmix = \frac{a\ kfet\ kI\ kuI}{a\ i\ kI + i\ kA\ kuI + a\ kI\ kuI + kA\ kI\ kuI}$$

$$\frac{a\ kfet\ kI\ kuI}{a\ i\ kI + i\ kA\ kuI + a\ kI\ kuI + kA\ kI\ kuI}$$

There are 4 kinetic parameters, and so 4 velocities have to be determined to estimate the kinetic parameters from the minimum number of measurements of velocities. The following values of the kinetic parameters are used for a test.

```
vmix /. kfet → 1 /. kA → 10 /. kI → 20 /. kuI → 30
```

$$\frac{600\ a}{6000 + 600\ a + 300\ i + 20\ a\ i}$$

The characteristics of vmix are shown by a 3D plot.

```
plot3a = Plot3D[Evaluate[vmix /. kfet → 1 /. kA → 10 /. kI → 20 /. kuI → 30],
    {a, 0.00001, 50}, {i, 0.00001, 100}, ViewPoint → {-2, -2, 1}, PlotRange -> {0, 1},
    Lighting -> {{"Ambient", GrayLevel[1]}}, AxesLabel → {"[A]", "[I]"}, PlotLabel → "vmix"];

plot3b = Plot3D[Evaluate[D[vmix /. kfet → 1 /. kA → 10 /. kI → 20 /. kuI → 30, a]],
    {a, 0.00001, 50}, {i, 0.00001, 100}, ViewPoint → {-2, -2, 1},
    PlotRange -> {0, .1}, Lighting -> {{"Ambient", GrayLevel[1]}},
    AxesLabel → {"[A]", "[I]"}, PlotLabel → "dvmix/d[A]"];

plot3c = Plot3D[Evaluate[D[vmix /. kfet → 1 /. kA → 10 /. kI → 20 /. kuI → 30, i]],
    {a, 0.00001, 50}, {i, 0.00001, 100}, ViewPoint → {-2, -2, 1},
    PlotRange -> {-.03, 0}, Lighting -> {{"Ambient", GrayLevel[1]}},
    AxesLabel → {"[A]", "[I]"}, PlotLabel → "dvmix/d[I]"];

plot3d = Plot3D[Evaluate[D[vmix /. kfet → 1 /. kA → 10 /. kI → 20 /. kuI → 30, a, i]],
    {a, 0.00001, 50}, {i, 0.00001, 100}, ViewPoint → {-2, -2, 1},
    PlotRange -> {-.005, 0}, Lighting -> {{"Ambient", GrayLevel[1]}},
    AxesLabel → {"[A]", "[I]"}, PlotLabel → "d(dvmix/d[A])/d[I]"];
```

fig3vmix = GraphicsArray[{{plot3a, plot3b}, {plot3c, plot3d}}]

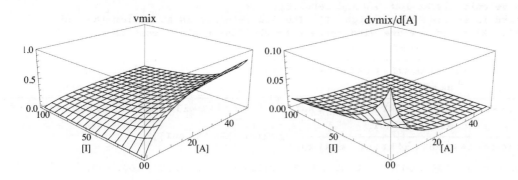

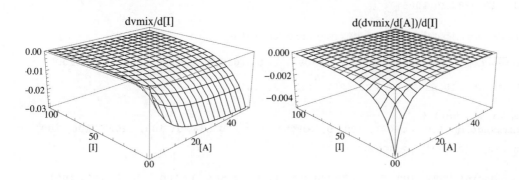

Fig. 9.3 Plots of vmix, $\partial\text{vmix}/\partial[A]$, $\partial\text{vmix}/\partial[I]$, and $\partial(\partial\text{vmix}/\partial[A])/\partial[I]$ versus [A] and [I] for mixed inhibition of A → products when kfet = 1, kA = 10, kI = 20, and kuI = 30.

Four velocities can be measured at {a,i} = {100,0},{1,0},{1,100}, {100,100}

vmix /. kfet → 1 /. kA → 10 /. kI → 20 /. kuI → 30 /. a → 100 /. i → 0 // N

0.909091

vmix /. kfet → 1 /. kA → 10 /. kI → 20 /. kuI → 30 /. a → 1 /. i → 0 // N

0.0909091

vmix /. kfet → 1 /. kA → 10 /. kI → 20 /. kuI → 30 /. a → 1 /. i → 100 // N

0.015544

vmix /. kfet → 1 /. kA → 10 /. kI → 20 /. kuI → 30 /. a → 100 /. i → 100 // N

0.202703

The following program was written to estimate the kinetic parameters for mixed inhibition.

```
calckinparsmixedcomp[v1_, a1_, i1_, v2_, a2_, i2_, v3_, a3_, i3_, v4_, a4_, i4_] := Module[{},
   (*This program calculates kfet, kA,kI,and kuI from 4 experimental velocities for A →
    products with a mixed inhibitor I.  The first velocity is at high [A] and zero [I],
    the second velocity is at low [A] and zero [I],
    and the third is at low [A] and high [I]. The 4th velocity is at medium [A] and
    medium [I].  High and low are with respect to the dissociation constants.*)
   Solve[{v1 == (a1 kfet kI kuI)/(a1 i1 kI + i1 kA kuI + a1 kI kuI + kA kI kuI),
      v2 == (a2 kfet kI kuI)/(a2 i2 kI + i2 kA kuI + a2 kI kuI + kA kI kuI), v3 == (a3 kfet kI kuI)/(a3 i3 kI + i3 kA kuI + a3 kI kuI + kA kI kuI),
      v4 == (a4 kfet kI kuI)/(a4 i4 kI + i4 kA kuI + a4 kI kuI + kA kI kuI)}, {kfet, kA, kI, kuI}]]
```

```
calckinparsmixedcomp[.9091, 100, 0, .09091, 1, 0, .01554, 1, 100, .2027, 100, 100]
```

{{kfet → 1.00001, kA → 10., kI → 19.993, kuI → 30.0005}}

```
line1 = {kfet, kA, kI, kuI} /.
   calckinparsmixedcomp[.9091, 100, 0, .09091, 1, 0, .01554, 1, 100, .2027, 100, 100]
```

{{1.00001, 10., 19.993, 30.0005}}

These kinetic parameters are correct, but the effects of 5% errors are calculated as follows:

```
calckinparsmixedcomp[1.05 * .9091, 100, 0, .09091, 1, 0, .01554, 1, 100, .2027, 100, 100]
```

{{kfet → 1.05588, kA → 10.6145, kI → 20.1201, kuI → 27.9692}}

```
line2 = {kfet, kA, kI, kuI} /.
   calckinparsmixedcomp[1.05 * .9091, 100, 0, .09091, 1, 0, .01554, 1, 100, .2027, 100, 100]
```

{{1.05588, 10.6145, 20.1201, 27.9692}}

```
calckinparsmixedcomp[.9091, 100, 0, 1.05 * .09091, 1, 0, .01554, 1, 100, .2027, 100, 100]
```

{{kfet → 0.994747, kA → 9.42105, kI → 18.737, kuI → 30.2072}}

```
line3 = {kfet, kA, kI, kuI} /.
   calckinparsmixedcomp[.9091, 100, 0, 1.05 * .09091, 1, 0, .01554, 1, 100, .2027, 100, 100]
```

{{0.994747, 9.42105, 18.737, 30.2072}}

```
calckinparsmixedcomp[.9091, 100, 0, .09091, 1, 0, 1.05 * .01554, 1, 100, .2027, 100, 100]
```

{{kfet → 1.00001, kA → 10., kI → 21.3119, kuI → 29.7245}}

```
line4 = {kfet, kA, kI, kuI} /.
   calckinparsmixedcomp[.9091, 100, 0, .09091, 1, 0, 1.05 * .01554, 1, 100, .2027, 100, 100]
```

{{1.00001, 10., 21.3119, 29.7245}}

```
calckinparsmixedcomp[.9091, 100, 0, .09091, 1, 0, .01554, 1, 100, 1.05 * .2027, 100, 100]
```

{{kfet → 1.00001, kA → 10., kI → 19.8986, kuI → 32.3}}

```
line5 = {kfet, kA, kI, kuI} /.
   calckinparsmixedcomp[.9091, 100, 0, .09091, 1, 0, .01554, 1, 100, 1.05 * .2027, 100, 100]
```

{{1.00001, 10., 19.8986, 32.3}}

Table 9.3 Estimation of four kinetic parameters for mixed inhibition of A → products using velocities at {a,i} = {100,0}, {1,0}, {1,100}, {100,100}

```
TableForm[Round[{line1[[1]], line2[[1]], line3[[1]], line4[[1]], line5[[1]]}, 0.01],
  TableHeadings -> {{"No errors", "1.05*v1", "1.05*v2", "1.05*v3", "1.05*v4", "1.05*v5"},
    {"kfet", "kA", "kI", "kuI"}}]
```

	kfet	kA	kI	kuI
No errors	1.	10.	19.99	30.
1.05*v1	1.06	10.61	20.12	27.97
1.05*v2	0.99	9.42	18.74	30.21
1.05*v3	1.	10.	21.31	29.72
1.05*v4	1.	10.	19.9	32.3

The 3 reactions in the mechanism for mixed inhibition provide 3 sides of a thermodynamic cycle. This makes it possible to calculate a fourth kinetic parameter kuA.

$$
\begin{array}{l}
\quad\quad kA \\
E + A = EA \rightarrow products \\
+ \quad\quad\quad + \\
I \quad\quad\quad\; I \\
\parallel kI \quad\;\; \parallel kuI \\
EI + A = EAI \\
\quad\quad kuA
\end{array}
$$

This is the mechanism in Figure 5.3 of Cornish-Bowden [21]. This thermodynamic cycle shows that kA kuI = kuA kI, so that kuA = kA kuI/kI. When 3 of these reactions are a equilibrium, the fourth is also at equilibrium. When there are no errors, kuA is given by

kuA = 10*30/20 = 15

When there are errors, kuA is given by

$$
\frac{\begin{array}{c}10.\grave{}\\10.61\grave{}\\9.42\grave{}\\10.\grave{}\\10.\grave{}\end{array}}{}\;\frac{\begin{array}{c}30.\grave{}\\27.97\grave{}\\30.21\grave{}\\29.72\grave{}\\32.3\grave{}\end{array}}{}\; \Big/ \;\frac{\begin{array}{c}19.990000000000002\grave{}\\20.12\grave{}\\18.740000000000002\grave{}\\21.31\grave{}\\19.900000000000002\grave{}\end{array}}{}
$$

{{15.0075}, {14.7496}, {15.1856}, {13.9465}, {16.2312}}

These values can be added to Table 9.3. The lines in Table 9.3 have to be augmented.

```
line1[[1]]
```

{1.00001, 10., 19.993, 30.0005}

```
line1a = {1.00001`, 10.`, 19.993028933775392`, 30.00053303245061`, 15.00}
```

{1.00001, 10., 19.993, 30.0005, 15.}

```
line2[[1]]
```

{1.05588, 10.6145, 20.1201, 27.9692}

```
line2a = {1.0558764804469274`, 10.614525139664805`,
  20.120095841271816`, 27.969239176861713`, 14.75}
```

{1.05588, 10.6145, 20.1201, 27.9692, 14.75}

```
line3[[1]]
```

{0.994747, 9.42105, 18.737, 30.2072}

```
line3a =
  {0.9947467894736842`, 9.421052631578945`, 18.736990789665754`, 30.20721482342275`, 15.19}
```

{0.994747, 9.42105, 18.737, 30.2072, 15.19}

```
line4[[1]]
```

{1.00001, 10., 21.3119, 29.7245}

```
line4a = {1.00001`, 10.`, 21.31189113878749`, 29.724511768546716`, 13.95}
```

{1.00001, 10., 21.3119, 29.7245, 13.95}

{1.00001`, 10.`, 21.31189113878749`, 29.724511768546716`, 13.95`}

{1.00001, 10., 21.3119, 29.7245, 13.95}

```
line5[[1]]
```

{1.00001, 10., 19.8986, 32.3}

```
line5a = {1.00001`, 10.`, 19.898623339432476`, 32.30000241199354`, 16.23}
```

{1.00001, 10., 19.8986, 32.3, 16.23}

Table 9.4 Estimation of five kinetic parameters for mixed inhibition of A → products using velocities at {a,i} = {100,0}, {1,0}, {1,100}, {100,100}. The fifth kinetic parameter has been calculated using a thermodynamic cycle.

```
TableForm[Round[{line1a, line2a, line3a, line4a, line5a}, 0.01],
  TableHeadings -> {{"No errors", "1.05*v1", "1.05*v2", "1.05*v3", "1.05*v4", "1.05*v5"},
    {"kfet", "kA", "kI", "kuI", "kuA"}}]
```

	kfet	kA	kI	kuI	kuA
No errors	1.	10.	19.99	30.	15.
1.05*v1	1.06	10.61	20.12	27.97	14.75
1.05*v2	0.99	9.42	18.74	30.21	15.19
1.05*v3	1.	10.	21.31	29.72	13.95
1.05*v4	1.	10.	19.9	32.3	16.23

Application of a more general program to velocity data from simpler mechanisms.

The program calckinparsmixedcomp can be used to determine whether the inhibition is competitive, uncompetitive, or mixed. To apply this program to data for a reaction with a competitive mechanism, one more velocity is needed (see Section 9.2.1). This fourth velocity is determined at {a,i} = {100,100}.

```
vcomp /. kfet → 1 /. kA → 10 /. kI → 20 /. a → 100 /. i → 100 // N
```

0.625

```
calckinparsmixedcomp[.9091, 100, 0, .09091, 1, 0, .01639, 1, 100, .625, 100, 100]
```

{{kfet → 1.00001, kA → 10., kI → 19.9946, kuI → -837 364.}}

Since kuI is negative, this shows that there is no kuI = ea*i/eai in the competitive mechanism. But correct values for the other 3 kinetic parameters are obtained.

When the inhibition is uncompetitive, one more velocity is needed when calckinparsmixedcomp is applied. This fourth velocity is determined at {a,i} = {100,100} (see Section 9.3).

```
vuncomp /. kfet → 1 /. kA → 10 /. kuI → 30 /. a → 100 /. i → 100 // N
```

0.225564

```
calckinparsmixedcomp[.9091, 100, 0, .09091, 1, 0, .06977, 1, 100, .2256, 100, 100]
```

$\{\{kfet \to 1.00001, kA \to 10., kI \to 3.50213 \times 10^6, kuI \to 30.006\}\}$

Since the value of kI is unreasonable, this shows that there is no kI = e*i/ei in the mechanism, and this calculation yields correct values for the other 3 kinetic parameters.

9.5 Essential activation

Cornish-Bowden [21] has represented essential activation by X in his Figure 5.8 with the following mechanism

```
      kAX    kfx
A + EX = EAX → products
   || kX
   E
   +
   X
```

This mechanism can also be written as

```
E + X = EX        kX = e*x/ex
EX + A = EAX      kAX = ex*a/eax
EAX → products    v = kfx*eax = kfx*et×(eax/et) = kfxet×(eax/et)
```

The rapid-equilibrium rate equation can be derived by use of Solve.

```
Solve[{kX == e * x / ex, kAX == ex * a / eax, et == e + ex + eax}, {eax}, {e, ex}]
```

$$\left\{\left\{eax \to \frac{a\,et\,x}{kAX\,kX + a\,x + kAX\,x}\right\}\right\}$$

The rapid-equilibrium velocity is given by

$$vessact = \frac{a\,kfet\,x}{kAX\,kX + a\,x + kAX\,x}$$

$$\frac{a\,kfet\,x}{kAX\,kX + a\,x + kAX\,x}$$

Notice that the velocity is zero when [X] = 0. There are 3 kinetic parameters, and so 3 velocities have to be measured to estimate the kinetic parameters from the minimum number of measurements of velocities. For a test calculation, the following values of the kinetic parameters are chosen:

```
vessact /. kfet → 1 /. kX → 20 /. kAX → 40
```

$$\frac{a\,x}{800 + 40\,x + a\,x}$$

The characteristics of vessact are shown by a 3D plot.

```
plot4a = Plot3D[Evaluate[vessact /. kfet → 1 /. kX → 20 /. kAX → 40],
    {a, 0.00001, 200}, {x, 0.00001, 100}, ViewPoint → {-2, -2, 1}, PlotRange -> {0, 1},
    Lighting -> {{"Ambient", GrayLevel[1]}}, AxesLabel → {"[A]", "[X]"}, PlotLabel → "vessact"];

plot4b = Plot3D[Evaluate[D[vessact /. kfet → 1 /. kX → 20 /. kAX → 40, a]],
    {a, 0.00001, 200}, {x, 0.00001, 100}, ViewPoint → {-2, -2, 1},
    PlotRange -> {0, .03}, Lighting -> {{"Ambient", GrayLevel[1]}},
    AxesLabel → {"[A]", "[X]"}, PlotLabel → "dvessact/d[A]"];

plot4c = Plot3D[Evaluate[D[vessact /. kfet → 1 /. kX → 20 /. kAX → 40, x]],
    {a, 0.00001, 200}, {x, 0.00001, 100}, ViewPoint → {-2, -2, 1},
    PlotRange -> {0, .2}, Lighting -> {{"Ambient", GrayLevel[1]}},
    AxesLabel → {"[A]", "[X]"}, PlotLabel → "dvessact/d[X]"];

plot4d = Plot3D[Evaluate[D[vessact /. kfet → 1 /. kX → 20 /. kAX → 40, a, x]],
    {a, 0.00001, 200}, {x, 0.00001, 100}, ViewPoint → {-2, -2, 1},
    PlotRange -> {-.0005, .001}, Lighting -> {{"Ambient", GrayLevel[1]}},
    AxesLabel → {"[A]", "[X]"}, PlotLabel → "d(dvessact/d[A])/d[X]"];
```

fig4vessact = GraphicsArray[{{plot4a, plot4b}, {plot4c, plot4d}}]

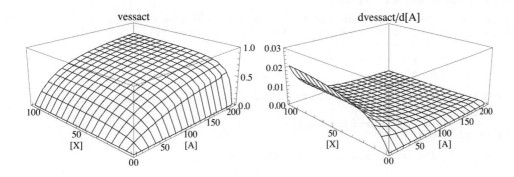

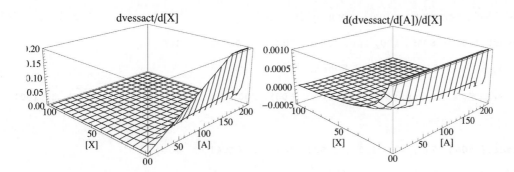

Fig. 9.4. Plots of vessact, $\partial\,\text{vessact}/\partial\,[A]$, $\partial\,\text{vessact}/\partial\,[X]$, and $\partial(\partial\,\text{vessact}/\partial\,[A])/\partial\,[X]$ versus [A] and [X] for mixed inhibition of A → products when kfet = 1, kX = 20, and kAX = 40.

The rate equation vessact can be used to estimate the 3 kinetic parameters from 3 measurements of velocities: {a,x} = {200,100}, {200,1}, and {1,100} are chosen.

vessact /. kfet → 1 /. kX → 20 /. kAX → 40 /. a → 200 /. x → 100 // N

0.806452

vessact /. kfet → 1 /. kX → 20 /. kAX → 40 /. a → 200 /. x → 1 // N

0.192308

vessact /. kfet → 1 /. kX → 20 /. kAX → 40 /. a → 1 /. x → 100 // N

0.0204082

In Section 9.7 (mixed activation) the velocity at {5,5} is going to be needed, and so it is calculated here.

vessact /. kfet → 1 /. kX → 20 /. kAX → 40 /. a → 5 /. x → 5 // N

0.0243902

The following program is written to estimate the kinetic parameters for essential activation:

```
calckinparsessact[v1_, a1_, x1_, v2_, a2_, x2_, v3_, a3_, x3_] :=
 Module[{}, (*This program calculates kfet,
   kX and kAX from 3 measured velocities for A → products with essential
     activation by X.  The first velocity is at high [A] and high [I],
    the second velocity is at high [A] and low [I], and the third is at low [A]
     and high [I].  High and low are with respect to the dissociation constants.*)
```

$$\text{Solve}\left[\left\{v1 == \frac{a1\ kfet\ x1}{kAX\ kX + a1\ x1 + kAX\ x1}, v2 == \frac{a2\ kfet\ x2}{kAX\ kX + a2\ x2 + kAX\ x2},\right.\right.$$
$$\left.\left. v3 == \frac{a3\ kfet\ x3}{kAX\ kX + a3\ x3 + kAX\ x3}\right\}, \{kfet, kX, kAX\}\right]\right]$$

```
calckinparsessact[.8065, 200, 100, .1923, 200, 1, .02041, 1, 100]
```

{{kfet → 1.00005, kX → 20.0039, kAX → 39.9972}}

```
r1 = {kfet, kX, kAX} /. calckinparsessact[.8065, 200, 100, .1923, 200, 1, .02041, 1, 100]
```

{{1.00005, 20.0039, 39.9972}}

Experimental errors in the velocities can be taken into account one at a time.

```
calckinparsessact[1.05 * .8065, 200, 100, .1923, 200, 1, .02041, 1, 100]
```

{{kfet → 1.06314, kX → 20.3326, kAX → 42.4568}}

```
r2 = {kfet, kX, kAX} /. calckinparsessact[1.05 * .8065, 200, 100, .1923, 200, 1, .02041, 1, 100]
```

{{1.06314, 20.3326, 42.4568}}

```
calckinparsessact[.8065, 200, 100, 1.05 * .1923, 200, 1, .02041, 1, 100]
```

{{kfet → 1.00005, kX → 18.5214, kAX → 40.4975}}

```
r3 = {kfet, kX, kAX} /. calckinparsessact[.8065, 200, 100, 1.05 * .1923, 200, 1, .02041, 1, 100]
```

{{1.00005, 18.5214, 40.4975}}

```
calckinparsessact[.8065, 200, 100, .1923, 200, 1, 1.05 * .02041, 1, 100]
```

{{kfet → 0.988463, kX → 21.2497, kAX → 37.2159}}

```
r4 = {kfet, kX, kAX} /. calckinparsessact[.8065, 200, 100, .1923, 200, 1, 1.05 * .02041, 1, 100]
```

{{0.988463, 21.2497, 37.2159}}

Table 9.5 Estimation of three kinetic parameters for essential activation of A → products using velocities at {a,i} = {200,100}, {{200,1}, and {1,100}.

```
TableForm[Round[{r1[[1]], r2[[1]], r3[[1]], r4[[1]]}, 0.01],
  TableHeadings -> {{"No errors", "1.05*v1", "1.05*v2", "1.05*v3"}, {"kfet", "kX", "kAX"}}]
```

	kfet	kX	kAX
No errors	1.	20.	40.
1.05*v1	1.06	20.33	42.46
1.05*v2	1.	18.52	40.5
1.05*v3	0.99	21.25	37.22

9.6 Essential activation when two, three, or four molecules of activator X are bound in a single reaction

When an activator is studied for the first time, the number molecules of the activator bound is unknown. The determination of this number is discussed here. When two or more molecules of an activator have to be bound, sigmoid plots of velocity versus activator concentration are obtained.

■ 9.6.1 Two molecules of activator X are bound

This mechanism can be written as

$E + 2X = EX_2$ kX2^2 = e*x^2/ex2
$EX_2 + A = EAX_2$ kA = ex2*a/eax2
$EAX_2 \rightarrow$ products   v = kf*eax2 = kf*et×(eax2/et) = kfet×(eax/et)

The rapid-equilibrium rate equation can be derived by use of Solve.

```
Solve[{kX2^2 == e * x^2 / ex2, kA == ex2 * a / eax2, et == e + ex2 + eax2}, {eax2}, {e, ex2}]
```

$$\left\{\left\{eax2 \rightarrow \frac{a\ et\ x^2}{kA\ kX2^2 + a\ x^2 + kA\ x^2}\right\}\right\}$$

$$vAX2 = \frac{a\ kfet\ x^2}{kA\ kX2^2 + a\ x^2 + kA\ x^2};$$

The following kinetic parameters are specified for test calculations.

```
vAX2 /. kfet → 1 /. kA → 10 /. kX2 → 30
```

$$\frac{a\ x^2}{9000 + 10\ x^2 + a\ x^2}$$

Construct 3D plots

```
plotAX2a = Plot3D[Evaluate[vAX2 /. kfet → 1 /. kA → 10 /. kX2 → 30],
  {a, 0.00001, 50}, {x, 0.00001, 100}, ViewPoint → {-2, -2, 1}, PlotRange -> {0, 1},
  Lighting -> {{"Ambient", GrayLevel[1]}}, AxesLabel → {"[A]", "[X]"}, PlotLabel → "vAX2"];

plotAX2b = Plot3D[Evaluate[D[vAX2 /. kfet → 1 /. kA → 10 /. kX2 → 30, a]],
  {a, 0.00001, 50}, {x, 0.00001, 100}, ViewPoint → {-2, -2, 1},
  PlotRange -> {0, .1}, Lighting -> {{"Ambient", GrayLevel[1]}},
  AxesLabel → {"[A]", "[X]"}, PlotLabel → "∂vAX2/∂[A]"];

plotAX2c = Plot3D[Evaluate[D[vAX2 /. kfet → 1 /. kA → 10 /. kX2 → 30, x]],
  {a, 0.00001, 50}, {x, 0.00001, 100}, ViewPoint → {-2, -2, 1},
  PlotRange -> {0, .05}, Lighting -> {{"Ambient", GrayLevel[1]}},
  AxesLabel → {"[A]", "[X]"}, PlotLabel → "∂vAX2/∂[X]"];

plotAX2d = Plot3D[Evaluate[D[vAX2 /. kfet → 1 /. kA → 10 /. kX2 → 30, a, x]],
  {a, 0.00001, 50}, {x, 0.00001, 100}, ViewPoint → {-2, -2, 1},
  PlotRange -> {-.001, .003}, Lighting -> {{"Ambient", GrayLevel[1]}},
  AxesLabel → {"[A]", "[X]"}, PlotLabel → "∂(∂vAX2/∂[A])/∂[X]"];
```

fig2AX2 = GraphicsArray[{{plotAX2a, plotAX2b}, {plotAX2c, plotAX2d}}]

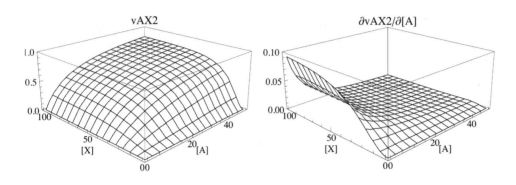

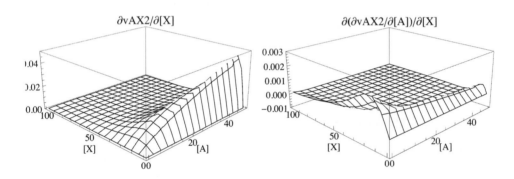

Fig. 9.5 Plots of vAX2, ∂vAX2/∂[A], ∂vAX2/∂[X], and $\partial(\partial$vAX2/∂[A])/∂[X] versus [A] and [X] when two molecules of X are bound in a single reaction. Notice the sigmoid plots of vX2 versus [X] at constant [A].

Since there are 3 kinetic parameters, 3 velocities have to be determined to estimate the kinetic parameters.

vAX2 /. kX2 → 30 /. kA → 10 /. kfet → 1 /. a → 100 /. x → 100 // N

0.901713

vAX2 /. kX2 → 30 /. kA → 10 /. kfet → 1 /. a → 1 /. x → 100 // N

0.0840336

vAX2 /. kX2 → 30 /. kA → 10 /. kfet → 1 /. a → 100 /. x → 1 // N

0.0109769

```
calckinparsAX2[v1_, a1_, x1_, v2_, a2_, x2_, v3_, a3_, x3_] :=
 Module[{}, (*This program calculates kfet, kA,
```

and kX2 from 3 experimental velocities for A → products with essential
  activation by 2X. The first velocity is at high [A] and high [X],
  the second velocity is at low [A] and high [X], and the third is at high [A]
  and medium[X]. High and low are with respect to the two dissociation constants.*)

$$\text{Solve}\left[\left\{v1 == \frac{a1\ kfet\ x1^2}{kA\ kX2^2 + a1\ x1^2 + kA\ x1^2}, \ v2 == \frac{a2\ kfet\ x2^2}{kA\ kX2^2 + a2\ x2^2 + kA\ x2^2},\right.\right.$$

$$\left.\left. v3 == \frac{a3\ kfet\ x3^2}{kA\ kX2^2 + a3\ x3^2 + kA\ x3^2}\right\}, \{kfet, kA, kX2\}\right]\right]$$

```
calckinparsAX2[.9017, 100, 100, .08403, 1, 100, .01098, 100, 1]
```

{{kfet → 0.999989, kA → 10.0006, kX2 → -29.9946}, {kfet → 0.999989, kA → 10.0006, kX2 → 29.9946}}

```
line1 = {kfet, kA, kX2} /. calckinparsAX2[.9017, 100, 100, .08403, 1, 100, .01098, 100, 1][[2]]
```

{0.999989, 10.0006, 29.9946}

These values are correct. Their sensitivity to experimental errors in measured velocities can be tested.

```
calckinparsAX2[1.05 * .9017, 100, 100, .08403, 1, 100, .01098, 100, 1]
```

{{kfet → 1.05634, kA → 10.62, kX2 → -29.9245}, {kfet → 1.05634, kA → 10.62, kX2 → 29.9245}}

```
line2 =
 {kfet, kA, kX2} /. calckinparsAX2[1.05 * .9017, 100, 100, .08403, 1, 100, .01098, 100, 1][[2]]
```

{1.05634, 10.62, 29.9245}

```
calckinparsAX2[.9017, 100, 100, 1.05 * .08403, 1, 100, .01098, 100, 1]
```

{{kfet → 0.994297, kA → 9.37457, kX2 → -30.8918}, {kfet → 0.994297, kA → 9.37457, kX2 → 30.8918}}

```
line3 =
 {kfet, kA, kX2} /. calckinparsAX2[.9017, 100, 100, 1.05 * .08403, 1, 100, .01098, 100, 1][[2]]
```

{0.994297, 9.37457, 30.8918}

```
calckinparsAX2[.9017, 100, 100, .08403, 1, 100, 1.05 * .01098, 100, 1]
```

{{kfet → 0.999989, kA → 10.044, kX2 → -29.1995}, {kfet → 0.999989, kA → 10.044, kX2 → 29.1995}}

```
line4 =
 {kfet, kA, kX2} /. calckinparsAX2[.9017, 100, 100, .08403, 1, 100, 1.05 * .01098, 100, 1][[2]]
```

{0.999989, 10.044, 29.1995}

Table 9.6 Estimation of kinetic constants for essential activation of A → products when 2X are bound using velocities at {100,0}, {1,100}, {100,1}

```
TableForm[Round[{line1, line2, line3, line4}, 0.01],
 TableHeadings -> {{"No errors", "1.05*v1", "1.05*v2", "1.05*v3"}, {"kfet", "kA", "kX2"}}]
```

	kfet	kA	kX2
No errors	1.	10.	29.99
1.05*v1	1.06	10.62	29.92
1.05*v2	0.99	9.37	30.89
1.05*v3	1.	10.04	29.2

9.6.2 Three molecules of activator X are bound

The mechanism is given by

$E + 3X = EX_3$ kX3^3 = e*x^3/ex3

$EX_3 + A = EAX_3$ kA = ex3*a/eax3

$EAX_3 \to$ products   v = kf*eax3 = kfet*eax3/et

The rapid-equilibrium rate equation can be derived using Solve.

```
Solve[{kX3^3 == e * x^3 / ex3, kA == ex3 * a / eax3, et == e + ex3 + eax3}, {eax3}, {e, ex3}]
```

$$\left\{\left\{eax3 \to \frac{a\ et\ x^3}{kA\ kX3^3 + a\ x^3 + kA\ x^3}\right\}\right\}$$

```
vAX3 =
```
$$\frac{a\ kfet\ x^3}{kA\ kX3^3 + a\ x^3 + kA\ x^3};$$

```
vAX3 /. kfet → 1 /. kA → 10 /. kX3 → 30
```

$$\frac{a\ x^3}{270\,000 + 10\ x^3 + a\ x^3}$$

Construct 3D plots.

```
plotAX3a = Plot3D[Evaluate[vAX3 /. kfet → 1 /. kA → 10 /. kX3 → 30],
    {a, 0.00001, 50}, {x, 0.00001, 100}, ViewPoint → {-2, -2, 1}, PlotRange -> {0, 1},
    Lighting -> {{"Ambient", GrayLevel[1]}}, AxesLabel → {"[A]", "[X]"}, PlotLabel → "vAX3"];

plotAX3b = Plot3D[Evaluate[D[vAX3 /. kfet → 1 /. kA → 10 /. kX3 → 30, a]],
    {a, 0.00001, 50}, {x, 0.00001, 100}, ViewPoint → {-2, -2, 1},
    PlotRange -> {0, .1}, Lighting -> {{"Ambient", GrayLevel[1]}},
    AxesLabel → {"[A]", "[X]"}, PlotLabel → "∂vAX3/∂[A]"];

plotAX3c = Plot3D[Evaluate[D[vAX3 /. kfet → 1 /. kA → 10 /. kX3 → 30, x]],
    {a, 0.00001, 50}, {x, 0.00001, 100}, ViewPoint → {-2, -2, 1},
    PlotRange -> {0, .05}, Lighting -> {{"Ambient", GrayLevel[1]}},
    AxesLabel → {"[A]", "[X]"}, PlotLabel → "∂vAX3/∂[X]"];

plotAX3d = Plot3D[Evaluate[D[vAX3 /. kfet → 1 /. kA → 10 /. kX3 → 30, a, x]],
    {a, 0.00001, 50}, {x, 0.00001, 100}, ViewPoint → {-2, -2, 1},
    PlotRange -> {-.001, .003}, Lighting -> {{"Ambient", GrayLevel[1]}},
    AxesLabel → {"[A]", "[X]"}, PlotLabel → "∂(∂vAX3/∂[A])/∂[X]"];
```

fig3AX3 = GraphicsArray[{{plotAX3a, plotAX3b}, {plotAX3c, plotAX3d}}]

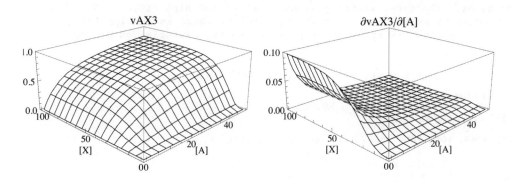

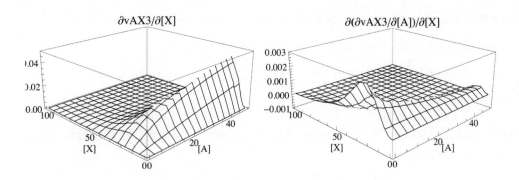

Fig. 9.6 Plots of vAX3, ∂vAX3/∂[A], ∂vAX3/∂[X], and $\partial(\partial$vAX3/∂[A])/∂[X] versus [A] and [X] when three molecules of X are bound in a single reaction. Note that the ridge has moved ot higher [X].

Three measured velocities can be used to estimate the kinetic parameters.

vAX3 /. kX3 → 30 /. kA → 10 /. kfet → 1 /. a → 100 /. x → 100 // N

0.906865

vAX3 /. kX3 → 30 /. kA → 10 /. kfet → 1 /. a → 1 /. x → 100 // N

0.0887311

vAX3 /. kX3 → 30 /. kA → 10 /. kfet → 1 /. a → 100 /. x → 4 // N

0.0231014

It is necessary to use x = 4 to avoid too low a velocity in the last calculated velocity.

```
calckinparsAX3[v1_, a1_, x1_, v2_, a2_, x2_, v3_, a3_, x3_] :=
 Module[{}, (*This program calculates kfet, kA,

   and kX3 from 3 experimental velocities for A → products with essential
     activation by 3X.  The first velocity is at high [A] and high [X],
   the second velocity is at low [A] and high [X], and the third is at high [A]
     and medium [X]. High and low are with respect to the two dissociation constants.*)
```

$$\text{Solve}\left[\left\{v1 == \frac{a1\ kfet\ x1^3}{kA\ kX3^3 + a1\ x1^3 + kA\ x1^3},\ v2 == \frac{a2\ kfet\ x2^3}{kA\ kX3^3 + a2\ x2^3 + kA\ x2^3},\right.\right.$$

$$\left.\left. v3 == \frac{a3\ kfet\ x3^3}{kA\ kX3^3 + a3\ x3^3 + kA\ x3^3}\right\},\ \{kfet, kA, kX3\}\right]\right]$$

```
calckinparsAX3[.9069, 100, 100, .08873, 1, 100, .02310, 100, 4]
```

```
{{kfet → 1.00004, kA → 10.0006, kX3 → -15.0002 - 25.9811 i},
 {kfet → 1.00004, kA → 10.0006, kX3 → -15.0002 + 25.9811 i},
 {kfet → 1.00004, kA → 10.0006, kX3 → 30.0004}}
```

```
line1 = {kfet, kA, kX3} /. calckinparsAX3[.9069, 100, 100, .08873, 1, 100, .02310, 100, 4][[3]]
```

```
{1.00004, 10.0006, 30.0004}
```

These values are correct, but the effects of experimental errors in the velocity measurements have to be considered.

```
calckinparsAX3[1.05 * .9069, 100, 100, .08873, 1, 100, .02310, 100, 4]
```

```
{{kfet → 1.05606, kA → 10.6164, kX3 → -14.9802 - 25.9464 i},
 {kfet → 1.05606, kA → 10.6164, kX3 → -14.9802 + 25.9464 i},
 {kfet → 1.05606, kA → 10.6164, kX3 → 29.9603}}
```

```
line2 =
 {kfet, kA, kX3} /. calckinparsAX3[1.05 * .9069, 100, 100, .08873, 1, 100, .02310, 100, 4][[3]]
```

```
{1.05606, 10.6164, 29.9603}
```

```
calckinparsAX3[.9069, 100, 100, 1.05 * .08873, 1, 100, .02310, 100, 4]
```

```
{{kfet → 0.994652, kA → 9.4075, kX3 → -15.2815 - 26.4683 i},
 {kfet → 0.994652, kA → 9.4075, kX3 → -15.2815 + 26.4683 i},
 {kfet → 0.994652, kA → 9.4075, kX3 → 30.563}}
```

```
line3 =
 {kfet, kA, kX3} /. calckinparsAX3[.9069, 100, 100, 1.05 * .08873, 1, 100, .02310, 100, 4][[3]]
```

```
{0.994652, 9.4075, 30.563}
```

```
calckinparsAX3[.9069, 100, 100, .08873, 1, 100, 1.05 * .02310, 100, 4]
```

```
{{kfet → 1.00004, kA → 10.0138, kX3 → -14.7453 - 25.5397 i},
 {kfet → 1.00004, kA → 10.0138, kX3 → -14.7453 + 25.5397 i},
 {kfet → 1.00004, kA → 10.0138, kX3 → 29.4906}}
```

```
line4 =
 {kfet, kA, kX3} /. calckinparsAX3[.9069, 100, 100, .08873, 1, 100, 1.05 * .02310, 100, 4][[3]]
```

```
{1.00004, 10.0138, 29.4906}
```

Table 9.7 Estimation of kinetic constants for essential activation of A $\rightarrow$ products when 3X are bound in a single reaction using velocities at {100,0}, {1,100}, {100,4}

```
TableForm[Round[{line1, line2, line3, line4}, 0.01],
  TableHeadings -> {{"No errors", "1.05*v1", "1.05*v2", "1.05*v3"}, {"kfet", "kA", "kX3"}}]
```

	kfet	kA	kX3
No errors	1.	10.	30.
1.05*v1	1.06	10.62	29.96
1.05*v2	0.99	9.41	30.56
1.05*v3	1.	10.01	29.49

■ 9.6.3 Four molecules of activator X are bound

The mechanism is given by

$E + 4X = EX_4$ kX4^4 = e*x^4/ex4

$EX_4 + A = EAX_4$ kA = ex4*a/eax4

$EAX_4 \rightarrow$ products   v = kf*eax4 = kfet*eax4/et

The rapid-equilibrium rate equation can be derived using Solve.

```
Solve[{kX4^4 == e * x^4 / ex4, kA == ex4 * a / eax4, et == e + ex4 + eax4}, {eax4}, {e, ex4}]
```

$$\left\{ \left\{ eax4 \rightarrow \frac{a \ et \ x^4}{kA \ kX4^4 + a \ x^4 + kA \ x^4} \right\} \right\}$$

```
        a kfet x^4
vAX4 = ─────────────────────────── ;
       kA kX4^4 + a x^4 + kA x^4
```

```
vAX4 /. kfet → 1 /. kA → 10 /. kX4 → 30
```

$$\frac{a \ x^4}{8\,100\,000 + 10 \ x^4 + a \ x^4}$$

```
plotAX4a = Plot3D[Evaluate[vAX4 /. kfet → 1 /. kA → 10 /. kX4 → 30],
    {a, 0.00001, 50}, {x, 0.00001, 100}, ViewPoint → {-2, -2, 1}, PlotRange -> {0, 1},
    Lighting -> {{"Ambient", GrayLevel[1]}}, AxesLabel → {"[A]", "[X]"}, PlotLabel → "vAX4"];
```

```
plotAX4b = Plot3D[Evaluate[D[vAX4 /. kfet → 1 /. kA → 10 /. kX4 → 30, a]],
    {a, 0.00001, 50}, {x, 0.00001, 100}, ViewPoint → {-2, -2, 1},
    PlotRange -> {0, .1}, Lighting -> {{"Ambient", GrayLevel[1]}},
    AxesLabel → {"[A]", "[X]"}, PlotLabel → "∂vAX4/∂[A]"];
```

```
plotAX4c = Plot3D[Evaluate[D[vAX4 /. kfet → 1 /. kA → 10 /. kX4 → 30, x]],
    {a, 0.00001, 50}, {x, 0.00001, 100}, ViewPoint → {-2, -2, 1},
    PlotRange -> {0, .05}, Lighting -> {{"Ambient", GrayLevel[1]}},
    AxesLabel → {"[A]", "[X]"}, PlotLabel → "∂vAX4/∂[X]"];
```

```
plotAX4d = Plot3D[Evaluate[D[vAX4 /. kfet → 1 /. kA → 10 /. kX4 → 30, a, x]],
    {a, 0.00001, 50}, {x, 0.00001, 100}, ViewPoint → {-2, -2, 1},
    PlotRange -> {-.001, .003}, Lighting -> {{"Ambient", GrayLevel[1]}},
    AxesLabel → {"[A]", "[X]"}, PlotLabel → "∂(∂vAX4/∂[A])/∂[X]"];
```

```
fig4AX4 = GraphicsArray[{{plotAX4a, plotAX4b}, {plotAX4c, plotAX4d}}]
```

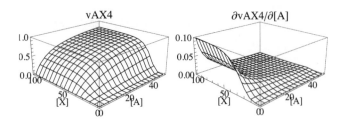

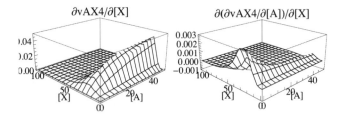

Fig. 9.7 Plots of vAX4, ∂vAX4/∂[A], ∂vAX4/∂[X], and $\partial(\partial$vAX4/∂[A])/∂[X] versus [A] and [X] when four molecules of X are bound. Note that the ridge has moved to higher [X].

Three measured velocities can be used to estimate the three kinetic parameters.

```
vAX4 /. kX4 → 30 /. kA → 10 /. kfet → 1 /. a → 100 /. x → 100 // N
```

0.908422

```
vAX4 /. kX4 → 30 /. kA → 10 /. kfet → 1 /. a → 1 /. x → 100 // N
```

0.0902446

```
vAX4 /. kX4 → 30 /. kA → 10 /. kfet → 1 /. a → 100 /. x → 6 // N
```

0.0157233

For reactions with smaller stoichiometric coefficients, x → 1 might be used in the third pair of concentrations, but that produces a very small velocity.

```
vAX4 /. kX4 → 30 /. kA → 10 /. kfet → 1 /. a → 100 /. x → 1 // N
```

0.0000123455

```
calckinparsAX4[v1_, a1_, x1_, v2_, a2_, x2_, v3_, a3_, x3_] :=
  Module[{}, (*This program calculates kfet, kA,
   and kX4 from 3 experimental velocities for A → products with essential
     activation by 4X.  The first velocity is at high [A] and high [X],
    the second velocity is at low [A] and high [X], and the third is at high [A]
     and medium [X]. High and low are with respect to the two dissociation constants.*)
```

$$\text{Solve}\left[\left\{v1 == \frac{a1\ kfet\ x1^4}{kA\ kX4^4 + a1\ x1^4 + kA\ x1^4}, v2 == \frac{a2\ kfet\ x2^4}{kA\ kX4^4 + a2\ x2^4 + kA\ x2^4},\right.\right.$$

$$\left.\left. v3 == \frac{a3\ kfet\ x3^4}{kA\ kX4^4 + a3\ x3^4 + kA\ x3^4}\right\}, \{kfet, kA, kX4\}\right]\right]$$

```
calckinparsAX4[.9084, 100, 100, .09024, 1, 100, .01572, 100, 6]
```

```
{{kfet → 0.999979, kA → 10.0003, kX4 → -30.0012},
 {kfet → 0.999979, kA → 10.0003, kX4 → 0. - 30.0012 i},
 {kfet → 0.999979, kA → 10.0003, kX4 → 0. + 30.0012 i},
 {kfet → 0.999979, kA → 10.0003, kX4 → 30.0012}}
```

```
line1 = {kfet, kA, kX4} /. calckinparsAX4[.9084, 100, 100, .09024, 1, 100, .01572, 100, 6][[4]]
```

```
{0.999979, 10.0003, 30.0012}
```

These values are correct. Make a table to show the effects of 5% errors in the velocities, one at a time.

```
calckinparsAX4[1.05 * .9084, 100, 100, .09024, 1, 100, .01572, 100, 6]
```

```
{{kfet → 1.05589, kA → 10.6153, kX4 → -29.9679},
 {kfet → 1.05589, kA → 10.6153, kX4 → 0. - 29.9679 i},
 {kfet → 1.05589, kA → 10.6153, kX4 → 0. + 29.9679 i},
 {kfet → 1.05589, kA → 10.6153, kX4 → 29.9679}}
```

```
line2 =
  {kfet, kA, kX4} /. calckinparsAX4[1.05 * .9084, 100, 100, .09024, 1, 100, .01572, 100, 6][[4]]
```

```
{1.05589, 10.6153, 29.9679}
```

```
calckinparsAX4[.9084, 100, 100, 1.05 * .09024, 1, 100, .01572, 100, 6]
```

```
{{kfet → 0.994677, kA → 9.4171, kX4 → -30.4148},
 {kfet → 0.994677, kA → 9.4171, kX4 → 0. - 30.4148 i},
 {kfet → 0.994677, kA → 9.4171, kX4 → 0. + 30.4148 i},
 {kfet → 0.994677, kA → 9.4171, kX4 → 30.4148}}
```

```
line3 =
  {kfet, kA, kX4} /. calckinparsAX4[.9084, 100, 100, 1.05 * .09024, 1, 100, .01572, 100, 6][[4]]
```

```
{0.994677, 9.4171, 30.4148}
```

```
calckinparsAX4[.9084, 100, 100, .09024, 1, 100, 1.05 * .01572, 100, 6]
```

```
{{kfet → 0.999979, kA → 10.0042, kX4 → -29.628},
 {kfet → 0.999979, kA → 10.0042, kX4 → 0. - 29.628 i},
 {kfet → 0.999979, kA → 10.0042, kX4 → 0. + 29.628 i},
 {kfet → 0.999979, kA → 10.0042, kX4 → 29.628}}
```

```
line4 =
  {kfet, kA, kX4} /. calckinparsAX4[.9084, 100, 100, .09024, 1, 100, 1.05 * .01572, 100, 6][[4]]
```

```
{0.999979, 10.0042, 29.628}
```

Table 9.8 Estimation of kinetic constants for essential activation of A $\rightarrow$ products by 4X using velocities at {100,0}, {1,100}, {100,6}

```
TableForm[Round[{line1, line2, line3, line4}, 0.01],
  TableHeadings -> {{"No errors", "1.05*v1", "1.05*v2", "1.05*v3"}, {"kfet", "kA", "kX4"}}]
```

	kfet	kA	kX4
No errors	1.	10.	30.
1.05*v1	1.06	10.62	29.97
1.05*v2	0.99	9.42	30.41
1.05*v3	1.	10.	29.63

The effect of an activator can be very dramatic because the number of activator molecules is in the exponent of $[X]^n$. Therefore, n can be determined by plotting the velocity at high [A] versus [X].

vAX2 /. kfet → 1 /. kA → 10 /. kX2 → 30 /. a → 50

$$\frac{50 \, x^2}{9000 + 60 \, x^2}$$

vAX3 /. kfet → 1 /. kA → 10 /. kX3 → 30 /. a → 50

$$\frac{50 \, x^3}{270\,000 + 60 \, x^3}$$

vAX4 /. kfet → 1 /. kA → 10 /. kX4 → 30 /. a → 50

$$\frac{50 \, x^4}{8\,100\,000 + 60 \, x^4}$$

```
Plot[{vAX2 /. kfet → 1 /. kA → 10 /. kX2 → 30 /. a → 50,
  vAX3 /. kfet → 1 /. kA → 10 /. kX3 → 30 /. a → 50,
  vAX4 /. kfet → 1 /. kA → 10 /. kX4 → 30 /. a → 50}, {x, 0, 100}, PlotStyle -> {Black}]
```

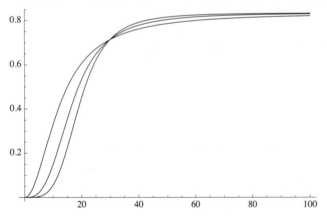

Fig. 9.8 Plots of velocities vAX2, vAX3, and vAX4 at [A] = 100 versus [X] to determine n in $[X]^n$.

9.7 Mixed activation when one X is bound

Cornish-Bowden [21] has represented mixed activation by X in his Figure 5.9 with the mechanism

```
      kAX     kfxet
A + EX = EAX → products
    ‖ kX  ‖ kXA
A + E  =  EA
  + kA +
  X     X
```

The four equilibria are not independent. Since kA kXA = kX kAX, kXA = kX kAX/kA. Therefore, EA + X = EAX and kXA are omitted from the derivation of the rapid-equilibrium rate equation. The mechanism used to derive the rapid-equilibrium velocity can be written as

```
E + X = EX        kX = e*x/ex
EX + A = EAX      kAX = ex*a/eax
E + A = EA        kA = e*a/ea
EAX → products    v = kfx*eax
```

The rapid-equilibrium rate equation can be derived by use of Solve.

```
Solve[{kX == e * x / ex, kAX == ex * a / eax, kA == e * a / ea, et == e + ea + ex + eax}, {eax}, {e, ea, ex}]
```

$$\left\{\left\{ eax \rightarrow \frac{a\ et\ kA\ x}{a\ kAX\ kX + kA\ kAX\ kX + a\ kA\ x + kA\ kAX\ x} \right\}\right\}$$

The rapid-equilibrium velocity is given by

$$vmixact = \frac{a\ kfxet\ kA\ x}{a\ kAX\ kX + kA\ kAX\ kX + a\ kA\ x + kA\ kAX\ x}$$

$$\frac{a\ kA\ kfxet\ x}{a\ kAX\ kX + kA\ kAX\ kX + a\ kA\ x + kA\ kAX\ x}$$

There are 4 kinetic parameters, and so 4 velocities have to be measured to estimate the kinetic parameters from the minimum number of velocity measurements. For a test calculation, the following values of the kinetic parameters are specified:

```
vmixact /. kfxet → 1 /. kX → 20 /. kAX → 40 /. kA → 60
```

$$\frac{60\ a\ x}{48\,000 + 800\ a + 2400\ x + 60\ a\ x}$$

The characteristics of vmixact are shown by a 3D plot.

```
plotmixa = Plot3D[Evaluate[vmixact /. kfxet → 1 /. kX → 20 /. kAX → 40 /. kA → 60],
    {a, 0.00001, 200}, {x, 0.00001, 100}, PlotLabel → "vmixact", ViewPoint → {-2, -2, 1},
    PlotRange -> {0, 1}, Lighting -> {{"Ambient", GrayLevel[1]}}, AxesLabel → {"[A]", "[X]"}];

plotmixb = Plot3D[Evaluate[D[vmixact /. kfxet → 1 /. kX → 20 /. kAX → 40 /. kA → 60, a]],
    {a, 0.00001, 50}, {x, 0.00001, 100}, PlotLabel → "∂vmixact/∂[A]", ViewPoint → {-2, -2, 1},
    PlotRange -> {0, .02}, Lighting -> {{"Ambient", GrayLevel[1]}}, AxesLabel → {"[A]", "[X]"}];

plotmixc = Plot3D[Evaluate[D[vmixact /. kfxet → 1 /. kX → 20 /. kAX → 40 /. kA → 60, x]],
    {a, 0.00001, 50}, {x, 0.00001, 100}, ViewPoint → {-2, -2, 1},
    PlotRange -> {0, .05}, Lighting -> {{"Ambient", GrayLevel[1]}},
    AxesLabel → {"[A]", "[X]"}, PlotLabel → "∂vmixact/∂[X]"];
```

```
plotmixd = Plot3D[Evaluate[D[vmixact /. kfxet → 1 /. kX → 20 /. kAX → 40 /. kA → 60, a, x]],
    {a, 0.00001, 50}, {x, 0.00001, 100}, ViewPoint → {-2, -2, 1},
    PlotRange -> {0, .0015}, Lighting -> {{"Ambient", GrayLevel[1]}},
    AxesLabel → {"[A]", "[X]"}, PlotLabel → "∂(∂vmixact/∂[A])/∂[X]"];

fig6mix = GraphicsArray[{{plotmixa, plotmixb}, {plotmixc, plotmixd}}]
```

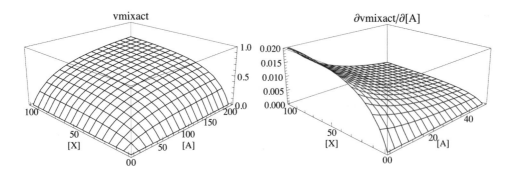

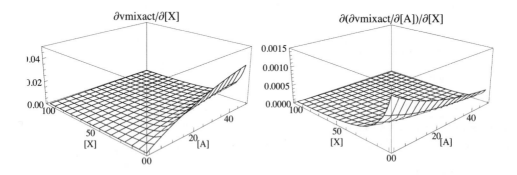

Fig. 9.9 Plots of vmixact, ∂vmixact/$\partial[A]$, ∂vmixact/$\partial[X]$, and $\partial(\partial$vmixact/$\partial[A])/\partial[X]$ versus [A] and [X] for mixed activation when one molecule of X is bound.

The rate equation vmixact can be used to estimate the 4 kinetic parameters from 4 measurements of velocities: {a,x} = {200,100}, {200,1}, {1,100}, and {5,5} are chosen.

```
vmixact /. kfxet → 1 /. kX → 20 /. kAX → 40 /. kA → 60 /. a → 200 /. x → 100 // N

0.728155

vmixact /. kfxet → 1 /. kX → 20 /. kAX → 40 /. kA → 60 /. a → 200 /. x → 1 // N

0.0539568

vmixact /. kfxet → 1 /. kX → 20 /. kAX → 40 /. kA → 60 /. a → 1 /. x → 100 // N

0.0203528
```

```
vmixact /. kfxet → 1 /. kX → 20 /. kAX → 40 /. kA → 60 /. a → 5 /. x → 5 // N
```

0.0229008

The following program is written to estimate the kinetic parameters:

```
calckinparsmixact[v1_, a1_, x1_, v2_, a2_, x2_, v3_, a3_, x3_, v4_, a4_, x4_] := Module[{},
```
(*This program calculates kfxet, kX, kAX, and kA from 4 measured velocities for A →
   products with mixed activation by X. The first velocity is at high [A] and high [X],
   the second velocity is at high [A] and low [X],
   and the third is at low [A] and high [X]. The fourth is at low [A] and
   low [X]. High and low are with respect to the dissociation constants.*)

$$\text{Solve}\Big[\Big\{v1 == \frac{a1\ kfxet\ kA\ x1}{a1\ kAX\ kX + kA\ kAX\ kX + a1\ kA\ x1 + kA\ kAX\ x1},$$

$$v2 == \frac{a2\ kfxet\ kA\ x2}{a2\ kAX\ kX + kA\ kAX\ kX + a2\ kA\ x2 + kA\ kAX\ x2},\ v3 == \frac{a3\ kfxet\ kA\ x3}{a3\ kAX\ kX + kA\ kAX\ kX + a3\ kA\ x3 + kA\ kAX\ x3},$$

$$v4 == \frac{a4\ kfxet\ kA\ x4}{a4\ kAX\ kX + kA\ kAX\ kX + a4\ kA\ x4 + kA\ kAX\ x4}\Big\},\ \{kfxet,\ kX,\ kAX,\ kA\}\Big]\Big]$$

```
calckinparsmixact[.7282, 200, 100, .05396, 200, 1, .02035, 1, 100, .02290, 5, 5]
```

{{kfxet → 1.00011, kX → 19.9969, kAX → 40.011, kA → 60.0056}}

```
rr1 = {kfxet, kX, kAX, kA} /.
   calckinparsmixact[.7282, 200, 100, .05396, 200, 1, .02035, 1, 100, .02290, 5, 5]
```

{{1.00011, 19.9969, 40.011, 60.0056}}

These values are correct, but test for errors in velocity measurements.

```
calckinparsmixact[1.05 * .7282, 200, 100, .05396, 200, 1, .02035, 1, 100, .02290, 5, 5]
```

{{kfxet → 1.07116, kX → 19.9965, kAX → 42.9126, kA → 59.8145}}

```
rr2 = {kfxet, kX, kAX, kA} /.
   calckinparsmixact[1.05 * .7282, 200, 100, .05396, 200, 1, .02035, 1, 100, .02290, 5, 5]
```

{{1.07116, 19.9965, 42.9126, 59.8145}}

```
calckinparsmixact[.7282, 200, 100, 1.05 * .05396, 200, 1, .02035, 1, 100, .02290, 5, 5]
```

{{kfxet → 0.991046, kX → 20.1342, kAX → 39.6032, kA → 64.7916}}

```
rr3 = {kfxet, kX, kAX, kA} /.
   calckinparsmixact[.7282, 200, 100, 1.05 * .05396, 200, 1, .02035, 1, 100, .02290, 5, 5]
```

{{0.991046, 20.1342, 39.6032, 64.7916}}

```
calckinparsmixact[.7282, 200, 100, .05396, 200, 1, 1.05 * .02035, 1, 100, .02290, 5, 5]
```

{{kfxet → 0.987879, kX → 21.6457, kAX → 37.0763, kA → 61.2182}}

```
rr4 = {kfxet, kX, kAX, kA} /.
   calckinparsmixact[.7282, 200, 100, .05396, 200, 1, 1.05 * .02035, 1, 100, .02290, 5, 5]
```

{{0.987879, 21.6457, 37.0763, 61.2182}}

```
calckinparsmixact[.7282, 200, 100, .05396, 200, 1, .02035, 1, 100, 1.05 * .02290, 5, 5]
```

{{kfxet → 1.00292, kX → 18.3368, kAX → 40.6865, kA → 54.6457}}

```
rr5 = {kfxet, kX, kAX, kA} /.
    calckinparsmixact[.7282, 200, 100, .05396, 200, 1, .02035, 1, 100, 1.05 * .02290, 5, 5]
```

```
{{1.00292, 18.3368, 40.6865, 54.6457}}
```

Table 9.9 Estimation of four kinetic parameters for mixed activation of A → products using velocities at {200,100}, {200,1}, {1,100}, and {5,5}

```
TableForm[Round[{rr1[[1]], rr2[[1]], rr3[[1]], rr4[[1]], rr5[[1]]}, 0.01],
  TableHeadings -> {{"No errors", "1.05*v1", "1.05*v2", "1.05*v3", "1.05*v4", "1.05*v5"},
    {"kfxet", "kX", "kAX", "kA"}}]
```

	kfxet	kX	kAX	kA
No errors	1.	20.	40.01	60.01
1.05*v1	1.07	20.	42.91	59.81
1.05*v2	0.99	20.13	39.6	64.79
1.05*v3	0.99	21.65	37.08	61.22
1.05*v4	1.	18.34	40.69	54.65

Since this mechanism yields 3 sides of a thermodynamic cycle, a fourth equilibrium constant can be calculated using kXA = kAX kX/kA.

40.01`	20.`	60.01`
42.910000000000004`	20.`	59.81`
39.6`	* 20.13`	64.79`
37.08`	21.650000000000002` /	61.22`
40.69`	18.34`	54.65`

```
{{13.3344}, {14.3488}, {12.3036}, {13.1131}, {13.6552}}
```

This column can be added to Table 9.9.

```
rr1 = {kfxet, kX, kAX, kA} /.
    calckinparsmixact[.7282, 200, 100, .05396, 200, 1, .02035, 1, 100, .02290, 5, 5]
```

```
{{1.00011, 19.9969, 40.011, 60.0056}}
```

```
rr1a = {1.0001077379196124`, 19.996907025631472`,
    40.01103657602468`, 60.00561126008378`, 13.33}
```

```
{1.00011, 19.9969, 40.011, 60.0056, 13.33}
```

```
rr2a = {1.0711634554074936`, 19.996461774761155`,
    42.91256928113075`, 59.81453377711739`, 14.35}
```

```
{1.07116, 19.9965, 42.9126, 59.8145, 14.35}
```

```
{1.0711634554074936`, 19.996461774761155`, 42.91256928113075`, 59.81453377711739`, 13.35`}
```

```
{1.07116, 19.9965, 42.9126, 59.8145, 13.35}
```

```
rr3a = {0.9910460401758454`, 20.134176040501433`,
    39.60320410646092`, 64.79155808822323`, 12.30}
```

```
{0.991046, 20.1342, 39.6032, 64.7916, 12.3}
```

```
rr4a = {0.9878787632108709`, 21.64569027106111`,
    37.076263866223904`, 61.21818748111276`, 13.11}
```

```
{0.987879, 21.6457, 37.0763, 61.2182, 13.11}
```

```
{0.9878787632108709`, 21.64569027106111`, 37.076263866223904`, 61.21818748111276`, 13.11}
```

```
{0.987879, 21.6457, 37.0763, 61.2182, 13.11}
```

```
{0.9878787632108709`, 21.64569027106111`, 37.076263866223904`, 61.21818748111276`, 13.66`}

{0.987879, 21.6457, 37.0763, 61.2182, 13.66}

rr5a = {1.0029224966545358`, 18.33677086487405`,
    40.686536960698916`, 54.64567843902394`, 13.66}

{1.00292, 18.3368, 40.6865, 54.6457, 13.66}
```

Table 9.10 Estimation of five kinetic parameters for mixed activation of A $\rightarrow$ products using velocities at {200,100}, {200,1}, {1,100}, and {5,5}

```
TableForm[Round[{rr1a, rr2a, rr3a, rr4a, rr5a}, 0.01],
  TableHeadings -> {{"No errors", "1.05*v1", "1.05*v2", "1.05*v3", "1.05*v4", "1.05*v5"},
    {"kfxet", "kX", "kAX", "kA", "kXA"}}]
```

	kfxet	kX	kAX	kA	kXA
No errors	1.	20.	40.01	60.01	13.33
1.05*v1	1.07	20.	42.91	59.81	14.35
1.05*v2	0.99	20.13	39.6	64.79	12.3
1.05*v3	0.99	21.65	37.08	61.22	13.11
1.05*v4	1.	18.34	40.69	54.65	13.66

The column for kXA was calculated using a thermodynamic cycle.

Since calckinparsmixedact is a more general program, it can be applied to velocities obtained from a reaction following the essential activation mechanism.

Using velocities calculated earlier yields the following kinetic parameters:

```
calckinparsmixact[.8065, 200, 100, .1923, 200, 1, .02041, 1, 100, .02439, 5, 5]
```

$$\{\{kfxet \rightarrow 1.00005, kX \rightarrow 20.0032, kAX \rightarrow 39.9974, kA \rightarrow 7.10877 \times 10^6\}\}$$

The fact that the value of kA is unreasonable shows that this data is not from a mixed activation mechanism. The other kinetic parameters are correct.

9.8 Mixed activation when two X are bound in a single reaction

The mechanism is

```
    kAX2    kf2xet
A + EX₂ = EAX₂ → products
   ‖ kX²  ‖ kXA²
A + E  =  EA
   + kA +
   2X    2X
```

The four equilibria are not independent. Since kA kXA^2 = kX^2 kAX2, kXA^2 = kX^2 kAX2/kA. Therefore, EA + 2X = EAX$_2$ and kXA^2 are omitted from the derivation of the rapid-equilibrium rate equation. The mechanism can also be written as

E + 2X = EX₂   kX^2 = e*x^2/ex2
EX₂ + A = EAX₂   kAX2 = ex2*a/eax2
E + A = EA   kA = e*a/ea
EAX₂ → products   v = kf2x*eax2

The rapid-equilibrium rate equation can be derived by use of Solve.

```
Solve[{kX^2 == e * x^2 / ex2, kAX2 == ex2 * a / eax2, kA == e * a / ea, et == e + ea + ex2 + eax2},
  {eax2}, {e, ea, ex2}]
```

$$\left\{\left\{eax2 \to \frac{a \, et \, kA \, x^2}{a \, kAX2 \, kX^2 + kA \, kAX2 \, kX^2 + a \, kA \, x^2 + kA \, kAX2 \, x^2}\right\}\right\}$$

The rapid-equilibrium velocity is given by

$$vmixact2 = \frac{\dfrac{a \, kfx2et \, kA \, x^2}{a \, kAX2 \, kX^2 + kA \, kAX2 \, kX^2 + a \, kA \, x^2 + kA \, kAX2 \, x^2}}{\dfrac{a \, kA \, kfx2et \, x^2}{a \, kAX2 \, kX^2 + kA \, kAX2 \, kX^2 + a \, kA \, x^2 + kA \, kAX2 \, x^2}}$$

There are 4 kinetic parameters, and so 4 velocities have to be measured to estimate the kinetic parameters from the minimum number of velocity measurements. For a test calculation, the following values of the kinetic parameters are specified:

```
vmixact2 /. kfx2et → 1 /. kX → 20 /. kAX2 → 40 /. kA → 60
```

$$\frac{60 \, a \, x^2}{960\,000 + 16\,000 \, a + 2400 \, x^2 + 60 \, a \, x^2}$$

The characteristics of vmixact2 are shown by a 3D plot.

```
plotmix2a = Plot3D[Evaluate[vmixact2 /. kfx2et → 1 /. kX → 20 /. kAX2 → 40 /. kA → 60],
   {a, 0.00001, 200}, {x, 0.00001, 100}, PlotLabel → "vmixact2", ViewPoint → {-2, -2, 1},
   PlotRange -> {0, 1}, Lighting -> {{"Ambient", GrayLevel[1]}}, AxesLabel → {"[A]", "[X]"}];
```

```
plotmix2b = Plot3D[Evaluate[D[vmixact2 /. kfx2et → 1 /. kX → 20 /. kAX2 → 40 /. kA → 60, a]],
   {a, 0.00001, 50}, {x, 0.00001, 100}, PlotLabel → "∂vmixact2/∂[A]", ViewPoint → {-2, -2, 1},
   PlotRange -> {0, .025}, Lighting -> {{"Ambient", GrayLevel[1]}}, AxesLabel → {"[A]", "[X]"}];
```

```
plotmix2c = Plot3D[Evaluate[D[vmixact2 /. kfx2et → 1 /. kX → 20 /. kAX2 → 40 /. kA → 60, x]],
   {a, 0.00001, 50}, {x, 0.00001, 100}, ViewPoint → {-2, -2, 1},
   PlotRange -> {0, .05}, Lighting -> {{"Ambient", GrayLevel[1]}},
   AxesLabel → {"[A]", "[X]"}, PlotLabel → "∂vmixact2/∂[X]"];
```

```
plotmix2d = Plot3D[Evaluate[D[vmixact2 /. kfx2et → 1 /. kX → 20 /. kAX2 → 40 /. kA → 60, a, x]],
   {a, 0.00001, 50}, {x, 0.00001, 100}, ViewPoint → {-2, -2, 1},
   PlotRange -> {0, .0015}, Lighting -> {{"Ambient", GrayLevel[1]}},
   AxesLabel → {"[A]", "[X]"}, PlotLabel → "∂(∂vmixact2/∂[A])/∂[X]"];
```

fig7mix = GraphicsArray[{{plotmix2a, plotmix2b}, {plotmix2c, plotmix2d}}]

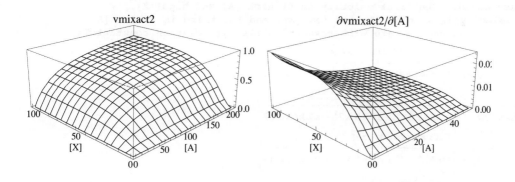

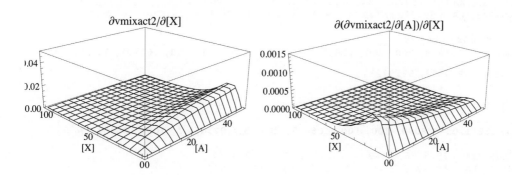

Fig. 9.10 Plots of vmixact2, ∂vmixact2/∂[A], ∂vmixact2/∂[X], and $\partial(\partial$vmixact2/∂[A])/∂[X] versus [A] and [X] for mixed activation when one molecule of X is bound.

The rate equation vmixact2 can be used to estimate the 4 kinetic parameters from 4 measurements of velocities: {a,x} = {200,100}, {200,5}, {1,100}, and {7,7} are chosen.

vmixact2 /. kfx2et → 1 /. kX → 20 /. kAX2 → 40 /. kA → 60 /. a → 200 /. x → 100 // N

0.809935

vmixact2 /. kfx2et → 1 /. kX → 20 /. kAX2 → 40 /. kA → 60 /. a → 200 /. x → 5 // N

0.0663717

vmixact2 /. kfx2et → 1 /. kX → 20 /. kAX2 → 40 /. kA → 60 /. a → 1 /. x → 100 // N

0.0234595

vmixact2 /. kfx2et → 1 /. kX → 20 /. kAX2 → 40 /. kA → 60 /. a → 7 /. x → 7 // N

0.0170057

The following program was written to estimate the kinetic parameters:

```
calckinparsmixact2[v1_, a1_, x1_, v2_, a2_, x2_, v3_, a3_, x3_, v4_, a4_, x4_] :=
  Module[{}, (*This program calculates kfx2et, kX, kAX2,
   and kA from 4 measured velocities for A → products with mixed activation
     when 2X are bound.  The first velocity is at high [A] and high [X],
    the second velocity is at high [A] and low [X], and the third is at low [A]
     and high [X].  High and low are with respect to the dissociation constants.*)
```

$$\text{Solve}\Big[\Big\{v1 == \frac{a1\ kfx2et\ kA\ x1^2}{a1\ kAX2\ kX^2 + kA\ kAX2\ kX^2 + a1\ kA\ x1^2 + kA\ kAX2\ x1^2},$$

$$v2 == \frac{a2\ kfx2et\ kA\ x2^2}{a2\ kAX2\ kX^2 + kA\ kAX2\ kX^2 + a2\ kA\ x2^2 + kA\ kAX2\ x2^2},$$

$$v3 == \frac{a3\ kfx2et\ kA\ x3^2}{a3\ kAX2\ kX^2 + kA\ kAX2\ kX^2 + a3\ kA\ x3^2 + kA\ kAX2\ x3^2},$$

$$v4 == \frac{a4\ kfx2et\ kA\ x4^2}{a4\ kAX2\ kX^2 + kA\ kAX2\ kX^2 + a4\ kA\ x4^2 + kA\ kAX2\ x4^2}\Big\}, \{kfx2et, kX, kAX2, kA\}\Big]\Big]$$

```
calckinparsmixact2[.8099, 200, 100, .06637, 200, 5, .02346, 1, 100, .01701, 7, 7]
```

```
{{kfx2et → 0.999945, kA → 59.9723, kAX2 → 39.9973, kX → -19.9968},
 {kfx2et → 0.999945, kA → 59.9723, kAX2 → 39.9973, kX → 19.9968}}
```

```
rrr1 = {kfx2et, kX, kAX2, kA} /.
   calckinparsmixact2[.8099, 200, 100, .06637, 200, 5, .02346, 1, 100, .01701, 7, 7][[2]]
```

```
{0.999945, 19.9968, 39.9973, 59.9723}
```

These values are correct, but test for errors in velocity measurements.

```
calckinparsmixact2[1.05 * .8099, 200, 100, .06637, 200, 5, .02346, 1, 100, .01701, 7, 7]
```

```
{{kfx2et → 1.0629, kA → 59.6786, kAX2 → 42.5777, kX → -19.9869},
 {kfx2et → 1.0629, kA → 59.6786, kAX2 → 42.5777, kX → 19.9869}}
```

```
rrr2 = {kfx2et, kX, kAX2, kA} /.
   calckinparsmixact2[1.05 * .8099, 200, 100, .06637, 200, 5, .02346, 1, 100, .01701, 7, 7][[2]]
```

```
{1.0629, 19.9869, 42.5777, 59.6786}
```

```
calckinparsmixact2[.8099, 200, 100, 1.05 * .06637, 200, 5, .02346, 1, 100, .01701, 7, 7]
```

```
{{kfx2et → 0.998085, kA → 65.0031, kAX2 → 39.9099, kX → -20.0814},
 {kfx2et → 0.998085, kA → 65.0031, kAX2 → 39.9099, kX → 20.0814}}
```

```
rrr3 = {kfx2et, kX, kAX2, kA} /.
   calckinparsmixact2[.8099, 200, 100, 1.05 * .06637, 200, 5, .02346, 1, 100, .01701, 7, 7][[2]]
```

```
{0.998085, 20.0814, 39.9099, 65.0031}
```

```
calckinparsmixact2[.8099, 200, 100, .06637, 200, 5, 1.05 * .02346, 1, 100, .01701, 7, 7]
```

```
{{kfx2et → 0.9898, kA → 60.4628, kAX2 → 37.5624, kX → -20.5942},
 {kfx2et → 0.9898, kA → 60.4628, kAX2 → 37.5624, kX → 20.5942}}
```

```
rrr4 = {kfx2et, kX, kAX2, kA} /.
   calckinparsmixact2[.8099, 200, 100, .06637, 200, 5, 1.05 * .02346, 1, 100, .01701, 7, 7][[2]]
```

```
{0.9898, 20.5942, 37.5624, 60.4628}
```

```
calckinparsmixact2[.8099, 200, 100, .06637, 200, 5, .02346, 1, 100, 1.05 * .01701, 7, 7]
```

```
{{kfx2et → 1.00045, kA → 55.1884, kAX2 → 40.1174, kX → -19.3375},
 {kfx2et → 1.00045, kA → 55.1884, kAX2 → 40.1174, kX → 19.3375}}
```

```
rrr5 = {kfx2et, kX, kAX2, kA} /.
   calckinparsmixact2[.8099, 200, 100, .06637, 200, 5, .02346, 1, 100, 1.05 * .01701, 7, 7][[2]]
```

{1.00045, 19.3375, 40.1174, 55.1884}

Table 9.11 Estimation of four kinetic parameters for mixed activation of A → products using velocities at {200,100}, {200,5}, {1,100}, and {7,7}

```
TableForm[Round[{rrr1, rrr2, rrr3, rrr4, rrr5}, 0.01],
  TableHeadings -> {{"No errors", "1.05*v1", "1.05*v2", "1.05*v3", "1.05*v4", "1.05*v5"},
    {"kfx2et", "kX", "kAX2", "kA"}}]
```

	kfx2et	kX	kAX2	kA
No errors	1.	20.	40.	59.97
1.05*v1	1.06	19.99	42.58	59.68
1.05*v2	1.	20.08	39.91	65.
1.05*v3	0.99	20.59	37.56	60.46
1.05*v4	1.	19.34	40.12	55.19

Since this mechanism involves 3 sides of a thermodynamic cycle, a fourth equilibrium constant can be calculated using $kXA^2 = kX^2\, kAX2/kA$ or $kXA = kX\,(kAX2/kA)^{(1/2)}$

$$
\begin{pmatrix} 20.` \\ 19.990000000000002` \\ 20.080000000000002` \\ 20.59` \\ 19.34` \end{pmatrix}
*
\begin{pmatrix} 40.` \\ 42.58` \\ 39.910000000000004` \\ 37.56` \\ 40.12` \end{pmatrix}
\Big/
\begin{pmatrix} 59.97` \\ 59.68` \\ 65.` \\ 60.46` \\ 55.19` \end{pmatrix}
\text{ ^ (1 / 2)}
$$

{{16.334}, {16.885}, {15.7343}, {16.2287}, {16.4895}}

This column can be added to Table 9.11.

```
rrr1
```

{0.999945, 19.9968, 39.9973, 59.9723}

```
rrr1a =
  {0.9999449082997035`, 19.99680119082004`, 39.99734681933998`, 59.9723287779935`, 16.33}
```

{0.999945, 19.9968, 39.9973, 59.9723, 16.33}

```
rrr2
```

{1.0629, 19.9869, 42.5777, 59.6786}

```
rrr2a = {1.062902831643799`, 19.986854866273916`,
   42.57765733962105`, 59.678627718448375`, 16.89}
```

{1.0629, 19.9869, 42.5777, 59.6786, 16.89}

```
rrr3
```

{0.998085, 20.0814, 39.9099, 65.0031}

```
rrr3a =
  {0.9980851677794321`, 20.081448373134545`, 39.90993913367678`, 65.00310062875417`, 15.75}
```

{0.998085, 20.0814, 39.9099, 65.0031, 15.75}

```
rrr4
```

{0.9898, 20.5942, 37.5624, 60.4628}

```
rrr4a = {0.989799808269654`, 20.594179934624393`,
    37.56241558440943`, 60.462840615359596`, 16.23}

{0.9898, 20.5942, 37.5624, 60.4628, 16.23}

rrr5

{1.00045, 19.3375, 40.1174, 55.1884}

rrr5a =
    {1.0004450761823749`, 19.33748498584726`, 40.1173923976604`, 55.18842762607657`, 16.49}

{1.00045, 19.3375, 40.1174, 55.1884, 16.49}
```

Table 9.12 Estimation of five kinetic parameters for mixed activation of A → products using velocities at {200,100}, {200,5}, {1,100}, and {7,7} including kXA from the thermodynamic cycle.

```
TableForm[Round[{rrr1a, rrr2a, rrr3a, rrr4a, rrr5a}, 0.01],
   TableHeadings -> {{"No errors", "1.05*v1", "1.05*v2", "1.05*v3", "1.05*v4", "1.05*v5"},
      {"kfx2et", "kX", "kAX2", "kA", "kXA"}}]
```

	kfx2et	kX	kAX2	kA	kXA
No errors	1.	20.	40.	59.97	16.33
1.05*v1	1.06	19.99	42.58	59.68	16.89
1.05*v2	1.	20.08	39.91	65.	15.75
1.05*v3	0.99	20.59	37.56	60.46	16.23
1.05*v4	1.	19.34	40.12	55.19	16.49

The last column in this table has been calculated using a thermodynamic cycle.

9.9 Derivation of the simplest rate equation that is sigmoid in both A and X.

This mechanism is like essential activation except that 2A and 2X are bound.

$$E + 2A = EA_2 \qquad kA^2 = e*a^\wedge 2/ea2$$
$$EA_2 + 2X = EA_2X_2 \quad kA2\,X2^\wedge 2 = ea2*x^\wedge 2/ea2x2$$
$$EA_2X_2 \rightarrow products \qquad v = kf*ea2x2$$

The rate equation can be derived by use of Solve.

```
Solve[{kA^2 == e * a^2 / ea2, kA2X2^2 == ea2 * x^2 / ea2x2, et == e + ea2 + ea2x2}, {ea2x2}, {e, ea2}]
```

$$\left\{\left\{ea2x2 \rightarrow \frac{a^2\ et\ x^2}{a^2\ kA2X2^2 + kA^2\ kA2X2^2 + a^2\ x^2}\right\}\right\}$$

The rate equation is given by

$$vA2X2 = \frac{a^2\ kfet\ x^2}{a^2\ kA2X2^2 + kA^2\ kA2X2^2 + a^2\ x^2}$$

$$\frac{a^2\ kfet\ x^2}{a^2\ kA2X2^2 + kA^2\ kA2X2^2 + a^2\ x^2}$$

The following kinetic parameters are arbitrarily assigned for test calculations:

```
vA2X2 /. kfet → 1 /. kA → 10 /. kA2X2 → 30
```

$$\frac{a^2 \, x^2}{90\,000 + 900\,a^2 + a^2\,x^2}$$

Construct 3D plots.

```
plotA2X2a = Plot3D[Evaluate[vA2X2 /. kfet → 1 /. kA → 10 /. kA2X2 → 30], {a, .00001, 30},
    {x, .00001, 100}, ViewPoint → {-2, -2, 1}, AxesLabel → {"[A]", "[X]"},
    PlotRange → {0, 1}, Lighting -> {{"Ambient", GrayLevel[1]}}, PlotLabel → "vA2X2"];

plotA2X2b = Plot3D[Evaluate[D[vA2X2 /. kfet → 1 /. kA → 10 /. kA2X2 → 30, a]],
    {a, .00001, 30}, {x, .00001, 100}, ViewPoint → {-2, -2, 1}, AxesLabel → {"[A]", "[X]"},
    PlotRange → {0, .3}, Lighting -> {{"Ambient", GrayLevel[1]}}, PlotLabel → "dvA2X2/d[A]"];

plotA2X2c = Plot3D[Evaluate[D[vA2X2 /. kfet → 1 /. kA → 10 /. kA2X2 → 30, x]],
    {a, .00001, 30}, {x, .00001, 100}, ViewPoint → {-2, -2, 1}, AxesLabel → {"[A]", "[X]"},
    PlotRange → {0, .03}, Lighting -> {{"Ambient", GrayLevel[1]}}, PlotLabel → "dvA2X2/d[X]"];

plotA2X2d = Plot3D[Evaluate[D[vA2X2 /. kfet → 1 /. kA → 10 /. kA2X2 → 30, a, x]],
    {a, .00001, 30}, {x, .00001, 100}, ViewPoint → {-2, -2, 1},
    AxesLabel → {"[A]", "[X]"}, PlotRange → {-.002, .005},
    Lighting -> {{"Ambient", GrayLevel[1]}}, PlotLabel → "d(dvA2X2/d[A])/d[X]"];
```

```
fig1A2X2 = GraphicsArray[{{plotA2X2a, plotA2X2b}, {plotA2X2c, plotA2X2d}}]
```

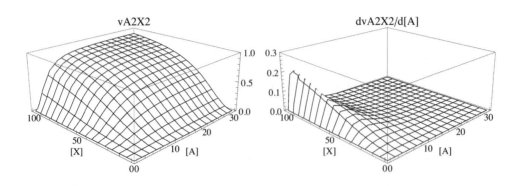

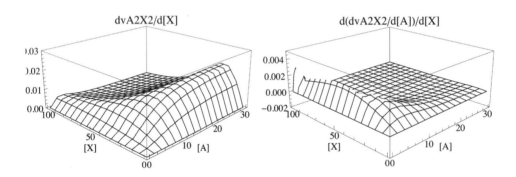

Fig. 9.11 Plots of vA2X2, $\partial vA2X2/\partial[A]$, $\partial vA2X2/\partial[X]$, and $\partial(\partial vA2X2/\partial[A])/\partial[X]$ versus [A] and [X] for the simplest mechanism for A → products that is sigmoid in both [A] and [X].

This figure shows that the plots of vA2X2 versus both [A] and [X] are sigmoid. Figure 9.11 shows that the greatest sensitivity of vA2X2 to [A] is at low [A] and high [X]. However, in the determination of kinetic parameters, [A] should not be too low. Figure 9.11c shows that the greatest sensitivity of vA2X2 to [X] is at low [X] and high [A]. However, in the determination of kinetic parameters, [X] should not be too low because there the sensitivity is very low.

Since there are three kinetic parameters, only three velocities have to be measured.

```
vA2X2 /. kfet → 1 /. kA → 10 /. kA2X2 → 30 /. a -> 100 /. x → 100 // N
```

0.916674

```
vA2X2 /. kfet → 1 /. kA → 10 /. kA2X2 → 30 /. a → 5 /. x → 100 // N
```

0.689655

```
vA2X2 /. kfet → 1 /. kA → 10 /. kA2X2 → 30 /. a -> 100 /. x → 20 // N
```

0.305577

A program is written to calculate the three kinetic parameters from three velocities.

```
calckinparsA2X2[v1_, a1_, x1_, v2_, a2_, x2_, v3_, a3_, x3_] := Module[{},
```

(*This program calculates kfet, kA, and kA2X2 from 3 experimental velocities for A →
products with the simplest mechanism that yields sigmoid plots
for both A and X The first velocity is at high [A] and high [X],
the second velocity is at low [A] and high [X], and the third is at high [A]
and medium [X]. High and low are with respect to the two dissociation constants.*)

$$\text{Solve}\Big[\Big\{v1 == \frac{a1^2\ kfet\ x1^2}{a1^2\ kA2X2^2 + kA^2\ kA2X2^2 + a1^2\ x1^2}, v2 == \frac{a2^2\ kfet\ x2^2}{a2^2\ kA2X2^2 + kA^2\ kA2X2^2 + a2^2\ x2^2},$$

$$v3 == \frac{a3^2\ kfet\ x3^2}{a3^2\ kA2X2^2 + kA^2\ kA2X2^2 + a3^2\ x3^2}\Big\}, \{kfet, kA, kA2X2\}\Big]\Big]$$

```
calckinparsA2X2[.9167, 100, 100, .6897, 5, 100, .3056, 100, 20]
```

{{kfet → 1.00002, kA → -9.99961, kA2X2 → -29.9988},
 {kfet → 1.00002, kA → -9.99961, kA2X2 → 29.9988},
 {kfet → 1.00002, kA → 9.99961, kA2X2 → -29.9988},
 {kfet → 1.00002, kA → 9.99961, kA2X2 → 29.9988}}

```
line1 = {kfet, kA, kA2X2} /. calckinparsA2X2[.9167, 100, 100, .6897, 5, 100, .3056, 100, 20][[4]]
```

{1.00002, 9.99961, 29.9988}

This yields the correct values of the kinetic parameters, but their sensitivity to experimental errors in velocities needs to be tested.

```
calckinparsA2X2[1.05 * .9167, 100, 100, .6897, 5, 100, .3056, 100, 20]
```

{{kfet → 1.05723, kA → -10.5797, kA2X2 → -31.1917},
 {kfet → 1.05723, kA → -10.5797, kA2X2 → 31.1917},
 {kfet → 1.05723, kA → 10.5797, kA2X2 → -31.1917},
 {kfet → 1.05723, kA → 10.5797, kA2X2 → 31.1917}}

```
line2 =
 {kfet, kA, kA2X2} /. calckinparsA2X2[1.05 * .9167, 100, 100, .6897, 5, 100, .3056, 100, 20][[4]]
```

{1.05723, 10.5797, 31.1917}

```
calckinparsA2X2[.9167, 100, 100, 1.05 * .6897, 5, 100, .3056, 100, 20]
```

{{kfet → 1.00002, kA → -8.97823, kA2X2 → -30.0277},
 {kfet → 1.00002, kA → -8.97823, kA2X2 → 30.0277},
 {kfet → 1.00002, kA → 8.97823, kA2X2 → -30.0277},
 {kfet → 1.00002, kA → 8.97823, kA2X2 → 30.0277}}

```
line3 =
 {kfet, kA, kA2X2} /. calckinparsA2X2[.9167, 100, 100, 1.05 * .6897, 5, 100, .3056, 100, 20][[4]]
```

{1.00002, 8.97823, 30.0277}

```
calckinparsA2X2[.9167, 100, 100, .6897, 5, 100, 1.05 * .3056, 100, 20]
```

{{kfet → 0.993571, kA → -10.3811, kA2X2 → -28.8031},
 {kfet → 0.993571, kA → -10.3811, kA2X2 → 28.8031},
 {kfet → 0.993571, kA → 10.3811, kA2X2 → -28.8031},
 {kfet → 0.993571, kA → 10.3811, kA2X2 → 28.8031}}

```
line4 =
 {kfet, kA, kA2X2} /. calckinparsA2X2[.9167, 100, 100, .6897, 5, 100, 1.05 * .3056, 100, 20][[4]]
```

{0.993571, 10.3811, 28.8031}

Table 9.13 Estimation of three kinetic parameters for the simplest mechanism of A → products that is sigmoid in both [A] and [X] using velocities at {100,100}, {5,100}, and {100,20}

```
TableForm[Round[{line1, line2, line3, line4}, 0.01], TableHeadings ->
  {{"No errors", "1.05*v1", "1.05*v2", "1.05*v3", "1.05*v4"}, {"kfet", "kA", "kA2X2"}}]
```

	kfet	kA	kA2X2
No errors	1.	10.	30.
1.05*v1	1.06	10.58	31.19
1.05*v2	1.	8.98	30.03
1.05*v3	0.99	10.38	28.8

9.10 Derivation of the rate equation when 2A is bound in two reactions and 2X is bound in one reaction

Since there are two different dissociation constants for A (kA1 and kA2) the effects of positive cooperativity (kA1 > kA2) and negative cooperativity (kA1 < kA2) can be studied [31]. The mechanism and the general rate equation provide for both kinds of cooperation. Hemoglobin has positive cooperativity because the second oxygen molecule is bound more tightly than the first.

The mechanism is given by

$E + A = EA$	$kA1 = e*a/ea$
$EA + A = EA_2$	$kA2 = ea*a/ea2$
$EA_2 + 2X = EA_2X_2$	$kA2\ X2^2 = ea2*x^2/ea2x2$
$EA_2X_2 \rightarrow$ products	$v = kf*ea2x2$

The general rate equation can be derived by use of Solve.

```
Solve[{kA1 == e * a / ea, kA2 == ea * a / ea2, kA2X2^2 == ea2 * x^2 / ea2x2, et == e + ea + ea2 + ea2x2},
  {ea2x2}, {e, ea, ea2}]
```

$$\left\{\left\{ea2x2 \rightarrow \frac{a^2\ et\ x^2}{a^2\ kA2X2^2 + a\ kA2\ kA2X2^2 + kA1\ kA2\ kA2X2^2 + a^2\ x^2}\right\}\right\}$$

The rate equation is given by

$$vA1A2X2 = \frac{a^2\ kfet\ x^2}{a^2\ kA2X2^2 + a\ kA2\ kA2X2^2 + kA1\ kA2\ kA2X2^2 + a^2\ x^2}$$

$$\frac{a^2\ kfet\ x^2}{a^2\ kA2X2^2 + a\ kA2\ kA2X2^2 + kA1\ kA2\ kA2X2^2 + a^2\ x^2}$$

▪ 9.10.1 Positive cooperativity with respect to A

The following values for the kinetic parameters are chosen arbitrarily for test calculations:

```
vA1A2X2 /. kA1 → 15 /. kA2 → 5 /. kA2X2 → 30 /. kfet → 1
```

$$\frac{a^2\ x^2}{67\,500 + 4500\ a + 900\ a^2 + a^2\ x^2}$$

This rate equation can be used to make 3D plots.

```
plotA1A2X2a = Plot3D[Evaluate[vA1A2X2 /. kA1 → 15 /. kA2 → 5 /. kA2X2 → 30 /. kfet → 1],
    {a, .00001, 30}, {x, .00001, 100}, ViewPoint → {-2, -2, 1}, AxesLabel → {"[A]", "[X]"},
    PlotRange → {0, 1}, Lighting -> {{"Ambient", GrayLevel[1]}}, PlotLabel → "vA1A2X2"];

plotA1A2X2b = Plot3D[Evaluate[D[vA1A2X2 /. kA1 → 15 /. kA2 → 5 /. kA2X2 → 30 /. kfet → 1, a]],
    {a, .00001, 30}, {x, .00001, 100}, ViewPoint → {-2, -2, 1}, AxesLabel → {"[A]", "[X]"},
    PlotRange → {0, .25}, Lighting -> {{"Ambient", GrayLevel[1]}}, PlotLabel → "dvA1A2X2/d[A]"];

plotA1A2X2c =
    Plot3D[Evaluate[D[vA1A2X2 /. kA1 → 15 /. kA2 → 5 /. kA2X2 → 30 /. kfet → 1, x]], {a, .00001, 30},
    {x, .00001, 100}, ViewPoint → {-2, -2, 1}, AxesLabel → {"[A]", "[X]"}, PlotRange → {0, .025},
    Lighting -> {{"Ambient", GrayLevel[1]}}, PlotLabel → "dvA1A2X2/d[X]"];

plotA1A2X2d = Plot3D[Evaluate[D[vA1A2X2 /. kA1 → 15 /. kA2 → 5 /. kA2X2 → 30 /. kfet → 1, a, x]],
    {a, .00001, 30}, {x, .00001, 100}, ViewPoint → {-2, -2, 1},
    AxesLabel → {"[A]", "[X]"}, PlotRange → {-.001, .005},
    Lighting -> {{"Ambient", GrayLevel[1]}}, PlotLabel → "d(dvA1A2X2/d[A])/d[X]"];
```

```
fig9A1A2X2poscoop = GraphicsArray[{{plotA1A2X2a, plotA1A2X2b}, {plotA1A2X2c, plotA1A2X2d}}]
```

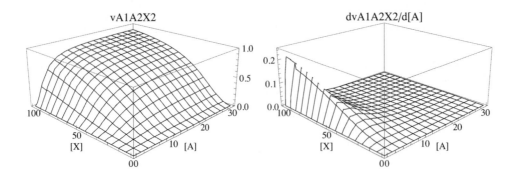

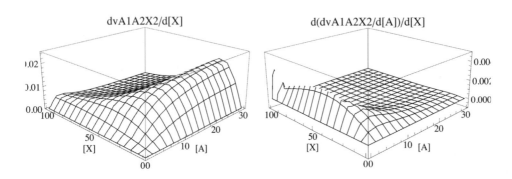

Fig. 9.12 Plots of vA1A2X2, ∂vA1A2X2/∂[A], ∂vA1A2X2/∂[X], and ∂(∂vA1A2X2/∂[A])/∂[X] versus [A] and [X] for A $\rightarrow$ products when the binding of A is positively cooperative (kA1 = 15 and kA2 = 5). These plots are not much different from Fig. 9.8.

Since there are kinetic parameters, only four velocities have to be measured.

```
vA1A2X2 /. kfet → 1 /. kA1 → 15 /. kA2 → 5 /. kA2X2 → 30 /. a -> 100 /. x → 100 // N
```

0.913096

```
vA1A2X2 /. kfet → 1 /. kA1 → 15 /. kA2 → 5 /. kA2X2 → 30 /. a → 1 /. x → 100 // N
```

0.120627

```
vA1A2X2 /. kfet → 1 /. kA1 → 15 /. kA2 → 5 /. kA2X2 → 30 /. a -> 100 /. x → 10 // N
```

0.0950796

```
vA1A2X2 /. kfet → 1 /. kA1 → 15 /. kA2 → 5 /. kA2X2 → 30 /. a → 5 /. x → 10 // N
```

0.0217391

A program is written to calculate the four kinetic parameters from four velocities.

```
calckinparsA1A2X2[v1_, a1_, x1_, v2_, a2_, x2_, v3_, a3_, x3_, v4_, a4_, x4_] :=
  Module[{}, (*This program calculates kfet, kA1,
   kA2,  and kA2X2 from 4 experimental velocities for A →
    products when 2A are bound in two reactions.*)
```

$$
\text{Solve}\Big[\Big\{v1 == \frac{a1^2\ kfet\ x1^2}{a1^2\ kA2X2^2 + a1\ kA2\ kA2X2^2 + kA1\ kA2\ kA2X2^2 + a1^2\ x1^2},
$$

$$
v2 == \frac{a2^2\ kfet\ x2^2}{a2^2\ kA2X2^2 + a2\ kA2\ kA2X2^2 + kA1\ kA2\ kA2X2^2 + a2^2\ x2^2},
$$

$$
v3 == \frac{a3^2\ kfet\ x3^2}{a3^2\ kA2X2^2 + a3\ kA2\ kA2X2^2 + kA1\ kA2\ kA2X2^2 + a3^2\ x3^2},
$$

$$
v4 == \frac{a4^2\ kfet\ x4^2}{a4^2\ kA2X2^2 + a4\ kA2\ kA2X2^2 + kA1\ kA2\ kA2X2^2 + a4^2\ x4^2}\Big\}, \{kfet, kA1, kA2, kA2X2\}\Big]\Big]
$$

```
calckinparsA1A2X2[.9131, 100, 100, .1206, 1, 100, .09508, 100, 10, .02174, 5, 10]

{{kfet → 1., kA1 → 15.0262, kA2 → 4.99283, kA2X2 → -30.001},
 {kfet → 1., kA1 → 15.0262, kA2 → 4.99283, kA2X2 → 30.001}}

ln1 = {kfet, kA1, kA2, kA2X2} /.
   calckinparsA1A2X2[.9131, 100, 100, .1206, 1, 100, .09508, 100, 10, .02174, 5, 10][[2]]

{1., 15.0262, 4.99283, 30.001}
```

This yields correct values, but the effects of experimental errors in velocity measurements have to be considered.

```
calckinparsA1A2X2[1.05 * .9131, 100, 100, .1206, 1, 100, .09508, 100, 10, .02174, 5, 10]

{{kfet → 1.05561, kA1 → 15.6544, kA2 → 4.80405, kA2X2 → -30.9364},
 {kfet → 1.05561, kA1 → 15.6544, kA2 → 4.80405, kA2X2 → 30.9364}}

ln2 = {kfet, kA1, kA2, kA2X2} /.
   calckinparsA1A2X2[1.05 * .9131, 100, 100, .1206, 1, 100, .09508, 100, 10, .02174, 5, 10][[2]]

{1.05561, 15.6544, 4.80405, 30.9364}

calckinparsA1A2X2[.9131, 100, 100, 1.05 * .1206, 1, 100, .09508, 100, 10, .02174, 5, 10]

{{kfet → 1., kA1 → 11.2871, kA2 → 6.225, kA2X2 → -29.8344},
 {kfet → 1., kA1 → 11.2871, kA2 → 6.225, kA2X2 → 29.8344}}

ln3 = {kfet, kA1, kA2, kA2X2} /.
   calckinparsA1A2X2[.9131, 100, 100, 1.05 * .1206, 1, 100, .09508, 100, 10, .02174, 5, 10][[2]]

{1., 11.2871, 6.225, 29.8344}

calckinparsA1A2X2[.9131, 100, 100, .1206, 1, 100, 1.05 * .09508, 100, 10, .02174, 5, 10]

{{kfet → 0.994971, kA1 → 13.9219, kA2 → 5.70406, kA2X2 → -29.0158},
 {kfet → 0.994971, kA1 → 13.9219, kA2 → 5.70406, kA2X2 → 29.0158}}

ln4 = {kfet, kA1, kA2, kA2X2} /.
   calckinparsA1A2X2[.9131, 100, 100, .1206, 1, 100, 1.05 * .09508, 100, 10, .02174, 5, 10][[2]]

{0.994971, 13.9219, 5.70406, 29.0158}

calckinparsA1A2X2[.9131, 100, 100, .1206, 1, 100, .09508, 100, 10, 1.05 * .02174, 5, 10]

{{kfet → 1., kA1 → 22.6984, kA2 → 3.32256, kA2X2 → -30.2402},
 {kfet → 1., kA1 → 22.6984, kA2 → 3.32256, kA2X2 → 30.2402}}
```

```
ln5 = {kfet, kA1, kA2, kA2X2} /.
    calckinparsA1A2X2[.9131, 100, 100, .1206, 1, 100, .09508, 100, 10, 1.05 * .02174, 5, 10][[2]]
```

```
{1., 22.6984, 3.32256, 30.2402}
```

Table 9.14 Estimation of four kinetic parameters for A $\rightarrow$ products when the binding of A is positively cooperative (kA1 = 15 and kA2 = 5) and {100,100}, {1,100}, {100,10}, and {5,10}

```
TableForm[Round[{ln1, ln2, ln3, ln4, ln5}, 0.01],
  TableHeadings -> {{"No errors", "1.05*v1", "1.05*v2", "1.05*v3", "1.05*v4", "1.05*v5"},
    {"kfet", "kA1", "kA2", "kA2X2"}}]
```

	kfet	kA1	kA2	kA2X2
No errors	1.	15.03	4.99	30.
1.05*v1	1.06	15.65	4.8	30.94
1.05*v2	1.	11.29	6.23	29.83
1.05*v3	0.99	13.92	5.7	29.02
1.05*v4	1.	22.7	3.32	30.24

▪ 9.10.2 Negative cooperativity with respect to A

The form of the general rate equation is unchanged and calckinparsA1A2X2 can be used to calculate the kinetic parameters. The only change is in the values of kA1 and kA2.

The following values for the kinetic parameters are chosen arbitrarily for test calculations:

```
vA1A2X2 /. kA1 → 5 /. kA2 → 15 /. kA2X2 → 30 /. kfet → 1
```

$$\frac{a^2\,x^2}{67\,500 + 13\,500\,a + 900\,a^2 + a^2\,x^2}$$

This rate equation can be used to make 3D plots.

```
plotA1A2X2nega = Plot3D[Evaluate[vA1A2X2 /. kA1 → 5 /. kA2 → 15 /. kA2X2 → 30 /. kfet → 1],
    {a, .00001, 30}, {x, .00001, 100}, ViewPoint → {-2, -2, 1}, AxesLabel → {"[A]", "[X]"},
    PlotRange → {0, 1}, Lighting -> {{"Ambient", GrayLevel[1]}}, PlotLabel → "vA1A2X2neg"];
```

```
plotA1A2X2negb =
  Plot3D[Evaluate[D[vA1A2X2 /. kA1 → 5 /. kA2 → 15 /. kA2X2 → 30 /. kfet → 1, a]], {a, .00001, 30},
    {x, .00001, 100}, ViewPoint → {-2, -2, 1}, AxesLabel → {"[A]", "[X]"}, PlotRange → {0, .25},
    Lighting -> {{"Ambient", GrayLevel[1]}}, PlotLabel → "dvA1A2X2neg/d[A]"];
```

```
plotA1A2X2negc =
  Plot3D[Evaluate[D[vA1A2X2 /. kA1 → 5 /. kA2 → 15 /. kA2X2 → 30 /. kfet → 1, x]], {a, .00001, 30},
    {x, .00001, 100}, ViewPoint → {-2, -2, 1}, AxesLabel → {"[A]", "[X]"}, PlotRange → {0, .025},
    Lighting -> {{"Ambient", GrayLevel[1]}}, PlotLabel → "dvA1A2X2neg/d[X]"];
```

```
plotA1A2X2negd =
  Plot3D[Evaluate[D[vA1A2X2 /. kA1 → 5 /. kA2 → 15 /. kA2X2 → 30 /. kfet → 1, a, x]],
    {a, .00001, 30}, {x, .00001, 100}, ViewPoint → {-2, -2, 1},
    AxesLabel → {"[A]", "[X]"}, PlotRange → {-.001, .005},
    Lighting -> {{"Ambient", GrayLevel[1]}}, PlotLabel → "d(dvA1A2X2neg/d[A])/d[X]"];
```

```
fig9A1A2X2negcoop =
  GraphicsArray[{{plotA1A2X2nega, plotA1A2X2negb}, {plotA1A2X2negc, plotA1A2X2negd}}]
```

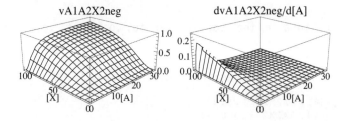

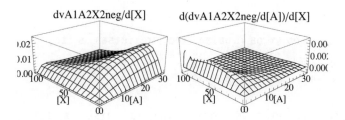

Fig. 9.13 Plots of vA1A2X2, ∂vA1A2X2/∂[A], ∂vA1A2X2/∂[X], and $\partial(\partial$vA1A2X2/∂[A])/∂[X] versus [A] and [X] for A $\rightarrow$ products when the binding of A is negatively cooperative (kA1 = 5 and kA2 = 15).

Since there are four kinetic parameters, only four velocities have to be measured.

```
vA1A2X2 /. kfet → 1 /. kA1 → 5 /. kA2 → 15 /. kA2X2 → 30 /. a -> 100 /. x → 100 // N
```

0.905654

```
vA1A2X2 /. kfet → 1 /. kA1 → 5 /. kA2 → 15 /. kA2X2 → 30 /. a → 1 /. x → 100 // N
```

0.108814

```
vA1A2X2 /. kfet → 1 /. kA1 → 5 /. kA2 → 15 /. kA2X2 → 30 /. a -> 100 /. x → 10 // N
```

0.0875848

```
vA1A2X2 /. kfet → 1 /. kA1 → 5 /. kA2 → 15 /. kA2X2 → 30 /. a → 5 /. x → 10 // N
```

0.015625

calckinparsA1A2X2 can be used with the changed velocities to estimate the values of the kinetic parameters.

```
calckinparsA1A2X2[.9057, 100, 100, .1088, 1, 100, .08758, 100, 10, .01563, 5, 10]
```

```
{{kfet → 1.00006, kA1 → 5.0086, kA2 → 14.9768, kA2X2 → -30.0049},
 {kfet → 1.00006, kA1 → 5.0086, kA2 → 14.9768, kA2X2 → 30.0049}}
```

```
lnn1 = {kfet, kA1, kA2, kA2X2} /.
  calckinparsA1A2X2[.9057, 100, 100, .1088, 1, 100, .08758, 100, 10, .01563, 5, 10][[2]]
```

{1.00006, 5.0086, 14.9768, 30.0049}

This yields correct values, but the effects of experimental errors in velocity measurements have to be considered. Make 5% changes in the velocities, one at a time.

```
calckinparsA1A2X2[1.05 * .9057, 100, 100, .1088, 1, 100, .08758, 100, 10, .01563, 5, 10]
```

{{kfet → 1.05616, kA1 → 5.11068, kA2 → 14.7134, kA2X2 → -30.9484},
 {kfet → 1.05616, kA1 → 5.11068, kA2 → 14.7134, kA2X2 → 30.9484}}

```
lnn2 = {kfet, kA1, kA2, kA2X2} /.
   calckinparsA1A2X2[1.05 * .9057, 100, 100, .1088, 1, 100, .08758, 100, 10, .01563, 5, 10][[2]]
```

{1.05616, 5.11068, 14.7134, 30.9484}

```
calckinparsA1A2X2[.9057, 100, 100, 1.05 * .1088, 1, 100, .08758, 100, 10, .01563, 5, 10]
```

{{kfet → 1.00006, kA1 → 4.23427, kA2 → 16.4681, kA2X2 → -29.8202},
 {kfet → 1.00006, kA1 → 4.23427, kA2 → 16.4681, kA2X2 → 29.8202}}

```
lnn3 = {kfet, kA1, kA2, kA2X2} /.
   calckinparsA1A2X2[.9057, 100, 100, 1.05 * .1088, 1, 100, .08758, 100, 10, .01563, 5, 10][[2]]
```

{1.00006, 4.23427, 16.4681, 29.8202}

```
calckinparsA1A2X2[.9057, 100, 100, .1088, 1, 100, 1.05 * .08758, 100, 10, .01563, 5, 10]
```

{{kfet → 0.994601, kA1 → 4.85205, kA2 → 16.4469, kA2X2 → -28.9343},
 {kfet → 0.994601, kA1 → 4.85205, kA2 → 16.4469, kA2X2 → 28.9343}}

```
lnn4 = {kfet, kA1, kA2, kA2X2} /.
   calckinparsA1A2X2[.9057, 100, 100, .1088, 1, 100, 1.05 * .08758, 100, 10, .01563, 5, 10][[2]]
```

{0.994601, 4.85205, 16.4469, 28.9343}

```
calckinparsA1A2X2[.9057, 100, 100, .1088, 1, 100, .08758, 100, 10, 1.05 * .01563, 5, 10]
```

{{kfet → 1.00006, kA1 → 6.06843, kA2 → 12.4508, kA2X2 → -30.3371},
 {kfet → 1.00006, kA1 → 6.06843, kA2 → 12.4508, kA2X2 → 30.3371}}

```
lnn5 = {kfet, kA1, kA2, kA2X2} /.
   calckinparsA1A2X2[.9057, 100, 100, .1088, 1, 100, .08758, 100, 10, 1.05 * .01563, 5, 10][[2]]
```

{1.00006, 6.06843, 12.4508, 30.3371}

Table 9.15 Estimation of four kinetic parameters for A → products when the binding of A is negatively cooperative (kA1 = 5 and kA2 = 15) and {100,100}, {1,100}, {100,10}, and {5,10}

```
TableForm[Round[{lnn1, lnn2, lnn3, lnn4, lnn5}, 0.01],
  TableHeadings -> {{"No errors", "1.05*v1", "1.05*v2", "1.05*v3", "1.05*v4", "1.05*v5"},
    {"kfx2et", "kA1", "kA2", "kA2X2"}}]
```

	kfx2et	kA1	kA2	kA2X2
No errors	1.	5.01	14.98	30.
1.05*v1	1.06	5.11	14.71	30.95
1.05*v2	1.	4.23	16.47	29.82
1.05*v3	0.99	4.85	16.45	28.93
1.05*v4	1.	6.07	12.45	30.34

9.11 Discussion

Inhibition and activation complicate the discussion of the kinetics of systems of biochemical reactions. Cornish-Bowden [21] has emphasized the importance of these in treating the kinetics of systems of enzyme-catalyzed reactions. Competitive inhibition is caused by the binding of molecules at the enzymatic site, and so molecules that are similar to the substrate may have significant effects on velocities. Uncompetitive inhibition is caused by the binding of molecules on enzyme-substrate complexes. This brings in the possibility that different types of molecules may be bound. Inhibitors reduce velocities, but activators can increase rates by binding at sites other than the enzymatic site. When activation occurs, there is a question as to how many activator molecules are bound. The determination of the number of activator molecules bound from velocity measurements is discussed in Sections 9.5 and 9.6.

Chapter 10 Modification of A -> products

10.1 General modifier mechanism for A → products with two paths to products

- 10.1.1 Mechanism

- 10.1.2 Estimation of kinetic parameters for the general modification mechanism

- 10.1.3 Use of a thermodynamic cycle to estimate the apparent equilibrium constant for the reaction EAX = EA + X

10.2 General modifier mechanism for A → products with the binding of 2X and two paths to products

- 10.2.1 Mechanism

- 10.2.2 Estimation of kinetic parameters for the general modifier mechanism with the binding of 2X and two paths to products

- 10.2.3 Estimation of the apparent equilibrium constant for the reaction EAX_2 = EA + 2X

- 10.2.4 Determination of the number of modifier molecules bound

---

10.3 General modifier mechanism for A → products when two X are bound in two reactions and there are three paths to products

- 10.3.1 Mechanism and derivation of the rate equation

- 10.3.2 Plots of cross sections

10.4 Discussion

10.1 General modifier mechanism for A → products with two paths to products

- ### 10.1.1 Mechanism

The following mechanism is the rapid-equilibrium version of Cornish-Bowden's [21] general modifier mechanism (Fig. 5.10). This mechanism involves the binding of one molecule of modifier X. In the following mechanism [36] the reaction EA + X = EAX has been omitted because it is redundant in calculating the equilibrium concentrations of EA and EAX. However, the equilibrium constant for EA + X = EAX can be estimated from kinetic measurements, as will be shown.

```
      kA   kf1
E + A = EA → products
+
X
‖ kX
EX + A = EAX → products
      kAX      kf2
```

The mechanism can also be written as

```
E + A = EA          kA=e*a/ea
E + X = EX          kX=e*x/ex
EX + A = EAX        kAX=ex*a/eax
v = kf1*ea+kf2*eax
```

The rapid-equilibrium rate equation vEAX can be derived by use of Solve.

```
Solve[{kA == e * a / ea, kX == e * x / ex, kAX == ex * a / eax, et == e + ea + ex + eax}, {ea, eax}, {e, ex}]
```

$$\left\{\left\{ea \rightarrow \frac{a\ et\ kAX\ kX}{a\ kAX\ kX + kA\ kAX\ kX + a\ kA\ x + kA\ kAX\ x}, \ eax \rightarrow \frac{a\ et\ kA\ x}{a\ kAX\ kX + kA\ kAX\ kX + a\ kA\ x + kA\ kAX\ x}\right\}\right\}$$

The rate equation is given by

$$vEAX = \frac{a\ kf1et\ kAX\ kX + a\ kf2et\ kA\ x}{a\ kAX\ kX + kA\ kAX\ kX + a\ kA\ x + kA\ kAX\ x}$$

$$\frac{a\ kAX\ kf1et\ kX + a\ kA\ kf2et\ x}{a\ kAX\ kX + kA\ kAX\ kX + a\ kA\ x + kA\ kAX\ x}$$

The kinetic parameters for test calculations are arbitrarily assigned as follows: kf1et=1, kf2et=2, kA=10, kX=20, and kAX=5. The rate equation with these values for the kinetic parameters is named velM1.

```
velM1 = vEAX /. kf1et -> 1 /. kf2et → 2 /. kA → 10 /. kX → 20 /. kAX → 5
```

$$\frac{100\ a + 20\ a\ x}{1000 + 100\ a + 50\ x + 10\ a\ x}$$

3D plots of velM1 and its derivatives versus [A] and [X] can be calculated.

```
plotvEAX1 = Plot3D[Evaluate[velM1], {a, .00001, 30}, {x, .00001, 100},
    ViewPoint → {-2, -2, 1}, AxesLabel → {"[A]", "[X]"}, PlotRange → {0, 2},
    Lighting -> {{"Ambient", GrayLevel[1]}}, PlotLabel → "velM1"];
```

```
plotvEAX2 = Plot3D[Evaluate[D[velM1, a]], {a, .00001, 30}, {x, .00001, 100},
    ViewPoint → {-2, -2, 1}, AxesLabel → {"[A]", "[X]"}, PlotRange → {0, .4},
    Lighting -> {{"Ambient", GrayLevel[1]}}, PlotLabel → "dvelM1/d[A]"];

plotvEAX3 = Plot3D[Evaluate[D[velM1, x]], {a, .00001, 30}, {x, .00001, 100},
    ViewPoint → {-2, -2, 1}, AxesLabel → {"[A]", "[X]"}, PlotRange → {0, .1},
    Lighting -> {{"Ambient", GrayLevel[1]}}, PlotLabel → "dvelM1/d[X]"];

plotvEAX4 = Plot3D[Evaluate[D[velM1, a, x]], {a, .00001, 30}, {x, .00001, 100},
    ViewPoint → {-2, -2, 1}, AxesLabel → {"[A]", "[X]"}, PlotRange → {-.001, .015},
    Lighting -> {{"Ambient", GrayLevel[1]}}, PlotLabel → "d(dvelM1/d[A])/d[X]"];

fig9generalmod = GraphicsArray[{{plotvEAX1, plotvEAX2}, {plotvEAX3, plotvEAX4}}]
```

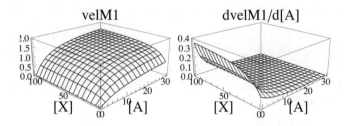

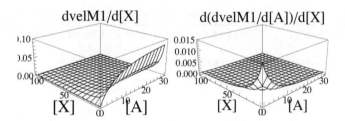

Fig. 10.1 Plots of (a) velM1, (b) ∂velM1$/\partial$[A], (c) ∂velM1$/\partial$[X], and (d) $\partial(\partial$velM1$/\partial$[A])$/\partial$[X] versus [A] and [X] for A → products for the general modifier mechanism.

There are 5 kinetic parameters, and so 5 velocities have to be measured. Figure 10.1(a) shows the Michaelis-Menten plot in the absence of modifier on the right face. Thus in the absence of modifier X, the determination of velocities at {[A],[X]} = {1,0} and {100,0} makes it possible to estimate kf1et and K_A. The use of {100,100} makes the predominant contribution to the estimation of kf1et + kf2et. Figure 10.1 (b) shows that the velocity is very sensitive to [A] at low [A] and high [X]. Figure 10.1 (c) shows that the velocity is very sensitive to modifier concentrations at high [A] and low [X]. Figure 10.1 (d) shows that the velocity is very sensitive to both [A] and [X] at very low concentrations of both.

Velocities at the following pairs of substrate concentrations are used for test calculations: {[A],[X]} = {1,0}, {100,0}, {100,100}, {1,100}, and {5,5}.

```
velM1 /. a → 1 /. x → 0 // N
```

0.0909091

```
velM1 /. a → 100 /. x → 0 // N
```

0.909091

```
velM1 /. a → 100 /. x → 100 // N
```

1.81034

```
velM1 /. a → 1 /. x → 100 // N
```

0.295775

```
velM1 /. a → 5 /. x → 5 // N
```

0.5

These velocities are treated as experimental data.

■ 10.1.2 Estimation of kinetic parameters for the general modification mechanism

A program is written for the general modifier mechanism with one X bound to estimate 5 kinetic parameters from 5 velocity measurements.

```
calckinparsgenmod[v1_, a1_, x1_, v2_, a2_, x2_, v3_, a3_, x3_, v4_, a4_, x4_, v5_, a5_, x5_] :=
 Module[{}, (*This program calculates kf1et, kf2et, kA, kX, and kAX from 5
   measured velocities for the general modifier mechanism for A → products.*)
  Solve[{v1 == (a1 kf1et kAX kX + a1 kf2et kA x1)/(a1 kAX kX + kA kAX kX + a1 kA x1 + kA kAX x1),
    v2 == (a2 kf1et kAX kX + a2 kf2et kA x2)/(a2 kAX kX + kA kAX kX + a2 kA x2 + kA kAX x2),
    v3 == (a3 kf1et kAX kX + a3 kf2et kA x3)/(a3 kAX kX + kA kAX kX + a3 kA x3 + kA kAX x3), v4 == (a4 kf1et kAX kX + a4 kf2et kA x4)/(a4 kAX kX + kA kAX kX + a4 kA x4 + kA kAX x4),
    v5 == (a5 kf1et kAX kX + a5 kf2et kA x5)/(a5 kAX kX + kA kAX kX + a5 kA x5 + kA kAX x5)}, {kf1et, kf2et, kA, kX, kAX}]]
```

The 5 velocities are used to estimate 5 kinetic parameters.

```
calckinparsgenmod[.09091, 1, 0, .9091, 100, 0, 1.8103, 100, 100, .2958, 1, 100, .5, 5, 5]
```

{{kf2et → 1.99994, kf1et → 1.00001, kX → 20.0042, kA → 10., kAX → 4.99915}}

```
row1 = {kf2et, kf1et, kX, kA, kAX} /.
   calckinparsgenmod[.09091, 1, 0, .9091, 100, 0, 1.8103, 100, 100, .2958, 1, 100, .5, 5, 5][[1]]
```

{1.99994, 1.00001, 20.0042, 10., 4.99915}

These values are correct, but it is necessary to test the sensitivity to errors in measured velocities.

```
calckinparsgenmod[1.05 * .09091, 1, 0, .9091, 100, 0, 1.8103, 100, 100, .2958, 1, 100, .5, 5, 5]
```

{{kf2et → 2.01244, kf1et → 0.994747, kX → 21.2888, kA → 9.42105, kAX → 5.00509}}

```
row2 = {kf2et, kf1et, kX, kA, kAX} /. calckinparsgenmod[1.05 * .09091,
    1, 0, .9091, 100, 0, 1.8103, 100, 100, .2958, 1, 100, .5, 5, 5][[1]]
```

{2.01244, 0.994747, 21.2888, 9.42105, 5.00509}

```
calckinparsgenmod[.09091, 1, 0, 1.05 * .9091, 100, 0, 1.8103, 100, 100, .2958, 1, 100, .5, 5, 5]
```

{{kf2et → 1.99331, kf1et → 1.05588, kX → 21.2099, kA → 10.6145, kAX → 4.94396}}

```
row3 = {kf2et, kf1et, kX, kA, kAX} /. calckinparsgenmod[.09091,
    1, 0, 1.05 * .9091, 100, 0, 1.8103, 100, 100, .2958, 1, 100, .5, 5, 5][[1]]
```

{1.99331, 1.05588, 21.2099, 10.6145, 4.94396}

```
calckinparsgenmod[.09091, 1, 0, .9091, 100, 0, 1.05 * 1.8103, 100, 100, .2958, 1, 100, .5, 5, 5]
```

{{kf2et → 2.12091, kf1et → 1.00001, kX → 20.3427, kA → 10., kAX → 5.34206}}

```
row4 = {kf2et, kf1et, kX, kA, kAX} /. calckinparsgenmod[.09091,
    1, 0, .9091, 100, 0, 1.05 * 1.8103, 100, 100, .2958, 1, 100, .5, 5, 5][[1]]
```

{2.12091, 1.00001, 20.3427, 10., 5.34206}

```
calckinparsgenmod[.09091, 1, 0, .9091, 100, 0, 1.8103, 100, 100, 1.05 * .2958, 1, 100, .5, 5, 5]
```

{{kf2et → 1.99652, kf1et → 1.00001, kX → 22.368, kA → 10., kAX → 4.62353}}

```
row5 = {kf2et, kf1et, kX, kA, kAX} /. calckinparsgenmod[.09091,
    1, 0, .9091, 100, 0, 1.8103, 100, 100, 1.05 * .2958, 1, 100, .5, 5, 5][[1]]
```

{1.99652, 1.00001, 22.368, 10., 4.62353}

```
calckinparsgenmod[.09091, 1, 0, .9091, 100, 0, 1.8103, 100, 100, .2958, 1, 100, 1.05 * .5, 5, 5]
```

{{kf2et → 1.98196, kf1et → 1.00001, kX → 15.7649, kA → 10., kAX → 5.08905}}

```
row6 = {kf2et, kf1et, kX, kA, kAX} /. calckinparsgenmod[.09091,
    1, 0, .9091, 100, 0, 1.8103, 100, 100, .2958, 1, 100, 1.05 * .5, 5, 5][[1]]
```

{1.98196, 1.00001, 15.7649, 10., 5.08905}

Table 10.1 Estimation of five kinetic parameters for the general modified mechanism for A → products when one molecule of modifier is bound using velocities at {1,0}, {100,0}, {100,100}, {1,100}, and {5,5}

```
TableForm[Round[{row1, row2, row3, row4, row5, row6}, 0.01],
  TableHeadings -> {{"No errors", "1.05*v(1,0)", "1.05*v(100,0)", "1.05*v(100,100)",
     "1.05*v(1,100)", "1.05*v(5,5)"}, {"kf2et", "kf1et", "kX", "kA", "kAX"}}]
```

	kf2et	kf1et	kX	kA	kAX
No errors	2.	1.	20.	10.	5.
1.05*v(1,0)	2.01	0.99	21.29	9.42	5.01
1.05*v(100,0)	1.99	1.06	21.21	10.61	4.94
1.05*v(100,100)	2.12	1.	20.34	10.	5.34
1.05*v(1,100)	2.	1.	22.37	10.	4.62
1.05*v(5,5)	1.98	1.	15.76	10.	5.09

More accurate values can be obtained by using wider ranges of substrate concentrations or different pairs of substrate concentrations.

■ **10.1.3 Use of a thermodynamic cycle to estimate the equilibrium constant for the reaction EAX = EA + X**

The general modifier mechanism when one X is bound (given in Section 10.1.1) can also be represented by

```
       kA
E + A = EA → products
+        +
X        X
∥ kX     ∥ kXA
EX + A = EAX → products
       kAX
```

When the three reactions in the general modifier mechanism are at equilibrium, the reaction EA + X = EAX (kXA = ea*x/eax) is also at equilibrium because it completes a thermodynamic cycle. This cycle yields the relation kX kAX = kA kXA. Thus kXA = kAX kX/kA. The values of kXA can be calculated from Table 10.1.

$$\begin{matrix} 5.\grave{} \\ \overline{5.01\grave{}} \\ 4.94\grave{} \\ 5.34\grave{} \\ 4.62\grave{} \\ 5.09\grave{} \end{matrix} * \begin{matrix} 20.\grave{} \\ \overline{21.29\grave{}} \\ 21.21\grave{} \\ 20.34\grave{} \\ 22.37\grave{} \\ 15.76\grave{} \end{matrix} / \begin{matrix} 10.\grave{} \\ \overline{9.42\grave{}} \\ 10.61\grave{} \\ 10.\grave{} \\ 10.\grave{} \\ 10.\grave{} \end{matrix}$$

{{10.}, {11.323}, {9.87534}, {10.8616}, {10.3349}, {8.02184}}

This list of Michaelis constants can be added to Table 10.1 as a column. The six rows in Table 10.1 can be augmented with kXA values.

row1

{1.99994, 1.00001, 20.0042, 10., 4.99915}

row1a = {1.9999356042230985`, 1.00001`, 20.004237357195688`, 10.`, 4.999145911247946`, 10}

{1.99994, 1.00001, 20.0042, 10., 4.99915, 10}

row2

{2.01244, 0.994747, 21.2888, 9.42105, 5.00509}

row2a = {2.012435500541701`, 0.9947467894736842`,
  21.288800915047794`, 9.421052631578945`, 5.005087254659858`, 11.32}

{2.01244, 0.994747, 21.2888, 9.42105, 5.00509, 11.32}

row3

{1.99331, 1.05588, 21.2099, 10.6145, 4.94396}

row3a = {1.993312925145528`, 1.0558764804469274`,
  21.20991774027018`, 10.614525139664805`, 4.943956459847041`, 9.88}

{1.99331, 1.05588, 21.2099, 10.6145, 4.94396, 9.88}

row4

{2.12091, 1.00001, 20.3427, 10., 5.34206}

row4a = {2.1209068023590314`, 1.00001`, 20.342740608275125`, 10.`, 5.342063717079896`, 10.86}

{2.12091, 1.00001, 20.3427, 10., 5.34206, 10.86}

row5

{1.99652, 1.00001, 22.368, 10., 4.62353}

row5a = {1.9965209238348298`, 1.00001`, 22.367965341764055`, 10.`, 4.623526971224438`, 10.33}

{1.99652, 1.00001, 22.368, 10., 4.62353, 10.33}

row6

{1.98196, 1.00001, 15.7649, 10., 5.08905}

row6a = {1.9819589404847533`, 1.00001`, 15.76486893485299`, 10.`, 5.089051272185144`, 8.02}

{1.98196, 1.00001, 15.7649, 10., 5.08905, 8.02}

Table 10.2 Estimation of six kinetic parameters for the general modified mechanism for A → products using velocities at {1,0}, {100,0}, {100,100}, {1,100}, {5,5}, and the thermodynamic cycle.

```
TableForm[Round[{row1a, row2a, row3a, row4a, row5a, row6a}, 0.01],
  TableHeadings -> {{"No errors", "1.05*v(1,0)", "1.05*v(100,0)", "1.05*v(100,100)",
    "1.05*v(1,100)", "1.05*v(5,5)"}, {"kf2et", "kf1et", "kX", "kA", "kAX", "kXA"}}]
```

	kf2et	kf1et	kX	kA	kAX	kXA
No errors	2.	1.	20.	10.	5.	10.
1.05*v(1,0)	2.01	0.99	21.29	9.42	5.01	11.32
1.05*v(100,0)	1.99	1.06	21.21	10.61	4.94	9.88
1.05*v(100,100)	2.12	1.	20.34	10.	5.34	10.86
1.05*v(1,100)	2.	1.	22.37	10.	4.62	10.33
1.05*v(5,5}	1.98	1.	15.76	10.	5.09	8.02

More accurate values of the kinetic parameters can be obtained by using a wider range of substrate concentrations or different pairs of concentrations.

10.2 General modifier mechanism for A → products with the binding of 2X and two paths to products

■ 10.2.1 Mechanism

This is an extension of the mechanism in the preceding section to include the binding of 2X in one reaction. The reaction EA + 2X = EAX$_2$ (kX2A^2 = ea*x^2/eax2) has been omitted because it is redundant in the calculation of the equilibrium concentrations of EA and EAX$_2$.

```
       kA    kf1
E + A = EA → products
+
2X
‖ kX^2
EX₂ + A = EAX₂ → products
       kAX2    kf2
```

The mechanism can also be written as follows:

```
E + A = EA              kA=e*a/ea
E + 2X = EX₂            kX^2=e*x^2/ex2
EX₂ + A = EAX₂          kAX2=ex2*a/eax2
v = kf1*ea + kf2*eax2 = kf1et*(ea/et) + kf2et*(eax2/et)
```

The rate equation can be derived by using Solve.

```
Solve[{kA == e * a / ea, kX^2 == e * x^2 / ex2,
  kAX2 == ex2 * a / eax2, et == e + ea + ex2 + eax2}, {ea, eax2}, {e, ex2}]
```

$$\left\{\left\{ea \to \frac{a\ et\ kAX2\ kX^2}{a\ kAX2\ kX^2 + kA\ kAX2\ kX^2 + a\ kA\ x^2 + kA\ kAX2\ x^2},\right.\right.$$

$$\left.\left.eax2 \to \frac{a\ et\ kA\ x^2}{a\ kAX2\ kX^2 + kA\ kAX2\ kX^2 + a\ kA\ x^2 + kA\ kAX2\ x^2}\right\}\right\}$$

The rate equation is given by

$$vEAX2 = \frac{a\ kf1et\ kAX2\ kX^2 + a\ kf2et\ kA\ x^2}{a\ kAX2\ kX^2 + kA\ kAX2\ kX^2 + a\ kA\ x^2 + kA\ kAX2\ x^2}$$

$$\frac{a\ kAX2\ kf1et\ kX^2 + a\ kA\ kf2et\ x^2}{a\ kAX2\ kX^2 + kA\ kAX2\ kX^2 + a\ kA\ x^2 + kA\ kAX2\ x^2}$$

The kinetic parameters are arbitrarily assigned as follows: $kf1et = 1, kf2et = 2, kA = 10, kX = 20,$ and $kAX2 = 5$.

```
velM2 = vEAX2 /. kf1et -> 1 /. kf2et → 2 /. kA → 10 /. kX → 20 /. kAX2 → 5
```

$$\frac{2000\ a + 20\ a\ x^2}{20\,000 + 2000\ a + 50\ x^2 + 10\ a\ x^2}$$

3D plots can be made to show the sensitivities of velM2 to [A] and [X].

```
plotvEAX2a = Plot3D[Evaluate[velM2], {a, .00001, 30},
   {x, .00001, 100}, ViewPoint → {-2, -2, 1}, AxesLabel → {"[A]", "[X]"},
   Lighting -> {{"Ambient", GrayLevel[1]}}, PlotRange → {0, 2}, PlotLabel → "velM2"];

plotvEAX2b =
  Plot3D[Evaluate[D[velM2, a]], {a, .00001, 30}, {x, .00001, 100}, ViewPoint → {-2, -2, 1},
   AxesLabel → {"[A]", "[X]"}, Lighting -> {{"Ambient", GrayLevel[1]}},
   PlotRange → {0, .5}, PlotLabel → "dvelM2/d[A]"];

plotvEAX2c = Plot3D[Evaluate[D[velM2, x]], {a, .00001, 30}, {x, .00001, 100},
   ViewPoint → {-2, -2, 1}, AxesLabel → {"[A]", "[X]"}, PlotRange → {0, .06},
   Lighting -> {{"Ambient", GrayLevel[1]}}, PlotLabel → "dvelM2/d[X]"];

plotvEAX2d = Plot3D[Evaluate[D[velM2, a, x]], {a, .00001, 30}, {x, .00001, 100},
   ViewPoint → {-2, -2, 1}, AxesLabel → {"[A]", "[X]"}, PlotRange → {-.003, .01},
   Lighting -> {{"Ambient", GrayLevel[1]}}, PlotLabel → "d(dvelM2/d[A])/d[X]"];

fig10generalmod = GraphicsArray[{{plotvEAX2a, plotvEAX2b}, {plotvEAX2c, plotvEAX2d}}]
```

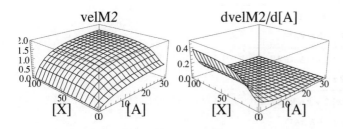

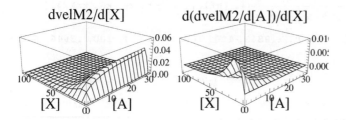

Fig. 10.2 Plots of (a) velM2, (b) $\partial \text{velM2}/\partial[A]$, (c) $\partial \text{velM2}/\partial[X]$, and (d) $\partial(\partial \text{velM2}/\partial[A])/\partial[X]$ versus [A] and [X] for A → products for the general modifier mechanism when 2X are bound in one reaction.

There are 5 kinetic parameters, and so 5 velocities have to be measured. Use the same 5 pairs of substrate concentrations used in the preceding section.

```
velM2 /. a → 1 /. x → 0 // N
```

0.0909091

```
velM2 /. a → 100 /. x → 0 // N
```

0.909091

```
velM2 /. a → 100 /. x → 100 // N
```

1.88433

```
velM2 /. a → 1 /. x → 100 // N
```

0.324759

```
velM2 /. a → 5 /. x → 5 // N
```

0.384615

■ **10.2.2 Estimation of kinetic parameters for the general modifier mechanism with the binding of 2X and two paths to products**

This program was written for the general modifier mechanism to estimate 5 kinetic parameters from 5 velocity measurements when 2X are bound in one reaction .

```
calckinparsgenmodX2[v1_, a1_, x1_, v2_, a2_, x2_, v3_, a3_, x3_, v4_, a4_, x4_, v5_, a5_, x5_] :=
Module[{}, (*This program calculates kf1et, kf2et, kA, kX,
   and kAX2 from 5 measured velocities for the general modifier mechanism for A →
     products when 2X are bound in one reaction.*)
```

$$\text{Solve}\left[\left\{v1 == \frac{a1\ kf1et\ kAX2\ kX^2 + a1\ kf2et\ kA\ x1^2}{a1\ kAX2\ kX^2 + kA\ kAX2\ kX^2 + a1\ kA\ x1^2 + kA\ kAX2\ x1^2},\right.\right.$$

$$v2 == \frac{a2\ kf1et\ kAX2\ kX^2 + a2\ kf2et\ kA\ x2^2}{a2\ kAX2\ kX^2 + kA\ kAX2\ kX^2 + a2\ kA\ x2^2 + kA\ kAX2\ x2^2},$$

$$v3 == \frac{a3\ kf1et\ kAX2\ kX^2 + a3\ kf2et\ kA\ x3^2}{a3\ kAX2\ kX^2 + kA\ kAX2\ kX^2 + a3\ kA\ x3^2 + kA\ kAX2\ x3^2},$$

$$v4 == \frac{a4\ kf1et\ kAX2\ kX^2 + a4\ kf2et\ kA\ x4^2}{a4\ kAX2\ kX^2 + kA\ kAX2\ kX^2 + a4\ kA\ x4^2 + kA\ kAX2\ x4^2},$$

$$\left.\left. v5 == \frac{a5\ kf1et\ kAX2\ kX^2 + a5\ kf2et\ kA\ x5^2}{a5\ kAX2\ kX^2 + kA\ kAX2\ kX^2 + a5\ kA\ x5^2 + kA\ kAX2\ x5^2}\right\}, \{kf1et, kf2et, kA, kX, kAX2\}\right]\right]$$

```
calckinparsgenmodX2[.09091, 1, 0, .9091, 100, 0, 1.8843, 100, 100, .3248, 1, 100, .3846, 5, 5]
```

{{kf2et → 1.99996, kf1et → 1.00001, kA → 10., kAX2 → 4.99902, kX → -20.0061},
 {kf2et → 1.99996, kf1et → 1.00001, kA → 10., kAX2 → 4.99902, kX → 20.0061}}

The negative value of kX is mathematically correct, but is impossible physically. The second set of kinetic parameters is correct, but it is necessary to test for errors in the measured velocities.

```
r1 = {kf2et, kf1et, kA, kAX2, kX} /. calckinparsgenmodX2[.09091,
      1, 0, .9091, 100, 0, 1.8843, 100, 100, .3248, 1, 100, .3846, 5, 5][[2]]
```

{1.99996, 1.00001, 10., 4.99902, 20.0061}

```
calckinparsgenmodX2[1.05 * .09091, 1, 0,
    .9091, 100, 0, 1.8843, 100, 100, .3248, 1, 100, .3846, 5, 5]
```

{{kf2et → 2.00701, kf1et → 0.994747, kA → 9.42105, kAX2 → 4.98014, kX → -22.6231},
 {kf2et → 2.00701, kf1et → 0.994747, kA → 9.42105, kAX2 → 4.98014, kX → 22.6231}}

```
r2 = {kf2et, kf1et, kA, kAX2, kX} /. calckinparsgenmodX2[1.05 * .09091, 1, 0,
      .9091, 100, 0, 1.8843, 100, 100, .3248, 1, 100, .3846, 5, 5][[2]]
```

{2.00701, 0.994747, 9.42105, 4.98014, 22.6231}

```
calckinparsgenmodX2[.09091, 1, 0,
    1.05 * .9091, 100, 0, 1.8843, 100, 100, .3248, 1, 100, .3846, 5, 5]
```

{{kf2et → 1.99996, kf1et → 1.05588, kA → 10.6145, kAX2 → 4.98005, kX → -21.2661},
 {kf2et → 1.99996, kf1et → 1.05588, kA → 10.6145, kAX2 → 4.98005, kX → 21.2661}}

```
r3 = {kf2et, kf1et, kA, kAX2, kX} /. calckinparsgenmodX2[.09091, 1, 0,
      1.05 * .9091, 100, 0, 1.8843, 100, 100, .3248, 1, 100, .3846, 5, 5][[2]]
```

{1.99996, 1.05588, 10.6145, 4.98005, 21.2661}

```
calckinparsgenmodX2[.09091, 1, 0,
    .9091, 100, 0, 1.05 * 1.8843, 100, 100, .3248, 1, 100, .3846, 5, 5]
```

{{kf2et → 2.10899, kf1et → 1.00001, kA → 10., kAX2 → 5.32389, kX → -20.0371},
 {kf2et → 2.10899, kf1et → 1.00001, kA → 10., kAX2 → 5.32389, kX → 20.0371}}

```
r4 = {kf2et, kf1et, kA, kAX2, kX} /. calckinparsgenmodX2[.09091, 1, 0,
      .9091, 100, 0, 1.05 * 1.8843, 100, 100, .3248, 1, 100, .3846, 5, 5][[2]]
```

{2.10899, 1.00001, 10., 5.32389, 20.0371}

```
calckinparsgenmodX2[.09091, 1, 0,
    .9091, 100, 0, 1.8843, 100, 100, 1.05 * .3248, 1, 100, .3846, 5, 5]
```

{{kf2et → 1.99435, kf1et → 1.00001, kA → 10., kAX2 → 4.68398, kX → -20.8238},
 {kf2et → 1.99435, kf1et → 1.00001, kA → 10., kAX2 → 4.68398, kX → 20.8238}}

```
r5 = {kf2et, kf1et, kA, kAX2, kX} /. calckinparsgenmodX2[.09091, 1, 0,
      .9091, 100, 0, 1.8843, 100, 100, 1.05 * .3248, 1, 100, .3846, 5, 5][[2]]
```

{1.99435, 1.00001, 10., 4.68398, 20.8238}

```
calckinparsgenmodX2[.09091, 1, 0,
    .9091, 100, 0, 1.8843, 100, 100, .3248, 1, 100, 1.05 * .3846, 5, 5]
```

{{kf2et → 1.99407, kf1et → 1.00001, kA → 10., kAX2 → 5.02849, kX → -16.6844},
 {kf2et → 1.99407, kf1et → 1.00001, kA → 10., kAX2 → 5.02849, kX → 16.6844}}

```
r6 = {kf2et, kf1et, kA, kAX2, kX} /. calckinparsgenmodX2[.09091, 1, 0,
      .9091, 100, 0, 1.8843, 100, 100, .3248, 1, 100, 1.05 * .3846, 5, 5][[2]]
```

{1.99407, 1.00001, 10., 5.02849, 16.6844}

Table 10.3 Estimation of five kinetic parameters for the general modifier mechanism for A → products when 2X are bound in one reaction using velocities at {1,0}, {100,0}, {100,100}, {1,100}, and {5,5}.

```
TableForm[Round[{r1, r2, r3, r4, r5, r6}, 0.01],
  TableHeadings -> {{"No errors", "1.05*v(1,0)", "1.05*v(100,0)", "1.05*v(100,100)",
     "1.05*v(1,100)", "1.05*v(5,5)"}, {"kf2et", "kf1et", "kA", "kAX2", "kX"}}]
```

	kf2et	kf1et	kA	kAX2	kX
No errors	2.	1.	10.	5.	20.01
1.05*v(1,0)	2.01	0.99	9.42	4.98	22.62
1.05*v(100,0)	2.	1.06	10.61	4.98	21.27
1.05*v(100,100)	2.11	1.	10.	5.32	20.04
1.05*v(1,100)	1.99	1.	10.	4.68	20.82
1.05*v(5,5}	1.99	1.	10.	5.03	16.68

■ **10.2.3 Estimation of the apparent equilibrium constant for the reaction EAX$_2$ = EA + 2X**

The reaction EAX$_2$ = EA + 2X can be added to the mechanism to complete a thermodynamic cycle.

```
         kA
E + A = EA → products
+          +
2X        2X
|| kX^2   || kX2A^2
EX₂ + A = EAX₂ → products
        kAX2
```

When the three reactions in the general modifier mechanism are at equilibrium, the reaction EA + 2X = EAX$_2$ (kX2A^2 = ea*x^2/eax2) is also at equilibrium because it completes a thermodynamic cycle. This cycle yields the relation kA*kX2A^2=kX^2*kAX2. Thus kX2A = (kX^2*kAX2/kA)^(1/2). The values of kX2A can be calculated from Table 10.3.

```
(20^2 * 5 / 10) ^ .5

14.1421
```

$$\left(\left(\begin{matrix}5.`\\4.98`\\4.98`\\5.32`\\4.68`\\5.03`\end{matrix}\right) * \left(\begin{matrix}20.01`\\22.62`\\21.27`\\20.04`\\20.82`\\16.68`\end{matrix}\right) \wedge 2 \bigg/ \left(\begin{matrix}10.`\\9.42`\\10.61`\\10.`\\10.`\\10.`\end{matrix}\right)\right) \wedge .5$$

```
{{14.1492}, {16.4468}, {14.5722}, {14.6168}, {14.2431}, {11.8299}}
```

This list can be added to Table 10.3 as a column.

```
r1

{1.99996, 1.00001, 10., 4.99902, 20.0061}

r1a = {1.999959970737145`, 1.00001`, 10.`, 4.999022864278886`, 20.00614150477403`, 14.14}

{1.99996, 1.00001, 10., 4.99902, 20.0061, 14.14}

r2

{2.00701, 0.994747, 9.42105, 4.98014, 22.6231}
```

r2a = {2.0070103447526844`, 0.9947467894736842`,
9.421052631578945`, 4.980137870483581`, 22.623103189058046`, 16.45}

{2.00701, 0.994747, 9.42105, 4.98014, 22.6231, 16.45}

r3

{1.99996, 1.05588, 10.6145, 4.98005, 21.2661}

r3a = {1.9999606703178803`, 1.0558764804469274`,
10.614525139664805`, 4.980051813380657`, 21.266116395091327`, 14.57}

{1.99996, 1.05588, 10.6145, 4.98005, 21.2661, 14.57}

r4

{2.10899, 1.00001, 10., 5.32389, 20.0371}

r4a = {2.108993183504071`, 1.00001`, 10.`, 5.323894254346965`, 20.03706833714209`, 14.62}

{2.10899, 1.00001, 10., 5.32389, 20.0371, 14.62}

r5

{1.99435, 1.00001, 10., 4.68398, 20.8238}

r5a = {1.9943483002388414`, 1.00001`, 10.`, 4.683977745783234`, 20.82376025327442`, 14.24}

{1.99435, 1.00001, 10., 4.68398, 20.8238, 14.24}

r6

{1.99407, 1.00001, 10., 5.02849, 16.6844}

r6a = {1.994067647533775`, 1.00001`, 10.`, 5.028491712199957`, 16.68444882157551`, 11.83}

{1.99407, 1.00001, 10., 5.02849, 16.6844, 11.83}

Table 10.4 Estimation of 6 kinetic parameters for the general modified mechanism for A → products when 2X are bound in one reaction using velocities at {1,0}, {100,0}, {100,100}, {1,100}, {5,5}, and the thermodynamic cycle.

```
TableForm[Round[{r1a, r2a, r3a, r4a, r5a, r6a}, 0.01],
  TableHeadings -> {{"No errors", "1.05*v(1,0)", "1.05*v(100,0)", "1.05*v(100,100)",
     "1.05*v(1,100)", "1.05*v(5,5)"}, {"kf2et", "kf1et", "kA", "kAX2", "kX", "kX2A"}}]
```

	kf2et	kf1et	kA	kAX2	kX	kX2A
No errors	2.	1.	10.	5.	20.01	14.14
1.05*v(1,0)	2.01	0.99	9.42	4.98	22.62	16.45
1.05*v(100,0)	2.	1.06	10.61	4.98	21.27	14.57
1.05*v(100,100)	2.11	1.	10.	5.32	20.04	14.62
1.05*v(1,100)	1.99	1.	10.	4.68	20.82	14.24
1.05*v(5,5}	1.99	1.	10.	5.03	16.68	11.83

This mechanism can be extended to provide for positive and negative cooperativity by binding 2X in two reactions, rather than one.

■ 10.2.4 Determination of the number of modifier molecules bound

The effects of increasing the number of modifier molecules bound is striking. The general modifier mechanism with the binding of 3X and two paths to products can be written as

E + A = EA   kA=e*a/ea
E + 3X = EX$_3$ kX^3=e*x^3/ex3
EX$_3$ + A = EAX$_3$ kAX3=ex3*a/eax3
v = kf1*ea+kf2*eax3

The rate equation can be derived by using Solve.

```
Solve[{kA == e * a / ea, kX^3 == e * x^3 / ex3,
   kAX3 == ex3 * a / eax3, et == e + ea + ex3 + eax3}, {ea, eax3}, {e, ex3}]
```

$$\left\{\left\{ea \rightarrow \frac{a\ et\ kAX3\ kX^3}{a\ kAX3\ kX^3 + kA\ kAX3\ kX^3 + a\ kA\ x^3 + kA\ kAX3\ x^3},\right.\right.$$
$$\left.\left.eax3 \rightarrow \frac{a\ et\ kA\ x^3}{a\ kAX3\ kX^3 + kA\ kAX3\ kX^3 + a\ kA\ x^3 + kA\ kAX3\ x^3}\right\}\right\}$$

The rate equation is given by

$$vEAX3 = \frac{a\ kf1et\ kAX3\ kX^3 + a\ kf2et\ kA\ x^3}{a\ kAX3\ kX^3 + kA\ kAX3\ kX^3 + a\ kA\ x^3 + kA\ kAX3\ x^3}$$
$$\frac{a\ kAX3\ kf1et\ kX^3 + a\ kA\ kf2et\ x^3}{a\ kAX3\ kX^3 + kA\ kAX3\ kX^3 + a\ kA\ x^3 + kA\ kAX3\ x^3}$$

The kinetic parameters are arbitrarily assigned as follows: kf1et = 1, kf2et = 2, kA = 10, kX = 20, and kAX3 = 5.

```
velM3 = vEAX3 /. kf1et -> 1 /. kf2et → 2 /. kA → 10 /. kX → 20 /. kAX3 → 5
```

$$\frac{40\,000\ a + 20\ a\ x^3}{400\,000 + 40\,000\ a + 50\ x^3 + 10\ a\ x^3}$$

```
plotvel =
  Plot[{velM1 /. a -> 100, velM2 /. a → 100, velM3 /. a → 100}, {x, 0, 50}, PlotRange → {0, 2},
   AxesLabel → {"[X]", "vel"}, PlotStyle → {Black, Black, Black}, PlotLabel → "(a)"];
```

Figure 10.3 shows that plots of measured velocities at high [A] over a range of [X] can be used to determine the number of modifiers bound. This is even clearer when derivatives of the velocity plot can be obtained.

```
plotdervel =
  Plot[Evaluate[{D[velM1 /. a -> 100, x], D[velM2 /. a → 100, x], D[velM3 /. a → 100, x]}],
   {x, 0, 50}, PlotRange → {0, .06}, AxesLabel → {"[X]", "dvel/d[X]"},
   PlotStyle → {Black, Black, Black}, PlotLabel → "(b)"];

fig3modNX = GraphicsArray[{{plotvel, plotdervel}}]
```

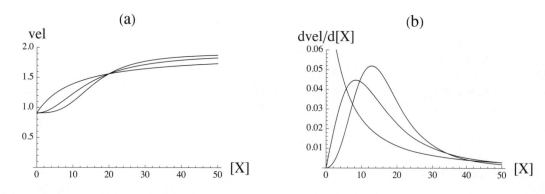

Fig. 10.3 (a) Velocities at [A] = 100 and various concentrations of [X]. (b) Plots of derivatives of velocities with respect to [X] at [A] = 100 and various concentrations of [X].

10.3 General modifier mechanism for A → products when two X are bound in two reactions and there are three paths to products

▪ 10.3.1 Mechanism and derivation of the rate equation

This is an extension of the mechanism with two paths to products to a mechanism with three paths to products. Since 2X are bound in two reactions, there is an opportunnity to discuss positive and negative cooperativity in the binding of X. Cooperativity is positive when the second molecule of X is bound more tightly than the first [31]. This section treats an example of positive cooperativity.

kA=20

E + A = EA → products   kf1et=1

```
+
X
‖  kX= 20
EX + A = EAX → products    kf2et=2
+       kAX=20
X
‖  kX2 = 5
EX₂ + A = EAX₂ → products   kf3et=3
     kAX2=10
```

These values of the parameters are for positive cooperativity because the second X is bound more strongly.

This mechanism can be written as

		Positive Coop.	Negative Coop. (see next section)
E + A = EA	kA=e*a/ea	20	20
E + X = EX	kX = e*x/ex	20	5
E X + A = EAX	kAX=ex*a/eax	20	20
E X + X = EX₂	kX2=ex*x/ex2	5	20
EX₂ + A = EAX₂	kAX2=ex2*a/eax2	10	10

v = kf1*ea + kf2*eax + kf3*eax2 = kf1et*(ea/et) + kf2et*(eax/et) + kf3et*(eax2/et)

where kf1et = 1, kf2et = 2, and kf3et = 3.

The equilibrium concentrations of EA, EAX, and EAX₂ can be derived by using Solve.

Derive the expression for ea:

```
Solve[{kA == e * a / ea, kX == e * x / ex, kAX == ex * a / eax, kX2 == ex * x / ex2,
  kAX2 == ex2 * a / eax2, et == e + ea + ex + eax + ex2 + eax2}, {ea}, {e, ex, eax, ex2, eax2}]
```

$$\{\{ea \to (a\ et\ kAX\ kAX2\ kX\ kX2) \big/ (a\ kAX\ kAX2\ kX\ kX2 +$$
$$kA\ kAX\ kAX2\ kX\ kX2 + a\ kA\ kAX2\ kX2\ x + kA\ kAX\ kAX2\ kX2\ x + a\ kA\ kAX\ x^2 + kA\ kAX\ kAX2\ x^2)\}\}$$

Derive the expression for eax:

```
Solve[{kA == e * a / ea, kX == e * x / ex, kAX == ex * a / eax, kX2 == ex * x / ex2,
  kAX2 == ex2 * a / eax2, et == e + ea + ex + eax + ex2 + eax2}, {eax}, {e, ea, ex, ex2, eax2}]
```

$$\{\{eax \to (a\ et\ kA\ kAX2\ kX2\ x) \big/ (a\ kAX\ kAX2\ kX\ kX2 +$$
$$kA\ kAX\ kAX2\ kX\ kX2 + a\ kA\ kAX2\ kX2\ x + kA\ kAX\ kAX2\ kX2\ x + a\ kA\ kAX\ x^2 + kA\ kAX\ kAX2\ x^2)\}\}$$

Derive the expression for eax2:

```
Solve[{kA == e * a / ea, kX == e * x / ex, kAX == ex * a / eax, kX2 == ex * x / ex2,
  kAX2 == ex2 * a / eax2, et == e + ea + ex + eax + ex2 + eax2}, {eax2}, {e, ea, ex, eax, ex2}]
```

$$\{\{eax2 \to (a\ et\ kA\ kAX\ x^2) \big/ (a\ kAX\ kAX2\ kX\ kX2 + kA\ kAX\ kAX2\ kX\ kX2 +$$
$$a\ kA\ kAX2\ kX2\ x + kA\ kAX\ kAX2\ kX2\ x + a\ kA\ kAX\ x^2 + kA\ kAX\ kAX2\ x^2)\}\}$$

These three equilibrium concentrations have the same six-term denominator, and so the numerator terms can be combined. The first numerator term is divided by et and multiplied by kf1et to obtain the contribution of ea to the velocity. The second numerator term is divided by et and multiplied by kf2et. The third numerator term is divided by et and multiplied by kf3et. The expression for the modified velocity (vM) is given by

$$vM = \left(a\ kf1et\ kAX\ kAX2\ kX\ kX2 + a\ kf2et\ kA\ kAX2\ kX2\ x + a\ kf3et\ kA\ kAX\ x^2\right) / \left(a\ kAX\ kAX2\ kX\ kX2 + \right.$$
$$\left. kA\ kAX\ kAX2\ kX\ kX2 + a\ kA\ kAX2\ kX2\ x + kA\ kAX\ kAX2\ kX2\ x + a\ kA\ kAX\ x^2 + kA\ kAX\ kAX2\ x^2\right)$$

$$\left(a\ kAX\ kAX2\ kf1et\ kX\ kX2 + a\ kA\ kAX2\ kf2et\ kX2\ x + a\ kA\ kAX\ kf3et\ x^2\right) /$$
$$\left(a\ kAX\ kAX2\ kX\ kX2 + kA\ kAX\ kAX2\ kX\ kX2 + a\ kA\ kAX2\ kX2\ x + kA\ kAX\ kAX2\ kX2\ x + a\ kA\ kAX\ x^2 + kA\ kAX\ kAX2\ x^2\right)$$

This equation applies to both positive and negative cooperativity.

When the X binding is positively cooperative, the velocity vMP (velocityModifierPositive) is given by

vMP =
vM /. kA → 20 /. kX → 20 /. kAX → 20 /. kX2 → 5 /. kAX2 → 10 /. kf1et → 1 /. kf2et → 2 /. kf3et → 3

$$\frac{20\,000\ a + 2000\ a\ x + 1200\ a\ x^2}{400\,000 + 20\,000\ a + 20\,000\ x + 1000\ a\ x + 4000\ x^2 + 400\ a\ x^2}$$

vMP can be plotted as a function of [A] and [X].

```
plotvMP = Plot3D[Evaluate[vMP], {a, .00001, 30}, {x, .00001, 100},
    ViewPoint → {-2, -2, 1}, AxesLabel → {"[A]", "[X]"}, PlotRange → {0, 3},
    Lighting -> {{"Ambient", GrayLevel[1]}}, PlotLabel → "vMP"];
```

As an example of negative cooperativity, the velocity is given by vMN (velocityModifierNegative).

vMN =
vM /. kA → 20 /. kX → 5 /. kAX → 20 /. kX2 → 20 /. kAX2 → 10 /. kf1et → 1 /. kf2et → 2 /. kf3et → 3

$$\frac{20\,000\ a + 8000\ a\ x + 1200\ a\ x^2}{400\,000 + 20\,000\ a + 80\,000\ x + 4000\ a\ x + 4000\ x^2 + 400\ a\ x^2}$$

This velocity is plotted versus [A] and [X].

```
plotvMN = Plot3D[Evaluate[vMN], {a, .00001, 30}, {x, .00001, 100},
    ViewPoint → {-2, -2, 1}, AxesLabel → {"[A]", "[X]"}, PlotRange → {0, 3},
    Lighting -> {{"Ambient", GrayLevel[1]}}, PlotLabel → "vMN"];
```

```
GraphicsArray[{{plotvMP, plotvMN}}]
```

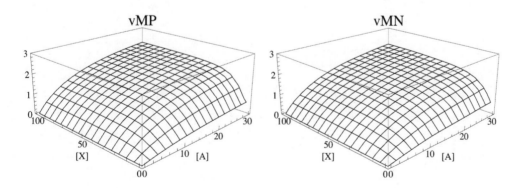

Fig. 10.4 (a) Velocity when the cooperativity is positive as a function of [A] and [X]. (b) Velocity when the cooperativity is negative as a function of [A] and [X].

10.3.2 Plots of cross sections

In order to see the differences between positive cooperativity and negative cooperativity more clearly, it is necessary to plot cross sections. Plot velocity vMN versus [X] at [A] = {10, 20, 40, 80, 160}. For vMN, kX = 5 and kX2 = 20; this means that the second molecule of modifier is bound less strongly than the first. This decreases the binding of X and the sigmoid character of the plot of the velocity versus [X].

```
plotA = Plot[vMP /. a → {10, 20, 40, 80, 160}, {x, 0, 40}, PlotRange → {0, 3},
    AxesLabel → {"[X]", "vMP"}, PlotStyle → {Black}, PlotLabel → "(a)"];

plotB = Plot[vMN /. a → {10, 20, 40, 80, 160}, {x, 0, 40}, PlotRange → {0, 3},
    AxesLabel → {"[X]", "vMN"}, PlotStyle → {Black}, PlotLabel → "(b)"];

plotC = Plot[vMP /. x → {10, 20, 40, 80, 160}, {a, 0, 40}, PlotRange → {0, 3},
    AxesLabel → {"[A]", "vMP"}, PlotStyle → {Black}, PlotLabel → "(c)"];

plotD = Plot[vMN /. x → {10, 20, 40, 80, 160}, {a, 0, 40}, PlotRange → {0, 3},
    AxesLabel → {"[A]", "vMN"}, PlotStyle → {Black}, PlotLabel → "(d)"];

plotE = Plot[{vMP /. x → {1, 100}, vMN /. x → {1, 100}},
    {a, 0, 100}, PlotRange → {0, 3}, AxesLabel → {"[A]", "vMP,vMN"},
    PlotStyle → {Black, {Black, Dashing[Medium]}}, PlotLabel → "(e)"];

plotF = Plot[{vMP /. a → {5, 40}, vMN /. a → {5, 40}},
    {x, 0, 100}, PlotRange → {0, 3}, AxesLabel → {"[X]", "vMP,vMN"},
    PlotStyle → {Black, {Black, Dashing[Medium]}}, PlotLabel → "(f)"];
```

fig4coop = GraphicsArray[{{plotA, plotB}, {plotC, plotD}, {plotE, plotF}}]

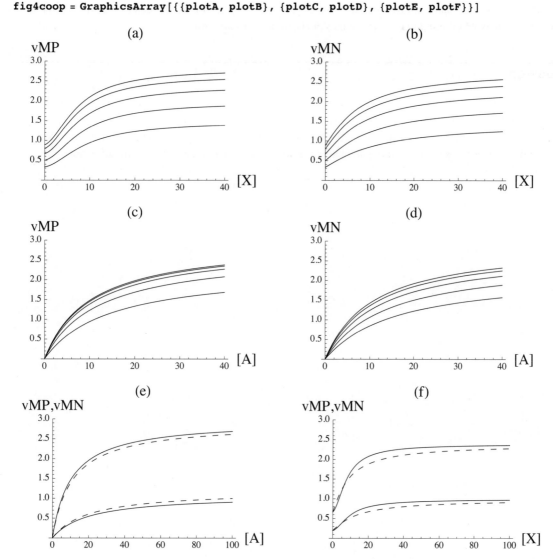

Fig. 10.5 (a) Plot of vMP (velModificationPostive) versus modifier concentration [X] when [A] = 10, 20, 40, 80, and 160. This shows sigmoid plots because K_X = 20 and K_{X2} = 5. (b) Plot of vMN (velModificationNegative) versus modifier concentration [X] when [A] = 10, 20, 40, 80, and 160. (c) Plots of vMP at [X] = 10, 20, 40, 80, and 160 versus [A]. (d) Plots of vMN at [X] = 10, 20, 40, 80, and 160 versus [A]. (e) The solid lines are for vMP at two concentrations of modifier X; the upper solid line is for [X] = 40 and the lower solid line is for [X] = 5. The dashed lines give the velocity vMN for the same two concentrations of modifier X. The top pair of lines shows that when K_X = 20 and K_{X2} = 5 the velocity is higher than when K_X = 5 and K_{X2} = 20. The bottom pair of lines shows that when K_X = 20 and K_{X2} = 5 the velocity is higher than when K_X = 5 and K_{X2} = 20. (f) Plots of vMP at [A] = 5 and 40 (solid line) and vMN at 5 and 40 (dashed line) as functions of [X]. In both cases, vMP is higher than vMN at higher concentrations of modifier X, but not at lower concentrations of modifier X.

Figure 10.6 confirms the statement by Cornish-Bowde*n* [21] that "the conditions that determine whether the modifier is an activator or inhibitor may be different in different ranges of substrate concentrations."

10.4 Discussion

In modification there can be activation or inhibition, but only activation is considered here. The determination of the number of modifier molecules bound from velocity measurements is discussed in Section 10.2.4. Cornish-Bowden [21] has emphasized the importance of these effects in treating the kinetics of systems of enzyme-catalyzed reactions. Deriving rate equations and calculating velocities with arbitrary parameters is a way to learn about the effects of modification.

Chapter 11 Inhibition, activation, and modification of A + B → products

11.1 Introduction

11.2 Ordered A + B → products

- 11.2.1 Competitive inhibition

- 11.2.2 Uncompetitive inhibition

- 11.2.3 Mixed inhibition

- 11.2.4 Essential activation

- 11.2.5 Mixed activation

- 11.2.6 Modification

11.3 Random A + B → products

- 11.3.1 Competitive inhibition

- 11.3.2 Uncompetitive inhibition

- 11.3.3 Mixed inhibition

11.4 Discussion

11.1 Introduction

Mechanisms and rapid-equilibrium rate equations can be complicated when only subsrates are present, but there are further complications when other molecules or species are present and affect the velocity. The effects of hydrogen ions and magnesium ions have been discussed earlier. The effects of inhibitors and activators on A → products have been discussed in Chapter 9. The effects of modifiers on A → products have been discussed in Chapter 10. These effects become more complicated for A + B → products because there are more enzyme-substrate complexes that can bind molecules or species that are not substrates in the biochemical reaction that is catalyzed. When an inhibitor I or activator X is present, the rapid-equilibrium rate equations become more complicated rather rapidly. The nomenclature of Cornish-Bowden is followed [21].

The effects of inhibitors, activators, and modifiers can be very large, and so they cannot be ignored in making kinetic calculations on systems of enzyme-catalyzed reactions. These effects lead to mechanisms with more reactions and more complicated rate equations. And so rapid-equilibrium rate equations are the place to start. As the number of kinetic parameters increases, it becomes impractical to use Lineweaver-Burk plots, and so the use of the minimum number of velocity measurements is recommended [32].

Solve can be used to derive rapid-equilibrium rate equations for very complicated mechanisms. Another advantage in using Solve is that the rapid-equilibrium rate equations are obtained in computer-readable form. This book uses Solve to estimate the kinetic parameters with the minimum number of velocity measurements, but Lineweaver-Burk plots can be used. However, that gets very complicated for mechanisms discussed here.

11.2 Ordered A + B → products

■ 11.2.1 Competitive inhibition

The mechanism for competitive inhibition is

E + A = EA   kA = e*a/ea
E + I = EI   kI = e*i/ei
EA + B = EAB   kB = ea*b/eab
EAB → products   v = kf*eab

The rapid-equilibrium rate equation can be derived by use of Solve.

Solve[{kA == e * a / ea, kI == e * i / ei, kB == ea * b / eab, et == e + ea + ei + eab}, {eab}, {e, ea, ei}]

$$\left\{\left\{eab \rightarrow \frac{a \, b \, et \, kI}{i \, kA \, kB + a \, b \, kI + a \, kB \, kI + kA \, kB \, kI}\right\}\right\}$$

$$\textbf{vcompordAB} = \frac{a \, b \, kfet \, kI}{i \, kA \, kB + a \, b \, kI + a \, kB \, kI + kA \, kB \, kI}$$

$$\frac{a \, b \, kfet \, kI}{i \, kA \, kB + a \, b \, kI + a \, kB \, kI + kA \, kB \, kI}$$

There are three independent variables, and so 3D plots will not be given.

For test calculations, the following kinetic parameters are chosen arbitrarily: kfet = 1, kA = 10, kB = 30 and kI = 20.

vcompordAB /. kfet → 1 /. kA → 10 /. kB → 30 /. kI → 20

$$\frac{20 \, a \, b}{6000 + 600 \, a + 20 \, a \, b + 300 \, i}$$

The rate equation vcompordAB can be used to write a program to estimate the 4 kinetic parameters from 4 velocity measurements.

```
vcompordAB /. kfet → 1 /. kA → 10 /. kB → 30 /. kI → 20 /. a → 100 /. b → 100 /. i → 0 // N
```

0.75188

```
vcompordAB /. kfet → 1 /. kA → 10 /. kB → 30 /. kI → 20 /. a → 1 /. b → 100 /. i → 0 // N
```

0.232558

```
vcompordAB /. kfet → 1 /. kA → 10 /. kB → 30 /. kI → 20 /. a → 5 /. b → 5 /. i → 0 // N
```

0.0526316

```
vcompordAB /. kfet → 1 /. kA → 10 /. kB → 30 /. kI → 20 /. a → 5 /. b → 5 /. i → 50 // N
```

0.0204082

In Section 11.2.3 the velocity at {100,100,100} will be needed, and so it is calculated here.

```
vcompordAB /. kfet → 1 /. kA → 10 /. kB → 30 /. kI → 20 /. a → 100 /. b → 100 /. i → 100 // N
```

0.675676

```
calckinparscompAB[v1_, a1_, b1_, i1_, v2_, a2_, b2_, i2_, v3_, a3_, b3_,
    i3_, v4_, a4_, b4_, i4_] := Module[{}, (*This program calculates kfet,
  kA,kB, and kI from 4 experimental velocities for ordered A + B →
   products with a competitive inhibitor I.  The first velocity is at high [A],
  high [B], and zero [I], the second velocity is at low [A], high [B],
  and zero [I], the third velocity is at medium [A],medium [B], and zero [I],
  and the fourth velocity is at medium [A], medium [B], and high [I].*)
```

$$\text{Solve}\left[\left\{v1 == \frac{a1\ b1\ kfet\ kI}{i1\ kA\ kB + a1\ b1\ kI + a1\ kB\ kI + kA\ kB\ kI},\right.\right.$$
$$v2 == \frac{a2\ b2\ kfet\ kI}{i2\ kA\ kB + a2\ b2\ kI + a2\ kB\ kI + kA\ kB\ kI}, v3 == \frac{a3\ b3\ kfet\ kI}{i3\ kA\ kB + a3\ b3\ kI + a3\ kB\ kI + kA\ kB\ kI},$$
$$\left.\left.v4 == \frac{a4\ b4\ kfet\ kI}{i4\ kA\ kB + a4\ b4\ kI + a4\ kB\ kI + kA\ kB\ kI}\right\}, \{kfet, kA, kB, kI\}\right]\right]$$

The kinetic parameters are estimated by use of this program.

```
calckinparscompAB[.7519, 100, 100, 0, .2326, 1, 100, 0, .05263, 5, 5, 0, .02041, 5, 5, 50]
```

{{kfet → 1.00022, kI → 19.9984, kA → 9.99124, kB → 30.0254}}

```
row5 = {kfet, kI, kA, kB} /.
   calckinparscompAB[.7519, 100, 100, 0, .2326, 1, 100, 0, .05263, 5, 5, 0, .02041, 5, 5, 50]
```

{{1.00022, 19.9984, 9.99124, 30.0254}}

These values are correct, but it is necessary to test the sensitivities of these values to experimental errors in velocity measurements.

```
calckinparscompAB[1.05 * .7519, 100, 100, 0, .2326, 1, 100, 0, .05263, 5, 5, 0, .02041, 5, 5, 50]
```

{{kfet → 1.05718, kI → 20.4249, kA → 10.5597, kB → 30.6675}}

```
row6 = {kfet, kI, kA, kB} /.
   calckinparscompAB[1.05 * .7519, 100, 100, 0, .2326, 1, 100, 0, .05263, 5, 5, 0, .02041, 5, 5, 50]
```

{{1.05718, 20.4249, 10.5597, 30.6675}}

```
calckinparscompAB[.7519, 100, 100, 0, 1.05 * .2326, 1, 100, 0, .05263, 5, 5, 0, .02041, 5, 5, 50]
```

{{kfet → 1.04338, kI → 18.6195, kA → 8.12672, kB → 35.8521}}

```
row7 = {kfet, kI, kA, kB} /.
   calckinparscompAB[.7519, 100, 100, 0, 1.05 * .2326, 1, 100, 0, .05263, 5, 5, 0, .02041, 5, 5, 50]
```

{{1.04338, 18.6195, 8.12672, 35.8521}}

```
calckinparscompAB[.7519, 100, 100, 0, .2326, 1, 100, 0, 1.05 * .05263, 5, 5, 0, .02041, 5, 5, 50]
```

{{kfet → 0.954742, kI → 19.4128, kA → 11.875, kB → 24.1137}}

```
row8 = {kfet, kI, kA, kB} /.
   calckinparscompAB[.7519, 100, 100, 0, .2326, 1, 100, 0, 1.05 * .05263, 5, 5, 0, .02041, 5, 5, 50]
```

{{0.954742, 19.4128, 11.875, 24.1137}}

```
calckinparscompAB[.7519, 100, 100, 0, .2326, 1, 100, 0, .05263, 5, 5, 0, 1.05 * .02041, 5, 5, 50]
```

{{kfet → 1.00022, kI → 21.6851, kA → 9.99124, kB → 30.0254}}

```
row9 = {kfet, kI, kA, kB} /.
   calckinparscompAB[.7519, 100, 100, 0, .2326, 1, 100, 0, .05263, 5, 5, 0, 1.05 * .02041, 5, 5, 50]
```

{{1.00022, 21.6851, 9.99124, 30.0254}}

Table 11.1 Estimation of kinetic parameters for competitive inhibition of ordered A + B → products using velocities at {a,b,i} = {100,100,0}, {1,100,0}, {5,5,0},{5,5,50}

```
TableForm[
 Round[{row5[[1]], row6[[1]], row7[[1]], row8[[1]], row9[[1]]}, 0.01], TableHeadings ->
  {{"No errors", "1.05*v1", "1.05*v2", "1.05*v3", "1.05*v4"}, {"kfet", "kI", "kA", "kB"}}]
```

	kfet	kI	kA	kB
No errors	1.	20.	9.99	30.03
1.05*v1	1.06	20.42	10.56	30.67
1.05*v2	1.04	18.62	8.13	35.85
1.05*v3	0.95	19.41	11.88	24.11
1.05*v4	1.	21.69	9.99	30.03

More accurate values can be obtained by using wider ranges of substrate and inhibitor concentrations.

■ 11.2.2 Uncompetitive inhibition

The inhibitor can be bound by EAB.

E + A = EA	kA = e*a/ea
EA + B = EAB	kB = ea*b/eab
EAB + I = EABI	kABI = eab*i/eabi
EAB → products	v = kf*eab

The inhibitor could also be bound by EA, but that case is not discussed.

The rapid-equilibrium rate equation can be derived by use of Solve.

```
Solve[{kA == e * a / ea, kB == ea * b / eab, kABI == eab * i / eabi, et == e + ea + eab + eabi},
  {eab}, {e, ea, eabi}]
```

$$\left\{\left\{eab \to \frac{a\ b\ et\ kABI}{a\ b\ i + a\ b\ kABI + a\ kABI\ kB + kA\ kABI\ kB}\right\}\right\}$$

```
vuncompAB =
```
$$\frac{a\ b\ kfet\ kABI}{a\ b\ i + a\ b\ kABI + a\ kABI\ kB + kA\ kABI\ kB}$$

$$\frac{a\ b\ kABI\ kfet}{a\ b\ i + a\ b\ kABI + a\ kABI\ kB + kA\ kABI\ kB}$$

For test calculations, kfet = 1, kA = 10, kB = 30, and kABI = 20.

```
vuncompAB /. kfet → 1 /. kA → 10 /. kB → 30 /. kABI → 20
```

$$\frac{20\ a\ b}{6000 + 600\ a + 20\ a\ b + a\ b\ i}$$

There are 4 kinetic parameters, and so 4 velocities have to be measured.

```
vuncompAB /. kfet → 1 /. kA → 10 /. kB → 30 /. kABI → 20 /. a → 100 /. b → 100 /. i → 0 // N
```

0.75188

```
vuncompAB /. kfet → 1 /. kA → 10 /. kB → 30 /. kABI → 20 /. a → 1 /. b → 100 /. i → 0 // N
```

0.232558

```
vuncompAB /. kfet → 1 /. kA → 10 /. kB → 30 /. kABI → 20 /. a → 100 /. b → 1 /. i → 0 // N
```

0.0294118

```
vuncompAB /. kfet → 1 /. kA → 10 /. kB → 30 /. kABI → 20 /. a → 5 /. b → 5 /. i → 100 // N
```

0.0416667

The velocity at {100,100,100} is needed in Section 11.2.3, and so it is calculated here.

```
vuncompAB /. kfet → 1 /. kA → 10 /. kB → 30 /. kABI → 20 /. a → 100 /. b → 100 /. i → 100 // N
```

0.157978

```
calckinparsuncompAB[v1_, a1_, b1_, i1_, v2_, a2_, b2_, i2_, v3_, a3_,
  b3_, i3_, v4_, a4_, b4_, i4_] := Module[{}, (*This program calculates kfet,
  kA,kB, and kABI from 4 experimental velocities for ordered A + B →
  products with an uncompetitive inhibitor I.  The first velocity is at high [A],
  high [B], and zero [I], the second velocity is at low [A], high [B],
  and zero [I], the third velocity is at high [A], high [B], and zero [I],
  and the fourth velocity is at medium [A], medium [B], and high [I].*)
```
$$Solve\left[\left\{v1 == \frac{a1\ b1\ kfet\ kABI}{a1\ b1\ i1 + a1\ b1\ kABI + a1\ kABI\ kB + kA\ kABI\ kB},\right.\right.$$
$$v2 == \frac{a2\ b2\ kfet\ kABI}{a2\ b2\ i2 + a2\ b2\ kABI + a2\ kABI\ kB + kA\ kABI\ kB},$$
$$v3 == \frac{a3\ b3\ kfet\ kABI}{a3\ b3\ i3 + a3\ b3\ kABI + a3\ kABI\ kB + kA\ kABI\ kB},$$
$$\left.\left.v4 == \frac{a4\ b4\ kfet\ kABI}{a4\ b4\ i4 + a4\ b4\ kABI + a4\ kABI\ kB + kA\ kABI\ kB}\right\}, \{kfet, kA, kB, kABI\}\right]\right]$$

```
calckinparsuncompAB[.7519, 100, 100, 0, .2326, 1, 100, 0, 0.02941, 100, 1, 0, .04167, 5, 5, 100]
```

{{kfet → 1.00006, kABI → 19.9967, kA → 9.99657, kB → 30.0045}}

```
r1 = {kfet, kABI, kA, kB} /.
   calckinparsuncompAB[.7519, 100, 100, 0, .2326, 1, 100, 0, 0.02941, 100, 1, 0, .04167, 5, 5, 100]
```

{{1.00006, 19.9967, 9.99657, 30.0045}}

These values are correct, but it is necessary to consider the errors in the velocity measurements.

```
calckinparsuncompAB[1.05 * .7519, 100, 100,
  0, .2326, 1, 100, 0, 0.02941, 100, 1, 0, .04167, 5, 5, 100]
```

{{kfet → 1.06841, kABI → 19.4644, kA → 10.2098, kB → 32.0553}}

```
r2 = {kfet, kABI, kA, kB} /. calckinparsuncompAB[1.05 * .7519,
    100, 100, 0, .2326, 1, 100, 0, 0.02941, 100, 1, 0, .04167, 5, 5, 100]
```

{{1.06841, 19.4644, 10.2098, 32.0553}}

```
calckinparsuncompAB[.7519, 100, 100, 0,
  1.05 * .2326, 1, 100, 0, 0.02941, 100, 1, 0, .04167, 5, 5, 100]
```

{{kfet → 1.00006, kABI → 17.281, kA → 9.24361, kB → 30.2114}}

```
r3 = {kfet, kABI, kA, kB} /. calckinparsuncompAB[.7519,
    100, 100, 0, 1.05 * .2326, 1, 100, 0, 0.02941, 100, 1, 0, .04167, 5, 5, 100]
```

{{1.00006, 17.281, 9.24361, 30.2114}}

```
calckinparsuncompAB[.7519, 100, 100, 0, .2326,
  1, 100, 0, 1.05 * 0.02941, 100, 1, 0, .04167, 5, 5, 100]
```

{{kfet → 0.983963, kABI → 19.1347, kA → 10.5729, kB → 27.9124}}

```
r4 = {kfet, kABI, kA, kB} /. calckinparsuncompAB[.7519,
    100, 100, 0, .2326, 1, 100, 0, 1.05 * 0.02941, 100, 1, 0, .04167, 5, 5, 100]
```

{{0.983963, 19.1347, 10.5729, 27.9124}}

```
calckinparsuncompAB[.7519, 100, 100, 0, .2326,
  1, 100, 0, 0.02941, 100, 1, 0, 1.05 * .04167, 5, 5, 100]
```

{{kfet → 1.00006, kABI → 25.9201, kA → 9.99657, kB → 30.0045}}

```
r5 = {kfet, kABI, kA, kB} /. calckinparsuncompAB[.7519,
    100, 100, 0, .2326, 1, 100, 0, 0.02941, 100, 1, 0, 1.05 * .04167, 5, 5, 100]
```

{{1.00006, 25.9201, 9.99657, 30.0045}}

Table 11.2 Estimation of kinetic parameters for uncompetitive inhibition of ordered A + B → products using velocities at {100,100,0}, {1,100,0}, {100,1,0},{5,5,100}

```
TableForm[Round[{r1[[1]], r2[[1]], r3[[1]], r4[[1]], r5[[1]]}, 0.01], TableHeadings ->
   {{"No errors", "1.05*v1", "1.05*v2", "1.05*v3", "1.05*v4"}, {"kfet", "kAI", "kA", "kB"}}]
```

	kfet	kAI	kA	kB
No errors	1.	20.	10.	30.
1.05*v1	1.07	19.46	10.21	32.06
1.05*v2	1.	17.28	9.24	30.21
1.05*v3	0.98	19.13	10.57	27.91
1.05*v4	1.	25.92	10.	30.

11.2.3 Mixed inhibition

A possible mechanism for mixed inhibition is

E + A = EA   kA = e*a/ea
E + I = EI   kI = e*i/ei
EA + B = EAB   kB = ea*b/eab
EAB + I = EABI   kABI = eab*i/eabi
EAB → products   v = kf*eab =kf*et*eab/et = kfet*eab/et

The inhibitor could also be bound by EA, but that case will not be treated here.

The rapid-equilibrium rate equation can be derived by use of Solve.

```
Solve[{kA == e * a / ea, kI == e * i / ei, kB == ea * b / eab,
  kABI == eab * i / eabi, et == e + ea + ei + eab + eabi}, {eab}, {e, ea, ei, eabi}]
```

$$\left\{\left\{eab \to \frac{a\,b\,et\,kABI\,kI}{i\,kA\,kABI\,kB + a\,b\,i\,kI + a\,b\,kABI\,kI + a\,kABI\,kB\,kI + kA\,kABI\,kB\,kI}\right\}\right\}$$

```
vmixinhibAB =
```
$$\frac{a\,b\,kfet\,kABI\,kI}{i\,kA\,kABI\,kB + a\,b\,i\,kI + a\,b\,kABI\,kI + a\,kABI\,kB\,kI + kA\,kABI\,kB\,kI}$$

$$\frac{a\,b\,kABI\,kfet\,kI}{i\,kA\,kABI\,kB + a\,b\,i\,kI + a\,b\,kABI\,kI + a\,kABI\,kB\,kI + kA\,kABI\,kB\,kI}$$

For test calculations, kfet = 1, kA = 10, kI = 20, kB = 30, and kABI = 20. kI and kABI do not need to be equal, but it could happen.

```
vmixinhibAB /. kfet -> 1 /. kA -> 10 /. kI -> 20 /. kB -> 30 /. kABI -> 20
```

$$\frac{400\,a\,b}{120\,000 + 12\,000\,a + 400\,a\,b + 6000\,i + 20\,a\,b\,i}$$

There are five kinetic parameters, and so five velocities are needed.

Calculate velocities at {a,b,i} = {100,100,0}, {1,100,0}, {100,1,0}, {5,5,50}, {5,5,100}, {100,100,100}:

```
vmixinhibAB /. kfet -> 1 /. kA -> 10 /. kI -> 20 /. kB -> 30 /. kABI -> 20 /. a → 100 /. b → 100 /.
  i → 0 // N
```

0.75188

```
vmixinhibAB /. kfet -> 1 /. kA -> 10 /. kI -> 20 /. kB -> 30 /. kABI -> 20 /. a → 1 /. b → 100 /.
  i → 0 // N
```

0.232558

```
vmixinhibAB /. kfet -> 1 /. kA -> 10 /. kI -> 20 /. kB -> 30 /. kABI -> 20 /. a → 100 /. b → 1 /.
  i → 0 // N
```

0.0294118

```
vmixinhibAB /. kfet -> 1 /. kA -> 10 /. kI -> 20 /. kB -> 30 /. kABI -> 20 /. a → 100 /. b → 100 /.
  i → 100 // N
```

0.154321

```
vmixinhibAB /. kfet -> 1 /. kA -> 10 /. kI -> 20 /. kB -> 30 /. kABI -> 20 /. a → 5 /. b → 5 /.
  i → 100 // N
```

0.0119048

The rate equation vmixinhibAB can be used to write a program to estimate the 5 kinetic parameters from 5 velocity measurements.

```
calckinparsmixinhibAB[v1_, a1_, b1_, i1_, v2_, a2_, b2_, i2_, v3_, a3_, b3_, i3_, v4_,
  a4_, b4_, i4_, v5_, a5_, b5_, i5_] := Module[{}, (*This program calculates kfet,
  kA, kuI, kB, and kI from 5 experimental velocities for ordered A + B →
   products with a mixed inhibitor I.  [The first velocity is at high [A], high [B],
    and zero [I], the second velocity is at low [A], high [B], and zero [I], the third
     velocity is at high [A], low [B], and zero [I], the fourth velocity is at high [A],
     high [B], and high [I], the fifth velocity is at low [A], low [B], and high [I]*)
```

$$\text{Solve}\left[\left\{v1 == \frac{a1\ b1\ kfet\ kABI\ kI}{i1\ kA\ kABI\ kB + a1\ b1\ i1\ kI + a1\ b1\ kABI\ kI + a1\ kABI\ kB\ kI + kA\ kABI\ kB\ kI},\right.\right.$$

$$v2 == \frac{a2\ b2\ kfet\ kABI\ kI}{i2\ kA\ kABI\ kB + a2\ b2\ i2\ kI + a2\ b2\ kABI\ kI + a2\ kABI\ kB\ kI + kA\ kABI\ kB\ kI},$$

$$v3 == \frac{a3\ b3\ kfet\ kABI\ kI}{i3\ kA\ kABI\ kB + a3\ b3\ i3\ kI + a3\ b3\ kABI\ kI + a3\ kABI\ kB\ kI + kA\ kABI\ kB\ kI},$$

$$v4 == \frac{a4\ b4\ kfet\ kABI\ kI}{i4\ kA\ kABI\ kB + a4\ b4\ i4\ kI + a4\ b4\ kABI\ kI + a4\ kABI\ kB\ kI + kA\ kABI\ kB\ kI},$$

$$\left.\left.v5 == \frac{a5\ b5\ kfet\ kABI\ kI}{i5\ kA\ kABI\ kB + a5\ b5\ i5\ kI + a5\ b5\ kABI\ kI + a5\ kABI\ kB\ kI + kA\ kABI\ kB\ kI}\right\},\right.$$

```
{kfet, kA, kB, kI, kABI}]]
```

```
calckinparsmixinhibAB[.7519, 100, 100, 0, .2326, 1,
  100, 0, .02941, 100, 1, 0, .1543, 100, 100, 100, .01190, 5, 5, 100]
```

{{kfet → 1.00006, kA → 9.99657, kI → 19.9833, kABI → 19.9955, kB → 30.0045}}

```
rr1 = {kfet, kA, kI, kABI, kB} /. calckinparsmixinhibAB[.7519, 100, 100,
    0, .2326, 1, 100, 0, .02941, 100, 1, 0, .1543, 100, 100, 100, .01190, 5, 5, 100]
```

{{1.00006, 9.99657, 19.9833, 19.9955, 30.0045}}

These values are correct, but it is necessary to take errors in velocity measurements into account.

```
calckinparsmixinhibAB[1.05 * .7519, 100, 100, 0, .2326, 1,
  100, 0, .02941, 100, 1, 0, .1543, 100, 100, 100, .01190, 5, 5, 100]
```

{{kfet → 1.06841, kA → 10.2098, kI → 20.4969, kABI → 18.4799, kB → 32.0553}}

```
rr2 = {kfet, kA, kI, kABI, kB} /. calckinparsmixinhibAB[1.05 * .7519, 100, 100,
    0, .2326, 1, 100, 0, .02941, 100, 1, 0, .1543, 100, 100, 100, .01190, 5, 5, 100]
```

{{1.06841, 10.2098, 20.4969, 18.4799, 32.0553}}

```
calckinparsmixinhibAB[.7519, 100, 100, 0, 1.05 * .2326, 1,
  100, 0, .02941, 100, 1, 0, .1543, 100, 100, 100, .01190, 5, 5, 100]
```

{{kfet → 1.00006, kA → 9.24361, kI → 18.3645, kABI → 20.0034, kB → 30.2114}}

```
rr3 = {kfet, kA, kI, kABI, kB} /. calckinparsmixinhibAB[.7519, 100, 100, 0,
    1.05 * .2326, 1, 100, 0, .02941, 100, 1, 0, .1543, 100, 100, 100, .01190, 5, 5, 100]
```

{{1.00006, 9.24361, 18.3645, 20.0034, 30.2114}}

```
calckinparsmixinhibAB[.7519, 100, 100, 0, .2326, 1, 100,
  0, 1.05 * .02941, 100, 1, 0, .1543, 100, 100, 100, .01190, 5, 5, 100]
```

{{kfet → 0.983963, kA → 10.5729, kI → 19.8801, kABI → 20.3258, kB → 27.9124}}

```
rr4 = {kfet, kA, kI, kABI, kB} /. calckinparsmixinhibAB[.7519, 100, 100, 0,
  .2326, 1, 100, 0, 1.05 * .02941, 100, 1, 0, .1543, 100, 100, 100, .01190, 5, 5, 100]
```

{{0.983963, 10.5729, 19.8801, 20.3258, 27.9124}}

```
calckinparsmixinhibAB[.7519, 100, 100, 0, .2326, 1, 100,
  0, .02941, 100, 1, 0, 1.05 * .1543, 100, 100, 100, .01190, 5, 5, 100]
```

{{kfet → 1.00006, kA → 9.99657, kI → 19.8808, kABI → 21.3142, kB → 30.0045}}

```
rr5 = {kfet, kA, kI, kABI, kB} /. calckinparsmixinhibAB[.7519, 100, 100, 0,
  .2326, 1, 100, 0, .02941, 100, 1, 0, 1.05 * .1543, 100, 100, 100, .01190, 5, 5, 100]
```

{{1.00006, 9.99657, 19.8808, 21.3142, 30.0045}}

```
calckinparsmixinhibAB[.7519, 100, 100, 0, .2326, 1, 100,
  0, .02941, 100, 1, 0, .1543, 100, 100, 100, 1.05 * .01190, 5, 5, 100]
```

{{kfet → 1.00006, kA → 9.99657, kI → 21.4142, kABI → 19.9555, kB → 30.0045}}

```
rr6 = {kfet, kA, kI, kABI, kB} /. calckinparsmixinhibAB[.7519, 100, 100, 0,
  .2326, 1, 100, 0, .02941, 100, 1, 0, .1543, 100, 100, 100, 1.05 * .01190, 5, 5, 100]
```

{{1.00006, 9.99657, 21.4142, 19.9555, 30.0045}}

Table 11.3 Estimation of kinetic parameters for mixed inhibition of ordered A + B → products using velocities at {a,b,i} = {100,100,0}, {1,100,0}, {100,1,0}, {100,100,100}, and {5,5,100},

```
TableForm[Round[{rr1[[1]], rr2[[1]], rr3[[1]], rr4[[1]], rr5[[1]], rr6[[1]]}, 0.01],
  TableHeadings -> {{"No errors", "1.05*v1", "1.05*v2", "1.05*v3", "1.05*v4", "1.05*v5"},
    {"kfet", "kA", "kI", "kABI", "kB"}}]
```

	kfet	kA	kI	kABI	kB
No errors	1.	10.	19.98	20.	30.
1.05*v1	1.07	10.21	20.5	18.48	32.06
1.05*v2	1.	9.24	18.36	20.	30.21
1.05*v3	0.98	10.57	19.88	20.33	27.91
1.05*v4	1.	10.	19.88	21.31	30.
1.05*v5	1.	10.	21.41	19.96	30.

Application of this more general program to velocities for the competitive mechanism in Section 11.2.1.

```
calckinparsmixinhibAB[.7519, 100, 100, 0, .2326, 1,
  100, 0, .05263, 5, 5, 0, .02041, 5, 5, 50, .6757, 100, 100, 100]
```

{{kfet → 1.00022, kA → 9.99124, kI → 19.9984, kABI → 1.32458×10^7, kB → 30.0254}}

The value of kABI is unreasonable, but the values of the other kinetic parameters are correct. This shows that the velocities are for the competitive inhibition of ordered A + B → products.

Application of this more general program to velocities for the uncompetitive mechanism in Section 11.2.2.

```
calckinparsmixinhibAB[.7519, 100, 100, 0, .2326, 1,
  100, 0, 0.02941, 100, 1, 0, .04167, 5, 5, 100, .1580, 100, 100, 100]
```

{{kfet → 1.00006, kA → 9.99657, kI → 854 579., kABI → 20.0023, kB → 30.0045}}

The value of kI is unreasonable, but the values of the other kinetic parameters are correct. This shows that the velocities are for the uncompetitive inhibition of ordered A + B → products. This shows that a more general program can be used to estimate the kinetic parameters for a less general program.

■ 11.2.4 Essential activation

In essential activation, the free enzyme without the activator has no activity and does not bind substrate. A possible mechanism for essential activation is

E + X = EX	kX = e*x/ex
EX + A = EAX	kAX = ex*a/eax
EAX + B = EABX	kABX = eax*b/eabx
EABX → products	v = kf*eabx

The rapid-equilibrium rate equation can be derived by use of Solve.

```
Solve[{kX == e * x / ex, kAX == ex * a / eax, kABX == eax * b / eabx, et == e + ex + eax + eabx},
  {eabx}, {e, ex, eax}]
```

$$\left\{\left\{eabx \rightarrow \frac{a\,b\,et\,x}{kABX\;kAX\;kX + a\,b\,x + a\,kABX\,x + kABX\,kAX\,x}\right\}\right\}$$

$$vessactAB = \frac{a\,b\,kfet\,x}{kABX\;kAX\;kX + a\,b\,x + a\,kABX\,x + kABX\,kAX\,x}$$

$$\frac{a\,b\,kfet\,x}{kABX\;kAX\;kX + a\,b\,x + a\,kABX\,x + kABX\,kAX\,x}$$

For trial calculations, assume that kX = 10, kAX = 30, kABX = 20, and kfet = 2.

```
vessactAB /. kX -> 10 /. kAX -> 30 /. kABX -> 20 /. kfet -> 2
```

$$\frac{2\,a\,b\,x}{6000 + 600\,x + 20\,a\,x + a\,b\,x}$$

Since there are 4 kinetic parameters, 4 velocities have to be measured. {a,b,x} = {100,100,100}, {1,100,100}, {100,1,100}, and {100,100,1}.

```
vessactAB /. kX -> 10 /. kAX -> 30 /. kABX -> 20 /. kfet -> 2 /. a → 100 /. b → 100 /. x → 100 // N
```

1.57978

```
vessactAB /. kX -> 10 /. kAX -> 30 /. kABX -> 20 /. kfet -> 2 /. a → 1 /. b → 100 /. x → 100 // N
```

0.25641

```
vessactAB /. kX -> 10 /. kAX -> 30 /. kABX -> 20 /. kfet -> 2 /. a → 100 /. b → 1 /. x → 100 // N
```

0.0724638

```
vessactAB /. kX -> 10 /. kAX -> 30 /. kABX -> 20 /. kfet -> 2 /. a → 100 /. b → 100 /. x → 1 // N
```

1.07527

```
calckinparsvessactAB[v1_, a1_, b1_, x1_, v2_, a2_, b2_, x2_, v3_, a3_,
   b3_, x3_, v4_, a4_, b4_, x4_] := Module[{}, (*This program calculates kfet,
  kX, kAX, and kABX from 4 experimental velocities for ordered A + B →
   products with essential activation.  The first velocity is at high [A],
   high [B], and high [X], the second velocity is at low [A], high [B],
   and high [X], the third velocity is at high [A], low[B], and high [X],
   the fourth velocity is at high [A], high[B], and low [X].*)
```

$$
\text{Solve}\Big[\Big\{v1 == \frac{a1\ b1\ kfet\ x1}{kABX\ kAX\ kX + a1\ b1\ x1 + a1\ kABX\ x1 + kABX\ kAX\ x1},
$$

$$
v2 == \frac{a2\ b2\ kfet\ x2}{kABX\ kAX\ kX + a2\ b2\ x2 + a2\ kABX\ x2 + kABX\ kAX\ x2},
$$

$$
v3 == \frac{a3\ b3\ kfet\ x3}{kABX\ kAX\ kX + a3\ b3\ x3 + a3\ kABX\ x3 + kABX\ kAX\ x3},
$$

$$
v4 == \frac{a4\ b4\ kfet\ x4}{kABX\ kAX\ kX + a4\ b4\ x4 + a4\ kABX\ x4 + kABX\ kAX\ x4}\Big\}, \{kfet, kX, kAX, kABX\}\Big]\Big]
$$

```
calckinparsvessactAB[1.580, 100, 100, 100,
 .2564, 1, 100, 100, .07246, 100, 1, 100, 1.075, 100, 100, 1]
```

{{kfet → 2.00039, kX → 10.0111, kAX → 29.9975, kABX → 20.005}}

```
rrr1 = {kfet, kX, kAX, kABX} /. calckinparsvessactAB[1.580,
    100, 100, 100, .2564, 1, 100, 100, .07246, 100, 1, 100, 1.075, 100, 100, 1]
```

{{2.00039, 10.0111, 29.9975, 20.005}}

These values are correct, but it is necessary to take errors in velocity measurments into account.

```
calckinparsvessactAB[1.05 * 1.580, 100, 100, 100,
 .2564, 1, 100, 100, .07246, 100, 1, 100, 1.075, 100, 100, 1]
```

{{kfet → 2.13011, kX → 11.0259, kAX → 29.9975, kABX → 21.3023}}

```
rrr2 = {kfet, kX, kAX, kABX} /. calckinparsvessactAB[1.05 * 1.580,
    100, 100, 100, .2564, 1, 100, 100, .07246, 100, 1, 100, 1.075, 100, 100, 1]
```

{{2.13011, 11.0259, 29.9975, 21.3023}}

```
calckinparsvessactAB[1.580, 100, 100, 100,
 1.05 * .2564, 1, 100, 100, .07246, 100, 1, 100, 1.075, 100, 100, 1]
```

{{kfet → 2.00039, kX → 10.6789, kAX → 27.6038, kABX → 20.3803}}

```
rrr3 = {kfet, kX, kAX, kABX} /. calckinparsvessactAB[1.580, 100,
    100, 100, 1.05 * .2564, 1, 100, 100, .07246, 100, 1, 100, 1.075, 100, 100, 1]
```

{{2.00039, 10.6789, 27.6038, 20.3803}}

```
calckinparsvessactAB[1.580, 100, 100, 100, .2564,
 1, 100, 100, 1.05 * .07246, 100, 1, 100, 1.075, 100, 100, 1]
```

{{kfet → 1.97417, kX → 10.0111, kAX → 32.1303, kABX → 18.4324}}

```
rrr4 = {kfet, kX, kAX, kABX} /. calckinparsvessactAB[1.580, 100,
    100, 100, .2564, 1, 100, 100, 1.05 * .07246, 100, 1, 100, 1.075, 100, 100, 1]
```

{{1.97417, 10.0111, 32.1303, 18.4324}}

```
calckinparsvessactAB[1.580, 100, 100, 100, .2564,
 1, 100, 100, .07246, 100, 1, 100, 1.05 * 1.075, 100, 100, 1]
```

{{kfet → 2.00039, kX → 8.39436, kAX → 30.4449, kABX → 20.005}}

```
rrr5 = {kfet, kX, kAX, kABX} /. calckinparsvessactAB[1.580, 100,
    100, 100, .2564, 1, 100, 100, .07246, 100, 1, 100, 1.05 * 1.075, 100, 100, 1]
```

$\{\{2.00039, 8.39436, 30.4449, 20.005\}\}$

Table 11.4 Estimation of kinetic parameters for essental activation of ordered A + B → products using velocities at {a,b,x} = {100,100,100}, {1,100,100}, {100,1,100},{100,100,1}

```
TableForm[
  Round[{rrr1[[1]], rrr2[[1]], rrr3[[1]], rrr4[[1]], rrr5[[1]]}, 0.01], TableHeadings ->
    {{"No errors", "1.05*v1", "1.05*v2", "1.05*v3", "1.05*v4"}, {"kfet", "kX", "kAX", "kABX"}}]
```

	kfet	kX	kAX	kABX
No errors	2.	10.01	30.	20.01
1.05*v1	2.13	11.03	30.	21.3
1.05*v2	2.	10.68	27.6	20.38
1.05*v3	1.97	10.01	32.13	18.43
1.05*v4	2.	8.39	30.44	20.01

■ 11.2.5 Mixed activation

A possible mechanism for mixed activation of ordered A + B → products is

```
      EABX→ products
        ‖ kABX
         B
    kAX  +
A + EX =  EAX
   ‖ kX   ‖ kEAX
A + E  =  EA
   + kA   +
    X      X
```

The 4 equilibrium constants in the cycle are not independent: kA kEAX = kX kAX. The rate equation must be derived using an independent set of reactions. The reaction EA + X = EAX is omitted in the derivation of the rate equation, and kEAX is calculated using kEAX = kX kAX/kA. The mechanism used to derive the rate equation is represented by

E + X = EX	kX = e*x/ex
E + A = EA	kA = e*a/ea
EX + A = EAX	kAX = ex*a/eax
EAX + B = EABX	kABX = eax*b/eabx
EABX → products	v = kf*eabx

These 4 equilibrium expressions are independent. The rapid-equilibrium velocity can be derived using Solve.

```
Solve[{kX == e * x / ex, kA == e * a / ea, kAX == ex * a / eax,
  kABX == eax * b / eabx, et == e + ex + ea + eax + eabx}, {eabx}, {e, ex, ea, eax}]
```

$$\left\{\left\{eabx \to \frac{a\,b\,et\,kA\,x}{a\,kABX\,kAX\,kX + kA\,kABX\,kAX\,kX + a\,b\,kA\,x + a\,kA\,kABX\,x + kA\,kABX\,kAX\,x}\right\}\right\}$$

The velocity is given by

$$\text{vmixactAB} = \frac{a\,b\,\text{kfet}\,\text{kA}\,x}{a\,\text{kABX}\,\text{kAX}\,\text{kX} + \text{kA}\,\text{kABX}\,\text{kAX}\,\text{kX} + a\,b\,\text{kA}\,x + a\,\text{kA}\,\text{kABX}\,x + \text{kA}\,\text{kABX}\,\text{kAX}\,x}$$

$$\frac{a\,b\,\text{kA}\,\text{kfet}\,x}{a\,\text{kABX}\,\text{kAX}\,\text{kX} + \text{kA}\,\text{kABX}\,\text{kAX}\,\text{kX} + a\,b\,\text{kA}\,x + a\,\text{kA}\,\text{kABX}\,x + \text{kA}\,\text{kABX}\,\text{kAX}\,x}$$

For trial calculations, the following values are used for the 5 kinetic parameters: kX = 10, kA = 30, kAX = 30, kABX = 20, and kfet = 2. It may seem strange to have two equilibrium constants with the same value, but it can happen.

```
vmixactAB /. kX -> 10 /. kA -> 30 /. kAX -> 30 /. kABX -> 20 /. kfet -> 2
```

$$\frac{60\,a\,b\,x}{180\,000 + 6000\,a + 18\,000\,x + 600\,a\,x + 30\,a\,b\,x}$$

Since there are 5 kinetic parameters, 5 velocities have to be measured. {a,b,x} = {100,100,100}, {1,100,100}, {100,1,100}, {100,100,1}, and {5,5,5}.

```
vmixactAB /. kX -> 10 /. kA -> 30 /. kAX -> 30 /. kABX -> 20 /. kfet -> 2 /. a → 100 /. b → 100 /.
  x → 100 // N
```

1.55521

```
vmixactAB /. kX -> 10 /. kA -> 30 /. kAX -> 30 /. kABX -> 20 /. kfet -> 2 /. a → 1 /. b → 100 /.
  x → 100 // N
```

0.255754

```
vmixactAB /. kX -> 10 /. kA -> 30 /. kAX -> 30 /. kABX -> 20 /. kfet -> 2 /. a → 100 /. b → 1 /.
  x → 100 // N
```

0.0675676

```
vmixactAB /. kX -> 10 /. kA -> 30 /. kAX -> 30 /. kABX -> 20 /. kfet -> 2 /. a → 100 /. b → 100 /.
  x → 1 // N
```

0.518135

```
vmixactAB /. kX -> 10 /. kA -> 30 /. kAX -> 30 /. kABX -> 20 /. kfet -> 2 /. a → 5 /. b → 5 /. x → 5 // N
```

0.0235294

```
calckinparsvmixactAB[v1_, a1_, b1_, x1_, v2_, a2_, b2_, x2_,
  v3_, a3_, b3_, x3_, v4_, a4_, b4_, x4_, v5_, a5_, b5_, x5_] := Module[{},
```

$$\text{Solve}\Big[\Big\{v1 == \frac{a1\,b1\,\text{kfet}\,\text{kA}\,x1}{a1\,\text{kABX}\,\text{kAX}\,\text{kX} + \text{kA}\,\text{kABX}\,\text{kAX}\,\text{kX} + a1\,b1\,\text{kA}\,x1 + a1\,\text{kA}\,\text{kABX}\,x1 + \text{kA}\,\text{kABX}\,\text{kAX}\,x1},$$

$$v2 == \frac{a2\,b2\,\text{kfet}\,\text{kA}\,x2}{a2\,\text{kABX}\,\text{kAX}\,\text{kX} + \text{kA}\,\text{kABX}\,\text{kAX}\,\text{kX} + a2\,b2\,\text{kA}\,x2 + a2\,\text{kA}\,\text{kABX}\,x2 + \text{kA}\,\text{kABX}\,\text{kAX}\,x2},$$

$$v3 == \frac{a3\,b3\,\text{kfet}\,\text{kA}\,x3}{a3\,\text{kABX}\,\text{kAX}\,\text{kX} + \text{kA}\,\text{kABX}\,\text{kAX}\,\text{kX} + a3\,b3\,\text{kA}\,x3 + a3\,\text{kA}\,\text{kABX}\,x3 + \text{kA}\,\text{kABX}\,\text{kAX}\,x3},$$

$$v4 == \frac{a4\,b4\,\text{kfet}\,\text{kA}\,x4}{a4\,\text{kABX}\,\text{kAX}\,\text{kX} + \text{kA}\,\text{kABX}\,\text{kAX}\,\text{kX} + a4\,b4\,\text{kA}\,x4 + a4\,\text{kA}\,\text{kABX}\,x4 + \text{kA}\,\text{kABX}\,\text{kAX}\,x4},$$

$$v5 == \frac{a5\,b5\,\text{kfet}\,\text{kA}\,x5}{a5\,\text{kABX}\,\text{kAX}\,\text{kX} + \text{kA}\,\text{kABX}\,\text{kAX}\,\text{kX} + a5\,b5\,\text{kA}\,x5 + a5\,\text{kA}\,\text{kABX}\,x5 + \text{kA}\,\text{kABX}\,\text{kAX}\,x5}\Big\},$$

```
  {kfet, kX, kA, kAX, kABX}]]
```

```
calckinparsvmixactAB[1.555, 100, 100, 100, .2558, 1,
   100, 100, .06757, 100, 1, 100, .5181, 100, 100, 1, .02353, 5, 5, 5]
```

{{kfet → 1.99963, kX → 10.0036, kAX → 29.9914, kA → 30.002, kABX → 19.9966}}

```
rrrr1 = {kfet, kX, kAX, kA, kABX} /. calckinparsvmixactAB[1.555, 100, 100,
   100, .2558, 1, 100, 100, .06757, 100, 1, 100, .5181, 100, 100, 1, .02353, 5, 5, 5]
```

{{1.99963, 10.0036, 29.9914, 30.002, 19.9966}}

These values are correct, but it is necessary to take errors in the velocity measurements in to account.

```
calckinparsvmixactAB[1.05 * 1.555, 100, 100, 100, .2558,
   1, 100, 100, .06757, 100, 1, 100, .5181, 100, 100, 1, .02353, 5, 5, 5]
```

{{kfet → 2.13147, kX → 9.79587, kAX → 30.4329, kA → 28.738, kABX → 21.2428}}

```
rrrr2 = {kfet, kX, kAX, kA, kABX} /. calckinparsvmixactAB[1.05 * 1.555, 100, 100,
   100, .2558, 1, 100, 100, .06757, 100, 1, 100, .5181, 100, 100, 1, .02353, 5, 5, 5]
```

{{2.13147, 9.79587, 30.4329, 28.738, 21.2428}}

```
calckinparsvmixactAB[1.555, 100, 100, 100, 1.05 * .2558,
   1, 100, 100, .06757, 100, 1, 100, .5181, 100, 100, 1, .02353, 5, 5, 5]
```

{{kfet → 1.99963, kX → 11.0637, kAX → 27.4684, kA → 31.3014, kABX → 20.3924}}

```
rrrr3 = {kfet, kX, kAX, kA, kABX} /. calckinparsvmixactAB[1.555, 100, 100, 100,
   1.05 * .2558, 1, 100, 100, .06757, 100, 1, 100, .5181, 100, 100, 1, .02353, 5, 5, 5]
```

{{1.99963, 11.0637, 27.4684, 31.3014, 20.3924}}

```
calckinparsvmixactAB[1.555, 100, 100, 100, .2558, 1, 100,
   100, 1.05 * .06757, 100, 1, 100, .5181, 100, 100, 1, .02353, 5, 5, 5]
```

{{kfet → 1.97156, kX → 10.0723, kAX → 32.2633, kA → 30.246, kABX → 18.3162}}

```
rrrr4 = {kfet, kX, kAX, kA, kABX} /. calckinparsvmixactAB[1.555, 100, 100, 100,
   .2558, 1, 100, 100, 1.05 * .06757, 100, 1, 100, .5181, 100, 100, 1, .02353, 5, 5, 5]
```

{{1.97156, 10.0723, 32.2633, 30.246, 18.3162}}

```
calckinparsvmixactAB[1.555, 100, 100, 100, .2558, 1, 100,
   100, .06757, 100, 1, 100, 1.05 * .5181, 100, 100, 1, .02353, 5, 5, 5]
```

{{kfet → 1.99963, kX → 10.1831, kAX → 29.6528, kA → 33.793, kABX → 20.192}}

```
rrrr5 = {kfet, kX, kAX, kA, kABX} /. calckinparsvmixactAB[1.555, 100, 100, 100,
   .2558, 1, 100, 100, .06757, 100, 1, 100, 1.05 * .5181, 100, 100, 1, .02353, 5, 5, 5]
```

{{1.99963, 10.1831, 29.6528, 33.793, 20.192}}

```
calckinparsvmixactAB[1.555, 100, 100, 100, .2558, 1, 100,
   100, .06757, 100, 1, 100, .5181, 100, 100, 1, 1.05 * .02353, 5, 5, 5]
```

{{kfet → 1.99963, kX → 8.985, kAX → 30.3568, kA → 26.4574, kABX → 19.9405}}

```
rrrr6 = {kfet, kX, kAX, kA, kABX} /. calckinparsvmixactAB[1.555, 100, 100, 100,
   .2558, 1, 100, 100, .06757, 100, 1, 100, .5181, 100, 100, 1, 1.05 * .02353, 5, 5, 5]
```

{{1.99963, 8.985, 30.3568, 26.4574, 19.9405}}

Table 11.5 Estimation of five kinetic parameters for mixed activation of ordered A + B → products using velocities at {a,b,x} = {100,100,100}, {1,100,100}, {100,1,100}, {100,100,1}, and {5,5,5}

```
TableForm[
  Round[{rrrr1[[1]], rrrr2[[1]], rrrr3[[1]], rrrr4[[1]], rrrr5[[1]], rrrr6[[1]]}, 0.01],
  TableHeadings -> {{"No errors", "1.05*v1", "1.05*v2", "1.05*v3", "1.05*v4", "1.05*v5"},
    {"kfet", "kX", "kAX", "kA", "kABX"}}]
```

	kfet	kX	kAX	kA	kABX
No errors	2.	10.	29.99	30.	20.
1.05*v1	2.13	9.8	30.43	28.74	21.24
1.05*v2	2.	11.06	27.47	31.3	20.39
1.05*v3	1.97	10.07	32.26	30.25	18.32
1.05*v4	2.	10.18	29.65	33.79	20.19
1.05*v5	2.	8.98	30.36	26.46	19.94

Since there is a thermodynamic cycle in this mechanism, kEAX is calculated using kEAX = kX kA/kAX.

```
10.`          29.990000000000002`        30.`
─────         ──────────────────        ──────────────────────
9.8`          30.43`                     28.740000000000002`

11.06`        27.47`                     31.3`
          *                         /
10.07`        32.26`                     30.25`

10.18`        29.650000000000002`        33.79`

8.98`         30.36`                     26.46`
```

{{9.99667}, {10.3763}, {9.70665}, {10.7391}, {8.93273}, {10.3036}}

This makes it possible to add a column to Table 11.5. Make a vector of a line by use of [[1]]. Insert a value for the last column and label it with an a for "augmented."

```
rrrr1[[1]]
```

{1.99963, 10.0036, 29.9914, 30.002, 19.9966}

```
rrrr1a = {1.99962776816609`, 10.003555781467792`,
   29.99140294278431`, 30.001972258871543`, 19.996571366976205`, 10.00}
```

{1.99963, 10.0036, 29.9914, 30.002, 19.9966, 10.}

```
rrrr2[[1]]
```

{2.13147, 9.79587, 30.4329, 28.738, 21.2428}

```
rrrr2a = {2.1314664258184126`, 9.795872513830684`,
   30.43285712970259`, 28.738002539573415`, 21.24283608254295`, 10.37}
```

{2.13147, 9.79587, 30.4329, 28.738, 21.2428, 10.37}

```
rrrr3[[1]]
```

{1.99963, 11.0637, 27.4684, 31.3014, 20.3924}

```
rrrr3a = {1.99962776816609`, 11.063696969849591`,
   27.4684013429761`, 31.30140253538949`, 20.392366568124228`, 9.71}
```

{1.99963, 11.0637, 27.4684, 31.3014, 20.3924, 9.71}

```
rrrr4[[1]]
```

{1.97156, 10.0723, 32.2633, 30.246, 18.3162}

```
rrrr4a = {1.971563640261018`, 10.072307396641152`,
   32.263269936357695`, 30.245958028197414`, 18.31615169903094`, 10.74}
```

{1.97156, 10.0723, 32.2633, 30.246, 18.3162, 10.74}

```
rrrr5[[1]]
```

{1.99963, 10.1831, 29.6528, 33.793, 20.192}

```
rrrr5a = {1.99962776816609`, 10.183066345577565`,
    29.652761748344847`, 33.79297062709186`, 20.191986175803972`, 11.60}
```

{1.99963, 10.1831, 29.6528, 33.793, 20.192, 11.6}

```
rrrr6[[1]]
```

{1.99963, 8.985, 30.3568, 26.4574, 19.9405}

```
rrrr6a = {1.99962776816609`, 8.984999231708663`,
    30.356786372084123`, 26.457445717556155`, 19.940521996447472`, 10.30}
```

{1.99963, 8.985, 30.3568, 26.4574, 19.9405, 10.3}

Table 11.6 Estimation of six kinetic parameters for mixed activation of ordered A + B → products using velocities at {a,b,x} = {100,100,100}, {1,100,100}, {100,1,100}, {100,100,1}, and {5,5,5}

```
TableForm[Round[{rrrr1a, rrrr2a, rrrr3a, rrrr4a, rrrr5a, rrrr6a}, 0.01],
  TableHeadings -> {{"No errors", "1.05*v1", "1.05*v2", "1.05*v3", "1.05*v4", "1.05*v5"},
    {"kfet", "kX", "kAX", "kA", "kABX", "kEAX"}}]
```

	kfet	kX	kAX	kA	kABX	kEAX
No errors	2.	10.	29.99	30.	20.	10.
1.05*v1	2.13	9.8	30.43	28.74	21.24	10.37
1.05*v2	2.	11.06	27.47	31.3	20.39	9.71
1.05*v3	1.97	10.07	32.26	30.25	18.32	10.74
1.05*v4	2.	10.18	29.65	33.79	20.19	11.6
1.05*v5	2.	8.98	30.36	26.46	19.94	10.3

The last column was calculated using kEAX = kX kAX/kA.

■ 11.2.6 Modification

A possible mechanism for modification of ordered A + B → products is [36]

E + X = EX	kX = e*x/ex	
E + A = EA	kA = e*a/ea	
EA + B = EAB	kAB = ea*b/eab	New reaction added to 11.2.5 Mixed activation
EAB → products	v1 = kf1*eab	New reaction added to 11.2.5 Mixed activation
EX + A = EAX	kAX = ex*a/eax	
EAX + B = EABX	kABX = eax*b/eabx	
EABX → products	v2 = kf2*eabx	
vEABX = v1 + v2		

These five equilibrium constants are independent. The rapid-equilibrium velocity v can be derived using Solve.

```
Solve[{kX == e * x / ex, kA == e * a / ea, kAB == ea * b / eab, kAX == ex * a / eax,
  kABX == eax * b / eabx, et == e + ex + ea + eab + eax + eabx}, {eab}, {e, ex, ea, eax, eabx}]
```

{{eab → (a b et kABX kAX kX) / (a b kABX kAX kX + a kAB kABX kAX kX +
    kA kAB kABX kAX kX + a b kA kAB x + a kA kAB kABX x + kA kAB kABX kAX x)}}

```
Solve[{kX == e * x / ex, kA == e * a / ea, kAB == ea * b / eab, kAX == ex * a / eax,
  kABX == eax * b / eabx, et == e + ex + ea + eab + eax + eabx}, {eabx}, {e, ex, ea, eax, eab}]
```

{{eabx → (a b et kA kAB x) / (a b kABX kAX kX + a kAB kABX kAX kX +
    kA kAB kABX kAX kX + a b kA kAB x + a kA kAB kABX x + kA kAB kABX kAX x)}}

These two equilibrium expressions have the same denominator, and so the rapid-equilibrium rate equation is given by.

```
vEABX = ((a b kf1et kABX kAX kX) + (a b kf2et kA kAB x)) / (a b kABX kAX kX +
    a kAB kABX kAX kX + kA kAB kABX kAX kX + a b kA kAB x + a kA kAB kABX x + kA kAB kABX kAX x)
```

(a b kABX kAX kf1et kX + a b kA kAB kf2et x) / (a b kABX kAX kX +
  a kAB kABX kAX kX + kA kAB kABX kAX kX + a b kA kAB x + a kA kAB kABX x + kA kAB kABX kAX x)

The following values for the kinetic parameters are chosen arbitrarily:

```
vEABX /. kf1et → 1 /. kf2et → 2 /. kX -> 10 /. kA → 30 /. kAX → 30 /. kABX → 20 /. kAB → 40
```

$$\frac{6000 \, a \, b + 2400 \, a \, b \, x}{7\,200\,000 + 240\,000 \, a + 6000 \, a \, b + 720\,000 \, x + 24\,000 \, a \, x + 1200 \, a \, b \, x}$$

This rapid-equilibrium rate equation involves seven parameters, and so seven velocities are needed to estimate the values of the kinetic parameters. A program can be written to do this, but when it comes to chosing seven triplets of substrate concentrations, it becomes very complicated. However, it is possible to calculate twelve 3D plots to show how the rapid-equilibrium velocity and its partial derivatives depend on [X] and [B] at constant [A], [X] and [A] at specified [B], and [A] and [B] at specified [X].

Make 3D plots at a = 50.

```
vEABX /. kf1et → 1 /. kf2et → 2 /. kX -> 10 /. kA → 30 /. kAX → 30 /. kABX → 20 /. kAB → 40 /. a → 50
```

$$\frac{300\,000\,b + 120\,000\,b\,x}{19\,200\,000 + 300\,000\,b + 1\,920\,000\,x + 60\,000\,b\,x}$$

```
plot101a = Plot3D[Evaluate[ (300 000 b + 120 000 b x) / (19 200 000 + 300 000 b + 1 920 000 x + 60 000 b x) ],
    {b, 0.00001, 150}, {x, 0.00001, 100}, ViewPoint → {-2, -2, 1},
    PlotRange -> {0, 2}, Lighting -> {{"Ambient", GrayLevel[1]}},
    AxesLabel → {"[B]", "[X]"}, PlotLabel → "vEABXaconst"];
```

```
plot101b = Plot3D[Evaluate[D[ (300 000 b + 120 000 b x) / (19 200 000 + 300 000 b + 1 920 000 x + 60 000 b x) , b]],
    {b, 0.00001, 150}, {x, 0.00001, 100}, ViewPoint → {-2, -2, 1},
    PlotRange -> {0, .1}, Lighting -> {{"Ambient", GrayLevel[1]}},
    AxesLabel → {"[B]", "[X]"}, PlotLabel → "dvEABXaconst/d[B]"];
```

```
plot101c = Plot3D[Evaluate[D[ (300 000 b + 120 000 b x) / (19 200 000 + 300 000 b + 1 920 000 x + 60 000 b x) , x]],
    {b, 0.00001, 150}, {x, 0.00001, 100}, ViewPoint → {-2, -2, 1},
    PlotRange -> {0, .2}, Lighting -> {{"Ambient", GrayLevel[1]}},
    AxesLabel → {"[B]", "[X]"}, PlotLabel → "dvEABXaconst/d[X]"];
```

```
plot101d = Plot3D[Evaluate[D[ (300 000 b + 120 000 b x) / (19 200 000 + 300 000 b + 1 920 000 x + 60 000 b x) , b, x]],
    {b, 0.00001, 150}, {x, 0.00001, 100}, ViewPoint → {-2, -2, 1},
    PlotRange -> {-.002, .004}, Lighting -> {{"Ambient", GrayLevel[1]}},
    AxesLabel → {"[B]", "[X]"}, PlotLabel → "d(dvEABXaconst/[B])/d[X]"];
```

This mechanism can be studied with X and without X.

```
fig101 = GraphicsArray[{{plot101a, plot101b}, {plot101c, plot101d}}]
```

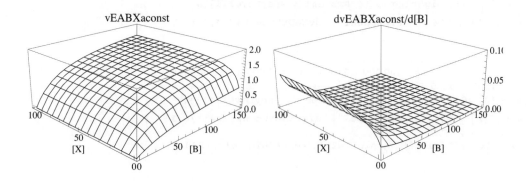

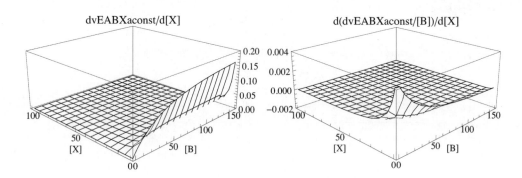

Fig. 11.1 3D plots for vEABX and partial derivatives for the modification mechanism of ordered A + B → products at [A] = 50. These plots are very similar to the plots in Fig. 9.14 for A → products.

Make 3D plots at b = 50.

```
vEABX /. kf1et → 1 /. kf2et → 2 /. kX -> 10 /. kA → 30 /. kAX → 30 /. kABX → 20 /. kAB → 40 /. b → 50
```

$$\frac{300\,000\,a + 120\,000\,a\,x}{7\,200\,000 + 540\,000\,a + 720\,000\,x + 84\,000\,a\,x}$$

```
plot102a = Plot3D[Evaluate[
```
$$\frac{300\,000\,a + 120\,000\,a\,x}{7\,200\,000 + 540\,000\,a + 720\,000\,x + 84\,000\,a\,x}$$
```
],
   {a, 0.00001, 150}, {x, 0.00001, 100}, ViewPoint → {-2, -2, 1},
   PlotRange -> {0, 2}, Lighting -> {{"Ambient", GrayLevel[1]}},
   AxesLabel → {"[A]", "[X]"}, PlotLabel → "vEABXbconst"];
```

```
plot102b = Plot3D[Evaluate[D[
```
$$\frac{300\,000\,a + 120\,000\,a\,x}{7\,200\,000 + 540\,000\,a + 720\,000\,x + 84\,000\,a\,x}
$$
```
, a]],
   {a, 0.00001, 150}, {x, 0.00001, 100}, ViewPoint → {-2, -2, 1},
   PlotRange -> {0, .2}, Lighting -> {{"Ambient", GrayLevel[1]}},
   AxesLabel → {"[A]", "[X]"}, PlotLabel → "dvEABXaconst/d[A]"];
```

```
plot102c = Plot3D[Evaluate[D[ (300 000 a + 120 000 a x) / (7 200 000 + 540 000 a + 720 000 x + 84 000 a x) , x]],
    {a, 0.00001, 150}, {x, 0.00001, 100}, ViewPoint → {-2, -2, 1},
    PlotRange -> {0, .2}, Lighting -> {{"Ambient", GrayLevel[1]}},
    AxesLabel → {"[A]", "[X]"}, PlotLabel → "dvEABXbconst/d[X]"];


plot102d = Plot3D[Evaluate[D[ (300 000 a + 120 000 a x) / (7 200 000 + 540 000 a + 720 000 x + 84 000 a x) , a, x]],
    {a, 0.00001, 150}, {x, 0.00001, 100}, ViewPoint → {-2, -2, 1},
    PlotRange -> {0, .01}, Lighting -> {{"Ambient", GrayLevel[1]}},
    AxesLabel → {"[A]", "[X]"}, PlotLabel → "d(dvEABXbconst/[A])/d[X]"];


fig102 = GraphicsArray[{{plot102a, plot102b}, {plot102c, plot102d}}]
```

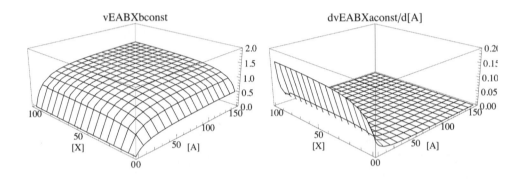

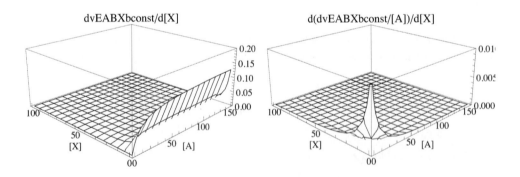

Fig. 11.2 3D plots for vEABX and partial derivatives for the modification mechanism of ordered A + B → products at [B] = 50. These plots are very similar to the plots in Fig. 10.1 for A → products.

Make 3D plots at x=50.

```
vEABX /. kf1et → 1 /. kf2et → 2 /. kX -> 10 /. kA → 30 /. kAX → 30 /. kABX → 20 /. kAB → 40 /. x → 50
```

$$\frac{126\,000\,a\,b}{43\,200\,000 + 1\,440\,000\,a + 66\,000\,a\,b}$$

It is perhaps a surprise to find only one term in the numerator.

```
plot103a =
    Plot3D[Evaluate[126 000 a b / (43 200 000 + 1 440 000 a + 66 000 a b)], {a, 0.00001, 150}, {b, 0.00001, 100},
        ViewPoint → {-2, -2, 1}, PlotRange -> {0, 2}, Lighting -> {{"Ambient", GrayLevel[1]}},
        AxesLabel → {"[A]", "[B]"}, PlotLabel → "vEABXxconst"];

plot103b =
    Plot3D[Evaluate[D[126 000 a b / (43 200 000 + 1 440 000 a + 66 000 a b), a]], {a, 0.00001, 150}, {b, 0.00001, 100},
        ViewPoint → {-2, -2, 1}, PlotRange -> {0, .3}, Lighting -> {{"Ambient", GrayLevel[1]}},
        AxesLabel → {"[A]", "[B]"}, PlotLabel → "dvEABXxconst/d[A]"];

plot103c =
    Plot3D[Evaluate[D[126 000 a b / (43 200 000 + 1 440 000 a + 66 000 a b), b]], {a, 0.00001, 150}, {b, 0.00001, 100},
        ViewPoint → {-2, -2, 1}, PlotRange -> {0, .1}, Lighting -> {{"Ambient", GrayLevel[1]}},
        AxesLabel → {"[A]", "[B]"}, PlotLabel → "dvEABXxconst/d[B]"];

plot103d = Plot3D[Evaluate[D[126 000 a b / (43 200 000 + 1 440 000 a + 66 000 a b), a, b]],
        {a, 0.00001, 150}, {b, 0.00001, 100}, ViewPoint → {-2, -2, 1},
        PlotRange -> {-.001, .005}, Lighting -> {{"Ambient", GrayLevel[1]}},
        AxesLabel → {"[A]", "[B]"}, PlotLabel → "d(dvEABXxconst/[A])/d[B]"];
```

```
fig103 = GraphicsArray[{{plot103a, plot103b}, {plot103c, plot103d}}]
```

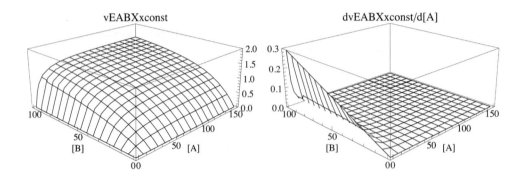

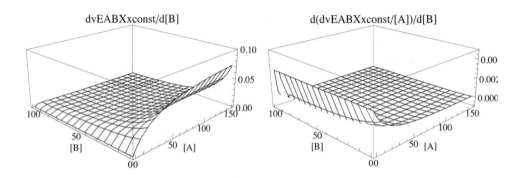

Fig. 11.3 3D plots for vEABX and partial derivatives for the modification mechanism of ordered A + B → products at [X] = 50.

11.3 Random A + B → products

■ 11.3.1 Competitive inhibition

The mechanism for competitive inhibition of ordered A + B → products involves a thermodynamic cycle.

```
EI
‖ kI
I
+       kIA
E +  A = EA
+        +
B        B
‖ kIB    ‖ kB
EB + A = EAB → products
        kA
```

The four reactions in the cycle are not independent because kIA kB = kA kIB. Since the rapid-equilibrium rate equation can only be derived using an independent set of reactions, the reaction EB + A = EAB is omitted, and the value of kA is calculated using kA = kIA kB/kIB.

The following mechanism with independent reactions is used to derive the rapid-equilibrium rate equation for competitive inhibition:

E + A = EA	kIA = e*a/ea	
E + B = EB	kIB = e*b/eb	Reaction added to the mechanism for ordered A + B → products in Section 10.2.1
E + I = EI	kI = e*i/ei	
EA + B = EAB	kB = ea*b/eab	
EAB → products	v = kf*eab	

The rapid-equilibrium rate equation can be derived by use of Solve.

```
Solve[{kIA == e * a / ea, kIB == e * b / eb, kI == e * i / ei,
   kB == ea * b / eab, et == e + ea + eb + ei + eab}, {eab}, {e, ea, eb, ei}]
```

$$\left\{\left\{eab \rightarrow \frac{a\,b\,et\,kI\,kIB}{b\,kB\,kI\,kIA + a\,b\,kI\,kIB + a\,kB\,kI\,kIB + i\,kB\,kIA\,kIB + kB\,kI\,kIA\,kIB}\right\}\right\}$$

$$\mathbf{vcomprandAB} = \frac{\mathbf{a\,b\,kfet\,kI\,kIB}}{\mathbf{b\,kB\,kI\,kIA + a\,b\,kI\,kIB + a\,kB\,kI\,kIB + i\,kB\,kIA\,kIB + kB\,kI\,kIA\,kIB}}$$

$$\frac{a\,b\,kfet\,kI\,kIB}{b\,kB\,kI\,kIA + a\,b\,kI\,kIB + a\,kB\,kI\,kIB + i\,kB\,kIA\,kIB + kB\,kI\,kIA\,kIB}$$

For trial calculations, the kinetic parameters have the following values:

```
vcomprandAB /. kfet → 1 /. kIA → 10 /. kIB → 15 /. kB → 30 /. kI → 20
```

$$\frac{300\,a\,b}{90\,000 + 9000\,a + 6000\,b + 300\,a\,b + 4500\,i}$$

Since there are 5 kinetic parameters, 5 velocities have to be measured. The following triplets of concentrations are chosen: {a,b,i} = {300,300,0}, {1,300,0}, {300,1,0}, {5,5,0}, and {10,10,100}.

```
vcomprandAB /. kfet → 1 /. kIA → 10 /. kIB → 15 /. kB → 30 /. kI → 20 /. a → 300 /. b → 300 /. i → 0 //
  N
```

0.854701

```
vcomprandAB /. kfet → 1 /. kIA → 10 /. kIB → 15 /. kB → 30 /. kI → 20 /. a → 1 /. b → 300 /. i → 0 // N
```

0.0452489

```
vcomprandAB /. kfet → 1 /. kIA → 10 /. kIB → 15 /. kB → 30 /. kI → 20 /. a → 300 /. b → 1 /. i → 0 // N
```

0.031185

```
vcomprandAB /. kfet → 1 /. kIA → 10 /. kIB → 15 /. kB → 30 /. kI → 20 /. a → 5 /. b → 5 /. i → 0 // N
```

0.0434783

```
vcomprandAB /. kfet → 1 /. kIA → 10 /. kIB → 15 /. kB → 30 /. kI → 20 /. a → 10 /. b → 10 /. i → 100 //
  N
```

0.0416667

```
calckinparscomprandAB[v1_, a1_, b1_, i1_, v2_, a2_, b2_, i2_, v3_, a3_, b3_, i3_, v4_,
  a4_, b4_, i4_, v5_, a5_, b5_, i5_] := Module[{}, (*This program calculates kfet,
  kIA,kIB,kB,  and kI from 5 experimental velocities for random A + B →
   products wth a competitive inhibitor I.*)
```

$$\text{Solve}\Big[\Big\{v1 == \frac{a1\ b1\ kfet\ kI\ kIB}{b1\ kB\ kI\ kIA + a1\ b1\ kI\ kIB + a1\ kB\ kI\ kIB + i1\ kB\ kIA\ kIB + kB\ kI\ kIA\ kIB},$$

$$v2 == \frac{a2\ b2\ kfet\ kI\ kIB}{b2\ kB\ kI\ kIA + a2\ b2\ kI\ kIB + a2\ kB\ kI\ kIB + i2\ kB\ kIA\ kIB + kB\ kI\ kIA\ kIB},$$

$$v3 == \frac{a3\ b3\ kfet\ kI\ kIB}{b3\ kB\ kI\ kIA + a3\ b3\ kI\ kIB + a3\ kB\ kI\ kIB + i3\ kB\ kIA\ kIB + kB\ kI\ kIA\ kIB},$$

$$v4 == \frac{a4\ b4\ kfet\ kI\ kIB}{b4\ kB\ kI\ kIA + a4\ b4\ kI\ kIB + a4\ kB\ kI\ kIB + i4\ kB\ kIA\ kIB + kB\ kI\ kIA\ kIB},$$

$$v5 == \frac{a5\ b5\ kfet\ kI\ kIB}{b5\ kB\ kI\ kIA + a5\ b5\ kI\ kIB + a5\ kB\ kI\ kIB + i5\ kB\ kIA\ kIB + kB\ kI\ kIA\ kIB}\Big\},$$

```
 {kfet, kIA, kIB, kB, kI}]]
```

```
calckinparscomprandAB[0.8547, 300, 300, 0, 0.04525, 1,
  300, 0, 0.03119, 300, 1, 0, 0.04348, 5, 5, 0, 0.04167, 10, 10, 100]
```

{{kfet → 0.99998, kI → 20.0022, kIA → 10.0019, kB → 29.9943, kIB → 15.0007}}

```
row10 = {kfet, kI, kIA, kB, kIB} /. calckinparscomprandAB[0.8547`, 300, 300, 0,
   0.04525`, 1, 300, 0, 0.03119`, 300, 1, 0, 0.04348`, 5, 5, 0, 0.04167`, 10, 10, 100]
```

{{0.99998, 20.0022, 10.0019, 29.9943, 15.0007}}

These values are correct, but the effects of errors in experimental velocities have to be taken into account.

```
calckinparscomprandAB[1.05 * 0.8547`, 300, 300, 0, 0.04525`, 1,
  300, 0, 0.03119`, 300, 1, 0, 0.04348`, 5, 5, 0, 0.04167`, 10, 10, 100]
```

{{kfet → 1.05939, kI → 20.0113, kIA → 10.0133, kB → 31.8323, kIB → 15.0045}}

```
row11 = {kfet, kI, kIA, kB, kIB} /. calckinparscomprandAB[1.05 * 0.8547`, 300, 300, 0,
   0.04525`, 1, 300, 0, 0.03119`, 300, 1, 0, 0.04348`, 5, 5, 0, 0.04167`, 10, 10, 100]
```

{{1.05939, 20.0113, 10.0133, 31.8323, 15.0045}}

```
calckinparscomprandAB[0.8547`, 300, 300, 0, 1.05 * 0.04525`, 1,
  300, 0, 0.03119`, 300, 1, 0, 0.04348`, 5, 5, 0, 0.04167`, 10, 10, 100]
```

{{kfet → 0.996414, kI → 20.29, kIA → 10.187, kB → 29.8694, kIB → 16.1354}}

```
row12 = {kfet, kI, kIA, kB, kIB} /. calckinparscomprandAB[0.8547`, 300, 300, 0,
  1.05 * 0.04525`, 1, 300, 0, 0.03119`, 300, 1, 0, 0.04348`, 5, 5, 0, 0.04167`, 10, 10, 100]
```

{{0.996414, 20.29, 10.187, 29.8694, 16.1354}}

```
calckinparscomprandAB[0.8547`, 300, 300, 0, 0.04525`, 1, 300,
  0, 1.05 * 0.03119`, 300, 1, 0, 0.04348`, 5, 5, 0, 0.04167`, 10, 10, 100]
```

{{kfet → 0.994814, kI → 20.419, kIA → 10.8237, kB → 28.2896, kIB → 15.4102}}

```
row13 = {kfet, kI, kIA, kB, kIB} /. calckinparscomprandAB[0.8547`, 300, 300, 0, 0.04525`,
  1, 300, 0, 1.05 * 0.03119`, 300, 1, 0, 0.04348`, 5, 5, 0, 0.04167`, 10, 10, 100]
```

{{0.994814, 20.419, 10.8237, 28.2896, 15.4102}}

```
calckinparscomprandAB[0.8547`, 300, 300, 0, 0.04525`, 1, 300,
  0, 0.03119`, 300, 1, 0, 1.05 * 0.04348`, 5, 5, 0, 0.04167`, 10, 10, 100]
```

{{kfet → 1.00029, kI → 17.8003, kIA → 9.02945, kB → 30.0981, kIB → 13.521}}

```
row14 = {kfet, kI, kIA, kB, kIB} /. calckinparscomprandAB[0.8547`, 300, 300, 0, 0.04525`,
  1, 300, 0, 0.03119`, 300, 1, 0, 1.05 * 0.04348`, 5, 5, 0, 0.04167`, 10, 10, 100]
```

{{1.00029, 17.8003, 9.02945, 30.0981, 13.521}}

```
calckinparscomprandAB[0.8547`, 300, 300, 0, 0.04525`, 1, 300,
  0, 0.03119`, 300, 1, 0, 0.04348`, 5, 5, 0, 1.05 * 0.04167`, 10, 10, 100]
```

{{kfet → 0.99998, kI → 21.6519, kIA → 10.0019, kB → 29.9943, kIB → 15.0007}}

```
row15 = {kfet, kI, kIA, kB, kIB} /. calckinparscomprandAB[0.8547`, 300, 300, 0, 0.04525`,
  1, 300, 0, 0.03119`, 300, 1, 0, 0.04348`, 5, 5, 0, 1.05 * 0.04167`, 10, 10, 100]
```

{{0.99998, 21.6519, 10.0019, 29.9943, 15.0007}}

Table 11.7 Estimation of kinetic parameters for competitive inhibition of random A + B → products using velocities at {300,300,0}, {1,300,0}, {300,1,0},{5,5,0},{10,10,100}

```
TableForm[
  Round[{row10[[1]], row11[[1]], row12[[1]], row13[[1]], row14[[1]], row15[[1]]}, 0.01],
  TableHeadings -> {{"No errors", "1.05*v1", "1.05*v2", "1.05*v3", "1.05*v4", "1.05*v5"},
    {"kfet", "kI", "kIA", "kB", "kIB"}}]
```

	kfet	kI	kIA	kB	kIB
No errors	1.	20.	10.	29.99	15.
1.05*v1	1.06	20.01	10.01	31.83	15.
1.05*v2	1.	20.29	10.19	29.87	16.14
1.05*v3	0.99	20.42	10.82	28.29	15.41
1.05*v4	1.	17.8	9.03	30.1	13.52
1.05*v5	1.	21.65	10.	29.99	15.

kA can be calculated using kA = kIA kB/kIB.

$$\begin{array}{ccc} \underline{10.^{\grave{}}} & \underline{29.990000000000002^{\grave{}}} & \underline{15.^{\grave{}}} \\ 10.01^{\grave{}} & 31.830000000000002^{\grave{}} & 15.^{\grave{}} \\ 10.19^{\grave{}} & 29.87^{\grave{}} & 16.14^{\grave{}} \\ 10.82^{\grave{}} & 28.29^{\grave{}} & 15.41^{\grave{}} \\ 9.03^{\grave{}} & 30.1^{\grave{}} & 13.52^{\grave{}} \\ 10.^{\grave{}} & 29.990000000000002^{\grave{}} & 15.^{\grave{}} \end{array} \; * \Big/ $$

{{19.9933}, {21.2412}, {18.8584}, {19.8636}, {20.1038}, {19.9933}}

Thse values can be added to Table 11.7.

row10[[1]]

{0.99998, 20.0022, 10.0019, 29.9943, 15.0007}

row10a = {0.9999798331858797`, 20.00223092570075`,
  10.001884115338173`, 29.99425144129188`, 15.000683202874837`, 20.0}

{0.99998, 20.0022, 10.0019, 29.9943, 15.0007, 20.}

row11[[1]]

{1.05939, 20.0113, 10.0133, 31.8323, 15.0045}

row11a = {1.0593861767906059`, 20.011325429816235`,
  10.013264382261317`, 31.832277331140308`, 15.00452543339831`, 21.24}

{1.05939, 20.0113, 10.0133, 31.8323, 15.0045, 21.24}

row12[[1]]

{0.996414, 20.29, 10.187, 29.8694, 16.1354}

row12a = {0.9964135073490427`, 20.28997977045844`,
  10.186954496321954`, 29.869448021348656`, 16.135446522348026`, 18.86}

{0.996414, 20.29, 10.187, 29.8694, 16.1354, 18.86}

row13[[1]]

{0.994814, 20.419, 10.8237, 28.2896, 15.4102}

row13a = {0.9948141640894599`, 20.419044861724878`,
  10.823702711720868`, 28.289575956751747`, 15.41015263724238`, 19.86}

{0.994814, 20.419, 10.8237, 28.2896, 15.4102, 19.86}

row14[[1]]

{1.00029, 17.8003, 9.02945, 30.0981, 13.521}

row14a = {1.0002945401014265`, 17.80030647030667`,
  9.029447103996693`, 30.098105008706703`, 13.521038224249715`, 20.10}

{1.00029, 17.8003, 9.02945, 30.0981, 13.521, 20.1}

row15[[1]]

{0.99998, 21.6519, 10.0019, 29.9943, 15.0007}

row15a = {0.9999798331858797`, 21.651925562166287`,
  10.001884115338173`, 29.99425144129188`, 15.000683202874837`, 20.00}

{0.99998, 21.6519, 10.0019, 29.9943, 15.0007, 20.}

Table 11.8 Estimation of kinetic parameters for competitive inhibition of random A + B → products using velocities at {300,300,0}, {1,300,0}, {300,1},{5,5,0},{10,10,100} and values of kA calculated using a thermodynamic cycle.

```
TableForm[Round[{row10a, row11a, row12a, row13a, row14a, row15a}, 0.01],
   TableHeadings -> {{"No errors", "1.05*v1", "1.05*v2", "1.05*v3", "1.05*v4", "1.05*v5"},
      {"kfet", "kI", "kIA", "kB", "kIB", "kA"}}]
```

	kfet	kI	kIA	kB	kIB	kA
No errors	1.	20.	10.	29.99	15.	20.
1.05*v1	1.06	20.01	10.01	31.83	15.	21.24
1.05*v2	1.	20.29	10.19	29.87	16.14	18.86
1.05*v3	0.99	20.42	10.82	28.29	15.41	19.86
1.05*v4	1.	17.8	9.03	30.1	13.52	20.1
1.05*v5	1.	21.65	10.	29.99	15.	20.

▪ 11.3.2 Uncompetitive inhibition

The following mechanism is used, but there are other mechanisms of uncompetitive inhibition when EA and EB also bind the uncompetitive inhibitor.

E + A = EA	kA = e*a/ea
E + B = EB	kB = e*b/eb
EA + B = EAB	kAB = ea*b/eab
EAB + I = EABI	kABI = eab*i/eabi
EAB → products	v = kf*eab

There are five kinetic parameters, but a sixth can be calculated using the thermodynamic cycle to form EB + A = EAB. The rapid-equilibrium rate equation can be derived using Solve.

```
Solve[{kA == e * a / ea, kB == e * b / eb, kAB == ea * b / eab,
   kABI == eab * i / eabi, et == e + ea + eb + eab + eabi}, {eab}, {e, ea, eb, eabi}]
```

$$\left\{\left\{eab \to \frac{a\ b\ et\ kABI\ kB}{b\ kA\ kAB\ kABI + a\ b\ i\ kB + a\ b\ kABI\ kB + a\ kAB\ kABI\ kB + kA\ kAB\ kABI\ kB}\right\}\right\}$$

$$\textbf{vranduncompAB} = \frac{a\ b\ kfet\ kABI\ kB}{b\ kA\ kAB\ kABI + a\ b\ i\ kB + a\ b\ kABI\ kB + a\ kAB\ kABI\ kB + kA\ kAB\ kABI\ kB}$$

$$\frac{a\ b\ kABI\ kB\ kfet}{b\ kA\ kAB\ kABI + a\ b\ i\ kB + a\ b\ kABI\ kB + a\ kAB\ kABI\ kB + kA\ kAB\ kABI\ kB}$$

For test calculations, the following values of the kinetic parameters are used.

vranduncompAB /. kA → 20 /. kB → 30 /. kAB → 40 /. kABI → 50 /. kfet → 1

$$\frac{1500\ a\ b}{1\,200\,000 + 60\,000\ a + 40\,000\ b + 1500\ a\ b + 30\ a\ b\ i}$$

The following triplets of substrate concentrations are used for test calculations: {a,b,i} = {300,300,0}, {1,300,0}, {300,1,0}, {5,5,5}, and {200,200,200}. The first four triplets predominate in determining kA, kB, and kfet. The fifth predominates in determining kABI.

vranduncompAB /. kA → 20 /. kB → 30 /. kAB → 40 /. kABI → 50 /. kfet → 1 /. a → 300 /. b → 300 /. i → 0 // N

0.812274

```
vranduncompAB /. kA → 20 /. kB → 30 /. kAB → 40 /. kABI → 50 /. kfet → 1 /. a → 1 /. b → 300 /. i → 0 //
  N
```

0.0328228

```
vranduncompAB /. kA → 20 /. kB → 30 /. kAB → 40 /. kABI → 50 /. kfet → 1 /. a → 300 /. b → 1 /. i → 0 //
  N
```

0.0228542

```
vranduncompAB /. kA → 20 /. kB → 30 /. kAB → 40 /. kABI → 50 /. kfet → 1 /. a → 5 /. b → 5 /. i → 5 // N
```

0.0215363

```
vranduncompAB /. kA → 20 /. kB → 30 /. kAB → 40 /. kABI → 50 /. kfet → 1 /. a → 200 /. b → 200 /.
  i → 200 // N
```

0.1868

```
calckinparsuncomprandAB[v1_, a1_, b1_, i1_, v2_, a2_, b2_, i2_, v3_, a3_, b3_, i3_, v4_,
  a4_, b4_, i4_, v5_, a5_, b5_, i5_] := Module[{}, (*This program calculates kfet,
  kA,kB,kAB,and kABI from 5 experimental velocities for random A + B →
  products wth an uncompetitive inhibitor I.*)
```

$$\text{Solve}\Big[\Big\{v1 == \frac{a1\ b1\ kABI\ kB\ kfet}{b1\ kA\ kAB\ kABI + a1\ b1\ i1\ kB + a1\ b1\ kABI\ kB + a1\ kAB\ kABI\ kB + kA\ kAB\ kABI\ kB},$$

$$v2 == \frac{a2\ b2\ kABI\ kB\ kfet}{b2\ kA\ kAB\ kABI + a2\ b2\ i2\ kB + a2\ b2\ kABI\ kB + a2\ kAB\ kABI\ kB + kA\ kAB\ kABI\ kB},$$

$$v3 == \frac{a3\ b3\ kABI\ kB\ kfet}{b3\ kA\ kAB\ kABI + a3\ b3\ i3\ kB + a3\ b3\ kABI\ kB + a3\ kAB\ kABI\ kB + kA\ kAB\ kABI\ kB},$$

$$v4 == \frac{a4\ b4\ kABI\ kB\ kfet}{b4\ kA\ kAB\ kABI + a4\ b4\ i4\ kB + a4\ b4\ kABI\ kB + a4\ kAB\ kABI\ kB + kA\ kAB\ kABI\ kB},$$

$$v5 == \frac{a5\ b5\ kABI\ kB\ kfet}{b5\ kA\ kAB\ kABI + a5\ b5\ i5\ kB + a5\ b5\ kABI\ kB + a5\ kAB\ kABI\ kB + kA\ kAB\ kABI\ kB}\Big\},$$

$$\{kfet, kA, kB, kAB, kABI\}\Big]\Big]$$

```
calckinparsuncomprandAB[.8123, 300, 300, 0, .03282, 1,
  300, 0, .02285, 300, 1, 0, .02154, 5, 5, 5, .1868, 200, 200, 200]
```

{{kfet → 1.00008, kABI → 49.996, kA → 19.9889, kAB → 40.0122, kB → 29.9862}}

```
line1 = {kfet, kABI, kA, kAB, kB} /. calckinparsuncomprandAB[.8123, 300, 300, 0,
  .03282, 1, 300, 0, .02285, 300, 1, 0, .02154, 5, 5, 5, .1868, 200, 200, 200][[1]]
```

{1.00008, 49.996, 19.9889, 40.0122, 29.9862}

{{1.0000777514535828`, 49.996021336736725`,
  19.988902826611753`, 40.01217328263129`, 29.98623107161352`}}

{{1.00008, 49.996, 19.9889, 40.0122, 29.9862}}

It is necessary to test the sensitivity of the kinetic parameters to experimental errors in the velocity measurements.

```
calckinparsuncomprandAB[1.05 * .8123, 300, 300, 0, .03282,
  1, 300, 0, .02285, 300, 1, 0, .02154, 5, 5, 5, .1868, 200, 200, 200]
```

{{kfet → 1.06279, kABI → 46.3685, kA → 19.983, kAB → 42.5807, kB → 29.9564}}

```
line2 = {kfet, kABI, kA, kAB, kB} /. calckinparsuncomprandAB[1.05 * .8123, 300, 300, 0,
    .03282, 1, 300, 0, .02285, 300, 1, 0, .02154, 5, 5, 5, .1868, 200, 200, 200][[1]]
```

{1.06279, 46.3685, 19.983, 42.5807, 29.9564}

```
calckinparsuncomprandAB[.8123, 300, 300, 0, 1.05 * .03282,
  1, 300, 0, .02285, 300, 1, 0, .02154, 5, 5, 5, .1868, 200, 200, 200]
```

{{kfet → 0.995166, kABI → 50.2126, kA → 20.1863, kAB → 39.7911, kB → 32.0424}}

```
line3 = {kfet, kABI, kA, kAB, kB} /. calckinparsuncomprandAB[.8123, 300, 300, 0,
    1.05 * .03282, 1, 300, 0, .02285, 300, 1, 0, .02154, 5, 5, 5, .1868, 200, 200, 200][[1]]
```

{0.995166, 50.2126, 20.1863, 39.7911, 32.0424}

```
calckinparsuncomprandAB[.8123, 300, 300, 0, .03282, 1, 300,
  0, 1.05 * .02285, 300, 1, 0, .02154, 5, 5, 5, .1868, 200, 200, 200]
```

{{kfet → 0.993039, kABI → 50.307, kA → 21.3915, kAB → 37.619, kB → 30.4252}}

```
line4 = {kfet, kABI, kA, kAB, kB} /. calckinparsuncomprandAB[.8123, 300, 300, 0, .03282,
    1, 300, 0, 1.05 * .02285, 300, 1, 0, .02154, 5, 5, 5, .1868, 200, 200, 200][[1]]
```

{0.993039, 50.307, 21.3915, 37.619, 30.4252}

```
calckinparsuncomprandAB[.8123, 300, 300, 0, .03282, 1, 300,
  0, .02285, 300, 1, 0, 1.05 * .02154, 5, 5, 5, .1868, 200, 200, 200]
```

{{kfet → 1.00071, kABI → 49.9623, kA → 18.4723, kAB → 40.2283, kB → 27.6456}}

```
line5 = {kfet, kABI, kA, kAB, kB} /. calckinparsuncomprandAB[.8123, 300, 300, 0, .03282,
    1, 300, 0, .02285, 300, 1, 0, 1.05 * .02154, 5, 5, 5, .1868, 200, 200, 200][[1]]
```

{1.00071, 49.9623, 18.4723, 40.2283, 27.6456}

```
calckinparsuncomprandAB[.8123, 300, 300, 0, .03282, 1, 300,
  0, .02285, 300, 1, 0, .02154, 5, 5, 5, 1.05 * .1868, 200, 200, 200]
```

{{kfet → 1.00008, kABI → 53.3992, kA → 19.9933, kAB → 40.0116, kB → 29.993}}

```
line6 = {kfet, kABI, kA, kAB, kB} /. calckinparsuncomprandAB[.8123, 300, 300, 0, .03282,
    1, 300, 0, .02285, 300, 1, 0, .02154, 5, 5, 5, 1.05 * .1868, 200, 200, 200][[1]]
```

{1.00008, 53.3992, 19.9933, 40.0116, 29.993}

Table 11.9 Estimation of kinetic parameters for uncompetitive inhibition of random A + B → products using velocities at {300,300,0}, {1,300,0}, {300,1,0}, {5,5,5}, and {200,200,200}

```
TableForm[Round[{line1, line2, line3, line4, line5, line6}, 0.01],
  TableHeadings -> {{"No errors", "1.05*v1", "1.05*v2", "1.05*v3", "1.05*v4", "1.05*v5"},
    {"kfet", "kABI", "kA", "kAB", "kB"}}]
```

	kfet	kABI	kA	kAB	kB
No errors	1.	50.	19.99	40.01	29.99
1.05*v1	1.06	46.37	19.98	42.58	29.96
1.05*v2	1.	50.21	20.19	39.79	32.04
1.05*v3	0.99	50.31	21.39	37.62	30.43
1.05*v4	1.	49.96	18.47	40.23	27.65
1.05*v5	1.	53.4	19.99	40.01	29.99

The equilibrium constant for EB + A = EAB can be calculated using a thermodynamic cycle, but that will not be done here.

■ 11.3.3 Mixed inhibition

The following mechanism is used, but there are other cases of mixed inhibition when EA and EB also bind the uncompetitive inhibitor.

E + A = EA	kA = e*a/ea	
E + B = EB	kB = e*b/eb	
E + I = EI	kI = e*i/ei	This is the new reaction.
EA + B = EAB	kAB = ea*b/eab	
EAB + I = EABI	kABI = eab*i/eabi	
EAB → products	v = kf*eab	

There are six kinetic parameters, but a seventh can be calculated using the thermodynamic cycle to estimate the equilibrium constant for EB + A = EAB.

The rapid-equilibrium rate equation can be derived using Solve.

```
Solve[{kA == e * a / ea, kB == e * b / eb, kI == e * i / ei, kAB == ea * b / eab,
   kABI == eab * i / eabi, et == e + ea + eb + ei + eab + eabi}, {eab}, {e, ea, eb, ei, eabi}]
```

```
{{eab → (a b et kABI kB kI) / (i kA kAB kABI kB +
       b kA kAB kABI kI + a b i kB kI + a b kABI kB kI + a kAB kABI kB kI + kA kAB kABI kB kI)}}
```

```
vrandmixAB = (a b kfet kABI kB kI) /
   (i kA kAB kABI kB + b kA kAB kABI kI + a b i kB kI + a b kABI kB kI + a kAB kABI kB kI + kA kAB kABI kB kI)
```

```
(a b kABI kB kfet kI) /
  (i kA kAB kABI kB + b kA kAB kABI kI + a b i kB kI + a b kABI kB kI + a kAB kABI kB kI + kA kAB kABI kB kI)
```

For test calculations, the following values of the kinetic parameters are used.

```
vrandmixAB /. kA → 20 /. kB → 30 /. kAB → 40 /. kABI → 50 /. kfet → 1 /. kI → 25
```

$$\frac{37\,500\;a\;b}{30\,000\,000 + 1\,500\,000\;a + 1\,000\,000\;b + 37\,500\;a\;b + 1\,200\,000\;i + 750\;a\;b\;i}$$

The following triplets of substrate concentrations are used for test calculations: {a,b,i} = {300,300,0}, {1,300,0}, {300,1,0}, {5,5,5}, {200,200,200}, and {10,10,100}. The first four triplets predominate in determining kA, kB, kAB, and kfet. The fifth and sixth predominate in determining kABI and kI .

```
vrandmixAB /. kA → 20 /. kB → 30 /. kAB → 40 /. kABI → 50 /. kfet → 1 /. kI → 25 /. a → 300 /.
   b → 300 /. i → 0 // N
```

```
0.812274
```

```
vrandmixAB /. kA → 20 /. kB → 30 /. kAB → 40 /. kABI → 50 /. kfet → 1 /. kI → 25 /. a → 1 /. b → 300 /.
   i → 0 // N
```

```
0.0328228
```

```
vrandmixAB /. kA → 20 /. kB → 30 /. kAB → 40 /. kABI → 50 /. kfet → 1 /. kI → 25 /. a → 300 /. b → 1 /.
   i → 0 // N
```

```
0.0228542
```

```
vrandmixAB /. kA → 20 /. kB → 30 /. kAB → 40 /. kABI → 50 /. kfet → 1 /. kI → 25 /. a → 5 /. b → 5 /.
   i → 5 // N
```

```
0.0189274
```

```
vrandmixAB /. kA → 20 /. kB → 30 /. kAB → 40 /. kABI → 50 /. kfet → 1 /. kI → 25 /. a → 200 /.
   b → 200 /. i → 200 // N
```

0.181378

```
vrandmixAB /. kA → 20 /. kB → 30 /. kAB → 40 /. kABI → 50 /. kfet → 1 /. kI → 25 /. a → 10 /. b → 10 /.
   i → 100 // N
```

0.0201342

```
calckinparsmixrandAB[v1_, a1_, b1_, i1_, v2_, a2_, b2_, i2_, v3_,
   a3_, b3_, i3_, v4_, a4_, b4_, i4_, v5_, a5_, b5_, i5_, v6_, a6_, b6_, i6_] :=
 Module[{}, (*This program calculates kfet, kA, kB, kI, kAB,
   and kABI from 6 experimental velocities for random A + B →
    products with mixed inhibition.*)
  Solve[{v1 == (a1 b1 kABI kB kfet kI) / (i1 kA kAB kABI kB + b1 kA kAB kABI kI +
        a1 b1 i1 kB kI + a1 b1 kABI kB kI + a1 kAB kABI kB kI + kA kAB kABI kB kI),
     v2 == (a2 b2 kABI kB kfet kI) / (i2 kA kAB kABI kB + b2 kA kAB kABI kI + a2 b2 i2 kB kI +
        a2 b2 kABI kB kI + a2 kAB kABI kB kI + kA kAB kABI kB kI),
     v3 == (a3 b3 kABI kB kfet kI) / (i3 kA kAB kABI kB + b3 kA kAB kABI kI + a3 b3 i3 kB kI +
        a3 b3 kABI kB kI + a3 kAB kABI kB kI + kA kAB kABI kB kI),
     v4 == (a4 b4 kABI kB kfet kI) / (i4 kA kAB kABI kB + b4 kA kAB kABI kI + a4 b4 i4 kB kI +
        a4 b4 kABI kB kI + a4 kAB kABI kB kI + kA kAB kABI kB kI),
     v5 == (a5 b5 kABI kB kfet kI) / (i5 kA kAB kABI kB + b5 kA kAB kABI kI + a5 b5 i5 kB kI +
        a5 b5 kABI kB kI + a5 kAB kABI kB kI + kA kAB kABI kB kI),
     v6 == (a6 b6 kABI kB kfet kI) / (i6 kA kAB kABI kB + b6 kA kAB kABI kI + a6 b6 i6 kB kI +
        a6 b6 kABI kB kI + a6 kAB kABI kB kI + kA kAB kABI kB kI)}, {kfet, kA, kB, kAB, kABI, kI}]]
```

```
calckinparsmixrandAB[.8123, 300, 300, 0, .03282, 1, 300, 0, .02285,
  300, 1, 0, .01893, 5, 5, 5, .1814, 200, 200, 200, .02013, 10, 10, 100]
```

{{kfet → 1.00008, kA → 19.9878, kI → 24.9807, kABI → 50.0048, kAB → 40.0123, kB → 29.9846}}

```
ln1 = {kfet, kA, kI, kABI, kAB, kB} /.
   calckinparsmixrandAB[.8123, 300, 300, 0, .03282, 1, 300, 0, .02285,
     300, 1, 0, .01893, 5, 5, 5, .1814, 200, 200, 200, .02013, 10, 10, 100][[1]]
```

{1.00008, 19.9878, 24.9807, 50.0048, 40.0123, 29.9846}

These values are correct, but it is necessary to test their sensitivity to errors in velocity measurements.

```
calckinparsmixrandAB[1.05 * .8123, 300, 300, 0, .03282, 1, 300, 0,
  .02285, 300, 1, 0, .01893, 5, 5, 5, .1814, 200, 200, 200, .02013, 10, 10, 100]
```

{{kfet → 1.06279, kA → 19.9805, kI → 24.9984, kABI → 46.3771, kAB → 42.5811, kB → 29.9526}}

```
ln2 = {kfet, kA, kI, kABI, kAB, kB} /.
   calckinparsmixrandAB[1.05 * .8123, 300, 300, 0, .03282, 1, 300, 0, .02285,
     300, 1, 0, .01893, 5, 5, 5, .1814, 200, 200, 200, .02013, 10, 10, 100][[1]]
```

{1.06279, 19.9805, 24.9984, 46.3771, 42.5811, 29.9526}

```
calckinparsmixrandAB[.8123, 300, 300, 0, 1.05 * .03282, 1, 300, 0,
  .02285, 300, 1, 0, .01893, 5, 5, 5, .1814, 200, 200, 200, .02013, 10, 10, 100]
```

{{kfet → 0.995171, kA → 20.175, kI → 25.1417, kABI → 50.2258, kAB → 39.7927, kB → 32.0238}}

```
ln3 = {kfet, kA, kI, kABI, kAB, kB} /.
   calckinparsmixrandAB[.8123, 300, 300, 0, 1.05 * .03282, 1, 300, 0, .02285,
     300, 1, 0, .01893, 5, 5, 5, .1814, 200, 200, 200, .02013, 10, 10, 100][[1]]
```

{0.995171, 20.175, 25.1417, 50.2258, 39.7927, 32.0238}

```
calckinparsmixrandAB[.8123, 300, 300, 0, .03282, 1, 300, 0, 1.05 * .02285,
  300, 1, 0, .01893, 5, 5, 5, .1814, 200, 200, 200, .02013, 10, 10, 100]
```

{{kfet → 0.993045, kA → 21.3748, kI → 25.2117, kABI → 50.3221, kAB → 37.6212, kB → 30.4008}}

```
ln4 = {kfet, kA, kI, kABI, kAB, kB} /.
  calckinparsmixrandAB[.8123, 300, 300, 0, .03282, 1, 300, 0, 1.05 * .02285,
    300, 1, 0, .01893, 5, 5, 5, .1814, 200, 200, 200, .02013, 10, 10, 100][[1]]
```

{0.993045, 21.3748, 25.2117, 50.3221, 37.6212, 30.4008}

```
calckinparsmixrandAB[.8123, 300, 300, 0, .03282, 1, 300, 0, .02285,
  300, 1, 0, 1.05 * .01893, 5, 5, 5, .1814, 200, 200, 200, .02013, 10, 10, 100]
```

{{kfet → 1.00084, kA → 18.1761, kI → 22.3978, kABI → 50.0045, kAB → 40.2707, kB → 27.1897}}

```
ln5 = {kfet, kA, kI, kABI, kAB, kB} /.
  calckinparsmixrandAB[.8123, 300, 300, 0, .03282, 1, 300, 0, .02285, 300,
    1, 0, 1.05 * .01893, 5, 5, 5, .1814, 200, 200, 200, .02013, 10, 10, 100][[1]]
```

{1.00084, 18.1761, 22.3978, 50.0045, 40.2707, 27.1897}

```
calckinparsmixrandAB[.8123, 300, 300, 0, .03282, 1, 300, 0, .02285,
  300, 1, 0, .01893, 5, 5, 5, 1.05 * .1814, 200, 200, 200, .02013, 10, 10, 100]
```

{{kfet → 1.00008, kA → 19.9735, kI → 24.8579, kABI → 53.527, kAB → 40.0144, kB → 29.9625}}

```
ln6 = {kfet, kA, kI, kABI, kAB, kB} /.
  calckinparsmixrandAB[.8123, 300, 300, 0, .03282, 1, 300, 0, .02285, 300,
    1, 0, .01893, 5, 5, 5, 1.05 * .1814, 200, 200, 200, .02013, 10, 10, 100][[1]]
```

{1.00008, 19.9735, 24.8579, 53.527, 40.0144, 29.9625}

```
calckinparsmixrandAB[.8123, 300, 300, 0, .03282, 1, 300, 0, .02285,
  300, 1, 0, .01893, 5, 5, 5, .1814, 200, 200, 200, 1.05 * .02013, 10, 10, 100]
```

{{kfet → 0.999935, kA → 20.3315, kI → 27.5258, kABI → 49.857, kAB → 39.9637, kB → 30.5166}}

```
ln7 = {kfet, kA, kI, kABI, kAB, kB} /.
  calckinparsmixrandAB[.8123, 300, 300, 0, .03282, 1, 300, 0, .02285, 300,
    1, 0, .01893, 5, 5, 5, .1814, 200, 200, 200, 1.05 * .02013, 10, 10, 100][[1]]
```

{0.999935, 20.3315, 27.5258, 49.857, 39.9637, 30.5166}

Table 11.10 Estimation of kinetic parameters for mixed inhibition of random A + B → products using velocities at {300,300,0}, {1,300,0}, {300,1,0}, {5,5,5}, {200,200,200}, and {10,10,100}

```
TableForm[Round[{ln1, ln2, ln3, ln4, ln5, ln6, ln7}, 0.01],
  TableHeadings -> {{"No errors", "1.05*v1", "1.05*v2", "1.05*v3",
    "1.05*v4", "1.05*v5", "1.05*v6"}, {"kfet", "kA", "kI", "kABI", "kAB", "kB"}}]
```

	kfet	kA	kI	kABI	kAB	kB
No errors	1.	19.99	24.98	50.	40.01	29.98
1.05*v1	1.06	19.98	25.	46.38	42.58	29.95
1.05*v2	1.	20.18	25.14	50.23	39.79	32.02
1.05*v3	0.99	21.37	25.21	50.32	37.62	30.4
1.05*v4	1.	18.18	22.4	50.	40.27	27.19
1.05*v5	1.	19.97	24.86	53.53	40.01	29.96
1.05*v6	1.	20.33	27.53	49.86	39.96	30.52

A seventh kinetic parameter for EB + A = EAB can be calculated using a thermodynamic cycle. Essential activation, mixed activation, and modification of random A + B → products can also be treated, but that is not done here.

11.4 Discussion

It is useful to make calculations of velocities for various mechanisms and various choices of kinetic parameters. This is the way to learn about the characteristics of these mechanisms. The following rate equations are provided:

vcompordAB	Section 11.2.1
vuncompAB	Section 11.2.2
vmixinhibAB	Section 11.2.3
vessactAB	Section 11.2.4
vmisactAB	Section 11.2.5
vEABX	Section 11.2.6
vcomprandAB	Section 11.3.1
vranduncompAB	Section 11.3.2
vrandmixAB	Section 11.3.3

This chapter provides eight *Mathematica* programs for calculating kinetic parameters using the minimum number of velocity measurements:

calckinparscompAB	Section 11.2.1
calckinparsuncompAB	Section 11.2.2
calckinparsmixinhibAB	Section 11.2.3
calckinparsvessactAB	Section 11.2.4
calckinparsmixactAB	Section 11.2.5
calckinparscomprandAB	Section 11.3.1
calckinparsuncomprandAB	Section 11.3.2
calckinparsmixrandAB	Section 11.3.3

The more general programs can be applied to data from less general mechanisms, as discussed in Section 11.2.3, but there are other cases that are not discussed. When mechanisms involve three sides of a thermodynamic cycle, an additional kinetic parameter can be calculated.

Chapter 12 Systems of Enzyme-Catalyzed Reactions

12.1 Simple Examples

- 12.1.1 Two consecutive first order reactions

- 12.1.2 Two consecutive reversible first order reactions

- 12.1.3 Chemical reaction $A + B = C + D$ in dilute aqueous solution

- 12.1.4 A chemical reaction with an intermediate

- 12.1.5 Two consecutive chemical reactions

12.2 Enzyme - catalyzed reactions

- 12.2.1 $A = P$

- 12.2.2 Ordered $A + B$ = ordered $P + Q$

12.3 Glyoxylate cycle

- 12.3.1 Introduction

- 12.3.2. Apparent equilibrium constants and rate constants

- 12.3.3 Calculation of equilibrium compositions

- 12.3.4 Calculation of reactant concentrations as functions of time with oxidizing conditions

- 12.3.5 Calculation of reactant concentrations as functions of time with more reducing conditions

- 12.3.6 Concentrations at longer times with oxidizing conditions

12.4 Discussion

12.1 Simple Examples

In dealing with the kinetics of enzyme-catalyzed reactions, Mathematica is very useful becuse it provides NDSolve that can be used to plot concentrations of reactants versus time for a given set of rate constants. The best way to see how this is done is to consider some simple examples.

■ 12.1.1 Two consecutive first order reactions

Two consecutive first order reactions are represented by

$$\begin{array}{cc} k_1 & k_2 \\ A_1 \rightarrow A_2 \rightarrow A_3 \end{array}$$

The time courses of the three concentrations are given by [22]

$$[A_1] = [A_1]_0\, e^{-k_1 t}$$

$$[A_2] = k_1 [A_1]_0 \left(e^{-k_1 t} - e^{-k_2 t}\right)/(k_2 - k_1)$$

$$[A_3] = [A_1]_0 [1 + \frac{1}{k_1 - k_2} (k_2 e^{-k_1 t} - k_1 e^{-k_2 t})]$$

It is not necessary to use these three equations because NDSolve can be used to obtain interpolating functions for these three concentrations. The input to NDSolve is as follows:

NDSolve [*eqns*, {y_1, y_2, ...}, {x, x_{min}, x_{max}}] finds numerical solutions for the functions y_i.

Assume that initially A_1 is at unit concentration and treat three cases: (*a*) $k_1 = 1$, $k_2 = 1$; (*b*) $k_1 = 1$, $k_2 = 5$; (*c*) $k_1 = 1$, $k_2 = 25$.

SOLUTION
(a) $k_1 = 1\,\text{s}^{-1}$, $k_2 = 1\,\text{s}^{-1}$

In NDSolve, a derivative is indicated by an apostrophe, and so the rate equations and initial conditions are given by

```
eqns1 = {c1'[t] == -c1[t], c2'[t] == c1[t] - c2[t],
         c3'[t] == c2[t], c1[0] == 1, c2[0] == 0, c3[0] == 0};
```

The list of variables to be calculated is given by

```
vars1 = {c1[t], c2[t], c3[t]};
```

NDSolve calculates interpolating functions that give the concentrations of the three reactants as functions of time.

```
concs1 = NDSolve[eqns1, vars1, {t, 0, 3}]

{{c1[t] → InterpolatingFunction[{{0., 3.}}, <>][t],
  c2[t] → InterpolatingFunction[{{0., 3.}}, <>][t],
  c3[t] → InterpolatingFunction[{{0., 3.}}, <>][t]}}
```

Plots are obtained by evaluating these interpolating functions. The ReplaceAll operation (/.) applies the interpolating functions to the three concentration variables.

```
Plot[Evaluate[vars1 /. concs1], {t, 0, 3}, AxesLabel → {"t", "[A_i] "}, PlotStyle → {Black}]
```

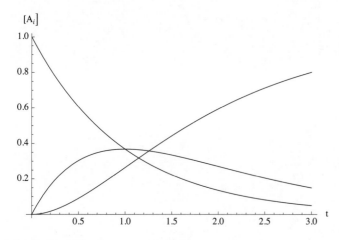

(b) $k_1 = 1$, $k_2 = 5$

```
eqns2 = {c1'[t] == -c1[t], c2'[t] == c1[t] - 5 c2[t],
        c3'[t] == 5 c2[t], c1[0] == 1, c2[0] == 0, c3[0] == 0};
```

```
vars2 = {c1[t], c2[t], c3[t]};
```

```
concs2 = NDSolve[eqns2, vars2, {t, 0, 3}];
```

```
Plot[Evaluate[vars2 /. concs2], {t, 0, 3}, AxesLabel → {"t", "[A_i] "}, PlotStyle → {Black}]
```

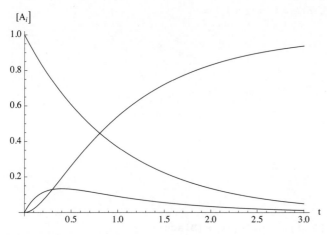

(c) $k_1 = 1$, $k_2 = 25$

```
eqns3 = {c1'[t] == -c1[t], c2'[t] == c1[t] - 25 c2[t],
        c3'[t] == 25 c2[t], c1[0] == 1, c2[0] == 0, c3[0] == 0};
```

```
vars3 = {c1[t], c2[t], c3[t]};
```

```
concs3 = NDSolve[eqns3, vars3, {t, 0, 3}];
```

```
Plot[Evaluate[vars3 /. concs3], {t, 0, 3}, AxesLabel → {"t", "[A_i] "}, PlotStyle → {Black}]
```

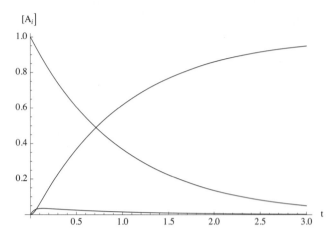

■ 12.1.2 Two consecutive reversible first order reactions

Two consecutive reversible first order reactions are represented by

$$\begin{array}{cc} k_1 & k_3 \\ A_1 \rightleftharpoons A_2 \rightleftharpoons A_3 \\ k_2 & k_4 \end{array}$$

Calculate the concentrations of A_1, A_2, and A_3 as functions of time. Assume that initially A_1 is at unit concentration and treat three cases: (*a*) $k_1 = k_2 = k_3 = k_4 = 1$, (*b*) $k_1 = k_2 = k_4 = 1$, $k_3 = 3$, (*c*) $k_1 = k_2 = k_4 = 1$, $k_3 = 9$,

SOLUTION

(a) $k_1 = k_2 = k_3 = k_4 = 1$

```
eqns4 = {c1'[t] == -c1[t] + c2[t],
         c2'[t] == c1[t] - 2 c2[t] + c3[t],
         c3'[t] == c2[t] - c3[t],
   c1[0] == 1, c2[0] == 0, c3[0] == 0};

vars4 = {c1[t], c2[t], c3[t]};

conc4 = NDSolve[eqns4, vars4, {t, 3}];

Plot[Evaluate[vars4 /. conc4], {t, 0, 3},
  AxesLabel → {"t", "[A_i] "}, PlotRange → {0, 1}, PlotStyle → {Black}]
```

(b) $k_1 = k_2 = k_4 = 1, k_3 = 3$

```
eqns5 = {c1'[t] == -c1[t] + c2[t],
        c2'[t] == c1[t] - 4 c2[t] + c3[t],
        c3'[t] == 3 c2[t] - c3[t],
  c1[0] == 1, c2[0] == 0, c3[0] == 0};

vars5 = {c1[t], c2[t], c3[t]};

conc5 = NDSolve[eqns5, vars5, {t, 3}];

Plot[Evaluate[vars5 /. conc5], {t, 0, 3},
  AxesLabel → {"t", "[A_i] "}, PlotRange → {0, 1}, PlotStyle → {Black}]
```

(c) $k_1 = k_2 = k_4 = 1, k_3 = 9$

```
eqns6 = {c1'[t] == -c1[t] + c2[t],
        c2'[t] == c1[t] - 10 c2[t] + c3[t],
        c3'[t] == 9 c2[t] - c3[t],
  c1[0] == 1, c2[0] == 0, c3[0] == 0};

vars6 = {c1[t], c2[t], c3[t]};

conc6 = NDSolve[eqns6, vars6, {t, 3}];

Plot[Evaluate[vars6 /. conc6], {t, 0, 3},
  AxesLabel → {"t", "[A_i] "}, PlotRange → {0, 1}, PlotStyle → {Black}]
```

12.1.3 Chemical reaction A + B = C + D in dilute aqueous solution

The reaction A + B = C + D is reversible and has an equilibrium constant equal to 2. (*a*) If the initial concentrations of A and B are 1 and 0.5, respectively, plot the four concentrations as a function of time, assuming the rate constant for the forward reaction is 1. The rate constant for the reverse reaction is 0.5. (*b*) To confirm the equilibrium concentrations of C and D, calculate these concentrations using the equilibrium constant expression.

SOLUTION

(a) The four rate equations and the initial conditions are given by

```
eqns7 = {cA'[t] == -cA[t] * cB[t] + .5 cC[t] * cD[t],
         cB'[t] == -cA[t] * cB[t] + .5 cC[t] * cD[t],
         cC'[t] == cA[t] * cB[t] - .5 cC[t] * cD[t],
         cD'[t] == cA[t] * cB[t] - .5 cC[t] * cD[t],
cA[0] == 1, cB[0] == .5, cC[0] == 0, cD[0] == 0};
```

Note that the rate equations for A and B are identical. The rate equations for C and D are also identical.

```
vars7 = {cA[t], cB[t], cC[t], cD[t]};

conc7 = NDSolve[eqns7, vars7, {t, 3}];

Plot[Evaluate[vars7 /. conc7], {t, 0, 3}, AxesLabel → {"t", "conc"}, PlotStyle → {Black}]
```

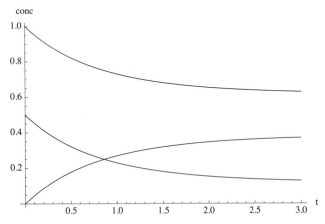

(b) Calculation of equilibrium concentrations of C and D:

The initial and equilibrium concentrations are given by

$$
\begin{array}{ccccc}
 & A & + \ B & = \ C + & D \\
\text{initial} & 1 & 0.5 & 0 & 0 \\
\text{equil.} & 1\text{-x} & 0.5\text{-x} & \text{x} & \text{x}
\end{array}
$$

The expression for the equilibrium constant is given by

$$
\frac{[C]_{eq}[D]_{eq}}{[A]_{eq}[B]_{eq}} = 2
$$

Substituting the equilibrium concentrations yields

$$
\frac{x^2}{(1-x)(0.5-x)} = 2
$$

```
FindRoot[2 * (1 - x) * (.5 - x) == x^2, {x, .5}]

{x → 0.381966}
```

This agrees with value that is being approached by the plot of [C] and [D].

- ### 12.1.4 A chemical reaction with an intermediate

   In the case of a reaction like that in the preceding problem, there may be a question as to whether there is an intermediate X:

$$
\begin{array}{cc}
1 & 10 \\
A + B \rightleftharpoons X \rightleftharpoons C + D \\
10 & 0.5
\end{array}
$$

The initial concentrations are [A] = 1 and [B] = 0.5. For the indicated values of the rate constants, explore the effects of intermediate X on the plots of concentration versus time.

SOLUTION
(a) The five rate equations and the initial conditions are given by

```
eqns8 = {cA'[t] == -cA[t] * cB[t] + 10 cX[t],
         cB'[t] == -cA[t] * cB[t] + 10 cX[t],
         cC'[t] == 10 cX[t] - .5 cC[t] * cD[t],
         cD'[t] == 10 cX[t] - .5 cC[t] * cD[t],
         cX'[t] == cA[t] * cB[t] + .5 cC[t] * cD[t] - 20 cX[t],
    cA[0] == 1, cB[0] == .5, cC[0] == 0, cD[0] == 0, cX[0] == 0};

vars8 = {cA[t], cB[t], cC[t], cD[t], cX[t]};

conc8 = NDSolve[eqns8, vars8, {t, 3}];
```

```
Plot[Evaluate[vars8 /. conc8], {t, 0, 3}, AxesLabel → {"t", "conc"}, PlotStyle → {Black}]
```

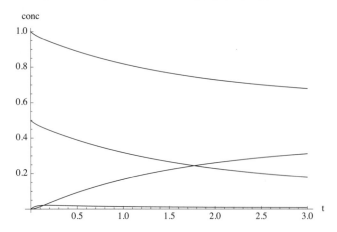

Note that C and D have equal concentrations at each time. The concentration of X barely shows on this plot, but its concentration goes through a maximum. This plot shows that intermediate X is in a nearly steady-state after the induction period (about $t = 0.1$) of its formation. Note that there is an induction period in the formation of C and D by comparing this plot with that in the preceeding problem. Thus even if X cannot be detected spectroscopically, the induction period in the formation of C and D will reveal the existence of the intermediate. This induction period can be seen more clearly by looking at short times.

```
Plot[Evaluate[vars8 /. conc8], {t, 0, 0.5`},
  AxesLabel → {"t", "conc"}, PlotRange → {0, 0.1`}, PlotStyle → {Black}]
```

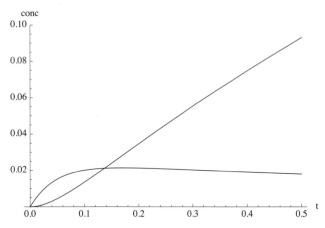

■ 12.1.5 Two consecutive chemical reactions

Consider two consecutive chemical reactions in which A, B, and D are partially converted to E.

$$A + B \underset{1}{\overset{2}{\rightleftharpoons}} C$$

$$C + D \underset{1}{\overset{2}{\rightleftharpoons}} E$$

The forward rate constants are both 2, and the reverse rate constants are both 1. When the initial concentrations are [A] = 1, [B] = 2, and D = 1, plot the concentrtions of the species as functions of time.

SOLUTION

```
eqns9 = {cA'[t] == -2 * cA[t] * cB[t] + cC[t],
        cB'[t] == -2 * cA[t] * cB[t] + cC[t],
    cC'[t] == 2 * cA[t] * cB[t] - cC[t] - 2 * cC[t] * cD[t] + cE[t], cD'[t] == -2 * cC[t] * cD[t] + cE[t],
    cE'[t] == 2 * cC[t] * cD[t] - cE[t],
cA[0] == 1, cB[0] == 2, cC[0] == 0, cD[0] == 1, cE[0] == 0};

vars9 = {cA[t], cB[t], cC[t], cD[t], cE[t]};

conc9 = NDSolve[eqns9, vars9, {t, 3}];

Plot[Evaluate[vars9 /. conc9], {t, 0, 3}, AxesLabel → {"t", "conc"}, PlotStyle → {Black}]
```

At long times, the order of the concentrations of chemical species starting with the lowest is [A], [C], [E], [D], and [B].

12.2 Enzyme - catalyzed reactions

■ 12.2.1 A = P

The rapid-equilibrium velocity for this reaction is given by vAP in Section 2.2.1, but, as a simplification, first consider experiments at concentrations much lower than the Michaelis constants. Under these conditions, the velocity is given by (kfet/kA)[A] - (kret/kP)[P]. Earlier calculations used kfet = 1, kret = 0.5, kA = 10, and kP = 20. With these kinetic parameters the velocity is given by 0.1*a - 0.025*p. In *Mathematica* the two rate equations and the initial conditions are given by

```
eqns20 = {cA'[t] == -.1 * cA[t] + .025 * cP[t],
    cP'[t] == .1 * cA[t] - .025 * cP[t], cA[0] == 1, cP[0] == 0};

vars20 = {cA[t], cP[t]};

conc20 = NDSolve[eqns20, vars20, {t, 40}];
```

```
Plot[Evaluate[vars20 /. conc20], {t, 0, 40}, AxesLabel → {"t", "conc"}, PlotStyle → {Black}]
```

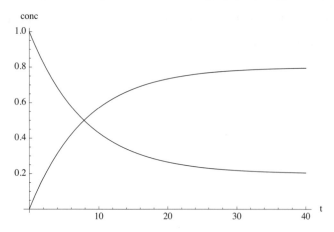

Fig. 12.1 Plots of the concentrations of A and P versus time at concentrations much lower than the Michaelis constants when $cA[0] = 0$.

The apparent equilibrium constant K' is equal to 4. Since units have not been given for the rate constants or Michaelis constants, this concentration scale is arbitrary.

However calculations can be made with the complete rapid-equilibrium rate equation given in Section 2.2.2.

```
(20 * a - 5 * p) / (200 + 20 * a + 10 * p)
```

$$\frac{20\,a - 5\,p}{200 + 20\,a + 10\,p}$$

At very high concentrations of A and in the absence of P, the velocity is given by 1

```
eqns21 = {cA'[t] == (-20 * cA[t] + 5 * cP[t]) / (200 + 20 * cA[t] + 10 * cP[t]),
    cP'[t] == (20 * cA[t] - 5 * cP[t]) / (200 + 20 * cA[t] + 10 * cP[t]), cA[0] == 1, cP[0] == 0};

vars21 = {cA[t], cP[t]};

conc21 = NDSolve[eqns21, vars21, {t, 40}];

Plot[Evaluate[vars21 /. conc21], {t, 0, 40}, AxesLabel → {"t", "conc"}, PlotStyle → {Black}]
```

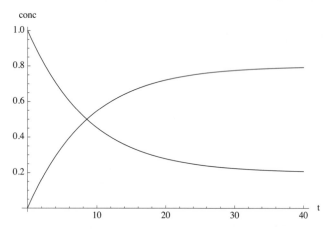

Fig. 12.2 Plots of concentrations of A and P when the whole rapid-equilibrium velocity is used and $cA[0] = 10$. This plot looks very much like Fig. 12.1, but it is different as shown by the next figure.

```
Plot[{Evaluate[vars20 /. conc20], Evaluate[vars21 /. conc21]},
 {t, 0, 40}, AxesLabel → {"t", "conc"}, PlotStyle → {Black}]
```

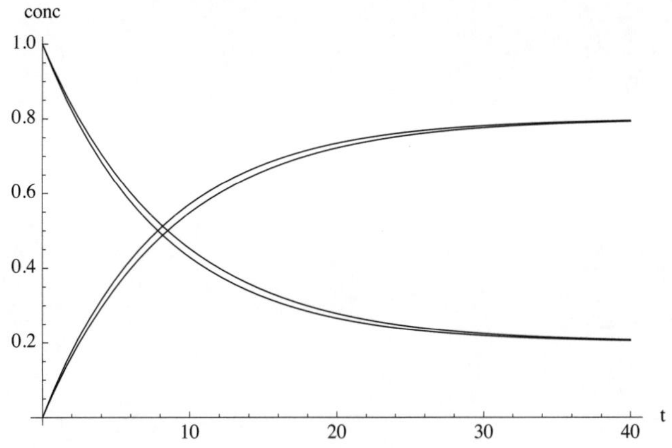

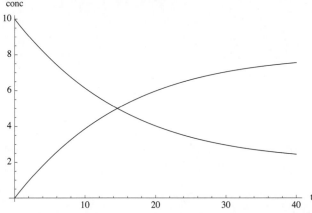

Fig. 12.3 Plots of Fig. 12.1 and 12.2 together.

Raise cA[0] for the calculation with the complete rate equation by a factor of 10.

```
eqns23 = {cA'[t] == (-20 * cA[t] + 5 * cP[t]) / (200 + 20 * cA[t] + 10 * cP[t]),
   cP'[t] == (20 * cA[t] - 5 * cP[t]) / (200 + 20 * cA[t] + 10 * cP[t]), cA[0] == 10, cP[0] == 0};
```

```
vars23 = {cA[t], cP[t]};
```

```
conc23 = NDSolve[eqns23, vars23, {t, 40}];
```

```
Plot[Evaluate[vars23 /. conc23], {t, 0, 40}, AxesLabel → {"t", "conc"}, PlotStyle → {Black}]
```

Fig. 12.4 Plot of concentrations of A and P when the whole rate equation is used and cA[0] = 10.

This calculation can be compared with Fig. 12.2.

```
Plot[{Evaluate[vars21 /. conc21], Evaluate[vars23 /. conc23]},
  {t, 0, 40}, AxesLabel → {"t", "conc"}, PlotStyle → {Black}]
```

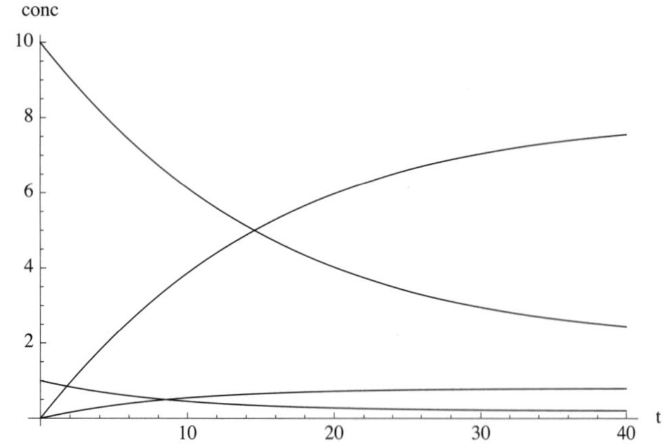

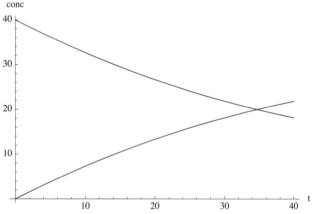

Fig. 12.5 Comparison of Fig. 12.2 with Fig. 12.4.

Raise cA[0] to 40 so that it is larger than the Michaelis constants

```
eqns24 = {cA'[t] == (-20 * cA[t] + 5 * cP[t]) / (200 + 20 * cA[t] + 10 * cP[t]),
    cP'[t] == (20 * cA[t] - 5 * cP[t]) / (200 + 20 * cA[t] + 10 * cP[t]), cA[0] == 40, cP[0] == 0};

vars24 = {cA[t], cP[t]};

conc24 = NDSolve[eqns24, vars24, {t, 40}];

Plot[{Evaluate[vars24 /. conc24]}, {t, 0, 40}, AxesLabel → {"t", "conc"}, PlotStyle → {Black}]
```

Fig. 12.6 Plot of concentrations of A and P when the whole rate equation is used and cA[0] = 40.

```
Plot[{Evaluate[vars23 /. conc23], Evaluate[vars24 /. conc24]},
 {t, 0, 40}, AxesLabel → {"t", "conc"}, PlotStyle → {Black}]
```

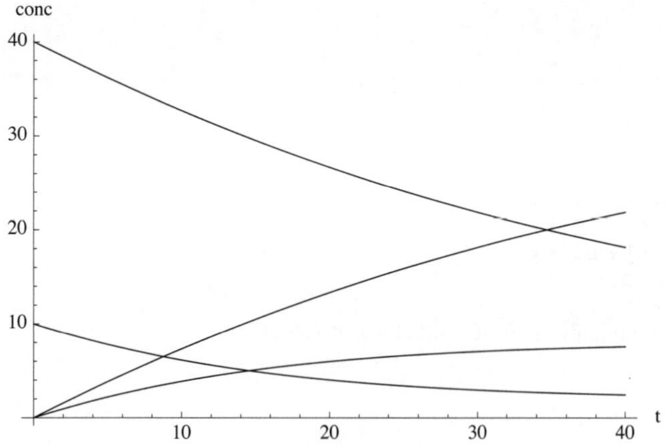

Fig. 12.7 Comparison of Fig. 12.4 with Fig. 12.6.

▪ 12.2.2 Ordered A + B = ordered P + Q

Section 5.1.1 shows that the rapid-equilibrium velocity of the forward reaction is given by

$$\frac{(a\,b\,vfexp\,kIQ\,kP - vrexp\,kB\,kIA\,p\,q)}{a\,b\,kIQ\,kP + a\,kB\,kIQ\,kP + kB\,kIA\,kIQ\,kP + kB\,kIA\,kP\,q + kB\,kIA\,p\,q}\,;$$

This shows that to calculate the concentrations of [A], [B], [P], and [Q] as functions of time it is necessary to have values for all 6 kinetic parameters. When the concentrations of the substrates are all significantly less than their Michaelis constants, the velocity in the forward reaction is given by

$$\frac{(a\,b\,vfexp\,kIQ\,kP - vrexp\,kB\,kIA\,p\,q)}{kB\,kIA\,kIQ\,kP}\,;$$

This rate equation can be rearranged to

(a b vfexp / (kB kIA) - vrexp p q / (kIQ kP))

$$\frac{a\,b\,vfexp}{kB\,kIA} - \frac{p\,q\,vrexp}{kIQ\,kP}$$

The values of these 6 kinetic parameters used in Section 5.1.1 can be inserted in this rate equation.

```
(a b vfexp / (kB kIA) - vrexp p q / (kIQ kP)) /. vfexp → 1 /. vrexp → 0.5 /. kIA → 5 /. kB → 20 /.
  kIQ → 10 /. kP -> 15
```

$$\frac{a\,b}{100} - 0.00333333\,p\,q$$

At equilibrium this rate is equal to zero, and so the apparent equilibrium constant K' is 3.

$$K' = \frac{[P]_{eq}[Q]_{eq}}{[A]_{eq}[P]_{eq}} = 3 = \frac{kff}{krr}$$

If kff = 1, krr = 0.33333. The calculation of the time course of this reaction is quite similar to the treatment of the chemical reaction in Section 12.1.3. In the previous calculation, the forward rate constant was 1 and the reverse rate constant was 0.5. The initial concentrations will be taken to be the same; that is [A] = 1 and [B] = 0.5.

Plot the concentrations of the 4 reactants as functions of time, assuming the rate constant for the forward reaction is 1, and the rate constant for the reverse reaction is 0.5. (*b*) To confirm the equilibrium concentrations of P and Q, calculate these concentrations using the equilibrium constant expression.

SOLUTION

(a) The four rate equations and the initial conditions are given by

```
eqns10 = {cA'[t] == -cA[t] * cB[t] + (1 / 3) cP[t] * cQ[t],
          cB'[t] == -cA[t] * cB[t] + (1 / 3) cP[t] * cQ[t],
          cP'[t] == cA[t] * cB[t] - (1 / 3) cP[t] * cQ[t],
          cQ'[t] == cA[t] * cB[t] - (1 / 3) cP[t] * cQ[t],
    cA[0] == 1, cB[0] == .5, cP[0] == 0, cQ[0] == 0};
```

Note that the rate equations for A and B are identical. The rate equations for C and D are also identical.

```
vars10 = {cA[t], cB[t], cP[t], cQ[t]};
```

```
conc10 = NDSolve[eqns10, vars10, {t, 3}];
```

```
Plot[Evaluate[vars10 /. conc10], {t, 0, 3}, AxesLabel → {"t", "conc"}, PlotStyle → {Black}]
```

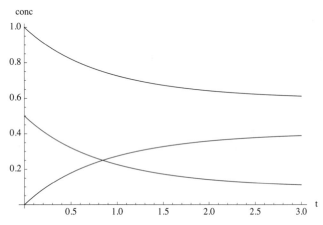

Fig. 12.8 Plots of [A], [B], [P], and [Q] for an enzyme-catalyzed reaction. The concentrations of P and Q are the same.

When the substrate concentrations are higher, their concentrations in the denominator of the rate equation have to be taken into account.

$$\frac{(a\ b\ vfexp\ kIQ\ kP - vrexp\ kB\ kIA\ p\ q)}{a\ b\ kIQ\ kP + a\ kB\ kIQ\ kP + kB\ kIA\ kIQ\ kP + kB\ kIA\ kP\ q + kB\ kIA\ p\ q};$$

Divide each denominator term by kB kIA kIQ kP.

```
{a b kIQ kP, a kB kIQ kP, kB kIA kIQ kP, kB kIA kP q, kB kIA p q} / (kB kIA kIQ kP)
```

$$\left\{ \frac{a\ b}{kB\ kIA},\ \frac{a}{kIA},\ 1,\ \frac{q}{kIQ},\ \frac{p\ q}{kIQ\ kP} \right\}$$

Thus the rate equation can be written as

$$\left(a\,b\,\text{vfexp}\,/\,(\text{kB kIA}) - \text{vrexp}\,p\,q\,/\,(\text{kIQ kP})\right) \Big/ \left(\frac{a\,b}{\text{kB kIA}} + \frac{a}{\text{kIA}} + 1 + \frac{q}{\text{kIQ}} + \frac{p\,q}{\text{kIQ kP}}\right)$$

$$\frac{\dfrac{a\,b\,\text{vfexp}}{\text{kB kIA}} - \dfrac{p\,q\,\text{vrexp}}{\text{kIQ kP}}}{1 + \dfrac{a}{\text{kIA}} + \dfrac{a\,b}{\text{kB kIA}} + \dfrac{q}{\text{kIQ}} + \dfrac{p\,q}{\text{kIQ kP}}}$$

Put in the kinetic parameters.

$$\left(a\,b\,\text{vfexp}\,/\,(\text{kB kIA}) - \text{vrexp}\,p\,q\,/\,(\text{kIQ kP})\right) \Big/ \left(\frac{a\,b}{\text{kB kIA}} + \frac{a}{\text{kIA}} + 1 + \frac{q}{\text{kIQ}} + \frac{p\,q}{\text{kIQ kP}}\right) /.\ \text{vfexp} \to 1\ /.$$

$$\text{vrexp} \to 0.5\ /.\ \text{kIA} \to 5\ /.\ \text{kB} \to 20\ /.\ \text{kIQ} \to 10\ /.\ \text{kP} \to 15$$

$$\frac{\dfrac{a\,b}{100} - 0.00333333\,p\,q}{1 + \dfrac{a}{5} + \dfrac{a\,b}{100} + \dfrac{q}{10} + \dfrac{p\,q}{150}}$$

This velocity will be multiplied by 100 to make it comparable to the preceeding plot.

$$\text{eqns11} = \Big\{ \text{cA}'[t] == -\text{cA}[t] * \text{cB}[t] \Big/ \left(1 + \frac{\text{cA}[t]}{5} + \frac{\text{cA}[t] * \text{cB}[t]}{100} + \frac{\text{cQ}[t]}{10} + \frac{\text{cP}[t] * \text{cQ}[t]}{150}\right) +$$

$$(1\,/\,3)\,\text{cP}[t] * \text{cQ}[t] \Big/ \left(1 + \frac{\text{cA}[t]}{5} + \frac{\text{cA}[t] * \text{cB}[t]}{100} + \frac{\text{cQ}[t]}{10} + \frac{\text{cP}[t] * \text{cQ}[t]}{150}\right),$$

$$\text{cB}'[t] == -\text{cA}[t] * \text{cB}[t] \Big/ \left(1 + \frac{\text{cA}[t]}{5} + \frac{\text{cA}[t] * \text{cB}[t]}{100} + \frac{\text{cQ}[t]}{10} + \frac{\text{cP}[t] * \text{cQ}[t]}{150}\right) +$$

$$(1\,/\,3)\,\text{cP}[t] * \text{cQ}[t] \Big/ \left(1 + \frac{\text{cA}[t]}{5} + \frac{\text{cA}[t] * \text{cB}[t]}{100} + \frac{\text{cQ}[t]}{10} + \frac{\text{cP}[t] * \text{cQ}[t]}{150}\right),$$

$$\text{cP}'[t] == \text{cA}[t] * \text{cB}[t] \Big/ \left(1 + \frac{\text{cA}[t]}{5} + \frac{\text{cA}[t] * \text{cB}[t]}{100} + \frac{\text{cQ}[t]}{10} + \frac{\text{cP}[t] * \text{cQ}[t]}{150}\right) -$$

$$(1\,/\,3)\,\text{cP}[t] * \text{cQ}[t] \Big/ \left(1 + \frac{\text{cA}[t]}{5} + \frac{\text{cA}[t] * \text{cB}[t]}{100} + \frac{\text{cQ}[t]}{10} + \frac{\text{cP}[t] * \text{cQ}[t]}{150}\right),$$

$$\text{cQ}'[t] == \text{cA}[t] * \text{cB}[t] \Big/ \left(1 + \frac{\text{cA}[t]}{5} + \frac{\text{cA}[t] * \text{cB}[t]}{100} + \frac{\text{cQ}[t]}{10} + \frac{\text{cP}[t] * \text{cQ}[t]}{150}\right) -$$

$$(1\,/\,3)\,\text{cP}[t] * \text{cQ}[t] \Big/ \left(1 + \frac{\text{cA}[t]}{5} + \frac{\text{cA}[t] * \text{cB}[t]}{100} + \frac{\text{cQ}[t]}{10} + \frac{\text{cP}[t] * \text{cQ}[t]}{150}\right),$$

$$\text{cA}[0] == 1,\ \text{cB}[0] == .5,\ \text{cP}[0] == 0,\ \text{cQ}[0] == 0\Big\};$$

Note that the rate equations for A and B are identical. The rate equations for P and Q are also identical.

```
vars11 = {cA[t], cB[t], cP[t], cQ[t]};
```

```
conc11 = NDSolve[eqns11, vars11, {t, 3}];
```

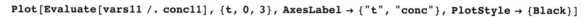

```
Plot[Evaluate[vars11 /. conc11], {t, 0, 3}, AxesLabel → {"t", "conc"}, PlotStyle → {Black}]
```

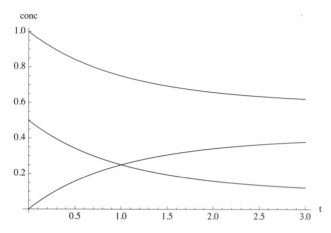

Fig. 12.9 Plots of [A], [B], [P], and [Q] for an enzyme-catalyzed reaction.

This plot is not much different from Fig. 12.8. This is because the Michaelis constants for A and B are not very different from the Michaelis constants for P and Q.

12.3 Glyoxylate cycle

▪ 12.3.1 Introduction

The thermodynamics and kinetics of systems of enzyme-catalyzed reactions are related because at long times the kinetic treatment must yield the equilibrium composition. The glyoxylate cycle is used as an example because of the complications it involves. Since it involves coenzymes that may have steady-state concentrations, oxidizing conditions can be compared with reducing conditions. There is a wide range of apparent equilibrium constants and rate constants in the glyoxylate cycle [24].

The reactions in the glyoxylate cycle are

EC 2.3.3.9 glyoxylate + acetyl-CoA + H_2 O = malate + CoA
EC 1.1.1.37 malate + NAD_{ox} = oxaloacetate + NAD_{red}
EC 2.3.1.1 oxaloacetate + acetyl-CoA + H_2 O = citrate + CoA
EC 4.2.1.3 citrate = isocitrate
EC 4.1.3.1 isocitrate = succinate + glyoxylate

The net reaction is

2acetyl-CoA + NAD_{ox} + 2H_2 O = succinate + 2CoA + NAD_{red}

▪ 12.3.2. Apparent equilibrium constants and rate constants

It is important to understand the thermodynamics of this cycle of reactions before making kinetic calculations.

The *Mathematica* package BasicBiochemData3 [*MathSource*1] can be used to calculate apparent equilibrium constants at 298.15 K, pHs 5 to 9, and ionic strengths zero to about 0.35 M for more than 200 enzyme-catalyzed reactions. Two files are available there for investigators with *Mathematica* in their computers: BasicBiochemData3.nb file and BasicBiochemData3.m file that were produced together. The BasicBiochemData3.m file that can be installed in your computer consists of the *Mathematica* input in the notebook file BasicBiochemData3.nb. The production and use of these files is described in Biochemical Thermodynamics: Applications of *Mathematica* (2006) [23].

The data needed to calculate the apparent equilibrium constants K' for the reactions in the glyoxylate cycle at 298.15 K and the desired pHs and ionic strengths are the species properties for the 11 reactants. Fortunately, these are known and are in BasicBiochemData3. These species properties have been used to calculate the following functions of pH and ionic strength for the reactants in the glyoxylate cycle at 298.15 K.

$$\text{acetylcoA} = -2.4789711565\ \text{Log}\left[e^{-0.4033931566238455\ \left(-180.36\ +\frac{8.74446\ is^{0.5}}{1+1.6\ is^{0.5}}-7.4369134695\ \text{Log}\left[10^{-pH}\right]\right)}\right]$$

$$-2.47897\ \text{Log}\left[e^{-0.403393\ \left(-180.36+\frac{8.74446\ is^{0.5}}{1+1.6\ is^{0.5}}-7.43691\ \text{Log}\left[10^{-pH}\right]\right)}\right]$$

$$\text{citrate} = -2.4789711565\ \text{Log}\left[e^{-0.4033931566238455\ \left(-1226.33\ +\frac{17.48892\ is^{0.5}}{1+1.6\ is^{0.5}}-17.3527980955\ \text{Log}\left[10^{-pH}\right]\right)} + \right.$$
$$e^{-0.4033931566238455\ \left(-1199.18\ +\frac{5.82964\ is^{0.5}}{1+1.6\ is^{0.5}}-14.873826939\ \text{Log}\left[10^{-pH}\right]\right)} +$$
$$\left. e^{-0.4033931566238455\ \left(-1162.69\ -\frac{11.65928\ is^{0.5}}{1+1.6\ is^{0.5}}-12.3948557825\ \text{Log}\left[10^{-pH}\right]\right)}\right]$$

$$-2.47897\ \text{Log}\left[e^{-0.403393\ \left(-1226.33+\frac{17.4889\ is^{0.5}}{1+1.6\ is^{0.5}}-17.3528\ \text{Log}\left[10^{-pH}\right]\right)} + \right.$$
$$e^{-0.403393\ \left(-1199.18+\frac{5.82964\ is^{0.5}}{1+1.6\ is^{0.5}}-14.8738\ \text{Log}\left[10^{-pH}\right]\right)} + e^{-0.403393\ \left(-1162.69-\frac{11.6593\ is^{0.5}}{1+1.6\ is^{0.5}}-12.3949\ \text{Log}\left[10^{-pH}\right]\right)}\left.\right]$$

$$\text{citrateiso} = -2.4789711565\ \text{Log}\left[e^{-0.4033931566238455\ \left(-1219.47\ +\frac{17.48892\ is^{0.5}}{1+1.6\ is^{0.5}}-17.3527980955\ \text{Log}\left[10^{-pH}\right]\right)} + \right.$$
$$e^{-0.4033931566238455\ \left(-1192.57\ +\frac{5.82964\ is^{0.5}}{1+1.6\ is^{0.5}}-14.873826939\ \text{Log}\left[10^{-pH}\right]\right)} +$$
$$\left. e^{-0.4033931566238455\ \left(-1156.04\ -\frac{11.65928\ is^{0.5}}{1+1.6\ is^{0.5}}-12.3948557825\ \text{Log}\left[10^{-pH}\right]\right)}\right]$$

$$-2.47897\ \text{Log}\left[e^{-0.403393\ \left(-1219.47+\frac{17.4889\ is^{0.5}}{1+1.6\ is^{0.5}}-17.3528\ \text{Log}\left[10^{-pH}\right]\right)} + \right.$$
$$e^{-0.403393\ \left(-1192.57+\frac{5.82964\ is^{0.5}}{1+1.6\ is^{0.5}}-14.8738\ \text{Log}\left[10^{-pH}\right]\right)} + e^{-0.403393\ \left(-1156.04-\frac{11.6593\ is^{0.5}}{1+1.6\ is^{0.5}}-12.3949\ \text{Log}\left[10^{-pH}\right]\right)}\left.\right]$$

$$\text{coA} = -2.4789711565\ \text{Log}\left[e^{\frac{1.1758184407903174\ is^{0.5}}{1+1.6\ is^{0.5}}} + e^{-0.4033931566238455\ \left(-47.83\ +\frac{2.91482\ is^{0.5}}{1+1.6\ is^{0.5}}-2.4789711565\ \text{Log}\left[10^{-pH}\right]\right)}\right]$$

$$-2.47897\ \text{Log}\left[e^{\frac{1.17582\ is^{0.5}}{1+1.6\ is^{0.5}}} + e^{-0.403393\ \left(-47.83+\frac{2.91482\ is^{0.5}}{1+1.6\ is^{0.5}}-2.47897\ \text{Log}\left[10^{-pH}\right]\right)}\right]$$

$$\text{glyoxylate} = -2.4789711565\ \text{Log}\left[e^{-0.4033931566238455\ \left(-468.6\ -2.4789711565\ \text{Log}\left[10^{-pH}\right]\right)}\right]$$

$$-2.47897\ \text{Log}\left[e^{-0.403393\ \left(-468.6-2.47897\ \text{Log}\left[10^{-pH}\right]\right)}\right]$$

$$\text{h2o} = -2.4789711565\ \text{Log}\left[e^{-0.4033931566238455\ \left(-237.19\ +\frac{5.82964\ is^{0.5}}{1+1.6\ is^{0.5}}-4.957942313\ \text{Log}\left[10^{-pH}\right]\right)}\right]$$

$$-2.47897\ \text{Log}\left[e^{-0.403393\ \left(-237.19+\frac{5.82964\ is^{0.5}}{1+1.6\ is^{0.5}}-4.95794\ \text{Log}\left[10^{-pH}\right]\right)}\right]$$

$$\text{malate} = -2.4789711565\grave{}\ \text{Log}\Big[e^{-0.4033931566238455\grave{}\left(-872.68\grave{}+\frac{11.65928\grave{}\ \text{is}^{0.5\grave{}}}{1+1.6\grave{}\ \text{is}^{0.5\grave{}}}-12.3948557825\grave{}\ \text{Log}\left[10^{-\text{pH}}\right]\right)} +$$

$$e^{-0.4033931566238455\grave{}\ \left(-842.66\grave{}-9.915884626\grave{}\ \text{Log}\left[10^{-\text{pH}}\right]\right)}\Big]$$

$$-2.47897\ \text{Log}\Big[e^{-0.403393\left(-872.68+\frac{11.6593\ \text{is}^{0.5}}{1+1.6\ \text{is}^{0.5}}-12.3949\ \text{Log}\left[10^{-\text{pH}}\right]\right)} + e^{-0.403393\left(-842.66-9.91588\ \text{Log}\left[10^{-\text{pH}}\right]\right)}\Big]$$

$$\text{nadox} = -2.4789711565\grave{}\ \text{Log}\Big[e^{-0.4033931566238455\grave{}\left(\frac{72.8705\grave{}\ \text{is}^{0.5}}{1+1.6\grave{}\ \text{is}^{0.5\grave{}}}-64.453250069\grave{}\ \text{Log}\left[10^{-\text{pH}}\right]\right)}\Big]$$

$$-2.47897\ \text{Log}\Big[e^{-0.403393\left(\frac{72.8705\ \text{is}^{0.5}}{1+1.6\ \text{is}^{0.5}}-64.4533\ \text{Log}\left[10^{-\text{pH}}\right]\right)}\Big]$$

$$\text{nadred} = -2.4789711565\grave{}\ \text{Log}\Big[e^{-0.4033931566238455\grave{}\left(22.65\grave{}+\frac{67.04086000000001\grave{}\ \text{is}^{0.5}}{1+1.6\grave{}\ \text{is}^{0.5\grave{}}}-66.9322212255\grave{}\ \text{Log}\left[10^{-\text{pH}}\right]\right)}\Big]$$

$$-2.47897\ \text{Log}\Big[e^{-0.403393\left(22.65+\frac{67.0409\ \text{is}^{0.5}}{1+1.6\ \text{is}^{0.5}}-66.9322\ \text{Log}\left[10^{-\text{pH}}\right]\right)}\Big]$$

$$\text{oxaloacetate} = -2.4789711565\grave{}\ \text{Log}\Big[e^{-0.4033931566238455\grave{}\left(-793.29\grave{}-\frac{5.82964\grave{}\ \text{is}^{0.5}}{1+1.6\grave{}\ \text{is}^{0.5\grave{}}}-4.957942313\grave{}\ \text{Log}\left[10^{-\text{pH}}\right]\right)}\Big]$$

$$-2.47897\ \text{Log}\Big[e^{-0.403393\left(-793.29-\frac{5.82964\ \text{is}^{0.5}}{1+1.6\ \text{is}^{0.5}}-4.95794\ \text{Log}\left[10^{-\text{pH}}\right]\right)}\Big]$$

$$\text{succinate} = -2.4789711565\grave{}\ \text{Log}\Big[e^{-0.4033931566238455\grave{}\left(-746.64\grave{}+\frac{17.48892\grave{}\ \text{is}^{0.5}}{1+1.6\grave{}\ \text{is}^{0.5\grave{}}}-14.873826939\grave{}\ \text{Log}\left[10^{-\text{pH}}\right]\right)} +$$

$$e^{-0.4033931566238455\grave{}\left(-722.62\grave{}+\frac{11.65928\grave{}\ \text{is}^{0.5}}{1+1.6\grave{}\ \text{is}^{0.5\grave{}}}-12.3948557825\grave{}\ \text{Log}\left[10^{-\text{pH}}\right]\right)} +$$

$$e^{-0.4033931566238455\grave{}\left(-690.44\grave{}-9.915884626\grave{}\ \text{Log}\left[10^{-\text{pH}}\right]\right)}\Big]$$

$$-2.47897\ \text{Log}\Big[e^{-0.403393\left(-746.64+\frac{17.4889\ \text{is}^{0.5}}{1+1.6\ \text{is}^{0.5}}-14.8738\ \text{Log}\left[10^{-\text{pH}}\right]\right)} +$$

$$e^{-0.403393\left(-722.62+\frac{11.6593\ \text{is}^{0.5}}{1+1.6\ \text{is}^{0.5}}-12.3949\ \text{Log}\left[10^{-\text{pH}}\right]\right)} + e^{-0.403393\left(-690.44-9.91588\ \text{Log}\left[10^{-\text{pH}}\right]\right)}\Big]$$

The following program is used to calculate the apparent equilibrium constants K' at 298.15 K, pHs 5-9, and 0.25 M ionic strength for the reactions in the glyoxylate cycle.

```
calckprime[eq_, pHlist_, islist_] := Module[{energy, dG},(*Calculates the apparent equili
    energy = Solve[eq, de];
    dG = energy[[1,1,2]] /. pH -> pHlist /. is -> islist;
    E^(-(dG/(8.31451*0.29815)))]
```

```
ec2339 = calckprime[glyoxylate + acetylcoA + h2o + de == malate + coA, {5, 6, 7, 8, 9}, .25];
```

```
ec11137 = calckprime[malate + nadox + de == oxaloacetate + nadred, {5, 6, 7, 8, 9}, .25];
```

```
ec2311 = calckprime[oxaloacetate + acetylcoA + h2o + de == citrate + coA, {5, 6, 7, 8, 9}, .25];
```

```
ec4213 = calckprime[citrate + de == citrateiso, {5, 6, 7, 8, 9}, .25];
```

```
ec4131 = calckprime[citrateiso + de == succinate + glyoxylate, {5, 6, 7, 8, 9}, .25];
```

```
glyoxnet =
    calckprime[2 * acetylcoA + nadox + 2 * h2o + de == succinate + 2 * coA + nadred, {5, 6, 7, 8, 9}, .25];
```

Table 12.1 Apparent equilibrium constants K_i' at 298.15 K, pHs 5-9, and 0.25 M ionic strength for the reactions in the glyoxylate cycle

```
TableForm[{ec2339, ec11137, ec2311, ec4213, ec4131, glyoxnet},
  TableHeadings → {{"ec2339", "ec11137", "ec2311", "ec4213", "ec4131", "net rx"},
   {"pH 5", "pH 6", "pH 7", "pH 7", "pH 9"}}, TableSpacing → {1, 1}]
```

	pH 5	pH 6	pH 7	pH 7	pH 9
ec2339	3.17543×10^6	2.24929×10^7	2.30869×10^8	3.83478×10^9	1.9189×10^{11}
ec11137	5.97546×10^{-8}	8.49672×10^{-7}	8.87102×10^{-6}	0.0000891027	0.000891421
ec2311	3.13288×10^6	8.51005×10^6	6.97298×10^7	1.12808×10^9	5.62969×10^{10}
ec4213	0.0685637	0.0686516	0.0684247	0.0683914	0.0683879
ec4131	0.410548	0.747702	0.887456	0.906067	0.90799
net rx	16 733.1	8.34846×10^6	8.67194×10^9	2.38854×10^{13}	5.97971×10^{17}

The apparent equilibrium constants are not known this accurately, but it is difficult to round exponential numbers in *Mathematica*. The product of the 5 apparent equilibrium constants of the five reactions at pH 7 is equal to the apparent equilibrium constant K_i' for the net reaction.

$$K' = K_1' K_2' K_3' K_4' K_5';$$

This equation is illustrated at pH 7.

```
(2.31 * 10^8) * (8.87 * 10^-6) * (6.97 * 10^7) * (.0684) * (.887)
```

8.66459×10^9

In treating thermodynamics and kinetics of enzyme-catalyzed reactions, it is often convenient to specify concentrations of NAD_{ox} and NAD_{red} because they are involved in many reactions and are therefor in steady states. Calculations are made here when $[NAD_{ox}] = 1000 \, [NAD_{red}]$ (oxidizing conditions) and $[NAD_{ox}] = 10 \, [NAD_{red}]$ (more reducing conditions).

Since $K' = k_f / k_r$ for each reaction, assuming all the forward rate constants are equal to unity, the reverse rate constants are calculated as follows:

k_r for EC 2.3.3.9:

```
1 / (2.31 * 10^8)
```

4.329×10^{-9}

k_r for EC1.1.1.37:

```
1 / (8.87 * 10^-3)
```

112.74

This apparent equilibrium constant for this reaction has been increased by 10^3 because $[NAD_{ox}] = 1000 \, [NAD_{red}]$ for the purpose of considering oxidizing conditions.

k_r for EC 2.3.1.1:

```
1 / (6.97 * 10^7)
```

1.43472×10^{-8}

k_r for EC 4.2.1.3:

1 / .0684

14.6199

k_r for EC 4.1.3.1

1 / .887 // N

1.1274

The fact that two of these rate constants are so low means that if the forward rate constants are equal to unity, it will take a very long time for this reaction system to reach equilibrium. Thus, the concentrations or rate constants of the enzymes for the first and third reactions are raised by a factor of 10^9.

When $[NAD_{ox}]/[NAD_{red}]$ is specified , the five reactions and the net reaction can be represented by

EC 2.3.3.9 glyoxylate(1) + acetyl-CoA(2) = malate(3) +CoA(4)
EC 1.1.1.37 malate(3) = oxaloacetate(5)
EC 2.3.1.1 oxaloacetate(5) + acetyl-CoA = citrate(6) + CoA(4)
EC 4.2.1.3 citrate(6) = isocitrate(7)
EC 4.1.3.1 isocitrate(7) = succinate(8) + glyoxylate(1)

The reactants are numbered because they are referred to by number later. The net reaction is

2acetyl-CoA = succinate + 2CoA

Note that H_2O has been deleted in two of the reactions and the net reaction because its concentration is not included in the expressions for apparent equilibrium constants or rate equations. Since concentrations of NAD_{ox} and NAD_{red} are specified, apparent equilibrium constants are represented by K'', rather than K'. More information on further transformed Gibbs energies of formation and apparent equilibrium constants are given in [20,23]. When $[NAD_{ox}] = 1000 [NAD_{red}]$, apparent equilibrium constants and temporary values of forward and reverse rate constants are given in Table 12.2.

Table 12.2 Apparent equilibrium constants K'' and rate constants k_f and k_r under oxidizing conditions specified as $[NAD_{ox}] = 1000 [NAD_{red}]$

```
TableForm[{{2.31 * 10^8, 1, 4.33 * 10^-9}, {8.87 * 10^-3, 1, 113},
  {6.97 * 10^7, 1, 1.43 * 10^-8}, {0.0684, 1, 14.6}, {0.887, 1, 1.13}},
 TableHeadings → {{"ec2339", "ec11137", "ec2311", "ec4213", "ec4131"}, {"K''", "k_f", "k_r"}}]
```

	K''	k_f	k_r
ec2339	2.31×10^8	1	4.33×10^{-9}
ec11137	0.00887	1	113
ec2311	6.97×10^7	1	1.43×10^{-8}
ec4213	0.0684	1	14.6
ec4131	0.887	1	1.13

This emphasizes the very wide range in reverse rate constants, which will make the approach to equilibrium very slow. In the first set of equilibrium calculations and rate calculations under oxidizing conditions, the forward rate constants will be taken to be 10^9, 1, 10^9, 1, and 1. The reverse rate constants will be taken to be 4.33, 113, 14.3, 14.6, and 1.13. This increase in forward rate constant does not affect the apparent equilibrium constants K''. Thus thermodynamics tells us that this set of reactions goes to equilibrium in two stages: In the first stage, glyoxylate and acetyl-coA are converted to malate and coA very quickly. and then the other reactions can occur.

When $[NAD_{ox}] = 10\ [NAD_{red}]$, apparent equilibrium constants and temporary values of forward and reverse rate constants are given in Table 12.3.

Table 12.3 Apparent equilibrium constants and rate constants when $[NAD_{ox}] = 10\ [NAD_{red}]$

```
TableForm[{{2.31 * 10^8, 1, 4.33 * 10^-9}, {8.87 * 10^-5, 1, 1.13 * 10^4},
    {6.97 * 10^7, 1, 1.43 * 10^-8}, {0.0684, 1, 14.6}, {0.887, 1, 1.13}},
   TableHeadings → {{"ec2339", "ec11137", "ec2311", "ec4213", "ec4131"}, {"K''", "k_f", "k_r"}}]
```

	K''	k_f	k_r
ec2339	2.31×10^8	1	4.33×10^{-9}
ec11137	0.0000887	1	11 300.
ec2311	6.97×10^7	1	1.43×10^{-8}
ec4213	0.0684	1	14.6
ec4131	0.887	1	1.13

Note that the apparent equilibrium constant for the second reaction has been reduced by a factor of 100. Again the concentrations of the enzymes for the first and third reactions are raised by a factor of 10^9. Thus, in the set of equilibrium calculations and rate calculations under more reducing conditions, the forward rate constants will be taken to be 10^9, 1, 10^9, 1, and 1. The reverse rate constants will be taken to be 4.33, $1.13 * 10^4$, 14.3, 14.6, and 1.13.

Note that the apparent equilibrium constant K_{net}'' for the net reaction is given by

$$K_1''\ K_2''\ K_3''\ K_4''\ K_5'' = K_{net}''$$

Under the oxidizing conditions defined here, K_{net}'' is 8.67×10^{12} and under the reducing conditions K_{net}'' is 8.67×10^{10}.

■ 12.3.3 Calculation of equilibrium compositions

Since the apparent equilibrium constants K'' are known for the five reactions in the glyoxylate cycle at two ratios of $[NAD_{ox}]/[NAD_{red}]$, equilibrium concentrations can be calculated for various initial amounts of substrates. The program equcalcc written by Fred Krambeck (Mobil Research and Development) [7] requires the conservation matrix for a reaction system, but $H_2 O$ in two of the reactions is a problem because $H_2 O$ is not included in the expression for the apparent equilibrium constant, even though its standard transformed Gibbs energy of formation is involved in the calculation of the apparent equilibrium constant. This problem is solved by using the program equcalcrx [18,20,23] that requires the transposed stoichiometric number matrix for the system. It calculates a suitable conservation matrix and calls on equcalcc to calculate the equilibrium composition.

```
equcalcc[as_, lnk_, no_] :=
 Module[{l, x, b, ac, m, n, e, k}, (*as=conservation matrix. lnk=
    -(1/RT)(Gibbs energy of formation vector at T). no=initial composition vector.*)
  (*Setup*)
  {m, n} = Dimensions[as];
  b = as.no;
  ac = as;
  (*Initialize*)l = LinearSolve[as.Transpose[as], -as.(lnk + Log[n])];
  (*Solve*)Do[e = b - ac.(x = E^(lnk + l.as));
   If[(10^-10) > Max[Abs[e]], Break[]];
   l = l + LinearSolve[ac.Transpose[as * Table[x, {m}]], e], {k, 100}];
  If[k = 100, Return["Algorithm Failed"]];
  Return[x]]

equcalcrx[nt_, lnkr_, no_] :=
 Module[{as, lnk}, (*nt=transposed stoichiometric number matrix. lnkr=
    ln of equilibrium constants of rxs (vector). no=initial composition vector.*)
  (*Setup*)
  lnk = LinearSolve[nt, lnkr];
  as = NullSpace[nt];
  equcalcc[as, lnk, no]]
```

The transposed stoichiometric number matrix nt is given by

```
nt = {{-1, -1, 1, 1, 0, 0, 0, 0}, {0, 0, -1, 0, 1, 0, 0, 0},
   {0, -1, 0, 1, -1, 1, 0, 0}, {0, 0, 0, 0, 0, -1, 1, 0}, {1, 0, 0, 0, 0, 0, -1, 1}};
```

Table 12.4 Transposed stoichiometric number matrix for the glyoxylate cycle

```
TableForm[nt, TableHeadings →
   {{"ec2339", "ec11137", "ec2311", "ec4213", "ec4131"}, {"glyoxylate", "acetyl-CoA", "malate",
     "CoA", "oxaloacetate", "citrate", "isocitrate", "succinate"}}, TableSpacing → {1, .5}]
```

	glyoxylate	acetyl-CoA	malate	CoA	oxaloacetate	citrate	isocitrate	succinate
ec2339	-1	-1	1	1	0	0	0	0
ec11137	0	0	-1	0	1	0	0	0
ec2311	0	-1	0	1	-1	1	0	0
ec4213	0	0	0	0	0	-1	1	0
ec4131	1	0	0	0	0	0	-1	1

This matrix applies under both oxidizing and more reducing conditions. In the first calculation of the equilibrium concentrations, all the eight reactants are initially present at 10^{-3} M. All reactants do not have to be present initially, but the initial reactants must contain atoms of all elements in the five reactions.

Calculation of the equilibrium composition when all reactants are initially present at 10^{-3} M under oxidizing conditions

The lnKprime required by equcalcrx under oxidizing conditions is given by:

```
lnKprime = Log[{2.31 * 10^8, 8.87 * 10^-3, 6.97 * 10^7, .0684, .887}]
```

$\{19.2579, -4.72508, 18.0597, -2.68238, -0.11991\}$

```
no1 = {10^-3, 10^-3, 10^-3, 10^-3, 10^-3, 10^-3, 10^-3, 10^-3}
```

$$\left\{ \frac{1}{1000}, \frac{1}{1000}, \frac{1}{1000}, \frac{1}{1000}, \frac{1}{1000}, \frac{1}{1000}, \frac{1}{1000}, \frac{1}{1000} \right\}$$

```
tab4 = equcalcrx[nt, lnKprime, no1]
```

$\{0.0010028, 3.38329 \times 10^{-11}, 0.00391865, 0.002,$
$0.0000347584, 0.0000409828, 2.80322 \times 10^{-6}, 0.00247951\}$

Table 12.5 Equilibrium concentrations in moles per liter for the glyoxylate cycle at 298.15 K, pH 7, and 0.25 M ionic strength when the eight reactants are initially present at 10^{-3} M and $[NAD_{ox}] = 1000[NAD_{red}]$

```
TableForm[tab4, TableHeadings → {{"glyoxylate", "acetylCoA", "malate",
    "CoA", "oxaloacetate", "citrate", "isocitrate", "succinate"}, None}]
```

glyoxylate	0.0010028
acetylCoA	3.38329×10^{-11}
malate	0.00391865
CoA	0.002
oxaloacetate	0.0000347584
citrate	0.0000409828
isocitrate	2.80322×10^{-6}
succinate	0.00247951

This equilibrium composition can be tested by using it to calculate the five apparent equilibrium constants involved. The following equation utilizes the dot product, lnc is the vector of logarithms of the equilibrium concentrations of the reactants (1x8), and v'' is the stoichiometric number matrix for the reaction system (8x5):

$$K'' = \exp[\text{lnc}.v'']$$

```
Exp[Log[tab4].Transpose[nt]]
```

$\{2.31 \times 10^8, 0.00887, 6.97 \times 10^7, 0.0684, 0.887\}$

This is in agreement with Table 12.2. Thus the calculated equilibrium composition satisfies this test.

Calculation of the equilibrium composition when all reactants are initially present at 10^{-3} M under more reducing conditions

The apparent equilibrium constant for the second reaction 8.87*10^-3 (see Table 12.2) is changed to 8.87*10^-5 for more reducing conditions.

```
lnKprime2 = Log[{2.31 * 10^8, 8.87 * 10^-5, 6.97 * 10^7, .0684, .887}]
```

$\{19.2579, -9.33025, 18.0597, -2.68238, -0.11991\}$

```
tab5 = equcalcrx[nt, lnKprime2, no1]
```

$\{0.000136216, 3.08803 \times 10^{-10}, 0.0048584, 0.002,$
$\quad 4.3094 \times 10^{-7}, 4.63769 \times 10^{-6}, 3.17218 \times 10^{-7}, 0.00206563\}$

Table 12.6 Equilibrium concentrations in moles per liter for the glyoxylate cycle at 298.15 K, pH 7, and 0.25 M ionic strength when the eight reactants are initially present at 10^{-3} M and $[NAD_{ox}] = 10[NAD_{red}]$

```
TableForm[tab5, TableHeadings → {{"glyoxylate", "acetylCoA", "malate",
    "CoA", "oxaloacetate", "citrate", "isocitrate", "succinate"}, None}]
```

glyoxylate	0.000136216
acetylCoA	3.08803×10^{-10}
malate	0.0048584
CoA	0.002
oxaloacetate	4.3094×10^{-7}
citrate	4.63769×10^{-6}
isocitrate	3.17218×10^{-7}
succinate	0.00206563

This equilibrium composition is tested by using it to calculate the five apparent equilibrium constants involved.

```
Exp[Log[tab5].Transpose[nt]]
```

$$\{2.31 \times 10^8, 0.0000887, 6.97 \times 10^7, 0.0684, 0.887\}$$

Thus the calculated equilibrium composition satisfies this test.

Calculation of the equilibrium composition when glyoxylate and acetyl-CoA are initially present at 10^{-3} M under oxidizing conditions

The lnKprime under oxidizing conditions is used.

```
lnKprime = Log[{2.31 * 10^8, 8.87 * 10^-3, 6.97 * 10^7, .0684, .887}]
```

$$\{19.2579, -4.72508, 18.0597, -2.68238, -0.11991\}$$

```
no2 = {10^-3, 10^-3, 0, 0, 0, 0, 0, 0}
```

$$\left\{\frac{1}{1000}, \frac{1}{1000}, 0, 0, 0, 0, 0, 0\right\}$$

```
tab6 = equcalcrx[nt, lnKprime, no2]
```

$$\{0.000455331, 5.11571 \times 10^{-12}, 0.000538078, 0.001,$$
$$4.77275 \times 10^{-6}, 1.7018 \times 10^{-6}, 1.16403 \times 10^{-7}, 0.000226756\}$$

Table 12.7 Equilibrium concentrations in moles per liter for the glyoxylate cycle at 298.15 K, pH 7, and 0.25 M ionic strength when glyoxylate and acetyl-CoA are initially present at 10^{-3} M under oxidizing conditions

```
TableForm[tab6, TableHeadings → {{"glyoxylate", "acetylCoA", "malate",
    "CoA", "oxaloacetate", "citrate", "isocitrate", "succinate"}, None}]
```

glyoxylate	0.000455331
acetylCoA	5.11571×10^{-12}
malate	0.000538078
CoA	0.001
oxaloacetate	4.77275×10^{-6}
citrate	1.7018×10^{-6}
isocitrate	1.16403×10^{-7}
succinate	0.000226756

This equilibrium composition is tested by using it to calculate the five apparent equilibrium constants involved.

```
Exp[Log[tab6].Transpose[nt]]
```

$\{2.31 \times 10^8, 0.00887, 6.97 \times 10^7, 0.0684, 0.887\}$

Thus the calculated equilibrium composition satisfies this test.

Calculation of the equilibrium composition when glyoxylate and acetyl-CoA are initially present at 10^{-3} M under more reducing conditions

The lnKprime under more reducing conditions is used.

```
lnKprime2 = Log[{2.31 * 10^8, 8.87 * 10^-5, 6.97 * 10^7, .0684, .887}]
```

$\{19.2579, -9.33025, 18.0597, -2.68238, -0.11991\}$

```
tab7 = equcalcrx[nt, lnKprime2, no2]
```

$\{0.000134519, 2.78449 \times 10^{-11}, 0.000865246, 0.001,$
$\;\; 7.67473 \times 10^{-8}, 1.4895 \times 10^{-7}, 1.01882 \times 10^{-8}, 0.0000671797\}$

Table 12.8 Equilibrium concentrations in moles per liter for the glyoxylate cycle at 298.15 K, pH 7, and 0.25 M ionic strength when glyoxylate and acetyl-CoA are initially present at 10^{-3} M under more reducing conditions

```
TableForm[tab7, TableHeadings → {{"glyoxylate", "acetylCoA", "malate",
    "CoA", "oxaloacetate", "citrate", "isocitrate", "succinate"}, None}]
```

glyoxylate	0.000134519
acetylCoA	2.78449×10^{-11}
malate	0.000865246
CoA	0.001
oxaloacetate	7.67473×10^{-8}
citrate	1.4895×10^{-7}
isocitrate	1.01882×10^{-8}
succinate	0.0000671797

This equilibrium composition is tested by using it to calculate the five apparent equilibrium constants involved.

```
Exp[Log[tab7].Transpose[nt]]
```

$\{2.31 \times 10^8, 0.0000887, 6.97 \times 10^7, 0.0684, 0.887\}$

Thus the calculated equilibrium composition satisfies this test.

▪ 12.3.4 Calculation of reactant concentrations as functions of time with oxidizing conditions

The thermodynamic and kinetic information on the glyoxylate cycle is included in the rate equations for the five enzyme-catalyzed reactions. As described in Sections 12.1 and 12.2, the rapid-equilibrium rate equations used here apply at substrate concentrations significantly below the Michaelis constant, but they do include the reverse reactions. The *Mathematica* built-in object NDSolve makes it possible to obtain interpolation functions for the eight reactants that can be plotted or tabulated. These solutions depend on the initial concentrations and whether there are oxidizing conditions or more reducing conditions. Kinetics are calculated here only for the case that glyoxylate and acetyl-CoA are present initially at 0.001 M. Since the rate constants k_f and k_r for a reaction are proportional to the concentration of the enzyme, the enzyme concentrations can be adjusted so that the calculations do not have to cover too long a time. In order to accomplish this, the forward rate constants for EC 2.3.3.9 and EC 2.3.1.1 in Tables 12.2 and 12.3 are multiplied by 10^9 so that there will not be any rate constants in the system less than unity. Thus for oxidizing conditions the forward rate constants are taken to be 10^9, 1, 10^9, 1, and 1. The corresponding reverse rate constants are taken to be 4.33, 113, 14.3, 14.6, and 1.13. This means that reaction EC 2.3.3.9 occurs very rapidly to make stoichiometric amounts of malate and CoA. Therefore the reaction occurs in two stages. In the first stage, glyoxylate and acetyl-CoA react to form malate and CoA. In the second stage, malate is essentially the initial reactant, and CoA remains at its equilibrium concentration since it is not involved in a forward reaction.

Calculations of kinetics at short times

NDSolve has three inputs: (1) rapid-equilibrium rate equations and the initial concentrations of reactants, (2) a list of the variables to be calculated, and (3) the time range.

```
eqns3 = {c1'[t] == - (10^9) * c1[t] * c2[t] + 4.33 * c3[t] * c4[t] + c7[t] - 1.13 * c8[t] * c1[t],
    c2'[t] ==
      - (10^9) * c1[t] * c2[t] + 4.33 * c3[t] * c4[t] - (10^9) * c5[t] * c2[t] + 14.3 * c4[t] * c6[t],
    c3'[t] == (10^9) * c1[t] * c2[t] - 4.33 * c3[t] * c4[t] - c3[t] + 113 * c5[t],
    c4'[t] ==
      (10^9) * c2[t] * c5[t] - 14.3 * c4[t] * c6[t] + (10^9) * c1[t] * c2[t] - 4.33 * c3[t] * c4[t],
    c5'[t] == c3[t] - 113 * c5[t] - (10^9) * c5[t] * c2[t] + 14.3 * c4[t] * c6[t],
    c6'[t] == (10^9) * c2[t] * c5[t] - 14.3 * c4[t] * c6[t] - c6[t] + 14.6 * c7[t],
    c7'[t] == c6[t] - 15.6 * c7[t] + 1.13 * c1[t] * c8[t],
    c8'[t] == c7[t] - 1.13 * c1[t] * c8[t],
    c1[0] == 1. * 10^-3, c2[0] == 1. * 10^-3,
    c3[0] == 0, c4[0] == 0, c5[0] == 0, c6[0] == 0, c7[0] == 0, c8[0] == 0};

vars = {c1[t], c2[t], c3[t], c4[t], c5[t], c6[t], c7[t], c8[t]};

solution16 = NDSolve[eqns3, vars, {t, 0, 5 * 10^-6}];
```

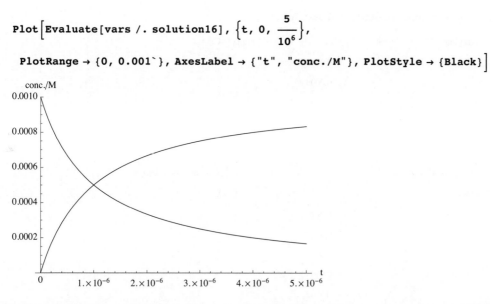

$$\text{Plot}\Big[\text{Evaluate[vars /. solution16]},\ \Big\{t,\ 0,\ \frac{5}{10^6}\Big\},$$

$$\text{PlotRange} \to \{0,\ 0.001\grave{}\},\ \text{AxesLabel} \to \{\text{"t"},\ \text{"conc./M"}\},\ \text{PlotStyle} \to \{\text{Black}\}\Big]$$

Fig. 12.10 Concentrations of glyoxylate, acetyl-CoA, malate, and CoA in the first phase as a function of time up to 5×10^{-6} units of arbitrary time

$$\text{plotA} = \text{Plot}\Big[\text{Evaluate[c1[t] /. solution16]},\ \Big\{t,\ 0,\ \frac{5}{10^6}\Big\},\ \text{PlotRange} \to \{0,\ 0.001\grave{}\},$$

$$\text{AxesLabel} \to \{\text{"t"},\ \text{"conc./M"}\},\ \text{PlotLabel} \to \text{"glyoxylate"},\ \text{PlotStyle} \to \{\text{Black}\}\Big];$$

$$\text{plotB} = \text{Plot}\Big[\text{Evaluate[c2[t] /. solution16]},\ \Big\{t,\ 0,\ \frac{5}{10^6}\Big\},\ \text{PlotRange} \to \{0,\ 0.001\grave{}\},$$

$$\text{AxesLabel} \to \{\text{"t"},\ \text{"conc./M"}\},\ \text{PlotLabel} \to \text{"acetyl-CoA"},\ \text{PlotStyle} \to \{\text{Black}\}\Big];$$

$$\text{plotC} = \text{Plot}\Big[\text{Evaluate[c3[t] /. solution16]},\ \Big\{t,\ 0,\ \frac{5}{10^6}\Big\},\ \text{PlotRange} \to \{0,\ 0.001\grave{}\},$$

$$\text{AxesLabel} \to \{\text{"t"},\ \text{"conc./M"}\},\ \text{PlotLabel} \to \text{"malate"},\ \text{PlotStyle} \to \{\text{Black}\}\Big];$$

$$\text{plotD} = \text{Plot}\Big[\text{Evaluate[c4[t] /. solution16]},\ \Big\{t,\ 0,\ \frac{5}{10^6}\Big\},\ \text{PlotRange} \to \{0,\ 0.001\grave{}\},$$

$$\text{AxesLabel} \to \{\text{"t"},\ \text{"conc./M"}\},\ \text{PlotLabel} \to \text{"CoA"},\ \text{PlotStyle} \to \{\text{Black}\}\Big];$$

```
fig1 = GraphicsArray[{{plotA, plotB}, {plotC, plotD}}]
```

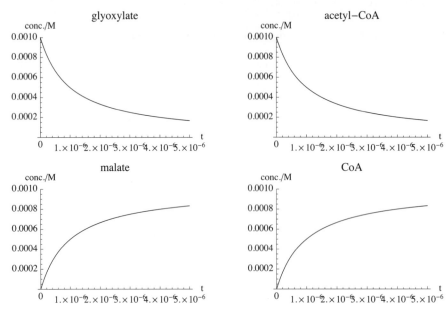

Fig. 12.11 Concentrations of four reactants in the first phase of the reaction as a function of time up to $t = 5 \times 10^{-6}$ units of arbitrary time

In 10^{-6} units of arbitrary time, glyoxylate and acetyl-coA are converted almost completely to malate and CoA.

Calculations of kinetic properties at longer times with oxidizing conditions

```
solution15 = NDSolve[eqns3, vars, {t, 10, 5000}];
```

```
figTOCgr = Plot[Evaluate[vars /. solution15], {t, 10, 5000},
    PlotRange → {0, 0.001`}, AxesLabel → {"t", "conc./M"}, PlotStyle → {Black}]
```

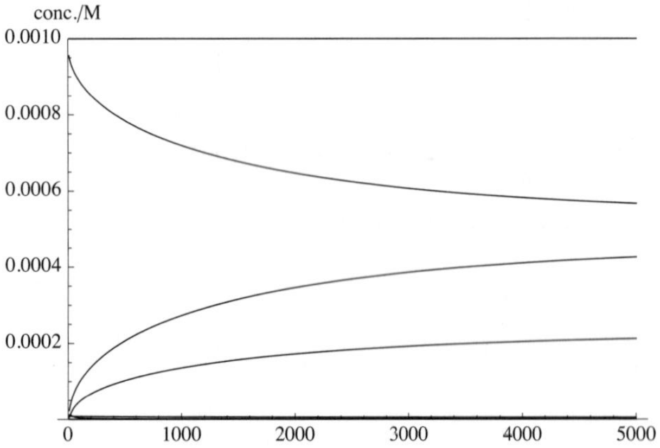

Fig. 12.12 Concentrations of reactants at longer times.

```
plota = Plot[Evaluate[c1[t] /. solution15], {t, 10, 5000}, PlotRange → {0, 0.001`},
    AxesLabel → {"t", "conc./M"}, PlotLabel → "glyoxylate", PlotStyle → {Black}];

plotb = Plot[Evaluate[c4[t] /. solution15], {t, 100, 5000}, PlotRange → {0, 0.001`},
    AxesLabel → {"t", "conc./M"}, PlotLabel → "CoA", PlotStyle → {Black}];

plotc = Plot[Evaluate[c3[t] /. solution15], {t, 10, 5000}, PlotRange → {0, 0.001`},
    AxesLabel → {"t", "conc./M"}, PlotLabel → "malate", PlotStyle → {Black}];

plotd = Plot[Evaluate[c8[t] /. solution15], {t, 10, 5000}, PlotRange → {0, 0.001`},
    AxesLabel → {"t", "conc./M"}, PlotLabel → "succinate", PlotStyle → {Black}];

GraphicsArray[{{plota, plotb}, {plotc, plotd}}]
```

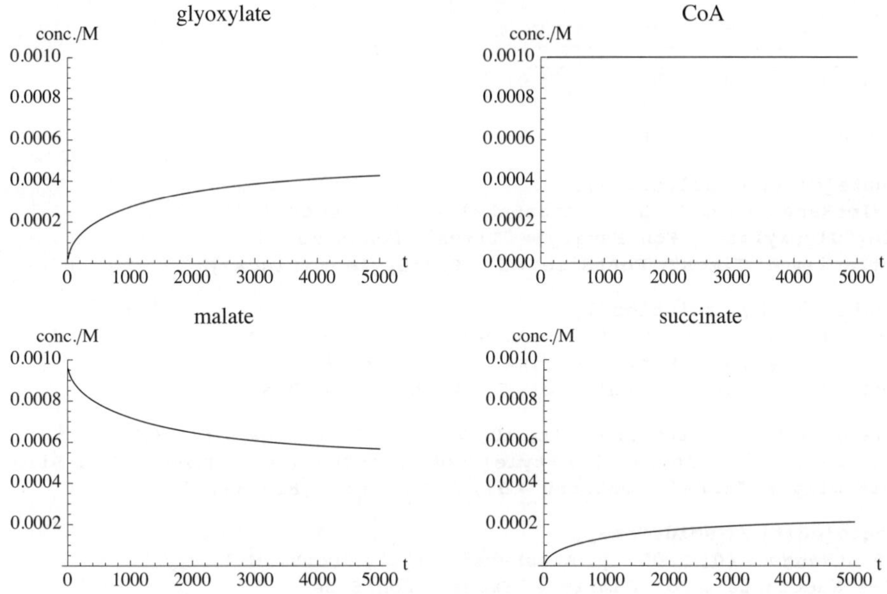

Fig. 12.13 Concentrations of four reactants with oxidizing conditions as a function of time up to t = 5000 units of arbitrary time. Notice that glyoxylate is produced.

■ 12.3.5 Calculation of reactant concentrations as functions of time with more reducing conditions

```
eqns4 = {c1'[t] == - (10^9) * c1[t] * c2[t] + 4.33 * c3[t] * c4[t] + c7[t] - 1.13 * c8[t] * c1[t],
    c2'[t] ==
      - (10^9) * c1[t] * c2[t] + 4.33 * c3[t] * c4[t] - (10^9) * c5[t] * c2[t] + 14.3 * c4[t] * c6[t],
    c3'[t] == (10^9) * c1[t] * c2[t] - 4.33 * c3[t] * c4[t] - c3[t] + (1.13 * 10^4) * c5[t],
    c4'[t] ==
      (10^9) * c2[t] * c5[t] - 14.3 * c4[t] * c6[t] + (10^9) * c1[t] * c2[t] - 4.33 * c3[t] * c4[t],
    c5'[t] == c3[t] - (1.13 * 10^4) * c5[t] - (10^9) * c5[t] * c2[t] + 14.3 * c4[t] * c6[t],
    c6'[t] == (10^9) * c2[t] * c5[t] - 14.3 * c4[t] * c6[t] - c6[t] + 14.6 * c7[t],
    c7'[t] == c6[t] - 15.6 * c7[t] + 1.13 * c1[t] * c8[t],
    c8'[t] == c7[t] - 1.13 * c1[t] * c8[t],
    c1[0] == 1. * 10^-3, c2[0] == 1. * 10^-3,
    c3[0] == 0, c4[0] == 0, c5[0] == 0, c6[0] == 0, c7[0] == 0, c8[0] == 0};

vars = {c1[t], c2[t], c3[t], c4[t], c5[t], c6[t], c7[t], c8[t]};

solution4 = NDSolve[eqns4, vars, {t, 10, 25 000}];
```

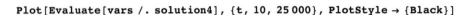

```
Plot[Evaluate[vars /. solution4], {t, 10, 25 000}, PlotStyle → {Black}]
```

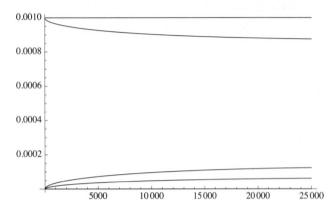

Fig. 12.14 Concentations of reactants as functions of time.

```
plotaa1 = Plot[Evaluate[c1[t] /. solution4],
    {t, 10, 25 000}, PlotRange → {0, 0.001`}, AxesLabel → {"t", "conc./M"},
    PlotLabel → Style["Glyoxylate", FontFamily → "Times", FontSize → 7],
    BaseStyle → {FontFamily → "Times", FontSize → 5}, PlotStyle → {Black}];

plotcc1 = Plot[Evaluate[c3[t] /. solution4],
    {t, 10, 25 000}, PlotRange → {0, 0.001`}, AxesLabel → {"t", "conc./M"},
    PlotLabel → Style["Malate", FontFamily → "Times", FontSize → 7],
    BaseStyle → {FontFamily → "Times", FontSize → 5}, PlotStyle → {Black}];

plotbb1 = Plot[Evaluate[c4[t] /. solution4], {t, 10, 25 000}, PlotRange → {0, 0.001`},
    AxesLabel → {"t", "conc./M"}, PlotLabel → Style["CoA", FontFamily → "Times", FontSize → 7],
    BaseStyle → {FontFamily → "Times", FontSize → 5}, PlotStyle → {Black}];

plotdd1 = Plot[Evaluate[c8[t] /. solution4],
    {t, 10, 25 000}, PlotRange → {0, 0.001`}, AxesLabel → {"t", "conc./M"},
    PlotLabel → Style["Succinate", FontFamily → "Times", FontSize → 7],
    BaseStyle → {FontFamily → "Times", FontSize → 5}, PlotStyle → {Black}];
```

fig5 = GraphicsArray[{{plotaa1, plotbb1}, {plotcc1, plotdd1}}]

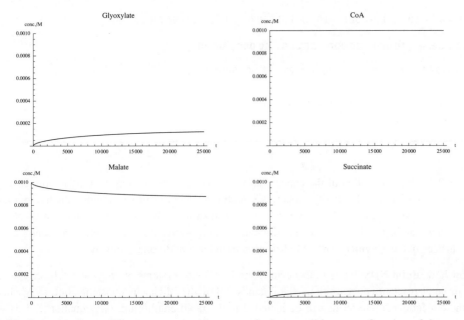

Fig. 12.15 Concentrations of four reactants with more reducing conditions as a function of time up to t = 5000 units of arbitrary time. Notice that less glyoxylate is produced under reducing conditions.

▪ 12.3.6 Concentrations at long times with oxidizing conditions

These calculations show that at very long times, the concentrations of the eight reactants reach the values calculated using equcalcrx.

solution20 = NDSolve[eqns3, vars, {t, 10, 50 000}];

c1[t] /. solution20 /. t → {10 000, 20 000, 30 000, 40 000, 50 000}

{{0.000451602, 0.000455299, 0.000455361, 0.000455361, 0.000455361}}

c2[t] /. solution20 /. t → {10 000, 20 000, 30 000, 40 000, 50 000}

$\left\{\left\{5.19313 \times 10^{-12},\ 5.11761 \times 10^{-12},\ 5.11636 \times 10^{-12},\ 5.11635 \times 10^{-12},\ 5.11635 \times 10^{-12}\right\}\right\}$

c3[t] /. solution20 /. t → {10 000, 20 000, 30 000, 40 000, 50 000}

{{0.0005418, 0.000538119, 0.000538057, 0.000538057, 0.000538057}}

c4[t] /. solution20 /. t → {10 000, 20 000, 30 000, 40 000, 50 000}

{{0.001, 0.001, 0.001, 0.001, 0.001}}

c5[t] /. solution20 /. t → {10 000, 20 000, 30 000, 40 000, 50 000}

$\left\{\left\{4.79469 \times 10^{-6},\ 4.76211 \times 10^{-6},\ 4.76157 \times 10^{-6},\ 4.76157 \times 10^{-6},\ 4.76157 \times 10^{-6}\right\}\right\}$

c6[t] /. solution20 /. t → {10 000, 20 000, 30 000, 40 000, 50 000}

$\left\{\left\{1.6875 \times 10^{-6},\ 1.70335 \times 10^{-6},\ 1.70362 \times 10^{-6},\ 1.70362 \times 10^{-6},\ 1.70362 \times 10^{-6}\right\}\right\}$

c7[t] /. solution20 /. t → {10 000, 20 000, 30 000, 40 000, 50 000}

$\{\{1.1553 \times 10^{-7}, 1.16667 \times 10^{-7}, 1.16686 \times 10^{-7}, 1.16687 \times 10^{-7}, 1.16687 \times 10^{-7}\}\}$

c8[t] /. solution20 /. t → {10 000, 20 000, 30 000, 40 000, 50 000}

$\{\{0.000224899, 0.00022674, 0.00022677, 0.00022677, 0.00022677\}\}$

These concentrations reach the equilibrium concentration to three digits by t = 50,000.

12.4 Discussion

As discussed in Chapters 9-11, the calculation of the concentrations of the reactants in a system of enzyme-catalayzed reactions as a function of time may involve the effects of competitive inhibitors, uncompetitive inhibitors, activators, and modifiers. Competitive inhibition involves binding at the enzymatic site and so structurally related molecules are likely to be involved, but other types of molecules can be involved in uncompetitive inhibition. Activation occurs when molecules are bound at neighboring sites that cause changes in the enzymatic site. Modification introduces additional pathways.

These calculations show how useful NDSolve is in calculating the kinetics of systems of enzyme-catalyzed reactions. These calculations have been made at reactant concentrations significantly below the Michaelis constants, but, as Michaelis constants are determined for these reactions, these terms can be included in the rapid-equilibrium rate equations. This will reduce the rapid-equilibrium velocities v.

These tables and figures can be calculated for different initial concentrations of reactants, different pHs and ionic strength, and different concentrations of coenzymes. These effects are too complicated to be summarized, but with rate equations in the computer, it is easy to change these variables.

References

1951

[1] R. A. Alberty, R. H. Smith, and R. M. Bock, The apparent ionization constants of adenosine phosphates and related compounds, J. Biol. Chem. 193, 425 (1951).

1953

[2] K. Burton and H. A. Krebs, The free energy changes associated with the individual steps of the tricarboxylic acid cycle, glycolysis, alcohol fermentation, and the hydrolysis of the phosphate group of adenosine triphosphate, Biochem. J. 54, 94-107 (1953).

1956

[3] R. N. Smith and R. A. Alberty, The apparent stability constants of ionic complexes of various adenosine phosphates and divalent cations, J. Am. Chem. Soc., 78, 2376-2380 (1956).

1957

[4] H. A. Krebs and H. L. Kornberg, Energy Transformations in Living Matter, Springer-Verlog, Berlin (1957) with Appendix by K. Burton.

1969

[5] R. C. Wilhoit, Thermodynamic properties of biochemical substances, in Biochemical Microcalorimetry, H. D. Brown (ed), Academic Press, New York, 1969.

1975

[6] I. H. Segel, 1975. Enzyme Kinetics: Behavoir and Analysis of Rapid-equilibrium and steady-state enzyme systems, Wiley-Interscience, Hoboken, NJ, 1975.

1978

[7] F. J. Krambeck, Presented at the 71st Annual Meeting of the AIChE, Miami Beach, FL, Nov. 16, 1978.

1979

[8] J. A. Beattie and I. Oppenheim, Principles of Thermodynamics, Elsevier, Amsterdam, 1979.

[9] R. G. Duggleby, Experimental designs for estimating kinetic parameters for enzyme-catalyzed reactions. J. Theor. Biol. 81, 672-684 (1979).

1982

[10] D. D. Wagman, W. H. Evans, V. B. Parker, R. H. Schumm, I. Halow, S. M. Bailey, K. L. Churney, and R. L. Nuttall, The NBS tables of chemical thermodynamic properties, J. Phys. Chem. Ref. Data, 11, Supplement 2 (1982).

[11] W. R. Smith and R. W. Missen, Chemical Equilibrium Reaction Analysis: Theory and Algorithms, Wiley, Hoboken, NJ (1982).

1985

[12] H. B. Callen, Thermodynamics and an Introduction to Thermostatistics, Wiley, Hoboken, NJ (1985).

1992

[13] R. A. Alberty, Equilibrium calculations on systems of biochemical reactions, Biophys. Chem. 42, 117-131 (1992).

[14] R. A. Alberty and R. N. Goldberg, Calculation of thermodynamic formation properties for the ATP series at specified pH and pMg. Biochem. 31, 10610-10615 (1992).

[15] R. A. Alberty, Calculation of transformed thermodynamic properties of biochemical reactants at specified pH and pMg, Biophys. Chem. 43, 239-254 (1992).

1993

[16] R. A. Alberty and A. Cornish-Bowden, On the pH dependence of the apparent equilibrium constant K' of a biochemical reaction, Trends Biochem. Sci. 18, 288-291 (1993).

1994

[17] R. A. Alberty, Legendre transforms in chemical thermodynamics, Chem. Rev. 94, 1457-1482 (1994).

2000

[18] R. A. Alberty, Calculation of equilibrium compositions of large systems of biochemical reactions, J. Phys. Chem. 104 B 4807-4814 (2000).

2001

[19] R. A. Alberty J. M. G. Barthel, E. R. Cohen, M. B. Ewing, R. N. Goldberg, and E. Wilhelm, Use of Legendre transforms in chemical thermodynamics (an IUPAC technical report), Pure. Appl. Chem. 73, No. 8 (2001).

2003

[20] R. A. Alberty, Thermodynamics of Biochemical Reactions, Wiley, Hoboken, NJ (2003).

2004

[21] A. Cornish-Bowden, Fundamentals of Enzyme Kinetics, 3rd ed. Portland Press, London, 2004.

2005

[22] R. J. Silbey, R. A. Alberty, and M. Bawendi, Physical Chemistry, Wiley, Hoboken, NJ (2005) and Solutions Manual.

2006

[23] R. A. Alberty, Biochemical Thermodynamics : Applications of Mathematica, Hoboken, NJ (2006).

[24] R. A. Alberty, Thermodynamics and kinetics of the glyoxylate cycle, Biochem., 45, 15838-15843 (2006).

2007

[25] P. F. Cook, W. W. Cleland, Enzyme Kinetics and Mechanism, Garland Science, London and New York, 2007.

[26] R. A. Alberty, Change in binding of hydrogen ions in biochemical reactions, Biophys. Chem. 125, 328-333 (2007).

[27] R. A. Alberty, Effects of pH in rapid-equilibrium enzyme kinetics, J. Phys. Chem. B, 111, 14064-14068 (2007).

[28] R. A. Alberty, Two different ways that hydrogen ions are involved in the thermodynamics and rapid-equilibrium kinetics of the enzyme-catalysis of S = P and S + H_2O = P, Biophys. Chem. 128, 204-209 (2007).

[29] R. A. Alberty, Three mechanisms and rapid-equilibrium rate equations for a type of reductase reaction, Biophys. Chem. 131, 71-91 (2007).

[30] R. A. Alberty, Effects of pH in biochemical thermodynamics and enzyme kinetics, Experimental standard conditions of enzyme characterization, Proceedings of the 3rd International Beilstein Workshop, Eds. M. G. Hicks and C. Kettner (2007).

2008

[31] Q. Cui, M. Karplus, Allostery and cooperativity revisited, Protein Science 17, 1295-1307 (2008).

[32] R. A. Alberty, Determination of kinetic parameters of enzyme-catalyzed reactions with a minimum number of velocity measurements, J. Theor. Biol. 254, 156-163 (2008).
http://library.wwolfram.com/infocenter/*MathSource*/7152

[33] R. A. Alberty, Rapid-equilibrium rate equations for the enzymatic catalysis of A + B = P + Q over a range of pH, Biophys. Chem. 132, 114-126 (2008).

2009

[34] R. A. Alberty, Determination of rapid-equilibrium kinetic parameters of enzyme-catalyzed reaction A + B + C -> with the minimum number of velocity measurements, J. Phys. Chem. B 113, 1225-1231 (2009).

[35] R. A. Alberty, Determination of Rapid-Equilibrium Kinetic Parameters of Enzyme-Catalyzed Reaction A + B = P + Q, J. Phys. Chem. B, 113, 10043-10048 (2009) .

2010

[36] R. A. Alberty, Estimation of Kinetic Parameters When Modifiers are Bound in Enzyme-Catalyzed Reactions, J. Phys. Chem. B, 114, 1684-1689 (2010).

[37] R. A. Alberty, Consumption of Hydrogen Ions in Rapid-Equilibrium Enzyme Kinetics, J. Phys. Chem. B, in press.

Web

[Web1] R. A. Alberty, A. Cornish-Bowden, Q. H. Gibson, R. N. Goldberg, G. G. Hammes, W. Jencks, K. F. Tipton, R. Veech, H. V. Westerhof, and E. C. Webb, Recommendations for nomenclature and tables for biochemical thermodynamics, Pure Appl. Chem. 66, 1641-1666 (1994).
http://www.chem.qmw.ac.uk/iubmb/thermod/

[Web2] Recommendations of the Nomenclature Committee of the International Union of Biochemistry and Molecular Biology on the Nomenclature and Classification of Enzymes by the Reactions they Catalyze
http : // www.chem.qmul.ac.uk/iubmb/enzyme

[Web3] R. N. Goldberg, Y. B. Tewari, and T. N. Bhat, "Thermodynamics of Enzyme-Catalyzed Reactions -a Database for Quantitative Biochemistry", Bioinformatics 2004;20(16):2874-2877.
http://xpdb.nist.gov/enzyme_thermodynamics/

[Web4] R. A. Alberty, A. Cornish-Bowden, R. N. Goldberg, G. G. Hammes, K. Tipton, and H. V. Westerhoff, Recommendations for Terminology and Databases for Biochemical Thermodynamics (2009).
http:www.chem.qmul.ac.uk/iubmb/thermod2/

MathSource

[MathSource1] R. A. Alberty, Basic Data for Biochemistry
http://library.wolfram.com/infocenter/*MathSource*/797

[MathSource2] R. A. Alberty, ProteinLigandProg
http://library.wolfram.com/infocenter/*MathSource*/4808

[MathSource3] R. A. Alberty, Mathematical Functions for Thermodynamic Properties of Biochemical Reactants
http://library.wolfram.com/infocenter/*MathSource*/5704

[MathSource4] R. A. Alberty, Changes in the Binding of Hydrogen Ions in Enzyme-Catalyzed Reactions
http://library.wolfram.com/infocenter/*MathSource*/6386

[MathSource5] R. A. Alberty, Rapid-equilibrium Rate Equations for A + B -> Products and Determination of Kinetic Parameters
http://library.wolfram.com/infocenter/*MathSource*/7151

[MathSource6] R. A. Alberty, Determination of the Rapid-Equilibrium Kinetic Parameters for the Enzyme-Catalyzed Reaction
A + mB -> Products with the Minimum Number of Velocity Measurements
http://library.wolfram.com/infocenter/*MathSource*/7152

Index

The page numbers give the Chapter numbers and the page numbers within the Chapter.

Alternate rate equation for ordered A + B = random P + Q, 5.32

Apparent equilibrium constant

    for a biochemical reaction, 1.6

    for EA = E + A as a function of temperature, 2.43

    for the hydrolysis of ATP as a function of pH, 1.26

    for the hydrolysis of ATP, 1.11

    for the net reaction for the glyoxylate cycle, 12.19

Apparent equilibrium constants for reactions in the glyoxylate cycle, 12.19

Average number of hydrogen atoms in a biochemical reactant, 7.3

Calculation of kinetic parameters from bell - shaped plots, 2.35

Calculation of the change in binding of hydrogen ions using pKs, 1.13

Change in binding of hydrogen ions in a biochemical reaction, 1.8, 7.2

Competitive inhibition of ordered A + B → products, 11.2

Competitive inhibition of random A + B → products, 11.22

Competitive inhibition, 9.2

Components in A + B + C → products, 6.5

Components in biochemical reactions, 1.8

Consecutive first order reactions, 12.2

Consecutive reversible first order reactions, 12.4

Cooperativity, 10.15

Criteria for equilibrium in chemical thermodynamics, 1.2

Database on species in biochemical reactions, 1.9

Dead - end enzyme - substrate complexes, 5.13

Derivation of the rate equation for A → products using Solve, 2.5

Derivation of the rate equation for A = P, 2.19

Determination of hydrogen ions consumed in ordered A + B → products, 3.45

Determination of the number of activator molecules bound, 9.27

Distinguishing the ordered mechanism from the random mechanism, 4.15

Effect of pH on standard transformed Gibbs energies of biochemical reactants, 1.10

Effect of pH on the distribution of species in a biochemical reactant, 1.18

Effect of temperature on kinetic properties for A → products, 2.37

Effects of changes in kinetic parameters for ordered A + B → products, 3.12

Effects of pH (factor method) on the kinetics of random A + B → products, 4.17

Effects of pH on ordered A + B → products

    when n hydrogen ions are consumed, 3.43

    when one hydrogen ion is consumed, 3.33

Effects of pH on ordered O + mR → products when mR is bound first, 7.8
Effects of pH on ordered O + mR → products when O is bound first, 7.18
Effects of pH on random A + B → products when hydrogen ions are consumed, 4.46
Effects of pH on the kinetics of ordered A + B → products, 3.12
Effects of pH on the kinetics of random A + B → products, 4.17
Equilibrium compositions for the glyoxylate cycle, 12.21
Essential activation of ordered A + B → products, 11.10
Essential activation when several molecules of activator are bound, 9.18
Essential activation, 9.14
Estimation of kinetic parameters for A → products, 2.3
        using two velocity measurements, 2.4
        when there are two pKs, 2.6
Estimation of kinetic parameters for A = P, 2.20
Estimation of kinetic parameters for ordered A + B → products, 3.8
Estimation of kinetic parameters for ordered mechanism for A + B + C → products, 6.7
Estimation of kinetic parameters for random A + B → products, 4.7
Estimation of kinetic parameters mechanism II for A + B + C → products, 6.9
Estimation of kinetic parameters mechanism III for A + B + C → products, 6.12
Estimation of kinetic parameters mechanism IV for A + B + C → products, 6.15
Estimation of kinetic parameters mechanism V for A + B + C → products, 6.20
Estimation of Michaelis constants of reactions not in the mechanism, 6.39
Estimation of parameters for ordered O + 2 R → products
        when 2 R is bound first, 7.5
        when O is bound first, 7.16
Estimation of parameters for ordered O + mR → products when mR is bound first, 7.7
Estimation of parameters for random O + 2 R → products, 8.8
Estimation of parameters for random O + 3 R → products, 8.18
Estimation of pH - dependent parameters for ordered A + B = ordered P + Q, 5.4
Estimation of pH - dependent parameters for ordered A + B = random P + Q, 5.15
Estimation of pH - dependent parameters for random A + B = random P + Q, 5.20
Estimation of pH - independent kinetic parameters for random A + B → products, 4.36
Estimation of the competitive inhibition constant, 9.5
Expression of the limiting velocity as a function of temperature, 2.39
Fundamental equation for the Gibbs energy in chemical thermodynamics, 1.3
Fundamental equation for the transformed Gibbs energy in biochemistry, 1.5
General modifier mechanism for A → products, 10.2
Glyoxylate cycle, 12.15
Haldane relation
        for A = P when the apparent equilibrium constant is very large, 29
        for A = P, 2.22
        for ordered A + B = ordered P + Q, 5.3, 5.6

for ordered A + B = random P + Q, 5.18

for random A + B = random P + Q, 5.25

including pH effects for A = P, 2.26

Independent chemical reactions, 1.4

Independent reactions in A + B + C → products, 6.3

Kinetic parameters

for ordered A + B → products as functions of pH, 3.18

for random A + B → products using the factor method, 4.34

Kinetics of a reversible chemical reaction with an intermediate, 12.7

Kinetics of a reversible chemical reaction, 12.6

Linear algebra of the reactions in A + B + C → products, 6.3

Lineweaver - Burk plots, 2.3

Mechanism

for A → products, 2.2

for competitive inhibition, 9.2

for essential activation when several molecules of activator are bound, 9.18

for essential activation, 9.14

for mixed activation when 2 X are bound in a single reaction, 9.28

for mixed activation, 9.28

for ordered O + mR → products when mR is bound first, 7.3

for ordered O + mR → products when O is bound first, 7.15

uncompetitive inhibition, 9.6

Mechanisms for random O + mR → products, 8.2

Michaelis - Menten plots for A → products when there are two pKs, 2.13

Mixed activation, 9.28

of ordered A + B → products, 11.13

when 2 X are bound in a single reaction, 9.28

when 2 X are bound in two reactions, 9.42

Mixed inhibition

of ordered A + B → products, 11.7

of random A + B → products, 11.29

Mixed inhibition, 9.9

Modification of ordered A + B → products, 11.17

Negative cooperativity, 10.15

with respect to A, 9.46

Net reaction, glyoxylate cycle, 12.15

Number of modifier molecules bound, 10.13

pH dependencies for random O + 2 R → products, 8.15

pH effects for random O + mR → products, 8.12

pH effects on the rate equation for A → products, 2.6

when the substrate has two pKs, 2.16

pH effects on the velocity of A = P, 2.22

pH independent kinetic parameters for A + B → products, 3.24

pH - dependencies of parameters for ordered O + mR → products when O is bound first, 7.21

pH - dependent parameters for random A + B → products, 4.22

Plots of kinetic parameters for a nitrate reductase reaction, 7.10

Plots of velocities

> for ordered A + B → products, 3.7

> for random A + B → products, 4.6

Positive cooperativity, 10.15

Rate equation for A → products, 2.2

> when the apparent equilibrium constant is very large, 2.27

Rate equation

> for ordered A + B → products, 3.2

> for ordered A + B = ordered P + Q, 5.2

> for ordered A + B = random P + Q, 5.13

> for ordered O + mR → products when mR is bound first, 7.3

> for random A + B = random P + Q, 5.19

> for random O + mR → products, 8.2

> for the completely ordered - mechanism for A + B + C → products, 6.6

> for the general modifier mechanism, 10.2

> for random A + B → products, 4.3

Rates for ordered A + B = ordered P + Q

> when A, B, and Q are present, 5.7

> when A, B, P, and Q are present, 5.9

Reactants in A + B + C → products, 6.2

Reductase reactions, 7.2

Sigmoid plots in both A and X, 9.38

Specificity constant, 2.44

Standard Gibbs energy of reaction in chemical thermodynamics, 1.4

Standard transformed Gibbs energies

> for the hydrolysis of ATP, 1.32

> of formation of reactants in the glyoxylate cycle, 12.17

Standard transformed Gibbs energy of a biochemical reactant (sum of species), 1.5, 1.7, 1.9

Systems of enzyme - catalyzed reactions, 12.1

Temperature effects on ordered A + B → products, 3.46

Thermodynamic cycle

> for random A + B → products, 4.2, 4.14

> in mixed inhibition, 9.12

> in the mechanism for ordered A + B = random P + Q, 5.17

> in the mechanism for random A + B = random P + Q, 5.24

Thermodynamic data on enzyme - catalyzed reactions on the web, 1.29

Thermodynamics of EA = E + A at two temperatures, 2.40
Time course of chemical reaction A = P, 12.9
Time course of ordered A + B = P + Q, 12.13
Titration curve for ATP, 1.14
Titration curves for diprotic acids, 2.29
Uncompetitive inhibition, 9.6
      of ordered A + B → products, 11.5
      of random A + B → products, 11.26
Velocities for random O + mR → products, 8.2-8.7